I0137855

WILDFLOWERS
OF FLORIDA AND THE
SOUTHEAST

DAVID W. HALL
WILLIAM J. WEBER
EDITED BY
JASON H. BYRD

Copyright 2011 DW Hall and JH Byrd
printed in China

To Barbara, my wife, my friend, my companion, and all those folks who wish to know more about our wildflowers.
W.J.W.

To Tiia for her patience and forbearance.
D.W.H.

Table of Contents

Acknowledgements

May 5, 2009

   I wish to thank a few of the many folks who helped me get the transparencies of wildflowers used in this book. Ben Burton, Leesburg, stimulated my interest in identifying wildflowers and started my taking pictures of wildflowers 50 years ago. Dr. David Hall took over my wildflower education - verifying the correct species from specimens collected when the images were made. Dr. Jason Byrd for converting all the film images to digital images and helping to organize the book.

   My wife, Barbara, who traveled with me with books, pad, and pens, writing down what I took, and where I took it. She was my support person. Edna Ferrell, my sister-in-law in Tallahassee, guided me to many North Florida spots rich in wildflowers. Angus Gholson, Chatahoochie, who spent many days with me on many trips in all seasons chasing all over the Panhandle while capturing many elusive blooms. Dayton Wild helped get many of the wildflowers in North Carolina. I'm proud to call Angus and Dayton my friends.

   Dick Workman, President of Coastplan Inc. of Fort Myers, for giving of his time, support, and friendship as we sought some of the rare wildflowers of the lower west coast. Roger Hammer, naturalist with the Metro Dade County Park System, for guiding me to some of the best South Florida wildflowers. Chuck Salter, of the Madison Tree Farm in Madison, who showed me some of the north Florida blooms. Paula Benshoff, Park Ranger at Myakka River State Park, Nancy Bissett from the Native Plant Nursery in Davenport, Jim Conner of the St. Johns River Water Management District, Bob Cook and Barbara Grigg from our local Native Plant society, Lavon Silvernell at Trout Lake Nature Center, and a whole host of friends who helped me in so many ways. Thank you all. This is your book.

   William J. Weber

Acknowledgements

Taking The Pictures For This Book

I've enjoyed the challenge of putting what I see onto film since I was in high school. Now, even after having authored several books, provided cover pictures for over 100 national magazines, written articles for numerous magazines with my pictures; I still find the challenge of getting what I see on to a chip of film exciting. Taking images of wildflowers that show you what you need to know to identify the flower is even more challenging.

I started taking images of Florida wildflowers after I met Ben Burton in the 1960's. He loved wildflowers, loved learning to know them, and he was willing to share his enthusiasm with anyone interested. I liked Ben and I liked what he was willing to teach me about wildflowers. The variety of sizes, shapes, colors, and the way light and shadow influence colors all added to the task of producing a decent photo.

In 1992 I met Dr. David Hall, a botanist at the University of Florida. After some time we came to the agreement we could do a service to our friends in Florida by doing a book of wildflowers. Over the years during the times we were both working to make a living, we enjoyed adding to our images and proper descriptions of wildflowers of Florida and the southeast.

It is not necessary to be a botanist to learn to identify our wildflowers. Certainly any knowledge of the science is helpful, but enthusiasm and willingness to study the good books available are as important as formal knowledge.

We have chosen to separate the wildflowers by color for identification. This is what you see first, and it is the easiest starting place. While some flowers do occur in several colors, we choose to put the bloom in the color it is most commonly seen and then in the text describe the other color variations. There are some blooms that have more than one color on the same bloom, or multiple colored blooms on the same plant. In these cases start with the predominant color category, and if you don't find your flower, check the less obvious colors you see.

While there is great satisfaction in doing this book, the greatest satisfaction has been the wonderful people I have met while chasing the seasons and the wildflowers all over the state. Their enthusiasm for knowledge of the world we live in has inspired me also. I look forward to sharing with you some of the wildflowers of Florida and the southeast I have always found so exciting.

William J. Weber

Introduction

Introduction

1 April 2010

Although the emphasis of this regional wildflower treatment is Florida, many of the featured plants occur through the Gulf and Eastern Coastal Plains, particularly from North Carolina west into eastern Texas. Many of the featured plants have ranges that extend far beyond the Coastal Plain. Approximately 4,000 seed plants, including about 164 ferns and fern allies, occur within Florida's boundaries. This treatment contains descriptions and photographs of 768 plants. Some are shown more than once depending on the dominant flower colors. In the U.S., Florida has the third highest number of species behind California at number one (about 5,900) and Texas at number two (nearly 5,000). Other than the few endemics (those species that only occur in Florida) most of these plants spread beyond the Florida borders. A number of the plants do not occur in Florida at all, but can be found in the Southeastern Coastal Plain. The climate of the Coastal Plain can be described as warm temperate. Winter temperatures usually range from a few degrees below freezing to slightly above 100 degrees Fahrenheit in the summers. The southernmost portions of the Florida peninsula can be subtropical, seldom enduring freezing temperatures. Rainfall varies greatly from year to year with the greatest amounts typically in the spring and mid and late summer.

Most of the species in this treatment are native, however, many are introduced as a result of escape from cultivation or inadvertent weeds that follow man. The flowers of numerous species in this treatment are showy or even spectacular. Conversely, many of the species have tiny flowers that will evade the notice of anyone not specifically looking for them. Some of the colors blend into the surrounding vegetation making them difficult to find. Most of the plants in this treatment are widespread, but some are quite rare. Several plants are so few in number that they are listed as threatened or endangered by either Florida or both Florida and the U.S. governments. Listed species cannot be harvested, moved, or propagation materials taken without a permit. There is no restriction on photography. A few of the listed species are quite desirable for cultivation and are legally available from nurseries that have acquired the proper permits.

The areas where the plant grows are listed, together with the types of habitats in which it typically occurs, the frequency of occurrence, and the season of flowering. Within the Coastal Plain range of this treatment the frequency of occurrence of the plant can be exceedingly variable. Even though a species is listed as common, it may not occur in any particular area. Lots of species, particularly those that are weedy, can be locally common, even though of very infrequent general occurrence.

A few species within this treatment that are quite frequent to common in other areas of the Coastal Plain rarely occur in Florida. This type of distribution usually is exhibited by dissemination in a few counties in the Florida panhandle. Flowering can vary greatly. The gentle climate of most of peninsula Florida allows us to find some wildfowers in bloom in all seasons.

More than one common name is often listed. The first name in the list is judged to be the most frequently used or the most proper. Many plants have numerous common names. Some of the common names are a result of regional differences. When a common name is needed, it is often made up on the spot and has no more relevance to the plant than the author's immediate impression. It has been remarkable how many common names have been carried along with people moving from one area to another.

There has been an onslaught of names brought from more northern areas of the U.S. as our population has moved southward, especially in retirement. Quite frequently a common name known to the large number of recent immigrants will overwhelm an established regional common name. Weeds commonly have several common names as a result of different interests, such as row crop farmers, foresters, home owners, pesticide operators, lawn care personnel, horticulturists, etc. As a result of all the interests, many common names are unknown to others or conflict because another or several other plants have the same name. The scientific name listed for each species is the one unchangeable constant by which we can separate all plants.

This treatment is catalogued by flower color. Colors of the flowers within this treatment rely on our best judgement. We certainly understand the limitless variations and subtle gradations of color. Flowers that are light pink or light blue often appear to be white, and, in fact, are sometimes white. Purple is included in blue. We have placed flowers that exhibit more than one color into the most common color for the species. When trying to identify a flower using color, do not forget that a range of variation can exist. Flowers frequently fade into different colors. When possible we have added an insert showing a color variation. If the flowers are multicolored, we have used the dominant color which occasionally has lead to treatment in more than one color. Separately, we have also added an insert for some species that shows the fruit, as, often, the fruit is the showiest and most notable feature of the plant.

Virtually all of Florida, as is the case throughout the Coastal Plain, has been disturbed by man's activities, logging, farming, buildings, roads, insect control, alteration of natural processes (fire, drainage), air, soil and water pollution, etc. The cosmopolitan nature of our times has allowed horticultural access to the entire world.

Modern transportation has greatly accelerated the movement of cultivated plants and weeds. Immigration has brought the inevitable treasured kitchen garden and medicinal species to new areas of settlement. Usually, we do not think about the numbers of plants that now move as a result of curiosity by tourists and transportation crews or by contaminants of food and other imports. The Coastal Plain and, in particular, Florida's climate with its warmer southern tip, has provided a wonderful opportunity for many of the cultivated species brought into the U.S. to escape. It can be estimated that one-third of the plants growing in the wild in Florida are native to some other place. Some of us have lived in Florida long enough to have noted the gradual movement of some of the weedy species from a very limited area of introduction as they spread throughout the state and northward on the Coastal Plain.

The terrain of the Coastal Plain is quite flat with some, but not great, changes in elevation. Unlike other regions of the U.S., a few inches of elevation often signals a change of habitat. The changes can be very subtle with a corresponding gradual change in vegetation. The zone between vegetation types is called an ecotone. Ecotones are very important to certain wildlife. Ecotonal vegetation can vary from a foot or two wide to hundreds of feet. Ecotones can make the determination of a habitat type problematical. Wildflower ranges include certain particular habitats and the accompanying ecotones No single example of community types will exactly match all the specific characteristics listed for them. Wildflowers will sometimes occur in habitats that are not at all suitable for their growth, but they have found a microclimatic area and taken advantage of it.

Introduction

Ecological Communities

Common ecological communities that occur throughout the coastal plain are reviewed as follows. Adapted from the 1990 *Guide to the Natural Communities of Florida* prepared by the Florida Natural Areas Inventory and the Florida Department of Natural Resources (now Florida Department of Environmental Protection) currently under revision.

Coastal

The substrate and vegetation of <u>Coastal Habitats</u> is influenced by coastal processes such as erosion, deposition, salt spray, and storms. Salt spray has, perhaps, the greatest affect on vegetation.

Coastal Dunes

<u>Beach Dunes</u> are characterized as wind- and wave-deposited and are sparsely to densely vegetated with pioneer species. This habitat is found along high energy shorelines which deposit sand. Onshore winds move the sand until it is slowed or stopped by vegetation. The plants must grow upward through the subsequent depositions of sand. New beach is deposited seaward of the older dune and new dunes develop on the sea side of the old dune. This procedure results in a dune and swale topography. Dune soils are unstable and nutrient poor. Species of plants usually have special adaptations which enable them to survive and reproduce in this environment. Plants need to be able to move about in salt water, root from fragments, root upwards on the stem to survive being buried, or spread by means of surface or subsurface runners. Typical plants are Sea Oats (*Uniola paniculata*), various Cord Gasses (*Spartina* species), Sand Spurs (*Cenchrus* species), Beach Grass (*Panicum amarum*), Railroad Vine (*Ipomoea pes-caprae*), Purslane (*Portulaca oleracea*), Beach Morning-glory (*Ipomoea imperati*), Seashore Paspalum (*Paspalum vaginatum*), Beach Elder (*Iva imbricata*), Dune Sunflower

(*Helianthus debilis*), Sea Purslane (*Sesuvium portulacastrum*), and Sea Rocket (*Cakile lanceolata*).

Coastal Berm communities usually occur along low-energy coastlines as a ridge, often storm-deposited. This ridge is parallel to the shore. This maritime habitat usually grades into other typical seashore communities. Several plant associations can occur on these ridges of sand or shells, such as maritime hammocks, or collections of xerophytic species, including Coco-plum (*Chrysobalanus icaco*), Seagrape (*Coccoloba uvifera*), Marsh Elder (*Iva frutescens*), Greenbrier (*Smilax* species), Prickly-pear Cactus (*Opuntia humifusa*), Coral Dropseed (*Sporobolus domingensis*), Marsh-hay Cord Grass (*Spartina patens*), Wax-myrtle (*Myrica cerifera*), Live Oak (*Quercus virginiana*), Sand Live Oak (*Quercus geminata*), Muhly Grass (*Muhlenbergia capillaris*), Sea Purslane (*Sesuvium portulacastrum*), Groundsel Bush (*Baccharis halimifolia*), Sea Oats (*Uniola paniculata*), Beach Morning-glory (*Ipomoea imperati*), Sea Oxeye (*Borrichia frutescens*), Love Vine (*Cassytha filiformis*), Snowberry (*Chiococca alba*), Stoppers (*Eugenia* species), and Cherokee-bean (*Erythrina herbacea*).

Salt Flats are sometimes known as Coastal Grasslands. This flat land is characterized by grasses, prostrate vines, low herbaceous plants, low shrubs, and a lack of trees. Species include Gulf Muhly Grass (*Muhlenbergia sericea*), Bluestem Grasses (*Schizacnyrium* species), Sea Oats (*Uniola paniculata*), Saltmarsh Cord Grass (*Spartina alterniflora*), Beach Grass (*Panicum amarum*), Beach Morning-glory (*Ipomoea imperati*), Sea Oxeye (*Borrichia frutescens*), Beach Elder (*Iva imbricata*), Sea Purslane (*Sesuvium portulacastrum*), Glasswort (*Salicornia* species), Beach Sandspur (*Cenchrus tribuloides*), Seaside Evening-primrose (*Oenothera humifusa*), Pennywort (*Hydrocotyle* species), Seaside Ground-cherry (*Physalis angustifolia*), Sedges (various genera), Coral Dropseed (*Sporobolus domingensis*), Seashore Dropseed (*Sporobolus virginicus*), Coastal Prickly-pear Cactus (*Opuntia stricta*), Rushes (*Juncus* species), Wax-myrtle (*Myrica cerifera*), Groundsel Bush (*Baccharis halimifolia*), and Saltwort (*Batis maritima*). Storms can wash major deposits of sand, salt, and debris into the flats.

Coastal Scrub, also known as Coastal Strand, is composed of wind-deposited coastal sandy dunes vegetated with salt-tolerant shrubs. The woody vegetation is shaped by the wind with the windward side pruned by salt spray into a smooth, upward-slanting canopy. This is the result of salt spray blown by winds retarding the growth of buds on the sides of the plant directly in the path of the spray. Buds on the opposite side of the plant will continue to grow normally leading to the upward-slanting growth. Coastal Scrub often progresses into scrub or, more typically, into maritime hammock. Typical plants found in this habitat are Sand Live Oak (*Quercus geminata*), Saw Palmetto (*Serenoa repens*), Cabbage Palm (*Sabal palmetto*), Myrtle Oak (*Quercus myrtifolia*), Yaupon Holly (*Ilex vomitoria*), Seagrape (*Coccoloba uvifera*), Cat's-claw (*Pithecellobium unguis-cati*), Nakec-wood (*Colubrina elliptica*), Lantana (*Lantana camara*), Greenbrier (*Smilax* species), Buckthorn (*Rhamnus caroliniana*), Coco-plum (*Chrysobalanus icaco*), Nicker Bean (*Caesalpinia* species), Coin Vine (*Dalbergia ecastaphyllum*), Beach Jacquemontia (*Jacquemontia reclinata*), Pinweeds (*Lechea* species), Bay-cedar (*Suriana maritima*), Necklace-pod (*Scphora tomentosa*), Sea-lavender (*Limonium carolinianum*), Spanish-bayonet (*Yucca aloifolia*), and Florida Rosemary (*Ceratiola ericoides*).

Shell Mounds are also called Shell Middens. Shell Mounds are piles of shells, shell fragments, and aboriginal garbage. The typical habitat which develops on these Mounds is that of a hardwood hammock with a closed canopy.

Introduction

Occasionally, the community is sparsely vegetated with shrubs and/or cactus. Drainage is rapid. Salt spray, storm surges, and high winds affect the vegetation. Tropical species are often found extending their range further north on these coastal shell mounds due to the warming influence of the water. Typical plants are Live Oak (*Quercus virginiana*), Sand Live Oak (*Quercus geminata*), Red Cedar (*Juniperus virginiana*), Florida Privet (*Forestiera segregata*), Cherokee-bean (*Erythrina herbacea*), Coontie (*Zamia floridana*), Saffron-plum (*Sideroxylon celastrinum*), Hackberry (*Celtis laevigata*), Marlberry (*Ardisia escallonioides*), Gumbo-limbo (*Bursera simaruba*), Cabbage Palm (*Sabal palmetto*), False Mastic (*Sideroxylon foetidissimum*), Climbing Buckthorn (*Sageretia minutiflora*), and several kinds of cactus (Cactaceae).

Disturbed and/or Cultivated

Disturbed and/or Cultivated habitats are those created or extensively altered by man. A large portion of the current landscape has been disturbed by farming, roads, dwellings, businesses, or industrial sites with the resulting changes to the former habitats. Habitats disturbed by any activities often react to this disturbance by an invasion of primary and secondary species. These first species to colonize open areas are frequently weedy by nature. Most weeds are able to move their propagules quickly and extensively throughout the landscape. While each type of disturbance (old fields, spoil piles, vacant crop land, pastures, pine plantations, orchard crops, lawns, gardens, etc.) can have plants that are more likely to occur, the vast majority of the first invaders are among the most common weeds. Many of the common weeds are not native having been introduced by accident or, all too frequently, intentionally. Weedy colonizers can be herbaceous or woody. So much of the landscape around urban areas where the majority of the current population lives has been cleared that most people think the woody weeds are a natural "terminal" habitat. These woody weeds often provide the better places to develop, but clearing is resisted because of the mistaken belief that the area is undisturbed.

Mesic Flatwoods

Flatwoods

Flatwoods are pine forests with herbaceous and/or evergreen shrub ground cover (often dense) on flat, poorly drained sandy soils with a mixture of organic material and often with a hard pan within three feet of the surface. The soil pH is often very low. Typically, flatwoods have been bedded and planted with slash pine. Before the intervention by man, flatwoods burned periodically. The plants that occur in flatwoods are adapted to these periodic fires. Without fires, flatwoods will succeed into different types of habitats.

Mesic Flatwoods are pine forests on land that rarely flood, but have saturated soil during wet periods and low soil moisture throughout the root zone during droughts. The natural fire cycle of moderate intensity fires is every 1 to 4 years. Characteristic plants are Longleaf Pine (*Pinus palustris*), Slash Pine (*Pinus elliottii*), Saw Palmetto (*Serenoa repens*), Gallberry (*Ilex glabra*), Shiny Blueberry (*Vaccinium myrsinites*), Dwarf Huckleberry (*Gaylussacia dumosa*), and Wire Grass (*Aristida stricta*). Slash Pine has become the most common tree on most areas of this community due to fire protection and planting.

Scrubby Flatwoods are pine forests growing with shrubs on a layer of well drained sand that is on top of poorly drained, flat subsoil. The fire cycle of moderate to intense fires varies from every 2 to perhaps every 10 years. Characteristic plants are Longleaf Pine (*Pinus palustris*), Slash Pine (*Pinus elliottii*), Sand Live Oak (*Quercus geminata*), Myrtle Oak (*Quercus myrtifolia*), Chapman Oak (*Quercus chapmanii*), Saw Palmetto (*Serenoa repens*), Shiny Lyonia (*Lyonia lucida*), Huckleberry (*Gaylussacia* species), Crooked Wood (*Lyonia ferruginea*), Tarflower (*Befaria racemosa*), Flatwoods Pawpaw (*Asimina reticulata*), Scrub Hedge-hyssop (*Gratiola hispida*), and Penny-royal (*Piloblephis rigida*).

Wet Flatwoods are pine forests on seasonally flooded, sandy soils. Fires are somewhat less frequent and of low to moderate intensity, perhaps every 2 to 5 years. Characteristic plants are Slash Pine (*Pinus elliottii*), Pond Pine (*Pinus serotina*), Loblolly-bay (*Gordonia lasianthus*), Swampbay (*Persea palustris*), Gallberry (*Ilex glabra*), Shiny Lyonia (*Lyonia lucida*), Wax-myrtle (*Myrica cerifera*), Redroot (*Lachnanthes caroliniana*), and Virginia Chain Fern (*Woodwardia virginica*).

Forest

Forest habitats have soils with moderate to good fertility and moisture holding capacity. The soils have varying amounts of clay, phosphatic rock, and/or limerock in the subsoil and usually a fair amount of organic matter in the topsoils. Depending on both the fire history and the soil type, the forest can be a pine forest similar to sandhill or may be a tall dense hardwood forest.

Bayheads are sometimes called Baygalls. These are seepage wetlands that may be on the side or base of a slope on either inorganic or organic soil. They can be at the head of or beside a stream or tributary that provides good outflow drainage, but often are at the edge of a swamp. The soil is kept saturated by downslope or lateral seepage. Flooding does not occur or is very mild and fire is rare or infrequent. Characteristic plants are Blackgum (*Nyssa sylvatica* var. *biflora*), Loblolly-bay (*Gordonia lasianthus*), Sweetbay (*Magnolia virginiana*), and Swampbay (*Persea palustris*), with evergreen shrubs and several kinds of ferns common in the understory. Characteristic soils are acidic peats. Damage to the peat layers through fires, drought, or commonly by tree plantations will result in a change to a different community.

Introduction

St. John's River, Hwy. 44 Bridge

Bottomland Forest is a tall, dense floodplain forest adjacent to a stream with a defined channel. The stream rarely floods and then for only short periods. The soil is always moist due to a high water table, lateral seepage, and abundant organic matter in the soil. Fire rarely, if ever, occurs. Characteristic plants are Water Hickory (*Carya aquatica*), Water Locust (*Gleditsia aquatica*), Water Oak (*Quercus nigra*), Swamp Laurel Oak (*Quercus laurifolia*), Swamp Chestnut Oak (*Quercus michauxii*), Overcup Oak (*Quercus lyrata*), Spruce Pine (*Pinus glabra*), Loblolly Pine (*Pinus taeda*), Red Maple (*Acer rubrum*), Sweetbay (*Magnolia virginiana*), Sweetgum (*Liquidambar styraciflua*), Blackgum (*Nyssa sylvatica* var. *biflora*), Florida Elm (*Ulmus americana* var. *floridana*), Swamp Dogwood (*Cornus foemina*), and Blue-beech (*Carpinus caroliniana*).

Seepage Slope Forests are characterized by forested slopes with downslope seepage saturating but rarely inundating the soils. They generally occur where water percolates down through the sand to hit an impermeable layer. Instead of being in discrete pockets like bayheads, these forests often occur along the entire slope of a stream. Fire is rare, as fire would have the result of changing the forest to a shrub or herbaceous system. Characteristic plants include Blackgum (*Nyssa sylvatica* var. *biflora*), Water Oak (*Quercus nigra*), Sweetgum (*Liquidambar styraciflua*), Red Maple (*Acer rubrum*), Sweetbay (*Magnolia virginiana*), Wax-myrtle (*Myrica cerifera*), Pond Pine (*Pinus serotina*), Large Gallberry (*Ilex coriacea*), Dahoon Holly (*Ilex cassine*), Cinnamon Fern (*Osmunda cinnamomea*), Virginia Chain Fern (*Woodwardia virginica*), Yellow-eyed-grass (*Xyris* species), and various other Grasses (Poaceae) and Sedges (Cyperaceae).

Slope Forests are densely shaded upland hardwood/pine forest types that occur on moderate to steep slopes with a clay subsoil underneath. The adjacent uplands have a clay layer underneath that slows or stops water allowing a continuous supply of water to seep down the hill within reach of the plant roots. Fire is very rare and never intense. Characteristic plants are Southern Magnolia (*Magnolia grandiflora*), Beech (*Fagus grandifolia*), Spruce Pine (*Pinus glabra*), Shumard Oak (*Quercus shumardii*), Water Oak (*Quercus nigra*), Florida Maple (*Acer saccharum* var. *floridanum*), Sweetgum (*Liquidambar styraciflua*), and Basswood (*Tilia americana*).

Upland Pine Forest, when managed with fire, is an open forest of pine, deciduous oaks and hickory with a ground cover of grasses and wild flowers. It is very similar in appearance and ecology to sandhill habitat. Sandhill and upland pine forest have historically been placed together in one main category called high pine.

However, the soil in an upland pine forest is more fertile. Typically upland pine has some clay in the subsoil, or throughout the soil profile, and often with some limerock near the surface. The natural fire frequency and intensity is the same as in sandhill or slightly more frequent and hotter. Upland pine has greater fuel loads produced by higher rates of vegetation on the more fertile soil. Characteristic plants are Longleaf Pine (*Pinus palustris*), Southern Red Oak (*Quercus falcata*), Post Oak (*Quercus stellata*), Mockernut Hickory (*Carya tomentosa*), Chinquapin (*Castanea pumila*), Sassafras (*Sassafras albidum*), New Jersey Tea (*Ceanothus americanus*), Summer Haw (*Crataegus michauxii*), Rusty Black Haw (*Viburnum rufidulum*), and Wire Grass (*Aristida stricta*). Other common plants include Bluejack Oak (*Quercus incana*), Sand Post Oak (*Quercus margaretta*), and a great many herbaceous plants. With protection from fire, this community quickly becomes invaded by Upland Laurel Oak (*Quercus hemisphaerica*), Live Oak (*Quercus virginiana*), Water Oak (*Quercus nigra*), Sweetgum (*Liquidambar styraciflua*), Loblolly Pine (*Pinus taeda*), and many other hammock species.

Hammock
A Hammock is a broad-leaved evergreen or mixed evergreen-deciduous climax forest. These communities have mature soils and some protection from fire.

Hydric Hammocks are mixed forests of hardwoods, pine, red cedar, and/or cabbage palm on low, flat land. Soils are sand, clay or organic, often over limestone. Conditions are mesic to mildly hydric with some occasional flooding. Fire is infrequent and generally mild, although hammocks dominated by cabbage palm can produce and withstand very hot fires. Characteristic plants are Live Oak (*Quercus virginiana*), Water Oak (*Quercus nigra*), Swamp Laurel Oak (*Quercus laurifolia*), Cabbage Palm (*Sabal palmetto*), Red Cedar (*Juniperus virginiana*), Loblolly Pine (*Pinus taeda*), Florida Elm (*Ulmus saccharum* var. *floridanum*), Sweetgum (*Liquidambar styraciflua*), Red Maple (*Acer rubrum*), Sugarberry (*Celtis laevigata*), Sweetbay (*Magnolia virginiana*), Persimmon (*Diospyros virginiana*), Blue-beech (*Carpinus caroliniana*), Walter Viburnum (*Viburnum obovatum*), Green Haw (*Crataegus viridis*), Rattan Vine (*Berchemia scandens*), Greenbrier (*Smilax* species), and Trumpet Vine (*Campsis radicans*).

Maritime Hammock is also known as a Coastal Hammock. A Maritime Hammock is a narrow band of hardwood forest just inland of the coastal strand (coastal scrub) dune vegetation. The forest is dominated by Live Oak (*Quercus virginiana*), Sand Live Oak (*Quercus geminata*), Cabbage Palm (*Sabal palmetto*), Redbay (*Persea borbonia*), Swampbay (*Persea palustris*), and Southern Red Cedar (*Juniperus virginiana*). The wind-pruned canopy is formed, just as with the coastal strand, by the salt spray slowing or stopping new growth and effecting a smooth, upward-slanting canopy becoming a dense barrier that helps prevent damage to inland areas. The upward-slanting canopy also helps protect the habitat from wind damage. Typical vegetation is American Holly (*Ilex opaca*), Southern Magnolia (*Magnolia grandiflora*), Sea Grape (*Coccoloba uvifera*), False Mastic (*Sideroxylon foetidissimum*), Paradise Tree (*Simarouba glauca*), Lancewood (*Ocotea coriacea*), Gumbo-limbo (*Bursera glauca*), Strangler Fig (*Ficus aurea*), Poison Wood (*Metopium toxiferum*), Wild Olive (*Osmanthus americana*), Saw Palmetto (*Serenoa repens*), French Mulberry (*Callicarpa americana*), Poison-ivy (*Toxicodendron radicans*), Coral Bean (*Erythrina herbacea*), Coontie (*Zamia floridana*), Hercules'-club (*Zanthoxylum clava-herculis*), Wild Coffee (*Psychotria* species), Snowberry (*Chiococca alba*), Myrsine (*Rapanea punctata*), Caper Tree (*Capparis* species), and Marlberry (*Ardisia escallonioides*). This habitat develops on old coastal dunes. Soils are well-drained and accumulate organic matter at the surface.

Introduction

Fire is rare, occurring at 25 to 100 year intervals. A Maritime Hammock is a terminal successional stage. It can grade into coastal strand, scrub, or hydric hammock. Most of this formerly continuous community has been developed, leaving discontinuous strips and pockets.

San Felasco Hammock State Park

Mesic Hammock can be called Upland Mixed Forest, Upland Hardwoods, and Pine-oak-hickory. A Mesic Hammock is a tall, dense, closed canopy hardwood forest on level to moderately sloping terrain. Soils can be somewhat fertile and are mostly well drained. There is no flooding. Fire is rare and never intense. Characteristic plants are Pignut Hickory (*Carya glabra*), Upland Laurel Oak (*Quercus hemisphaerica*), Water Oak (*Quercus nigra*), Sweetgum (*Liquidambar styraciflua*), Swamp Chestnut Oak (*Quercus michauxii*), White Ash (*Fraxinus americana*), Basswood (*Tilia americana*), and Spruce Pine (*Pinus glabra*) in the overstory and Hop-hornbeam (*Ostrya virginiana*) in the understory. Many other plant species are commonly present including several kinds of vines such as Wild Grape (*Vitis* species), Poison-ivy (*Toxicodendron radicans*), and Virginia Creeper (*Parthenocissus quinquefolia*)
along with many shade tolerant herbaceous plants such as Violets (*Viola* species), Spike Grass (*Chasmanthium* species), Woods Grass (*Oplismenus hirtellus*), and Partridge Berry (*Mitchella repens*) in the ground cover.

Rockland Hammocks are also known as Tropical Hammocks. A Rockland Hammock is a hardwood hammock located on upland rockland sites. As with pine rocklands, limestone is very near or at the surface and is often exposed. This community is the final successional stage of pine rockland. It takes twenty five to one hundred years of protection from fire, severe storms, and drought. The forest has high species diversity as is exhibited by these typical plants Gumbo-limbo (*Bursera simaruba*), Wild Tamarind (*Lysiloma latisiliquum*), Stoppers (*Eugenia* species), Pigeon Plum (*Coccoloba diversifolia*), False Mastic (*Sideroxylon foetidissimum*), Poison Wood (*Metopium toxiferum*), West

Introduction

Indian Mahogany (*Swietenia mahagoni*), Inkwood (*Exothea paniculata*), Marlberry (*Ardisia escallonioides*), Lancewood (*Ocotea coriacea*), Strangler Fig (*Ficus aurea*), Wild Coffee (*Psychotria* species), Bustic (*Sideroxylon salicifolium*), Black Ironwood (*Krugiodendron ferreum*), Paradise Tree (*Simarouba glauca*), Satin Leaf (*Chrysophyllum oliviforme*), Swampbay (*Persea palustris*), Cabbage Palm (*Sabal palmetto*), Swamp Laurel Oak (*Quercus laurifolia*), Tallow-wood (*Ximenia americana*), Hercules'-club (*Zanthoxylum clava-herculis*), Hackberry (*Celtis laevigata*), Live Oak (*Quercus virginiana*), Guiana-plum (*Drypetes lateriflora*), Cat's-claw (*Pithecellobium unguis-cati*), Soapberry (*Sapindus saponaria*), Sea Grape (*Coccoloba uvifera*), Coffee Colubrina (*Colubrina arborescens*), Naked-wood (*Colubrina elliptica*), Geiger Tree (*Cordia sebestena*), Wild-pine Air-plant (*Tillandsia* species), Spanish-moss (*Tillandsia usneoides*), Coontie (*Zamia floridana*), and Poison-ivy (*Toxicodendron radicans*). This habitat is frequently found on somewhat higher ground that rarely floods. Water levels must remain near the surface to provide the high humidity needed. It is frequently found surrounded by wetlands; the distinctively rounded shape provides protection from the desiccating effects of normal and storm winds. The surrounding wetlands also serve as fire protection.

Xeric Hammock - Lake Wales Ridge

Xeric Hammock is an advanced successional stage of scrub or sandhill. This is a closed canopy forest with only moderate shade and an understory containing the characteristic shrubs listed below. The canopy can be low or high and open or closed. The variability is mostly due to the community from which it is derived. Fire is infrequent or rare. Fires are catastrophic and regressive, usually resulting in the return of the community that existed before the hammock developed. Characteristic plants include Sand Live Oak (*Quercus geminata*), Saw Palmetto (*Serenoa repens*), Crooked Wood (*Lyonia ferruginea*), Sparkleberry (*Vaccinium arboreum*), French-mulberry (*Callicarpa americana*), Scrub Beakrush (*Rhynchospora megalocarpa*), and Bracken Fern (*Pteridium aquilinum*). Other plants often present include Live Oak (*Quercus virginiana*), Upland Laurel Oak (*Quercus hemisphaerica*), Pignut Hickory (*Carya glabra*), Southern Magnolia

Introduction

(*Magnolia grandiflora*), Huckleberry (*Gaylussacia dumosa*), and Deerberry (*Vaccinium stamineum*).

Marsh
A Marsh is a habitat dominated by herbaceous species with the water table usually at or above the soil surface for much of the year.

Basin Marsh is a large fresh water marsh in a large basin, such as the well known Paynes Prairie in Alachua County, Florida, with peat on top of sand or clay. Inundation is for the greater part of the year. Fire is occasional but can be intense and is an important part of the ecology. Normal burn cycles of 1 to 3 years are necessary to maintain herbaceous growth, and 3 to 10 years cycles will result in the growth/invasion of Button Bush (*Cephalanthus occidentalis*) and Willows (*Salix* species). Flooding for about half of the year is required to maintain the marsh. More frequent flooding will result in a lake and a less frequent hydroperiod will lead to successional growth of shrubs and eventually trees. Characteristic plants are Maidencane (*Panicum hemitomon*), Pickerel-weed (*Pontederia cordata*), Sawgrass (*Cladium jamaicense*), Cut Grass (*Leersia* species), Cat-tail (*Typha* species), Primrose-willow (*Ludwigia* species), Lotus (*Nelumbo lutea*), Water-lily (*Nymphaea* species), Spatter-dock (*Nuphar lutea*), etc. Redroot (*Lachnanthes caroliniana*), Maidencane (*Panicum hemitomon*), Pickerel-weed (*Pontederia cordata*), Spatter-dock (*Nuphar lutea*), St. John's-worts (*Hypericum* species), Willows (*Salix* species), Primrose-willow (*Ludwigia* species), Seedboxes (*Ludwigia* species), Arrowheads (*Sagittaria* species), and Yellow-eyed-grasses (*Xyris* species). Depression marshes are typical of limestone regions and are considered of critical importance to many wetland and upland animals.

Depression Marsh is a small freshwater marsh in a rounded depression in sandy soil with peat accumulating toward the center. It is seasonally inundated with still water, but usually goes completely dry every year or at least periodically. Fire is frequent to occasional. Characteristic plants are Virginia Chain Fern (*Woodwardia virginica*), Redroot (*Lachnanthes caroliniana*), Maidencane (*Panicum hemitomon*), Pickerel-weed (*Pontederia cordata*), Spatter-dock (*Nuphar lutea*), St. John's-worts (*Hypericum* species), Willows (*Salix* species), Primrose-willow (*Ludwigia* species), Seedboxes (*Ludwigia* species), Arrowheads (*Sagittaria* species), and Yellow-eyed-grasses (*Xyris* species). Depression marshes are typical of limestone regions and are considered of critical importance to many wetland and upland animals.

Saltwater and Brackish Marshes are also known as Tidal Marshes. Saltwater and Brackish Marshes occur along coastlines with low wave-energy and within mouths of rivers. These communities is found mostly north of the freeze line in Florida and along the Gulf and southeastern coasts. They are characterized by Grasses (Poaceae), Rushes (*Juncus* species), and Sedges (Cyperaceae). The topography is very gradual to flat. Soils are poorly drained, high in sulfur, and high in organic content. The plants must be able to tolerate high stress conditions of poor soil aeration, high soil salt content, frequent submersion and exposure, bright sunlight, and periodic fires. Typical vegetation is Black Needle Rush (*Juncus roemerianus*), Smooth Cord Grass (*Spartina alterniflora*), Marsh-hay Cord Grass (*Spartina patens*), Gulf Cord Grass (*Spartina spartinae*), Salt Grass (*Distichlis spicata*), Soft Rush (*Juncus effusus*), Marsh Elder (*Iva frutescens*), Saltwort (*Batis maritima*), Sea Oxeye (*Borrichia frutescens*), Bulrush (*Scirpus* species), Seashore Dropseed (*Sporobolus virginicus*), Seashore Paspalum (*Paspalum vaginatum*), Shore Grass (*Monanthochloe littoralis*), Glasswort (*Salicornia* species), Seablight (*Suaeda linearis*), and Saltmarsh Fleabane (*Pluchea odorata*).

Introduction

Gulf Coast

Pond and Lake

Ponds and Lakes contain non flowing, open water in natural depressions and are generally lacking emergent vegetation except around the perimeter. They can be with or without surface inflows or outflows. Some natural ponds and lakes occur within marshes or wooded wetlands, such as floodplains or flatwoods. Water chemistry is variable.

Man-made ponds are extremely common throughout. Often the ponds are created for agriculture, but just as commonly for enjoyment as a landscape feature. Development that has hardscape features such as buildings, parking lots, and roads are required to provide retention ponds to hold amounts of run off water to help prevent erosion, pollution, and flooding. Lakes can also be man made. They are usually created by constructing dams within streams or rivers. A few lakes are built to handle water at power plants or more frequently by digging large holes to provide fill soils and then allowing the hole to fill with water.

Prairie

Prairies are flat communities usually with a subsurface hard pan and soils having an organic component. They are often described as flatwoods without pines.

Dry Prairies are nearly treeless, contain a dense ground cover, and are maintained by frequent fire. The frequent fires limit the establishment of most trees. Typical plants are Wire Grass (*Aristida stricta*), Saw Palmetto (*Serenoa repens*), Broom Grass (*Andropogon* species), Carpet Grass (*Axonopus* species), Runner Oaks (*Quercus minima, Quercus pumila*), Indian Grass (*Sorghastrum* species), Love Grass (*Eragrostis* species), Blazing-star (*Liatris* species), Blackroot (*Pterocaulon pycnostachyum*), Catesby Lily (*Lilium catesbaei*), Marsh Pink (*Sabatia* species), Milkwort (*Polygala* species), Goldenrod (*Solidago* species), Musky Mint (*Hyptis alata*), Pawpaw (*Asimina* species),

Introduction

Dwarf Wax-myrtle (*Myrica cerifera* var. *pumila*), Gallberry (*Ilex glabra*), Stagger Bush (*Lyonia* species), and Shiny Lyonia (*Lyonia lucida*). Soils typically consist of a few inches or feet of sands above a hard pan layer of denser fine-particle soils. The hard pan slows water drainage and, during periods of heavy rains, will cause the habitat to be flooded for short periods. Dry Prairie, true to its name, is generally well-drained. This community differs from mesic flatwoods only by the lack of trees. Trees in Dry Prairie occur at a rate of less than one per acre. Wet prairie and mesic flatwoods are closely associated with one frequently grading into the other.

Marl Prairies occur in south Florida where limestone is near the surface. They are at the interface between deeper water and upland communities. Marl Prairies are seasonal freshwater marshes with marl that is derived from periphyton, masses of algae and other minute organisms forming a highly alkaline, fine, gray or white mud varying from a few inches to a few feet thick. Dominant plants are Sawgrass (*Cladium jamaicense*), Muhly Grass (*Muhlenbergia capillaris*), Spikerush (*Eleocharis* species), Bluestem (*Schizachyrium* species), Beakrush (*Rhynchospora* species), Shore Grass (*Monanthochloe littoralis*), and Pond-cypress (*Taxodium ascendens*). Other species are typical of shallow wetland habitats and include Fragrant Water-lily (*Nymphaea odorata*), Pickerel-weed (*Pontederia cordata*), Aster (*Aster* species), Cut Grass (*Leersia* species), Panic Grass (*Panicum* species), Black Sedge (*Schoenus nigricans*), Sand Cord Grass (*Spartina bakeri*), Colicroot (*Aletris* species), Milkwort (*Polygala* species), White-top Sedge (*Rhynchospora* species), Cowhorn Orchid (*Cyrtopodium punctatum*), and Fire Flag (*Thalia geniculata*). This unusual habitat is sparsely vegetated so much so that fire will not usually burn across it. The Pond-cypress trees are often very small, very old, and quite distorted and picturesque. The seasonally flooded soils have a variable hydroperiod and can be rock hard when dry and quite soft and slippery when wet.

Lake Kissimmee Prairie

Introduction

Wet Prairies are small to vast expanses of grassland or shallow marsh occupying flat basins on sandy soils that often have a substantial clay or organic component. Flat, shallow wetlands of herbaceous vegetation in pine flatwoods areas are also included in this category. Under natural conditions, both fire and flooding are frequent. The most characteristic plant is Maidencane (Panicum hemitomon), but most prairies have been so altered by drainage, fire protection, cattle grazing, mowing, fertilizing sewage effluent, etc. that a large assortment of native and exotic weedy plants often dominate. Karst prairies usually surround a basin marsh that occupies the center of the basin, whereas flatwoods prairies may or may not have deeper areas of marsh.

Rockland

Rocklands are biological communities having limerock exposed at the surface. Two types of rocklands are most frequently encountered.

Limestone Glade Rocklands are also known as Chalky Barrens or Upland Glaces. These limestone glades are herbaceous openings in forests dominated by grasses and sedges. The soils are calcareous with areas of exposed limestone. The glades can be located on dry uplands or on hillsides subject to water flow across the surface. The vegetation is typically composed of Florida Diamond-flower (*Hedyotis nigricans* var. *floridana*), Beakrush (*Rhynchospora* species), Black Sedge (*Schoenus nigricans*), Florida White-top Sedge (*Rhynchospora floridensis*), Panic Grass (*Panicum* species), Broom Grass (*Andropogon* species), and Wire Grass (*Aristida stricta*). Stunted woody species typical of secondary woods such as Wax-myrtle (*Myrica cerifera*), Black Cherry (*Prunus serotina*), Winged Sumac (*Rhus copallinum*), Red Cedar (*Juniperus virginiana*), Persimmon (*Diospyros virginiana*), and Sweetgum (*Liquidambar styraciflua*) are scattered in this habitat. Some quite rare species can be found in these glades.

Pine Rockland is a flat habitat with an open canopy of Slash Pines and a scattered understory of shrubs, palms, and herbaceous species. Many exposed patches of pitted and cracked oolitic limestone are at the surface. Typical plants are South Florida Slash Pine (*Pinus elliottii* var. *densa*), Saw Palmetto (*Serenoa repens*), Cabbage Palm (*Sabal palmetto*), Silver Palm (*Coccothrinax argentata*), Gallberry (*Ilex glabra*), Velvet Seed (*Guettarda* species), Blolly (*Pisonia aculeata*), Locust Berry (*Byrsonima lucida*), Myrsine (*Rapanea punctata*), Tetrazygia (*Tetrazygia bicolor*), Varnish Leaf (*Dodonaea viscosa*), Marlberry (*Ardisia escallonioides*), Indigo Berry (*Randia aculeata*), Poison Wood (*Metopium toxiferum*), Bustic (*Sideroxylon salicifolium*), Live Oak (*Quercus virginiana*), Stopper (*Eugenia* species), Winged Sumac (*Rhus copallinum*), Satin Leaf (*Chrysophyllum oliviforme*), Wild Tamarind (*Lysiloma latisiliquum*), Rubber Vine (*Echites umbellata*), Snowberry (*Chiococca alba*), Broom Grass (*Andropogon* species), Wire Grass (*Aristida stricta*), Muhly Grass (*Muhlenbergia capillaris*), Partridge Pea (*Chamaecrista* species), Coontie (*Zamia floridana*), and Pine Fern (*Anemia adiantifolia*). Pine rockland covers a limited area in the southeast peninsula. Soil is usually restricted to crevices and solution holes in the limestone. The community must be maintained by fire. Many warm temperate and tropical species are found in this habitat as well as many threatened and endangered plants. The very infrequent freezing temperatures can easily kill many of the plants in this open environment.

Sandhill

Sandhill is a fire climax community and when managed with fire, it is a very open, sunny forest of pine and deciduous oak with a grass and wild flower ground cover. The low intensity ground fires occurred naturally every 1 to 3 years before the advent of

Introduction

fire suppression. The community occurs on hilltops and slopes of rolling hills. The soils are composed of marine-deposited, well-drained, porous sands that function as important aquifer recharge areas. Characteristic plants are Longleaf Pine (*Pinus palustris*), Turkey Oak (*Quercus laevis*), Wire Grass (*Aristida stricta*), Pineywoods Dropseed (*Sporobolus junceus*), Blazing-star (*Liatris* species), and Bracken Fern (*Pteridium aquilinum*). There are several hundred other plant species that can occur in the ground cover of this community. Currently, the typical condition in Florida is represented by the conversion to a slash pine plantation with remnants of Sandhill ground cover flora, a few Turkey Oaks, and scattered gopher tortoises (*Gopherus polyphemus*).

Scrub

Scrub has very well-drained, open sands with poor water and nutrient holding capacity. It supports xeric-adapted vegetation. Scrub is a thicket of evergreen shrubs and small trees. It often has an overstory of sand pine and has little in the way of ground cover except for lichens. Fire is necessary to maintain this habitat. Fires often develop as very severe crown fires that occur on an erratic and variable cycle ranging from 5 to 100 years. Scrub occurs on sand ridges which are former shorelines. The loose sands are deep and drain rapidly creating xeric conditions. The sands are usually bright white and can be easily identified from aerials. Due to the sparse ground cover, little leaf litter occurs thus limiting fires. Characteristic plants include Sand Pine (*Pinus clausa*), Sand Live Oak (*Quercus geminata*), Myrtle Oak (*Quercus myrtifolia*), Chapman Oak (*Quercus chapmanii*), Crooked Wood (*Lyonia ferruginea*), Florida Rosemary (*Ceratiola ericoides*), Saw Palmetto (*Serenoa repens*), Scrub Palmetto (*Sabal etonia*), and deermoss (terrestrial lichens). Scrub sustains about 50 plant species that only grow in this habitat.

Seepage

Seepage Habitats are sloped or flat sands or peat with very stable, constant high moisture levels maintained by downslope or lateral seepage combined with good outlets for drainage. Vegetation can be dominated by either forest, shrubs, or herbs, depending on fire history and the volume per unit area of seepage.

Bogs occur on deep peat kept saturated by lateral seepage and capillary action from below. There is little fluctuation in water levels, and the plant growth is floating on the surface, so that it goes up and down to some extent with the water level. Soils are acidic. Fire is quite variable, from none to occasional, and can be severe. A severe or catastrophic fire burning into the peat may destroy the habitat. Characteristic plants are Sphagnum Moss (*Sphagnum* species), Sundews (*Drosera* species), Shiny Lyonia (*Lyonia lucida*), Maleberry (*Lyonia ligustrina*), Gallberry (*Ilex few glabra*), Highbush Gallberry (*Ilex coriacea*), Wax-myrtle (*Myrica cerifera*), Odorless Wax-myrtle (*Myrica inodora*), Bamboo Vine (*Smilax laurifolia*), Loblolly-bay (*Gordonia lasianthus*), White Cedar (*Chamaecyparis thyoides*), Cocoplum (*Chrysobalanus icaco*), Sweet Pepper Bush (*Clethra alnifolia*), Possum Haw (*Viburnum nudum*), Highbush Blueberry (*Vaccinium corymbosum*), Virginia-willow (*Itea virginica*), and sometimes Slash Pine (*Pinus elliottii*). Other plants often present are Pond-cypress (*Taxodium ascendens*), Blackgum (*Nyssa sylvatica* var. *biflora*), Pitcher Plant (*Sarracenia* species), and Highbush Blackberry (*Rubus argutus*).

Pitcher Plants in Bog or Savanna

Seepage Slopes are frequently called Shrub Bogs. They occur where massive downslope or lateral seepage continually comes to the surface on the side of a hill or at the head of a small stream. The soil is almost always very wet but never floods to the point that water stands above the surface. Fire can be frequent, occasional, or rare. Characteristic plants include Sphagnum Moss (*Sphagnum* species), various Ferns, Shiny Lyonia (*Lyonia lucida*), Gallberry (*Ilex glabra*), Titi (*Cyrilla racemiflora*), Poison Sumac (*Toxicodendron vernix*), Wild Azalea (*Rhododendron canescens*), Swamp Azalea (*Rhododendron viscosum*), Highbush Blueberry (*Vaccinium corymbosum*), Maleberry (*Lyonia ligustrina*), Virginia-willow (*Itea virginica*), Chokeberry (*Photinia pyrifolia*), Elderberry (*Sambucus canadensis*), Highbush Blackberry (*Rubus argutus*), Dahoon Holly (*Ilex cassine*), Sweet Pepper Bush (*Clethra alnifolia*), Wax-myrtle (*Myrica cerifera*), Odorless Wax-myrtle (*Myrica inodora*), Tulip Popular (*Liriodendron tulipifera*), and Bamboo Vine (*Smilax laurifolia*). Many other interesting plant species occur in this habitat in Clay County (north peninsula Florida) and in the Florida panhandle.

Streams and Rivers

Streams and Rivers are natural flowing waters, bounded by channel banks, and are located upstream of tidal influences.

Alluvial Streams and Rivers can be perennial, intermittent, or seasonal. They are characterized by turbid water with suspended silt, clay, sand, and small gravel. Generally, they have a distinct, sediment-derived (alluvial) floodplain and a sandy, elevated natural levee just inland from the bank. Emergent vegetation is scarce due to the flow rate and turbidity of the water. A few plants occur in quiet zones. Banks are lined with Willows (*Salix* species), Cottonwood (*Populus* species), River Birch (*Betula nigra*), Silver Maple (*Acer saccharinum*), and other wetland shrubs and trees.

Introduction

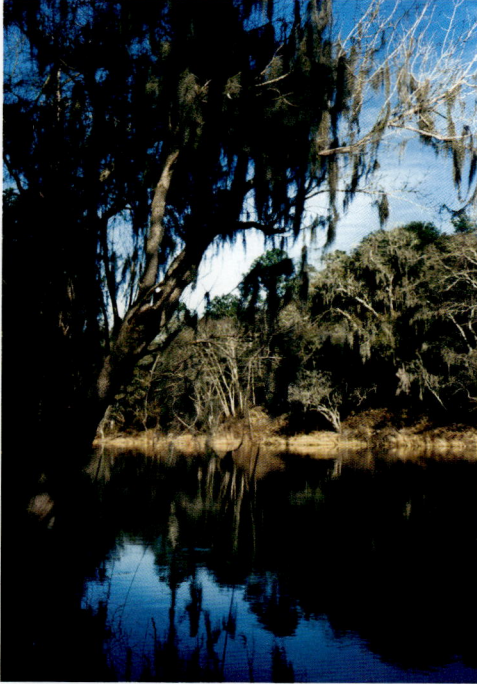

Suwannee River

<u>Blackwater Streams and Rivers</u> are perennial, intermittent, or seasonal water courses. They are characterized by tea-colored water with a high content of particulate and dissolved organic matter derived from drainage through swamps and marshes. Usually, they lack an alluvial floodplain. Most banks are steep leading to a general lack of vegetation. The dark water inhibits plant growth in the stream.

<u>Seepage Streams</u> are perennial, intermittent, or seasonal. They are characterized by clear to lightly colored water derived from shallow groundwater seepage. Most Seepage streams occur in areas of more topographic relief. These streams usually occur in densely shaded habitats which greatly limit the vegetation.

<u>Spring Run </u>is a perennial water course with deep aquifer headwaters. They are characterized by clear water, circumneutral pH and, frequently, a solid limestone bottom. Characteristic vegetation is Tape-grass (*Vallisneria americana*), Arrowheads (*Sagittaria* species), Wild Rice (*Zizania aquatica*), Giant Cut Grass (*Zizaniopsis miliacea*), Southern Naiad (*Najas* species), Pondweed (*Potamogeton* species), and Chara (*Chara* species).

Swamp

A <u>Swamp</u> habitat is dominated by trees with the water table usually at or above the soil surface for much of the year.

<u>Basin Swamps</u> are in large basins with peat substrate that have extended or seasonal hydroperiods of still water. Soils are generally acidic, poor in nutrients, and occur over an impervious zone, such as a layer of clay. Fire is occasional to rare. Characteristic plants are Pond-cypress (*Taxodium ascendens*), Blackgum (*Nyssa sylvatica* var. *biflora*), Red Maple (*Acer rubrum*), Virginia-willow (*Itea virginica*), and Shiny Lyonia (*Lyonia lucida*). Basin swamps strongly dominated by Blackgum are often called gum swamps. Green Ash (Fraxinus pennsylvanica) may also be present, and Slash Pine (Pinus elliottii) and various bay trees often occur in shallow areas or on the edge.

<u>Dome Swamps</u> are usually called Cypress Domes. These small, somewhat rounded depressions within pine Flatwoods forests often occupy 20 to 30 percent of the pine flatwoods region. These swamps are dominated by Pond-cypress trees. The tallest trees are often in the center of the dome and tree height decreases gradually toward the edge.

There is usually a well developed understory of shrubs or a herbaceous ground cover of depression marsh plants or both. The soil is usually sand underlain with a clay lens, and often has some peat accumulating toward the center. These depressions remain inundated with still, soft, acidic water for several months to most of the year. Fire varies from rare to frequent. Characteristic trees are Pond-cypress (*Taxodium ascendens*), Blackgum (*Nyssa sylvatica* var. *biflora*), and Slash Pine (*Pinus elliottii*), with some Sweetbay (*Magnolia virginiana*), Swampbay (*Persea palustris*), and/or Loblolly-bay (*Gordonia lasianthus*) around the edge. Shiny Lyonia (*Lyonia lucida*) is usually the dominant shrub, with various concentrations of Dahoon Holly (*Ilex cassine*), Pond-apple (*Annona glabra*), Titi (*Cyrilla racemiflora*), Virginia-willow (*Itea virginica*), and Primrose-willows (*Ludwigia* species). Characteristic ground covers are Seedboxes (*Ludwigia* species), Virginia Chain Fern (*Woodwardia virginica*), Netted Chain Fern (*Woodwardia areolata*), Redroot (*Lachnanthes caroliniana*), and Maidencane (*Panicum hemitomon*).

Withlacoochee River - Central Florida

Floodplain Swamps occur in low areas along rivers and streams. Flooding duration is usually longer than six months. Fire occurs rarely or never. Characteristic plants are Bald-cypress (*Taxodium distichum*), Blackgum (*Nyssa sylvatica* var. *biflora*), Ogeechee Tupelo (*Nyssa ogeche*), Water Tupelo (*Nyssa aquatica*), Pumpkin Ash (*Fraxinus profunda*), Green Ash (*Fraxinus pennsylvanica*), Dahoon Holly (*Ilex cassine*), Myrtle Holly (*Ilex myrtifolia*), Possum Haw (*Viburnum nudum*), Alder (*Alnus serrulata*), Wax-myrtle (*Myrica cerifera*), Cabbage Palm (*Sabal palmetto*), Red Maple (*Acer rubrum*), Lizard's-tail (*Saururus cernuus*), and various ferns.

Introduction

Lake Shore Swamps consist of built up peat deposits along shores of lakes that often form distinct swamps dominated by Pond- or Bald-cypress. Fires seldom extend into these deposits. Characteristic plants include Pond-cypress (*Taxodium ascendens*) and Bald-cypress (*Taxodium distichum*), with Pennywort (*Hydrocotyle* species), Soft Rush (*Juncus effusus*), Wax-myrtle (*Myrica cerifera*), Button Bush (*Cephalanthus occidentalis*), low Panic Grasses (*Panicum* species), and Spikerushes (*Eleocharis* species).

Red Mangrove - Merritt Island National Wildlife Refuge

Mangrove Swamps are also known as Tidal Swamps. Mangrove Swamps are found along relatively flat, tidal shorelines of low wave energy in south Florida. Soils are usually saturated with brackish water and inundated at high tides. These dense, low forests are dominated by Black Mangrove (*Avicennia germinans*), Red Mangrove (*Rhizophora mangle*), White Mangrove (*Laguncularia racemosa*), and Button Wood (*Conocarpus erectus*). Red Mangrove occurs in the lowest zones, Black Mangrove in the intermediate zone, and White Mangrove and Button Wood in the highest zone. The unusual roots of the three mangroves (prop roots of red, pneumatophores of black, dense root mats of white) serve to trap sediments. Mangrove swamps require an average water temperature of 66 degrees Fahrenheit. Low or widely fluctuating temperatures will decrease survival. Mangrove populations occur from Cedar Key south on the west coast of Florida and from Daytona Beach south on the east coast. A few Mangroves will germinate and occasionally grow for a short while north of these limits when winter temperatures are warmer than normal.

Introduction

<u>Shrub Swamps</u> are found in basins with peat substrate that is seasonally inundated with still water. The vegetation is shrubs that can withstand an extended hydroperiod. Fire is occasional to rare. Characteristic plants are Elderberry (*Sambucus canadensis*), Wax-myrtle (*Myrica cerifera*), Button Bush (*Cephalanthus occidentalis*), and/or Willow (*Salix* species).

<u>Strand Swamps</u> are linear cypress/hardwood swamps in an elongated depression that has flowing water most of the year, but no definite channel. The largest trees are found in the center where the most water occurs. Soils are usually deep peats, but also can be peats and sands over limestone. Fire is essential to prevent hardwood invasion and conversion to a bottomland forest. Characteristic plants are Bald-cypress (*Taxodium ascendens*) or Pond-cypress (*Taxodium distichum*), Blackgum (*Nyssa sylvatica*), Green Ash (*Fraxinus pennsylvanica*), Pumpkin Ash (*Fraxinus profunda*), Red Maple (*Acer rubrum*), Sweetbay (*Magnolia virginiana*), Swamp Laurel Oak (*Quercus laurifolia*), Coastal Plain Willow (*Salix caroliniana*), Button Bush (*Cephalanthus occidentalis*), Swamp Dogwood (*Cornus foemina*), Wax-myrtle (*Myrica cerifera*), Cabbage Palm (Sabal palmetto), Strangler Fig (Ficus aurea), Sweetbay (Magnolia virginiana), Myrsine (*Rapanea punctata*), Swampbay (*Persea palustris*), Poison-ivy (*Toxicodendron radicans*), and Leather Fern (*Acrostichum danaeifolium*).

Red

Acer rubrum L.
Red Maple
Aceraceae, Maple Family

Habit: Tree to 28 m.

Leaves: Opposite, papery, 3- to 5-lobed, green on upper surface and white underneath, pointed at tips, with teeth along margins. Leaf veination and stalks usually reddish.

Flowers: In clusters on slender stalks, seen before leaves appear in spring. Flowers small and reddish or sometimes yellow, but very showy in massed clusters with 5 short petals (1.5 to 2 mm long).

Fruit: Dry, usually reddish brown but with green, yellow, and brown variants, with 2 papery wings at an angle, each wing 1.5 -2.5 cm long.

Habitat and Distribution: Common in swamps and wet woods, less frequent in upland woods, throughout eastern North America, from eastern Canada to eastern Texas and southern Florida.

Comment: This eye catching deciduous tree is an excellent cultivated specimen. The showy flowers and fruits appear in the very early spring often during the first warm period of a day or two. The medium size tree is often multiple trunked and the light bark and red veined leaves are attractive during the year. Leaves produce bright red to yellow colors in the autumn.

Aesculus pavia L.
Red Buckeye
Hippocastanaceae, Horse Chestnut Family

Habit: Shrub or small tree, 1-12 m tall.

Leaves: The opposite, palmately compound leaves have 5-7 leaflets which are 6-17 cm long and 3-6 cm wide with fine marginal teeth.

Flowers: Spreading clusters of scarlet flowers, each 2-4 cm long, are borne at the tip.

Fruit: Large, round capsule, 3-6 mm in diameter.

Habitat and Distribution: Infrequent in woods and along swamp margins, from central Florida west to Texas, and north to the Carolinas, and Virginia.

Comment: The distinctive deciduous, opposite, palmately compound leaves make this shrub attractive all year, but not much noticed. As a wildflower this plant usually blends into the background except when in bloom. The flowers in early to mid spring are showy. Best flowering is attained in full sun, but it is adapted to shady areas. When cultivated it is better placed along a border or mixed with low shrubs. The hard fruit capsules usually contain one to three, hard, brown seeds. The seeds are quite poisonous! These seeds are sometimes polished and called "lucky nuts."

Acer rubrum L. - flowers

Acer rubrum L.

Aesculus pavia L.

Red

Aguilegia canadensis L.
Wild Columbine, American Columbine, Canadian Columbine
Ranunculaceae, Crowfoot Family

Habit: Perennial herb, to 80 cm tall.

Leaves: Compound leaves smooth, light green, divided 2 to 3 times, with rounded lobes. Leaflets 5 cm long and wide.

Flowers: The pendulous flowers are scarlet, bronze, and yellow. Petals are yellow tipped, and tube-shaped, 20-35 mm long, with scarlet spurs.

Fruit: Five-parted pod with small, oval, shiny, black seeds.

Habitat and Distribution: Found in rich often rocky woods throughout eastern North America, from Canada to Texas and north Florida.

Comment: As a wildflower Wild Columbine is a favorite with many homeowners. The lacy foliage is very attractive, especially when the plants are placed in drifts. The quite showy flowers appear in spring. The seeds can be spread among and around existing plants to achieve the effect of drifts or scattered into suitable habitats for surprise wildflowers.

Asclepias lanceolata Walter
Red Milkweed
Asclepiadaceae, Milkweed Family

Habit: Perennial herb with a single stem 0.4-1.6 m tall.

Leaves: A few pairs of opposite leaves, 1-25 cm long and 0.5 to 1 cm wide.

Flowers: Borne in a few, flat-topped clusters toward ends of branches that are 2-5 cm across, each with 3-8 flowers. The flowers are red-orange with petals that curve down and outwards, each 8-11 mm long. Red or orange hood-like petals occur in the center of flower, with horn-like protrusions.

Fruit: Smooth, erect pods, 7-10 cm long, 1 cm wide.

Habitat and Distribution: Occurs in fresh and brackish wet sites and swamps, from southern Florida west to Texas and north to New Jersey.

Comment: Red Milkweed makes an outstanding garden plant as well as an eye catching wildflower. It usually becomes somewhat leggy and is best mixed with plants shorter in stature that will soften the effect of the bare lower stems. The bright flowers can be seen all the warm months and attract butterflies. The juice of this species is milky and thought to be poisonous.

Aguilegia canadensis L.

Asclepias lanceolata Walter

Red

Asimina parviflora (Michx.) Dunal
Hammock Pawpaw
Annonaceae, Custard-apple Family

Habit: Perennial shrub or small tree, to 6 m tall.

Leaves: Thin, alternate, and oval. Blades to 18 cm long and 10 cm wide, with rust colored hair below.

Flowers: The maroon flowers smell fetid and are composed of 3 pointed, fleshy outer petals 1 cm long, and 3 smaller, fleshy, inner petals.

Fruit: 3-6 cm long, yellowish and smooth when ripe.

Habitat and Distribution: Found in wooded areas of the Coastal Plain, from central Florida west to Mississippi and north to Virginia.

Comment: As a wildflower this deciduous plant has inconspicuous, bad smelling flowers in the spring, but the plant is attractive for understory use. The ripe fruit pulp in late summer is edible both raw and cooked. Eating the fruit is toxic to some individuals causing vomiting, gastro intestinal pains, headaches, or dermatitis.

Calamintha coccinea (Nutt.) Benth.
Red Basil
Lamiaceae (Labiatae), Mint Family

Habit: Perennial shrub to 1 m tall.

Leaves: The small, opposite, narrow leaves are 0.5-2 cm long and hairy below, with smooth often rolled edges.

Flowers: Emerge singly towards ends of branches. Funnel shaped, scarlet or rarely yellow flowers are 3-5 cm long. Seeds occur in a cluster of 4 little nuts.

Habitat and Distribution: Found in dry sites, pine ridges, and scrub from central Florida to southern Georgia and Mississippi.

Comment: Flowers of Red Basil are usually easily seen as this shrub grows in open dry habitats. Leaves have a minty odor when crushed. The bright red flowers occurring from spring into autumn are showy, especially as the tardily deciduous leaves fall from the plant. This species can be cultivated in dry upland sites.

Asimina parviflora (Michx.)

Calamintha coccinea (Nutt.) Benth.

Red

Calycanthus floridus L.
Sweet Shrub
Calycanthaceae, Calycanthus Family

Habit: Deciduous shrub, to 3 m tall.

Leaves: The opposite leaves are dark green on the upper surface and white underneath, oval, smooth-sided, and aromatic.

Flowers: Distinctive red, fragrant flowers, 3 cm wide, with numerous petals are borne on ends of branches.

Fruit: A cluster of cup shaped carpels each having one seed.

Habitat and Distribution: Found in rich woodlands, on ravine slopes and streamsides, from northern Florida to Virginia and Kentucky, and west to Mississippi primarily in the coastal plain and piedmont.

Comment: Sweet Shrub is as good as its name in that the dried flowers were used to "sweeten" closets and drawers. The dark red, multi-petaled, sweet smelling flowers in the spring and opposite, oval leaves make this species easy to identify. The shrub is clonal, spreading by root shoots, making dense clumps. It is an attractive cultivated specimen, but requires room to spread. This species is listed as Endangered by the state of Florida.

Castilleja coccinea (L.) Spreng.
Indian Paintbrush
Scrophulariaceae, Figwort or Snapdragon Family

Habit: Biennial or sometimes annual herb, 20-60 cm tall.

Leaves: The 3- to 5-lobed, elongated leaves form a rosette at the base then are arranged sparsely along simple, ascending, hairy stems.

Flowers: Bright, showy, scarlet to yellow bracts are in clusters at ends of the stems and are to 4 cm long, often with 3 lobes. The actual flowers surrounded by these bracts are smaller, tubular, and pale yellow.

Fruit: Straw colored, angular seeds are found in a 1 cm long capsule.

Habitat and Distribution: Occurs in meadows, bogs, grassy woods, pastures and along roadsides and ledges. Uncommon in northern Florida, more common from Georgia to Manitoba and New Hampshire, and west to Oklahoma.

Comment: Indian Paintbrush is one of the most spectacular wildflower plants in the United States. Occurring in large to small populations, the bright shows of color in spring often spread into the distance and sparkle from distant slopes and rock ledges.

Calycanthus floridus L.

Castilleja coccinea (L.) Spreng

Red

Clerodendrum thomsoniae Balf.
Bleeding heart
[*Clerodendrum thompsoniae* Balf. f.]
Verbenaceae, Verbena Family

Habit: A twining, smooth, evergreen shrub, with vine like growth.

Leaves: The oblong to ovate, opposite leaves are to 15 cm long, with smooth margins.

Flowers: The red and white flowers are in clusters.

Fruit: A drupe enclosed by the remains of the flower.

Habitat and Distribution: Escaped from cultivation and found in mostly disturbed areas in central and south Florida.

Comment: Native to West Tropical Africa this warm weather vine can twine without supports, but most frequently is placed along a trellis or other backing. The large clusters of showy red and white flowers occur mostly in the summer. It prefers well drained fertile soils in light or partial shade.

Dicliptera sexangularis (L.) Juss.
Crimson Dicliptera
Acanthaceae, Acanthus Family

Habit: Perennial, erect and usually diffusely branching herb to 1 m tall.

Leaves: The opposite, smooth margined, elliptic to lanceolate leaves are 2 -10 cm long.

Flowers: The 2-3 cm long, narrow, red to crimson flowers are two-lipped and occur in spikes at the tip of the stem or in the axils of the leaves.

Fruit: The fruits are rounded capsules with four seeds.

Habitat and Distribution: Occurring in hammocks and disturbed habitats, it is found in peninsula Florida, through the Florida Keys into the West Indies.

Comment: The bright, red, distinctively two-lipped flowers of Crimson Dicliptera occurring throughout the year are occasionally seen in open disturbed areas. Seldom used, it can make a good wildflower.

Clerodendrum thomsoniae Balf.

Dicliptera sexangularis (L.) Juss.

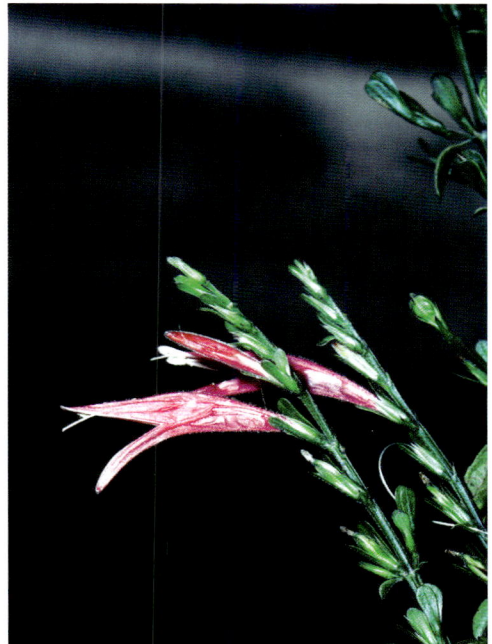

Red

Erythrina herbacea L.
Cherokee Bean
[*Erythrina arborea* (Chapm.) Small]
Fabaceae, Bean Family

Habit: Perennial herb or tree, usually 0.6-2 m tall, frequently 3-8 m tall shrub or tree in central and south peninsula Florida. Branchlets and leaves with prickles or spines.

Leaves: With long stalks and 3 leaflets, leaflets triangular or slightly 3- lobed.

Flowers: Many, bright, scarlet flowers, very slender and tube-like in shape, 3-5 cm long. Flowers clustered in at the tip of a long raceme on a blackish stem.

Fruits: Long narrow black pods, 7-15 cm long, that contain red seeds.

Habitat and Distribution: Found in open and sandy woods, common in Florida, but less common north of Florida to the Carolinas and west to Texas.

Comment: The plant freezes to the ground in colder climates, but becomes a shrub or tree when able to grow throughout the year. Growth is upright and often unkempt when shruby or herbaceous. Herbaceous growth is from an underground stem. Cherokee Bean can be a low maintenance native in the landscape. When planted along a border or against a tree the flowers in late spring are bright and attractive. In colder climates the foliage freezes in fall and can be mowed or cut to the ground. The mature pods in mid summer can be attractive when they open and expose the scarlet seeds. Seeds are poisonous and should be kept away from children. Plants do best in broken shade and loamy soils.

Euphorbia inundata Torr. ex Chapm.
Florida Pine Spurge
Euphorbiaceae, Spurge Family

Habit: Perennial herb, from a brown, gnarled rootstock. Several stems 10-50 cm tall, reddish green below and green or reddish green above.

Leaves: Narrow and oval, 3-10 cm long and 4-10 mm wide, sometimes tinged with red or purple, alternate below and opposite above in the inflorescence.

Flowers: Tiny flowers held in maroon, cup shaped structure 2-3 mm long. Cup with wavy margins.

Fruit: A 3-lobed, smooth capsule, 6-10 mm wide. Seed olive or gray, 2-3 mm in diameter.

Habitat and Distribution: Occurs in pine savannahs, flatwoods and bogs, central peninsula Florida into southern Alabama.

Comment: The reddish-green plant with its multiple stems and cup-shaped flower structures which occur in spring and early summer is attractive and well worth trying in a native garden.

Erythrina herbacea L.

Euphorbia inundata Torr. ex Chapm.

Red

Gaillardia pulchella Foug.
Blanket Flower
Asteraceae (Compositae), Aster Family or Sunflower Family

Habit: Annual or short-lived, hairy, perennial, to 60 cm tall. Stems ribbed, spreading and often branching from base, erect or lying flat.

Leaves: Oblong, some with many lobes, dissected, entire or toothed.

Flowers: Rays variously with reds and yellows, 1.5-2.5 cm long, distinctly toothed. Center disc usually yellow below and variously red and or yellow above.

Fruit: A hairy nutlet, 2 mm long, with 5-7 stiff scales at top.

Habitat and Distribution: Occurs naturally in dry and disturbed sites, often along the coast on beaches, from Florida west to Texas and north to the Carolinas.

Comment: Blanket Flower is a native success story. This plant has been cultivated and selected for its bright flowers. Various colors are now available from seed sources. Flowering occurs during warm months. It is often planted along roadsides and provides a colorful wildflower for home gardens. It is self seeding once established.

Hamelia patens Jacq.
Firebush
Rubiaceae, Coffee Family

Habit: Perennial shrub or small tree, to 5 m tall.

Leaves: Large, oval, 7-15 cm long, and sometimes folded.

Flowers: The red and orange flowers are in clusters, narrow and tubular in shape, and 1-2 cm long with 5 lobes at the tip.

Fruit: Berries are dark red or black, oval, and 6 mm long.

Habitat and Distribution: Occurring at coastal sites and edges of hammocks, it is found in southern and central Florida.

Comment: Blooms occur throughout the year, but are most common in the summer. Butterflies and hummingbirds are commonly attracted to the nectar of these very colorful flowers. Leaves in full sun are suffused with red. Leaves in shade are a velvet green with red petioles. This shrub will grow in sandy, calcareous, or loamy soils. It does best when exposed to full sun at least part of the day. Firebush will freeze, but will usually regrow from the rootstock.

Gaillardia pulchella Foug.

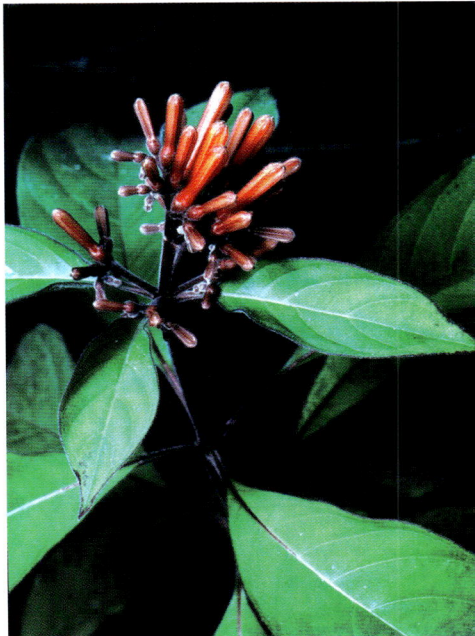

Hamelia patens Jacq.

Red

Helianthus radula (Pursh) Torr. & Gray
Rayless Sunflower
Asteraceae (Compositae), Aster or Sunflower Family

Habit: A perennial herb 0.5 -1 m tall. Stems are single or few and hairy.

Leaves: Form a basal rosette, with a few spaced along the lower half of the ascending stems.

Flowers: Heads occur singly at the ends of stems and are comprised of a dark red or purple disc, 3 cm wide, lacking typical sunflower rays.

Fruit: A slender, blackish-purple nutlet 4 mm long.

Habitat and Distribution: Found in moist pinelands, flatwoods, and prairies throughout Florida, north to Virginia and west to Louisiana.

Comment: Rayless Sunflower is the most distinct species of sunflower. It is a small sunflower without rays that flowers in the summer and fall. The maroon disc florets are accented by yellow stamens. The basal leaves are notable when the plants are placed together. Although preferring extra moisture, it will survive well if mulched.

Hibiscus coccineus Walt.
Red Hibiscus
Malvaceae, Mallow Family

Habit: Perennial shrub, 1-3 m tall.

Leaves: Deeply divided with 5 or 7 slender lobes, toothed edges, and pointed tips.

Flowers: 15 to 30 cm wide and a bright irredescent red with 5 large petals, each 7-10 cm long.

Fruit: An oval capsule, about 2-3 cm long and wide, with 5 chambers, each chamber containing 2 or more seeds. Seeds round, 3-4 mm in diameter and fuzzy brown hairy.

Habitat and Distribution: Occurs in swamps and marshy sites in Florida, southern Alabama, and southern Georgia.

Comment: Red Hibiscus has been touted as among the half dozen showiest plants native to North America. One color variation seen is an irridescent pink. Flowering is from June into winter. Plants in colder areas freeze back. This terrific wildflower can be cultivated in any rich, moist soil. It does best with full sun but can do well with at least a significant period of the day in full sun.

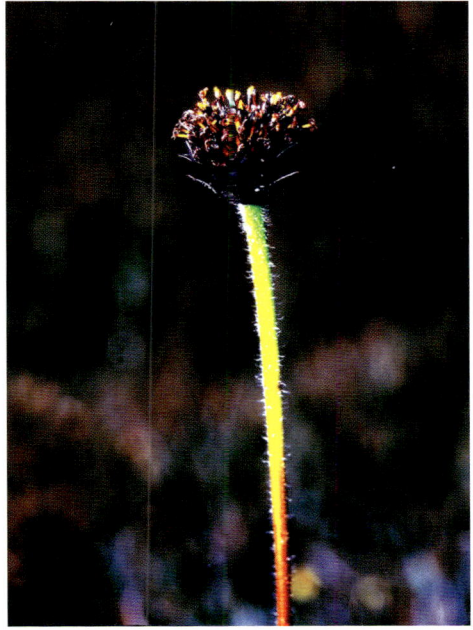

Helianthus radula (Pursh) Torr. & Gray

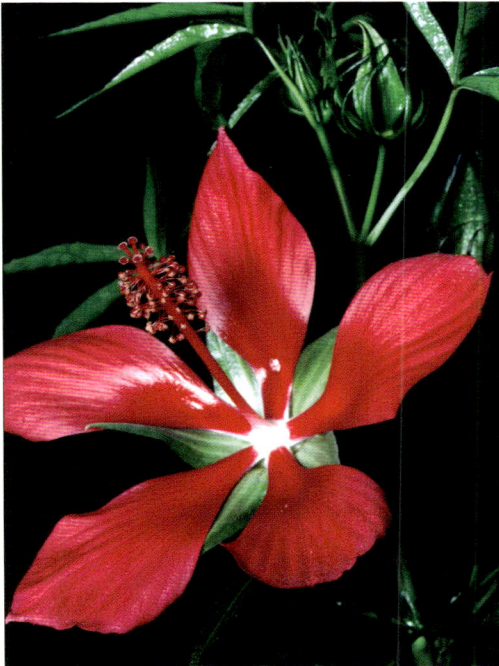

Hibiscus coccineus Walt.

Red

Hibiscus tiliaceus L.
Mahoe or Sea Hibiscus
[*Hibiscus pernambucensis* Arruda is closely related to this species, but is a native of the New World]
Malvaceae, Mallow Family

Habit: An evergreen perennial shrub or small tree, to 13 m tall.

Leaves: The long-petioled leaves are rounded or ovate with a heart-shaped base, 8-30 cm long, covered with sparse, star-shaped hairs, green above and white on the lower side.

Flowers: Solitary or few in leaf axils, large, on short stalks, with an 8- to 14-toothed cup. The 4-8 cm long, yellow or almost white petals fade through orange-yellow and turn red as they age during the day. The flower has a red center.

Fruit: A dry, hairy capsule, 1-3 cm long.

Habitat and Distribution: Occurring in dry to moist coastal habitats, this Asian species is widely distributed in the subtropics and tropics of the West Indies, and Mexico south into South America, and in Africa. It has escaped from cultivation and is frequently found in south Florida.

Comment: Flowering of this extremely attractive shrub is during the spring. Flowering is best when the plants are exposed to direct sunlight for at least a portion of the day. This large shrub is very desirable as a hedge plant due to the dense foliage. The hedge is useful as a visual barrier and is good for abating noise. It requires adequate moisture, readily wilting when dry, but recovering promptly when watered.

Hibiscus pernambucensis Arruda, also known as Mahoe or Sea Hibiscus, is a very similar species native to south peninsula Florida. Its yellow flowers lack the red center. Also, *Hibiscus pernambucensis* has a dense mat of star-shaped hairs on the lower surface of the leaves, as opposed to a sparse layer, and is a somewhat smaller plant.

Illicium floridanum Ellis
Florida Anise, Red Anise, Stink bush
Illiciaceae, Anise tree Family

Habit: Perennial shrub, 2-3 m tall.

Leaves: Aromatic, evergreen, smooth, oval, 6-15 cm long with smooth edges.

Flowers: Deep maroon, with 15-30 strap-like petals, 1.5-2 cm long, arranged like a star. Flowers have fetid smell, like rotten fish.

Fruit: Star-shaped, hard, 2.5-3 cm diameter, each segment containing one seed. Seeds slender, oval, slightly flattened, and glossy brown.

Habitat and Distribution: Grows on seepage slopes, in swampy areas and low, wet hammocks in the Florida panhandle and southwest Georgia, west to Alabama and Louisiana.

Comment: Flowering of this extremely attractive shrub is during the spring. Flowering is best when the plants are exposed to direct sunlight for at least a portion of the day. This large shrub is very desirable as a hedge plant. It requires adequate moisture and will wilt readily when dry but recovers promptly when watered. The foliage is so dense the hedge is a good large visual barrier and is useful for abating noise.

Hibiscus tiliaceus L.

Illicium floridanum Ellis

Red

Indigofera hirsuta Harvey
Hairy Indigo
Fabaceae (Leguminosae), Bean Family

Habit: Annual or biennial herb, 30-90 cm tall. Stems woody at base with long reddish brown hair.

Leaves: Compound, with 5-9 oval leaflets, leaflets 3 cm long, hairy below.

Flowers: Numerous, aggregated into a spike, small, salmon to maroon, pea-like, densely clustered, 4-5 mm long.

Fruit: A hairy, 4-angled pod, to 2 cm long, drooping downward in clusters.

Habitat and Distribution: An introduced plant from Africa, now pantropical, it occurs in disturbed areas, particularly abundant in old fields, throughout Florida and the southeast.

Comment: Flowering occurs throughout the year in south peninsula Florida and during the warm months in most of the temperate range. Hairy Indigo has an attractive flower, but to be effective as a wildflower the old stems need to be removed. The large somewhat woody stems from last season will last well into the next flowering period in such profusion that they distract and spoil the show. Additionally, seeds are quite resistant in the soil, lasting for several years. Originally, it was planted for soil improvement and forage. Once planted it will return for many years.

Ipomoea hederifolia L.
Scarlet Morning-glory
Convolvulaceae, Morning-glory Family

Habit: Annual twining vine. Stems smooth.

Leaves: 2-10 cm long, smooth, heart-shaped or usually 3- to sometimes 5- or 7-lobed.

Flowers: One to several flowers borne on long, hairy stalks. Sepals 4-4.5 mm long, acute, the outer 2 with a tail-like appendage. Flowers small and scarlet, with a tube, 2.5-4.5 cm long, and a flared mouth.

Fruit: A capsule, to 8 mm wide, on a straight stalk.

Habitat and Distribution: Found infrequently in disturbed areas, from Florida north into southern Georgia and west to Texas and Mexico. Native to South America, through cultivation this species now occurs throughout the tropics.

Comment: Blooming in the summer this vine is attractive but not showy. It is often seen climbing over shrubs. Best displays are obtained by using a trellis.

Indigofera hirsuta Harvey

Ipomoea hederifolia L.

Red

Ipomoea quamoclit L.
Cypress-vine Morning-glory
Convolvulaceae, Morning-glory Family

Habit: Annual twining vine. Stems smooth.

Leaves: 3-10 cm long, oval and finely divided into needle-like segments 3-5 cm long and 1 mm wide, resembling cypress tree leaves.

Flowers: Flower stalks gradually enlarging from the base to apex, bearing 1 to 3 flowers. Sepals are oblong, with a short, sharp, flexible point at the apex. Flowers scarlet, sometimes white, with tube 2.7-4 cm long, flaring at the mouth. Free lobes at the mouth triangular, thus entire mouth star shaped, 2 cm wide.

Fruit: An oval capsule, 5-8 mm long, usually 4-seeded, borne on an erect stalk. Seeds dull brown, 5-6 mm long, angular.

Habitat and Distribution: An escaped tropical American native, it can be found in disturbed sites, old fields, and on fences, from Florida to Virginia and Texas. It is extensively naturalized in warm temperate and tropical climates.

Comment: Cypress-vine Morning-glory, flowering in the summer and fall, can be a spectacular wildflower and a very problematic weed for agriculture. It is commonly found on fences and will reseed on open sandy soils.

Ipomopsis rubra (L.) Wherry
Standing-cypress
[*Gilia rubra* (L.) Heller]
Polmoniaceae, Phlox Family

Habit: Biennial herb, 1-1.7 m tall. Plant produces large basal rosette during first year, then erect stem the second year. Stem usually simple to the inflorescence.

Leaves: To 4 cm long, finely divided giving a feathery appearance.

Flowers: Red, usually with a yellow throat inside, borne in elongated simple or branched clusters, at top of stem. Flowers tubular, 2-4 cm long, with 5 free lobes at the mouth. Lobes longer than wide, 1 cm long.

Fruit: An oval capsule, with pointed tip, 8-10 mm long.

Habitat and Distribution: Occurs in dry open sites, dry pinelands, and sandhills, Virginia west to Kansas, south to Texas, and south in Florida to Lake Okeechobee.

Comment: Standing cypress is a startlingly bright red wildflower that blooms in the summer and fall. The tall stems fit well in wildflower gardens among shorter species. Cultivation requires dry, well drained soils.

Ipomoea quamoclit L.

Ipomopsis rubra (L.) Wherry

Red

Krameria lanceolata Torr.
Sandbur
Krameriaceae, Krameria or Ratany Family

Habit: Perennial herb, to 0.1-1.8 m long, prostrate or trailing. Stems covered with silky hairs.

Leaves: To 2.5 cm long, hairy, narrow and needle-like.

Flowers: Reddish or purplish, 5 narrow, pointed petals forming a star shape, 2-3 cm across.

Fruit: A rounded pod, woolly with many barbless spines, up to 1.2 cm in diameter.

Habitat and Distribution: Found in dry sites from central Florida north to Georgia and west to Kansas and Arizona and northern Mexico.

Comment: These prostrate herbaceous plants from woody rootstocks are difficult to see in their native habitats unless flowering. The flowers of a brick red color are showy in the spring.

Lobelia cardinalis L.
Cardinal Flower
Campanulaceae, Bell flower Family

Habit: Perennial herb, 0.5-2 m tall, from short basal offshoots. Stems simple, erect, smooth or hairy.

Leaves: Oval, with subtle teeth along edges, 5-20 cm long, 2- 6 cm wide.

Flowers: Bright red, many borne in raceme on top of stem. Flowers 3-4.5 cm long, with two strap-like upper lobes, and 3 lower lobes.

Fruit: An oval to rounded capsule, 0.8-1 cm wide and long. Seeds yellow brown, rough, oblong, 0.5-0.8 mm long.

Habitat and Distribution: Found at wet sites, streamsides, bogs, and on floating mats, central peninsula Florida, north to New Brunswick and Ontario west to Minnesota and south to Texas.

Comment: Cardinal Flower is a wonderful, extraordinarily showy wildflower blooming in most of the summer and all fall. Cultivation can be in any soil kept moist with mulch and irrigation. The simple stem that grows the first year is replaced with multiple stems from the basal offshoots the following years. These multiple stems produce even showier displays in succeeding years. Cardinal Flower is listed as Threatened by the State of Florida.

Krameria lanceolata Torr.

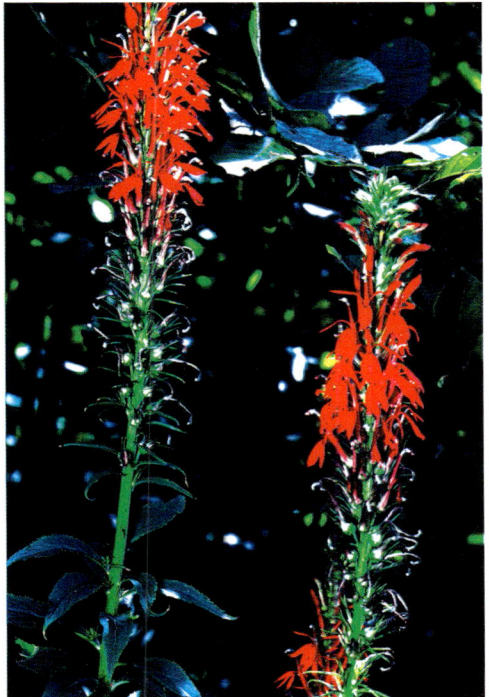

Lobelia cardinalis L.

Red

Lonicera sempervirens L.
Coral Honeysuckle
Caprifoliaceae, Honeysuckle Family

Habit: Perennial trailing or twining woody vine to 5 m long.

Leaves: Rounded to oval, green above, whitish below, in opposite pairs, and the pairs are fused around the stem below flowers.

Flowers: Red or red orange and are borne in clusters of a few to several flowers. Flowers shaped as long narrow tubes, 3-6 cm long, with 5 small lobes at the mouth. The throat sometimes yellow.

Fruit: Berries can be red or orange.

Habitat and Distribution: Found in woodlands, thickets, and along fences, from central Florida north to Maine and west to Texas and Nebraska.

Comment: This vine is an excellent ornamental. It is low climbing and flowers best in full sun. It can be used on fences, trellises, and mail boxes. Flowering somewhat irregularly in spring through summer, both the flowers and fruits are showy and attract butterflies and hummingbirds. The foliage is attractive due to the green and white colors. The leaves are semi-evergreen and last well into the winter or through the winter in warmer climates.

Malvaviscus arboreus Cav.
Turk's-cap
Malvaceae, Mallow Family

Habit: Perennial shrub to 3 m tall. Stems finely hairy.

Leaves: Stalked, evergreen, smooth to hairy, blades narrowly elliptic to rounded, often spade-shaped, 3-21 cm long, with toothed edges and sometimes 3 to 5 lobes.

Flowers: Red, solitary, pendulent, with 5 overlapping petals, never opening fully, 3-7 cm long.

Fruit: 5-parted, 1-2 cm wide, red and fleshy, becoming brown and dry, splitting apart, each falling separately.

Habitat and Distribution: Found in disturbed sites or as an ornamental throughout Florida, west to Texas, California, and Cuba. Native from Mexico south to Peru and Brazil.

Comment: Turk's cap is frequently used as a foundation plant or hedge. Flowering spring through fall it is a bright addition to the landscape. It can be cultivated in any soil, but will freeze back.

Lonicera sempervirens L.

Malvaviscus arboreus Cav.

Red

Poinsettia cyathophora (Murr.) Kl. & Gke.
Painted Leaf, Wild Poinsettia
[*Euphorbia cyathophora* Murr.]
Euphorbiaceae, Spurge Family

Habit: Annual herb, to 0.5 m tall. Stems slightly hairy, erect. Branches few to many.

Leaves: Stalked, opposite or alternate, blades linear to linear lanceolate or narrowly obovate, pointed, entire or lobed to create violin shapes, with toothed edges. Floral bracts (modified leaves below flowers) partly or entirely red at base. Flowers tiny and inconspicuous, yellow green, in clusters above floral bracts. Flower cups having glands on upper margins with mouth squeezed shut.

Fruit: A smooth capsule, 3-4 mm wide and 5-6 mm long.

Habitat and Distribution: Occurs in hammocks, old fields, and disturbed sites throughout Florida. Cultivated and escaped north to Virginia, west to Wisconsin, Minnesota and Kansas, south into South America, West Indies.

Comment: Painted Leaf flowers all year in warm climates and during the summer and fall elsewhere. These plants can be especially showy in mass and are most effective when sown along an open area.

Pueraria montana (Lour.) Merr.
Kudzu
[*Pueraria lobata* (Willd.) Ohwi]
Fabaceae (Leguminosae), Pea or Bean Family

Habit: Perennial herbaceous or semi-woody vine, to 30 m long, high climbing or trailing. Young stems hairy.

Leaves: Alternate, long stalked, with 3 broad, oval, 2- or 3-lobed leaflets, each 5-20 cm long, densely hairy underneath.

Flowers: In many-flowered clusters, 5-30 (usually 8-15) cm long, in the axils of the leaves. Flowers red-purple with a yellow patch, 1-2.5 cm long, with hood shaped upper lobes and 3 lower lobes.

Fruit: A pod, 4-10 cm long and 1 cm wide, covered with tan to reddish brown hairs.

Habitat and Distribution: Found in woods, old fields, vacant lots, woods borders and along roadsides throughout Florida, west to eastern Texas and eastern Oklahoma, north to Virginia and occasionally New York, Massachusetts, and Ohio.

Comment: Flowers can be found from the summer into the fall. Flowers are usually not noticed due to the rampant foliage. Originally native to Asia, Kudzu was introduced for forage and for soil stabilization. It has escaped to become a prolific weed perhaps causing more harm than it did good.

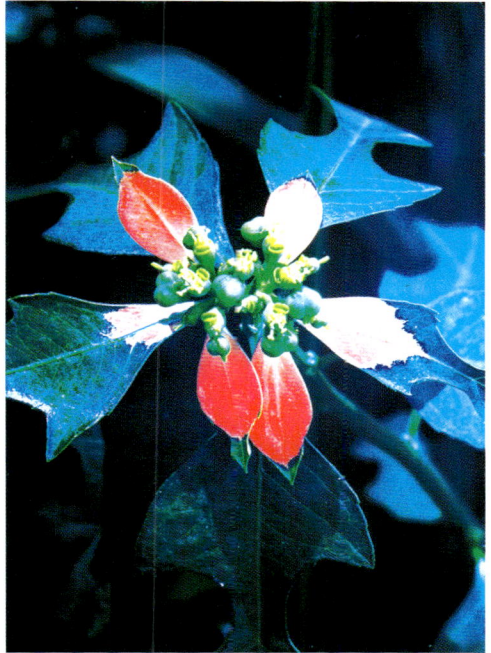

Poinsettia cyathophora
(Murr.) Kl. & Gke.

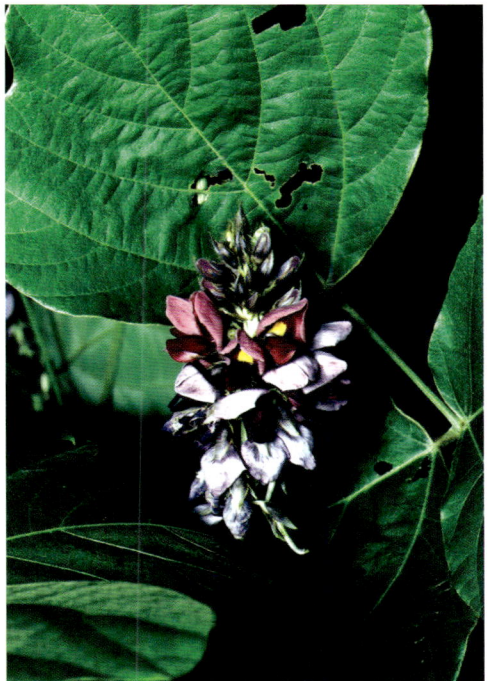

Pueraria montana (Lour.) Merr.

Red

Rudbeckia graminifolia (Torr. & Gray) Boyn. & Beadle
Grass-leaf Cone Flower
Asteraceae (Compositae), Aster or Sunflower Family

Habit: Perennial herb, 40-90 cm tall. Stems few, slender, usually unbranched, often hairy towards top of stem.

Leaves: Long, narrow, usually 10-20 cm long and not more than 1 cm wide, resembling blades of grass, midrib visible below, margins rolled under.

Flowers: Heads deep maroon, rarely orange-red, borne singly at ends of stems. Rays 5-12, slender and oblong, 0.8-2 cm long, often drooping, velvety and darker above, duller and lighter below. Disc dome-shaped, brownish red, 1-2 cm high.

Fruit: A nutlet, dry, cylindrical, 2-3 long.

Habitat and Distribution: Found in wet pinelands, ponds, and depressions, often in shallow water, central Florida panhandle.

Comment: Flowering in all warm months, Grass-leaf Cone Flower could be a wonderful addition to a landscape.

Rumex hastatulus Bald. ex Ell.
Heart-wing Sorrel or Wild Sorrel
Polygonaceae, Smartweed Family

Habit: Winter annual or short lived perennial herb, 0.2-1.3 m tall, from a slender taproot.

Leaves: To 8 cm long, 3-lobed, with the center lobe being long and sword shaped, and the outer lobes being slender, smaller, emerging from the base of the leaf, and perpendicular to the center lobe. Lower leaves with a petiole as long as blades.

Flowers: Borne in dense clusters on long slender flowering stems, 30- 40 cm long, male and female flowers on separate plants. Flowers very small, 1-1.5 mm long and 3-5 mm across, sometimes greenish, pink, or purple, becoming red as fruits on female plants mature.

Fruit: Dry, red to reddish brown, with 3 membranous wings, 3 mm across, containing a nutlet.

Habitat and Distribution: Occurs in sandy soil, old fields, pastures, and along roadsides, from central Florida north to Massachusetts, and west to Illinois, Kansas, and Texas.

Comment: Flowering in the spring and early summer, separate plants are attractive, but masses of flowers are spectacular. Fields of Heart-wing Sorrel flowers are sensational as they expand into the distance. Often used in salads and as cooked greens, plants do best in full sun. *Rumex hastatulus* is easily confused with *Rumex acetosella* L., Sheep Sorrel or Sour-grass. Separation is best accomplished by checking the root systems, *R. acetosella* has rhizomes and *R. hastatulus* has a taproot.

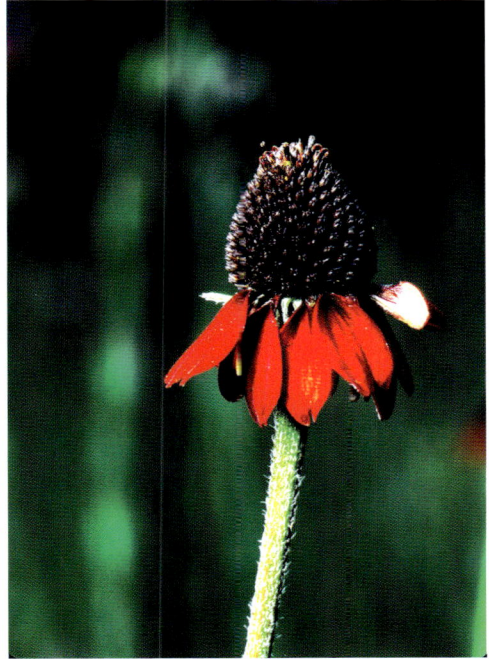

Rudbeckia graminifolia
(Torr. & Gray) Boyn. & Beadle

Rumex hastatulus Bald. ex Ell.

Red

Sacoila lanceolata (Aubl.) Garay
Leafless Beaked Orchid
[*Spiranthes lanceolata* (Aubl.) Leon, *Stenorrhynchos lanceolatus* (Aubl.) L.C. Rich. ex Spreng.]
Orchidaceae, Orchid Family

Habit: Perennial, terrestrial herb, to 60 cm tall, from a thick cluster of large fleshy roots.

Leaves: Thick and fleshy, narrowly oval, 10-40 cm long, appearing as a rosette after flowering.

Flowers: Red, orange red, coral, or brick, 25-30, arranged along end of thick stem. Flowers 2-3 cm long, somewhat tubular, with petals opening gracefully at mouth. Highest flowers open first.

Fruit: A capsule to 1.5 cm long.

Habitat and Distribution: Occurs infrequently in wet flatwoods, woods, pastures, and roadsides from southern Florida into the Florida panhandle and north into North Carolina, Mexico south through Central America into South America, West Indies.

Comment: Flowering from April into July the flowers are spectacular although seldom seen. Leafless Beaked Orchid is listed as a Threatened species by the state of Florida.

Salvia coccinea P.J. Buchoz ex Etlinger
Tropical Sage
Lamiaceae (Labiatae), Mint Family

Habit: Perennial herb, 30-70 cm tall. Stems branching, 4-sided in cross-section, rough hairy.

Leaves: Opposite, triangular to heart-shaped, 1-8 cm long, with blunt toothed edges.

Flowers: Inflorescences in panicles at the ends of stems. Flowers many, scarlet red, 2.5-3 cm long, short tube with lips at mouth, upper lip smaller and entire, lower lip larger with 2 broad lobes.

Fruit: Small, nut-like fruits oval, 2.5 mm long.

Habitat and Distribution: Occurs in disturbed areas and hammocks, from Florida to South Carolina and Texas, south through Mexico and the West Indies into tropical America.

Comment: Flowering can be all the warm months of the year. Tropical Sage can be an excellent landscape plant. The bright green foliage and the many bright red flowers make this a showy plant. It does best with full sun, but can do well with light shade. Best results are obtained with well drained soils. The plants can become rank in cultivation and should be pruned during the winter months.

*Sacoila lanceolata (*Aubl.) Garay

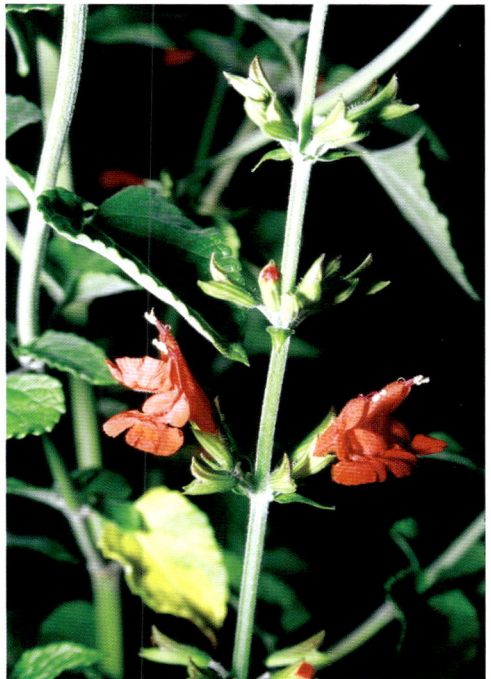

Salvia coccinea
P.J. Buchoz ex Etlinger

Red

Sarracenia leucophylla Raf.
White-top Pitcher Plant
Sarraceniaceae, Pitcher Plant Family

Habit: Perennial, carnivorous herb, to 1.2 m tall, from rhizomes.

Leaves: Aerial leaves deciduous in winter, erect, hollow, green at base, widening toward the top, opening at the top, top shaped like a hood. Leaf mouth white, with green, pink, or purple-red veination.

Flowers: Borne on leafless slender stalks, just taller than mature leaves. Flowers nodding, 7-9 cm broad, bracts red, sepals maroon, petals purplish red, fiddle shaped and 4-7 cm long.

Fruit: A capsule 2.5-3.5 cm in diameter, seeds wedge-shaped, 2.5 mm long.

Habitat and Distribution: Found in bogs, in northwest Florida, southwest Georgia, southern Alabama, to southern Mississippi.

Comment: Flowering March through May, the flowers are only one of the attractive features of White-top Pitcher Plant. The long whitish leaves, taller than the surrounding vegetation, are dramatic both in form and color and the frilled top adds to the effect. White-top Pitcher Plant is listed as Endangered by the State of Florida.

Sarracenia psittacina Michx.
Parrot Pitcher Plant
Sarraceniaceae, Pitcher Plant Family

Habit: Perennial, carnivorous herb, to 0.3 m tall, from rhizomes.

Leaves: Aerial leaves overwinter, prostrate, ascending at tips, 8-30 cm long, emerging as basal rosette. Leaves hollow, green at base, with hooded mouth. Mouth and hood white with purple patches and veination, hood arching upwards, curling to surround the mouth.

Flowers: Flower borne on leafless erect stalk, 15-40 cm tall. Flower nodding, 3-5 cm broad, bracts red, sepals maroon, petals maroon, broadly oval and 3-4 cm long.

Fruit: A capsule 9-14 mm diameter, seeds rough, almost triangular, 1.5 mm long.

Habitat and Distribution: Occurs in bogs, seeps and wet lowlands from northern Florida and southern Georgia west to southern Louisiana and southern Mississippi.

Comment: The flowers occur among the clusters of leaves in April through May. When surrounding vegetation is high one must look down to see the pretty flowers. Parrot Pitcher Plant is listed as Threatened by the State of Florida.

Sarracenia leucophylla Raf.

Sarracenia psittacina Michx.

Red

Sarracenia purpurea L.
Side-saddle Pitcher Plant
Sarraceniaceae, Pitcher Plant Family

Habit: Perennial, carnivorous herb, to 0.4 m tall, from rhizomes.

Leaves: Aerial leaves overwinter, spreading from a basal rosette, prostrate at base then arching upwards, 5-45 cm long. Leaves hollow, stout, urn shaped, with upward opening mouth bordered by an erect 2-lobed hood with undulating edges. Mouth and hood variable in color, from green to white to maroon, often variegated with these colors.

Flower: Nodding, borne on erect leafless stalk, 15-70 cm tall, no odor. Flower 5-6 cm broad, sepals purplish red, petals deep maroon to rose, fiddle-shaped and 3-7 cm long.

Fruit: A capsule 10-20 mm diameter, seeds almost drop shaped, 2 mm long.

Habitat and Distribution: Found in bogs, swamps, and wet flatwoods and woodlands, from Florida panhandle west to Mississippi, north to Ohio, Indiana, Illinois to Canada and west to Louisiana.

Comment: The flowering season for Side-saddle Pitcher Plant is usually April through May. These impressive flowers are quite surprising when seen, but not easily found. Side-saddle Pitcher Plant is listed as Threatened by the State of Florida.

Sarracenia rubra Walt.
Red-flowered Pitcher Plant
Sarraceniaceae, Pitcher Plant Family

Habit: Perennial, carnivorous herb, 10-50 cm tall, from rhizomes.

Leaves: Aerial leaves erect, hollow, upper half nearly tubular, green to reddish with variegation towards mouth, mouth opening upwards, hood almost erect or slightly arched over mouth.

Flowers: Flower nodding, borne on leafless erect stalk, 15-75 cm tall, usually just taller than leaves, sweet odor. Flower nodding, 3-5 cm broad, sepals purplish, petals maroon to red, broadly fiddle-shaped and 2-4 cm long.

Fruit: A capsule 5-15 mm diameter, seeds wedge-shaped and up to 2 mm long.

Habitat and Distribution: Occurs in bogs, seeps, wet pine savannahs, and along boggy streams in mountains, from the Florida panhandle to North Carolina and Mississippi.

Comment: The showy flowers occur from April through May. As a wildflower Red-flowered Pitcher Plant is very colorful. Red-flowered Pitcher Plant is listed as Threatened by the State of Florida.

Sarracenia purpurea L.

Sarracenia rubra Walt.

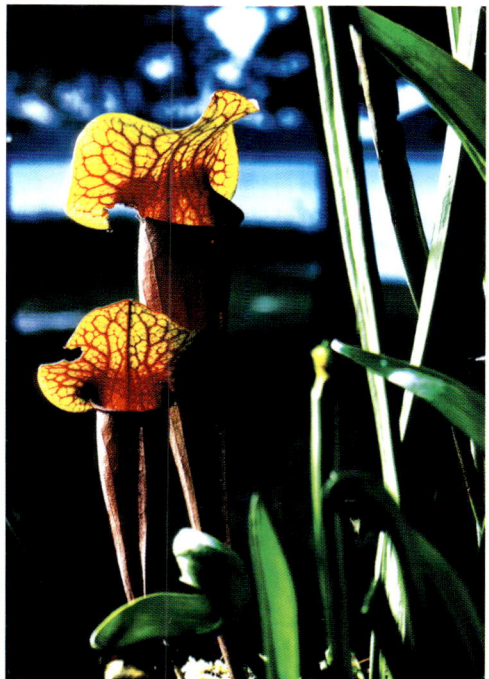

Red

Spigelia marilandica L.
Indian Pinkroot
Loganiaceae, Logania Family

Habit: Perennial herb, 15-70 cm tall. Stems with 4-7 pairs of leaves.

Leaves: Lanceolate to oval, 3-12 cm long, with very short leaf stalks.

Flowers: Borne in solitary clusters of 2-12 on ends of branches. Flower tubular, 3-4.5 cm long, red outside, pale yellow inside, 5 narrow triangular lobes at mouth, curling outwards.

Fruit: A firm capsule, 4-6 mm long, 6-10 mm broad, seeds dark brown, angular, 2-2.5 mm long.

Habitat and Distribution: Found in rich woodlands and hillsides, from the Florida panhandle north to Maryland, Missouri, and Indiana, west to Oklahoma and Texas.

Comment: Flowering in the spring the dark red flowers with clear yellow exposed lobes are dramatic against the dark green oval lance shaped leaves. This species contains a toxic alkaloid.

Tillandsia fasciculata Swartz
Common Air Plant, Quill-leaf, Stiff-leaved Wild-pine
Bromeliaceae, Pineapple Family

Habit: Perennial, epiphytic herb, 20-60 cm tall. Plant grows above the ground, on other trees.

Leaves: Gray-green with brown band at base, stiff, in a large rosette or whorled around a short stem, spreading outward and often curving downward. Leaves blade-shaped, up to 1 m long, tapering to the end, pointed.

Flowers: Flowering stem terminates with a cluster of spikes and has short erect leaves along lower portion of flower stem, equal to or below leaves. Spikes appearing solid and sword shaped, comprised of red or sometimes yellow overlapping bracts. Small inconspicuous flowers on stalks emerge between bracts.

Fruit: A capsule 3-3.5 cm long.

Habitat and Distribution: Found growing on cypress and hardwood trees, and occasionally on pines, in hammocks and swamps, central peninsula Florida south through the West Indies, Central and South America.

Comment: The spreading red bracts, most frequently seen in fall, of Common Air Plant can be ornamental when plants are placed onto lower tree branches or into containers at a home or commercial setting. This plant, although common in south peninsula Florida, is listed by the State of Florida as Endangered due to the spread of a moth which feeds on the species.

Spigelia marilandica L.

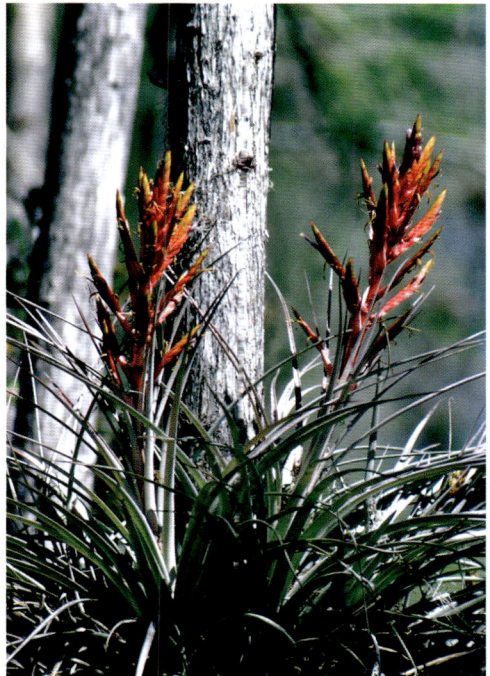

Tillandsia fasciculata Swartz

Red

Tillandsia setacea Swartz
Southern Needle-leaf or Red Needle-leaf
Bromeliaceae, Pineapple Family

Habit: Perennial, epiphytic herb, 10-30 cm tall. Plant grows on other tree trunks or large branches.

Leaves: Spreading in dense tufts, green or red, long and very slender, like pine needles, covered with minute scales.

Flowers: Flower spikes as long as leaves, with red tinged bracts, flowers emerging from between bracts small and purple.

Fruit: A capsule, 2.5-3 cm long.

Habitat and Distribution: Occurs on trees in hammocks and swamps of central and south peninsula Florida south into the West Indies.

Comment: These small densely tufted, needle-leaved plants usually flower during the summer. The reddish leaves and red tinged bracts are appealing and can be placed on low branches and in containers.

Trifolium incarnatum L.
Crimson Clover
Fabaceae (Leguminosae), Bean or Pea Family

Habit: Annual herb, 20-40 or occasionally to 80 cm tall. Stems and leaves hairy.

Leaves: Composed of 3 leaflets on a stalk, leaflets broadly oval, broader toward end than toward base, 1-4 cm long, almost as wide.

Flowers: Flower heads on a stalk up to 12 cm long, cylindrical, 3-8 cm long, 1-2.5 cm wide, comprised of many small scarlet flowers. Flowers tubular, about 1 cm long.

Fruit: An oval pod, 1-seeded.

Habitat and Distribution: A European native, now naturalized and found in fields, pastures and along roadsides, from central Florida north to Virginia and west to Texas.

Comment: The brilliant red flowers, so evident in spring, are among the most widespread and showiest wildflowers especially along roads. No roadside should be without them.

Tillandsia setacea Swartz

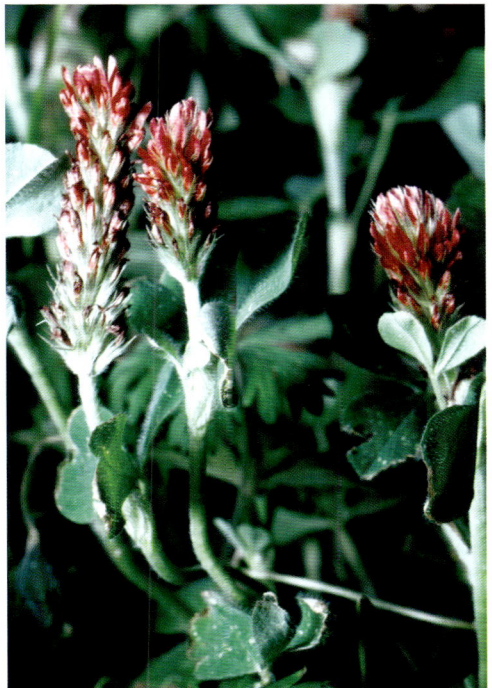

Trifolium incarnatum L.

Red

Trifolium pratense L.
Red Clover
Fabaceae (Leguminosae), Bean or Pea Family

Habit: Perennial herb, 20-50, or occasionally to 70 cm tall. Stems numerous and usually hairy.

Leaves: Composed of 3 leaflets on a stalk, leaflets oval to diamond shaped, 1-4 cm long and half as wide, hairy, often with light green chevron in center of leaflet.

Flowers: Flower heads lacking stalks or with short stalk, dome-shaped to spherical, 1-3 cm long. Flowers reddish to reddish-purple or rarely white, tubular, 1.2-1.8 cm long.

Fruit: An oval to almost spherical pod, smooth above, wrinkled below, 1-seeded.

Habitat and Distribution: A European native, it is naturalized and occurs in woods, fields, roadsides, and disturbed sites from southern Canada throughout most of the United States, south into central peninsula Florida.

Comment: One of the most widely distributed clovers in the United States, Red Clover contributes splendid shows of color in spring and to a lesser extent in fall.

Trillium decipiens J.D. Freeman
Chattahoochee River Wake-robin
Trilliaceae, Wake-robin Family

Habit: Perennial, erect herb, to 40 cm tall.

Leaves: Three, wide, ovate, mottled leaves, to 16 cm long.

Flowers: Solitary on each stem, parts in 3s; sepals straight; petals yellowish- to brownish-purple or sometimes greenish, to 8 cm long, 2 to 3.5 times as long as wide; ovary rounded, 6-ridged.

Fruit: A rounded berry, many-seeded.

Habitat and Distribution: Rich rocky slope forests and rocky hammocks, central panhandle Florida, north to southwest Alabama and southeast and central Georgia.

Comment: Flowering February through April, the spotted leaves and distinctive maroon flowers are a delightful wildflower find on the open floor of woods. This deceptive species strongly resembles *Trillium underwoodii*, Long-bract Wake-robin, but is consistently taller. Additionally, the sepals of *T. decipiens* are straight and never spread to a horizontal position with recurved tips as does *T. underwoodii*.

Trifolium pratense L.

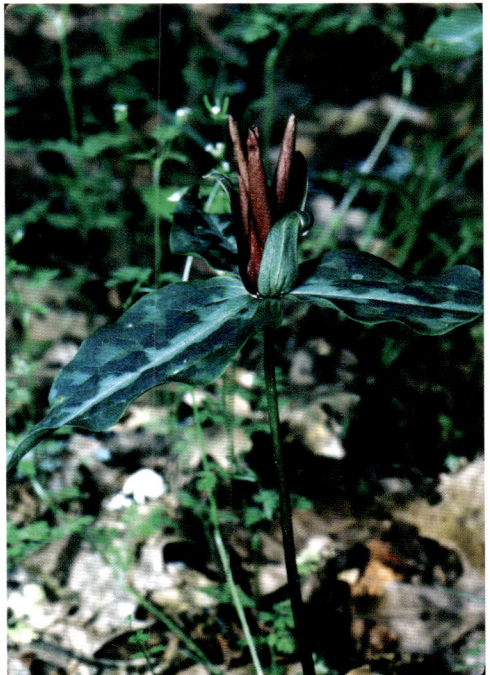

Trillium decipiens J.D. Freeman

Red

Trillium maculatum Raf.
Spotted Wake-robin
Trilliaceae, Wake-robin Family

Habit: Perennial herb to 40 cm tall.

Leaves: Three, wide, lanceolate to ovate, mottled leaves, to 20 cm long.

Flowers: Solitary on each stem, parts in 3s; sepals widely spreading with recurved tips; petals maroon or dark purple, rarely yellow, to 7 cm long, 3 to 4 times as long as wide; ovary 3-angled.

Fruit: A rounded berry, many-seeded.

Habitat and Distribution: Rocky hammocks and slope forests, central panhandle Florida, north into central Alabama and southwest South Carolina.

Comment: Like all Wake-robins the spotted leaves and maroon petals, seen flowering February through March, provide a sprinkling of color among the trees.

Trillium underwoodii Small
Long-bract Wake-robin
Trilliaceae, Wake-robin Family

Habit: Perennial herb to 25 cm tall.

Leaves: Three, wide, lanceolate to ovate, mottled leaves, to 18 cm long.

Flowers: Solitary on each stem, parts in 3s; sepals horizontal with recurved tips; petals maroon, to 5.5 cm long, more than 4 times as long as wide; ovary rounded.

Fruit: A rounded berry, many-seeded.

Habitat and Distribution: Hammocks, slope forests and pastures, panhandle Florida, north into southwest Alabama and central western Georgia.

Comment: Long-bract Wake-robin flowers January through March, frequently showing its distinctive three mottled leaves and maroon flowers before many other herbaceous species can make their statements in the winter or early spring. See *T. decipiens* for a discussion of the differences between these two similar species.

Trillium maculatum Raf.

Trillium underwoodii Small

Red

Verbesina chapmanii J.R. Coleman
Chapman's Crownbeard
Asteraceae (Compositae), Aster Family or Sunflower Family

Habit: Perennial herb to 1 m tall, from a short, thick rhizome.

Leaves: Without stalks (sessile), usually opposite or occasionally alternate, elliptic, sandpapery, margins are toothed; usually 1 or sometimes 2 heads at the tips of stems.

Flowers: Disc flowers only - no ray flowers; flowers small, yellowish- orange.

Fruit: Purplish, flattened nutlet, narrow wing along each edge.

Habitat and Distribution: Flatwoods and Wire Grass openings only in Florida (endemic) in Bay, Gulf, and Liberty Counties in the Florida panhandle. This area is known as the Apalachicola River lowlands. The plant occurs mostly on the floodplain of the river.

Comment: Flowering from July into August, Chapman's Crownbeard is not showy as a wildflower. This rare plant is listed as Threatened by the State of Florida.

Verbesina chapmanii J.R. Coleman

Pink

Abrus precatorius L.
Rosary Pea
Fabaceae (Leguminosae), Pea or Bean Family

Habit: Perennial, high climbing or trailing, woody vine, to 5 m or more long. Stems woody on lower parts of plant, herbaceous upwards on plant including branches.

Leaves: Alternate; stalked; blades once-divided, with 8-15 pairs of leaflets, each 9-18 mm long.

Inflorescences: Long-stalked crowded clusters from leaf axils.

Flowers: Pink, rose or white, typically shaped for pea flowers, petals 0.8-1 cm long.

Fruit: Pods to 2-4 cm long, flat, beaked. Seeds 5-7 mm long, bright red, with black base.

Habitat and Distribution: Frequent southward, very infrequent northward; dry thickets, margins, or other disturbed areas; south peninsula Florida north into northern Georgia and Alabama; West Indies. Native of tropical Asia and Africa, now pantropical.

Comment: Flowering in summer, Rosary Pea is most noted for the spectacular bright colored red and black seeds. The seeds are very POISONOUS if ingested. The common name comes from the practice of stringing the seeds together for necklaces, bracelets and other jewelry. However, poking the holes in the seeds to put them on a string also includes the risk of getting some of the POISONOUS material on needles, drills or hands and spreading it. This vine is quite weedy and is spreading throughout warmer regions in disturbed habitats.

Achillea millefolium L.
Common Yarrow
Asteraceae (Compositae), Aster or Sunflower or Daisy Family

Habit: Herbaceous perennial, 0.2-1.2 m tall, from underground runners. Stems erect, 1 to several, usually densely woolly hairy.

Leaves: Alternate, basal rosettes; stalked; blades 2-15 cm long, very finely once-divided, giving frilly appearance, densely hairy.

Inflorescences: One or more dense clusters of very small heads at ends of erect stems, heads 4-5 mm long.

Flowers: Ray flowers 5-12, 1-3 mm long, pink or usually white.

Fruit: A narrow, smooth achene, 2-3 mm long.

Habitat and Distribution: Occasional to common; meadows, old fields, disturbed and dry sites from central peninsula Florida north throughout the United States and into Canada.

Comment: Common Yarrow flowers in spring. It is a very complex variable species that is essentially circumboreal in distribution. It is very aromatic. Although an extremely weedy plant, many different forms of it are sold as garden ornamentals.

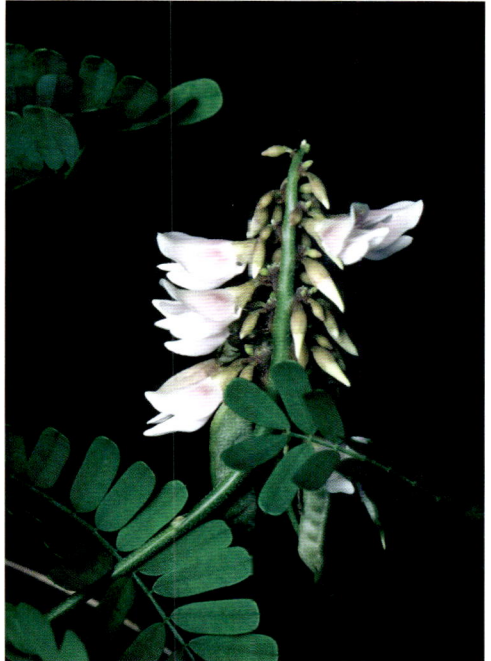

Abrus precatorius L.

Achillea millefolium L.

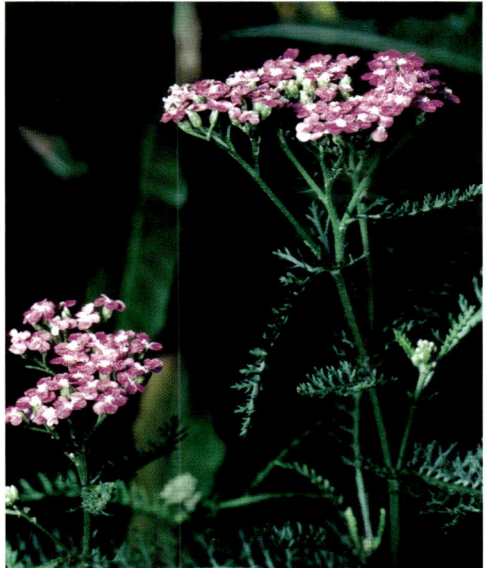

Pink

Agalinis fasciculata (Elliott) Raf.
Clusterleaf Gerardia
Scrophulariaceae, Figwort or Snapdragon Family

Habit: Herbaceous annual, 0.6-1.2 m tall. Stems sandpapery, with many erect branches.

Leaves: Opposite; stalkless; blades narrow, to 4 cm long and 1-2 mm wide, sandpapery on upper surface, younger leaves clustered in leaf axils.

Inflorescences: Solitary, short-stalked flowers in leaf axils at stem tips.

Flowers: Pinkish purple, 5-lobed tube, bell-shaped, 2.5-3.5 cm long, hairy, with yellow lines and purple dots inside throat of tube.

Fruit: A rounded capsule, 5-6 mm long.

Habitat and Distribution: Common; dry to moist to wet fields, open pinelands, flatwoods, sandhills, roadsides and other sandy sites; throughout Florida, west to Texas, and north to Arkansas, Missouri and Maryland.

Comment: The large showy flowers of Clusterleaf Gerardia are seen in summer and fall. Locally common, Gerardias can fill whole vistas with color.

Agalinis linifolia (Nutt.) Britton
Flaxleaf Gerardia

Scrophulariaceae, Figwort or Snapdragon Family

Habit: Herbaceous perennial, 0.8-1.5 m tall, from underground runners or cluster of fleshy roots. Stems smooth, rounded, with few erect branches.

Leaves: Opposite; stalkless; blades narrow, 3-5 cm long and 1-3 mm wide, smooth.

Inflorescences: Solitary, long-stalked flowers in leaf axils at stem tips.

Flowers: Pink to lavender, 5-lobed tube, cone-shaped, 3-4 cm long, outside of flower tube hairy, inside of tube with pale spots and lacking yellow lines.

Fruit: A rounded capsule, 6-9 mm long.

Habitat and Distribution: Occasional; wet pinelands, bogs and marsh margins; north peninsula Florida, west to Louisiana, and north to Delaware.

Comment: A fall flowering wildflower with a scattered distribution.

Agalinis fasciculata (Elliott) Raf.

Agalinis linifolia (Nutt.) Britton

Pink

Agalinis purpurea (L.) Pennell
Smooth Gerardia
Scrophulariaceae, Figwort or Snapdragon Family

Habit: Herbaceous annual, 0.4-1.2 m tall. Stems many-branched, smooth or only slightly rough.

Leaves: Opposite; stalkless; blades narrow, often curled, 2-4 cm long, 1-4 mm wide, often sandpapery on upper surface.

Inflorescences: Solitary, short-stalked flowers in leaf axils at stem tips.

Flowers: Pink to rose-pink to purple, 5-lobed tube, bell-shaped, 2-4 cm long, hairy, with yellow lines and purple dots inside throat of tube.

Fruit: A rounded capsule, 4-7 mm long.

Habitat and Distribution: Occasional; moist to wet meadows, woodlands, flatwoods, bogs, prairies, shores and marshy areas; panhandle Florida, west to Texas and north to South Dakota, Minnesota and Nova Scotia; Mexico; West Indies.

Comment: Summer and fall flowering, Smooth Gerardia can be separated from Clusterleaf Gerardia by the smooth stems. Like Clusterleaf Gerardia it is an eye-catching wildflower in bloom.

Albizia julibrissin Durazz.
Mimosa
Fabaceae (Leguminosae), Pea or Bean Family

Habit: Small deciduous tree, to 10 m or more tall, with short trunk, long upward reaching branches, and umbrella like crown.

Leaves: Alternate; long-stalked; blades 10-30 cm long and up to 15 cm broad, twice-divided into 4-12 pairs, each with 40 to 50 leaflets, leaflets 8-15 mm long and 3 mm wide.

Inflorescences: Panicles of stalked clusters occur in the axils of the uppermost leaves of the season, each cluster, 3-8 cm in diameter, contains 15-25 stalkless flowers.

Flowers: Sweetly fragrant, small, pale green flowers, with many long pink stamens radiating outward give the appearance of a pink silky puff.

Fruit: Pods long, narrow, and flat, 8-18 cm long, 2-3 cm broad.

Habitat and Distribution: Frequent; dry to moist native and disturbed sites; from central peninsula Florida, west to Texas, and north to New York; a native of Asia.

Comment: Mimosa flowers in spring and the sweet delicate fragrance spreads widely. The large canopies can be covered with blooms, making a spectacular display. It is extensively cultivated and naturalized throughout the southeast U.S. rapidly invading native habitats. The tree is short-lived, often succumbing to disease. Seeds are long-lived.

Agalinis purpurea (L.) Pennell

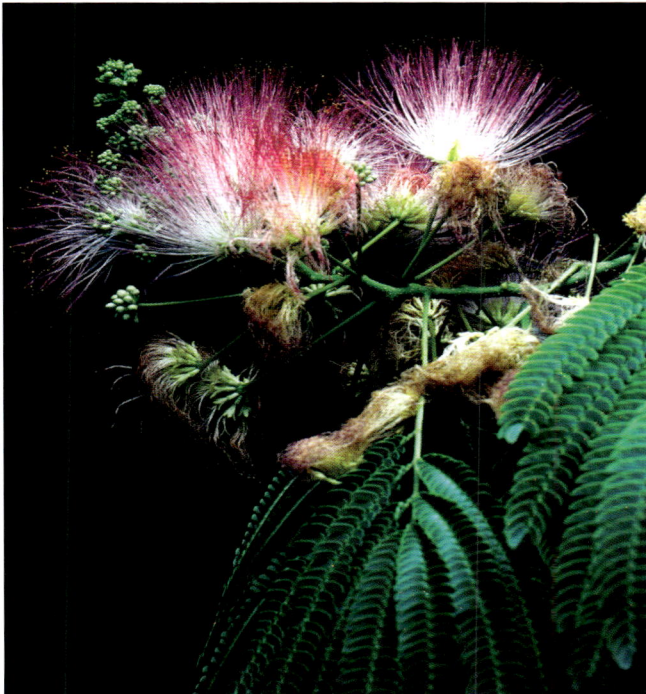

Albizia julibrissin Durazz.

55

Pink

Amaranthus spinosus L.
Spiny Amaranth
Amaranthaceae, Amaranth Family

Habit: Herbaceous annual, 0.2-1.2 m tall, from a taproot. Stems smooth, often reddish, with paired axillary spines and many branches.

Leaves: Alternate; stalked; blades ovate, 1.5-10 cm long, with smooth edges.

Inflorescences: Flowers dense on long, slender, terminal and axillary spikes, spikes often nodding.

Flowers: Pinkish-green to greenish-white, tiny, 2-3 mm long.

Fruits: Spherical and thin walled, to 2 mm diameter, containing one black seed, less than 1 mm long.

Habitat and Distribution: Frequent to common; disturbed sites and fields; throughout Florida, north to Maine, Manitoba, Indiana and Missouri, and west to Texas; a native of tropical America, now wide spreading in warmer parts of the world.

Comment: Flowers in all warm months. Extremely weedy, especially in crops, pastures and fence rows.

Antigonon leptopus Hook. & Arn.
Coral Vine
Polygonaceae, Buckwheat Family

Habit: Partly woody, perennial, high climbing vine, to 15 m long, from tuber-like roots.

Leaves: Alternate; stalked; blades triangular or arrowhead-shaped, to 13 cm long, margins slightly wavy.

Inflorescences: Flower spikes axillary or terminal, sometimes branched, with a tendril at tip.

Flowers: Pink, oval to heart-shaped petals and sepals, 1.5-2 cm long.

Fruit: Dry, 8-9 mm long, with 3 wings.

Habitat and Distribution: Frequent; hammocks, borders, roadsides, waste places, and cultivated; throughout Florida, Texas; a native of Mexico.

Comment: Coral Vine is widely cultivated and escaped into warm temperate and tropical climates. It is a tropical plant, primarily flowering in summer and fall, that persists only in lower peninsula Florida and southern Texas. It is easily cultivated, but is not cold tolerant. A wonderful shade and ornamental plant, growing rapidly on fences and trellises. Many color variations are sold at nurseries.

Amaranthus spinosus L.

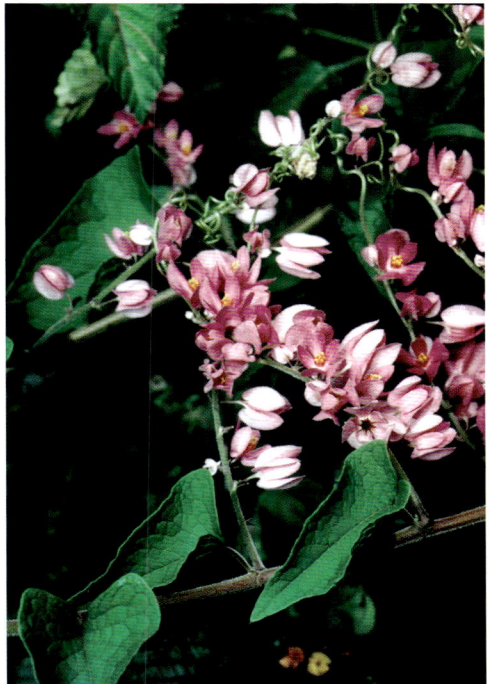

Antigonon leptopus Hook. & Arn.

Pink

Asclepias humistrata Walter
Sandhill Milkweed
Asclepiadaceae, Milkweed Family

Habit: Herbaceous perennial, 20-90 cm long or tall. Stems spreading, prostrate to ascending, stout, smooth, unbranched, rarely solitary.

Leaves: Opposite; stalkless; blades oval to heart-shaped with a wide base, 5-12 cm long, with pink to purple veins.

Inflorescences: Long-stalked, flat-topped clusters from the upper leaf axils.

Flowers: Rose, lavender, gray, of white, petals to 7 mm long, lobes reflexed, upright central crown white and 3-5 mm in diameter.

Fruit: An elongated capsule, broader in the middle and tapering towards both ends, 8-14 cm long, about 1.5 cm wide, erect.

Habitat and Distribution: Frequent; dry pinelands, sandhills, and scrub; central peninsula Florida, west to Mississippi, and north to North Carolina.

Comment: The leaves of this spring and summer bloomer are pretty. The flowers almost blend into the background, but in mass are attractive.

Asclepias michauxii Dcne.
Michaux's Milkweed
Asclepiadaceae, Milkweed Family

Habit: Herbaceous perennial, to 1 m tall. Stems several, simple or one branch from base, hairy.

Leaves: Opposite; stalkless; blades linear, 5-12 cm long.

Inflorescences: A stalked, solitary, terminal, disc-like cluster, usually with fewer than 20 flowers.

Flowers: 5 outer petals and hood shaped petals in center of flower, outer petals greenish white, tinged with rose, tips pink, 4-6 mm long.

Fruit: Long, slender capsules, 8-15 cm long, 5-8 cm wide, smooth, standing erect.

Habitat and Distribution: Occasional; pinelands, sandhills, savannahs, bogs, and marshes; northern peninsula Florida, along the coastal plain north to South Carolina, and west to Mississippi.

Comment: Spring and summer flowering, the pink-tipped greenish pink flowers and thin plant leave Michaux's Milkweed as a frequently overlooked wildflower.

Pink

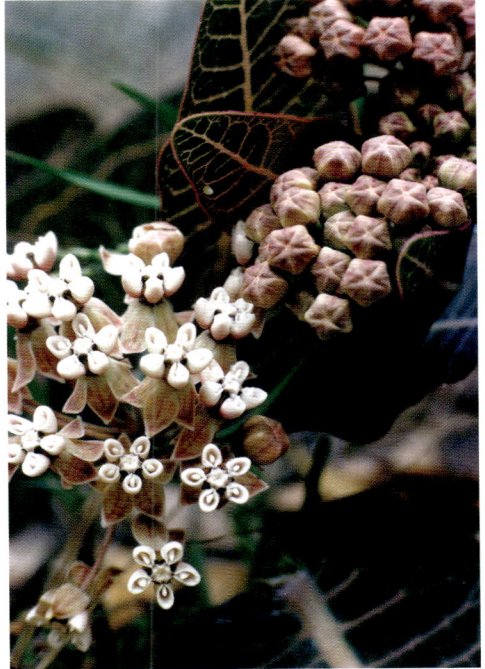

Asclepias humistrata Walter

Asclepias michauxii Dcne.

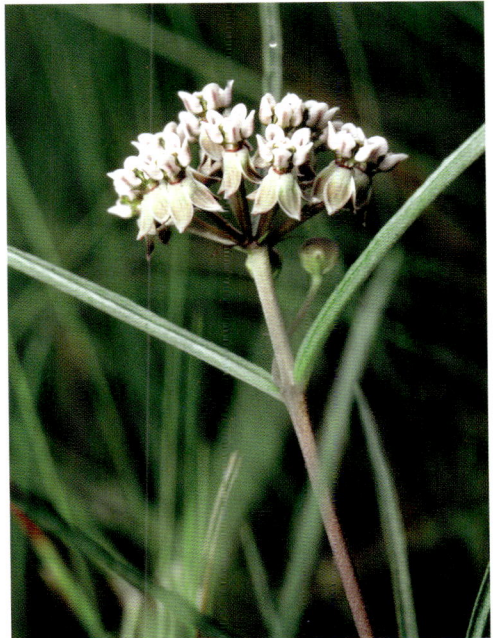

Pink

Asclepias viridula Chapm.
Southern Milkweed
Asclepiadaceae, Milkweed Family

Habit: Herbaceous perennial, to 0.7 m tall, from a woody, tuberous, rootstock. Stem usually 1, occasionally several, thin, purplish.

Leaves: Opposite; short-stalked; blades linear, 3-10 cm long, margins smooth.

Inflorescences: Flat-topped or dome-shaped, stalked clusters from upper leaf axils.

Flowers: Outwardly greenish, inwardly pinkish to brownish purple, 5 petals, 4-5 mm long, spreading upward.

Fruit: A smooth, narrow capsule, 8-10 cm long.

Habitat and Distribution: Occasional; flatwoods and bay and bog borders; northeast peninsula and eastern panhandle Florida (Gulf to Jefferson Counties).

Comment: The spring and summer flowers begin as green. The upward turned petals become a pinkish to brownish color, but are small and easily overlooked. The plant is thin with very skinny leaves which adds to the difficulty of finding it.

Asimina pygmaea (W. Bartram) Dunal
Dwarf Pawpaw
Annonaceae, Custard-apple Family

Habit: Small shrub, 20-60 cm tall. Stems 1 to several, ascending, arching or laying down, bark reddish brown with pale specks.

Leaves: Alternate; stalked; newly emerging blades - densely blonde to reddish brown underneath, soon become smooth, mature blades - dark green, leathery, lance-shaped to oblong, widest at the tip, 4-11 cm long.

Inflorescences: Solitary, nodding, long-stalked flowers from leaf axils of the current growth, fetid smelling.

Flowers: 3 outer petals, pink-striped with maroon or maroon, outward curving, bluntly triangular, 1.5-3 cm long, and 3 inner petals, deep maroon, shorter than outer petals.

Fruit: A berry, oblong, curved and bulging, 3-5 cm long, yellowish green when ripe, with 2 irregular rows of brown shiny seeds, up to 1 cm long.

Habitat and Distribution: Frequent; sandhills, flatwoods and savannas; central peninsula to eastern panhandle Florida, north into southeast Georgia.

Comment: Spring flowering or occasionally blooming in summer and fall. This low shrub has small, but striking dark maroon flowers. Although eaten by some people, the fruit pulp can be TOXIC to others.

Asclepias viridula Chapm.

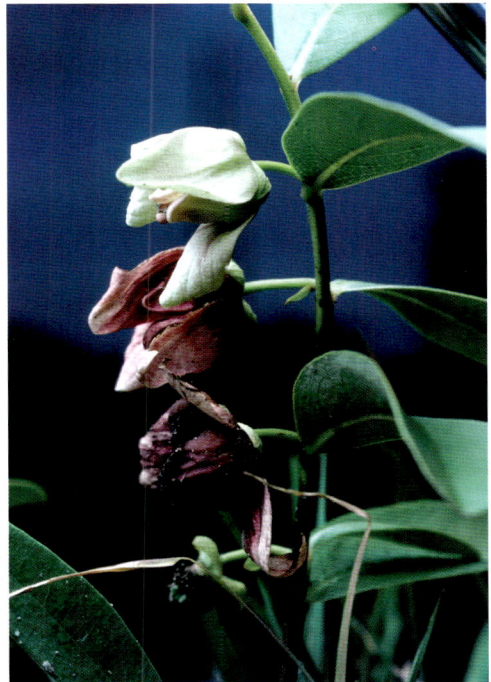

Asimina pygmaea
(W. Bartram) Dunal

Pink

Aster carolinianus Walter
Climbing Aster
[*Symphyotrichum carolinianum* (Walter) Wunderlin & B.F. Hansen]
Asteraceae (Compositae), Aster or Sunflower Family

Habit: Herbaceous or somewhat woody, sprawling or climbing perennial, 1-4 m tall. Stems herbaceous to woody, hairy, vinelike, climbing by twining or leaning into surrounding vegetation.

Leaves: Alternate; lacking a stalk; blades lance-shaped to broadly elliptical, 2-10 cm long, clasping stem at base.

Inflorescences: Heads 4-5 cm wide, stalked, in a diffuse, leafy panicle.

Flowers: Rays pink or light purple, numerous, narrow, and with yellow to reddish disc flowers.

Fruit: Nutlets to 3 mm long, smooth but with a tan tuft of bristles, 7-8 mm long on the top.

Habitat and Distribution: Frequent; swamps, wet hammocks, fresh and salt water marshes, ditches, and other wet areas; peninsula to the central panhandle of Florida; and north on the coastal plain to North Carolina.

Comment: This fall bloomer is a spectacular wildflower seen along wet roadsides and wetland borders. The large pink flowers can be easily seen as the plant scrambles towards the light. Climbing Aster is a good cultivated specimen growing in many soil types from sand to peat, and from dry (as long as irrigation is available) to wet.

Aster elliottii Torr. & A. Gray
Elliott's Aster
[*Aster puniceus* L. subsp. *elliottii* (Torr. & A. Gray) A.G. Jones, *Symphyotrichum elliottii* (Torr. & A. Gray) G.L. Nesom]
Asteraceae (Compositae), Aster or Sunflower Family

Habit: Herbaceous perennial, to 3 m tall, from underground runners (rhizomes). Stems erect, somewhat succulent and with lines of hairs at the top.

Leaves: Alternate; those at the bottom with a stalk, those at the top lacking a stalk; blades elliptic, to 25 cm long and 5 cm wide, upper surface sandpapery, margins with teeth.

Inflorescences: Somewhat flat-topped or elongated, branched cluster of many flowers at the tends of stems, heads 1.5-3 cm wide.

Flowers: 25-45 pink, blue, or lavender rays and yellow to red disc flowers.

Fruit: Nutlets brown, smooth or hairy, 2-3 mm long, and with a whitish tuft of bristles, 5-7 mm long at top.

Habitat and Distribution: Frequent; wet meadows, marshes, wet woods, swamps and ditches; from peninsula and central panhandle Florida, north into south Alabama and along the outer coastal plain into Virginia.

Comment: Elliott's Aster is a late summer and fall blooming wildflower with large displays in established colonies along roadside ditches and other wet habitats. It is cultivated and offered for sale by nurseries.

Aster carolinianus Walter

Aster elliottii Torr. & A. Gray

Pink

Aster eryngiifolius Torr. & A. Gray
Coyote-thistle Aster or Thistle-leaf Aster
[*Eurybia eryngiifolia* (Torr. & A. Gray) G.L. Nesom]
Asteraceae (Compositae), Aster or Sunflower or Daisy Family

Habit: Herbaceous perennial, 30-70 cm tall, from a crown or thick underground runner. Stem unbranched, erect, hairy.

Leaves: Alternate; stalkless; blades - basal, long, narrow, grasslike, to 35 cm long, 3-8 mm broad; stem - leaves also linear, smaller, some with spiny margins.

Inflorescences: Heads 1-6, terminal.

Flowers: About 20 pink to white, strap-like rays, 1-2 cm long, and yellow disk.

Fruit: Nutlets smooth, reddish, bristles at top.

Habitat and Distribution: Occasional; low flatwoods and swampy spots; from central panhandle Florida into adjacent Georgia and Alabama.

Comment: The strict stems, grass-like leaves, and large flower heads seen in spring and summer make this a very attractive wildflower that deserves to be more frequent in cultivation.

Bacopa monnieri (L.) Pennell
Smooth Water hyssop
Scrophulariaceae, Figwort or Snapdragon Family

Habit: Herbaceous, matted perennial, 10-30 cm tall. Stems smooth, glossy green, succulent, spreading or ascending, rooting at nodes.

Leaves: Opposite; stalkless; blades wider at tip, to 2 cm long, margins smooth.

Inflorescences: Flowers solitary from leaf axils, long-stalked.

Flowers: Pink to lilac to white tinged with blue to blue, bell-shaped, 7-10 mm long, with 5 petals.

Fruit: A somewhat rounded capsule, 5-7 mm long.

Habitat and Distribution: Common and practically ubiquitous; sandy shores of fresh and brackish waters; throughout Florida, along the Gulf Coast to Texas, and along the Coastal Plain to Virginia; widespread in the tropics.

Comment: Blooming all warm months, the flowers have a delicate color. In warm climates Smooth Water-hyssop is always in flower. It is useful as a ground cover on wet sandy sites. Sometimes in shallow water it occurs as a floating mat.

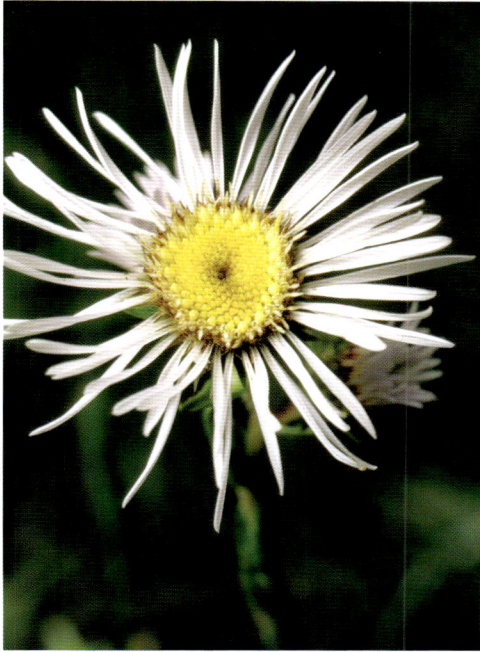

Aster eryngiifolius Torr. & A. Gray

Bacopa monnieri (L.) Pennell

Pink

Bauhinia variegata L.
Orchid Tree
Fabaceae (Leguminosae), Pea or Bean Family

Habit: Open, branched tree, to 15 m tall.

Leaves: Alternate; long-stalked; blades rounded, base with 2 rounded lobes, tip with deep cleft, 6-20 cm long.

Inflorescences: Short, simple, terminal, unbranched, spike-like.

Flowers: Pink to red to blue-purple to white, often mottled or streaked, 5 distinct petals, 4-6 cm long.

Fruit: A flattened, oblong, woody pod, 10-20 cm long.

Habitat and Distribution: Occasional; disturbed sites; central peninsula Florida, southern Texas and southern California; a native of tropical Asia.

Comment: Orchid Tree flowers in spring with large showy blooms that somewhat resemble a large orchid flower. It is frequently cultivated and is becoming more common as an escape.

Callicarpa americana L.
French-mulberry or American Beauty-berry
Verbenaceae, Vervain Family

Habit: Perennial shrub, 1-4 m tall. Twigs densely hairy with star-shaped hairs.

Leaves: Opposite, deciduous; stalked; blades broadly elliptical or oval, hairy, 7-23 cm long, margins toothed.

Inflorescences: Many-flowered clusters in leaf axils.

Flowers: Light pink, small, tubular, with 5 petals.

Fruit: Berry-like with a hard center, magenta or purple, rarely white, spherical, 3-6 mm diameter.

Habitat and Distribution: Common; open and closed canopy woody sites; throughout Florida, north to Maryland, North Carolina and Arkansas, west to Texas and Mexico; West Indies.

Comment: Flowering occurs mostly in spring. The pink clusters of blooms are attractive, but the fruits are spectacular. French-mulberry is a highly desired native ornamental shrub. It is almost weedy in its native habitats. The fruits are mostly gone about the time the leaves fall.

Bauhinia variegata L.

Callicarpa americana L.

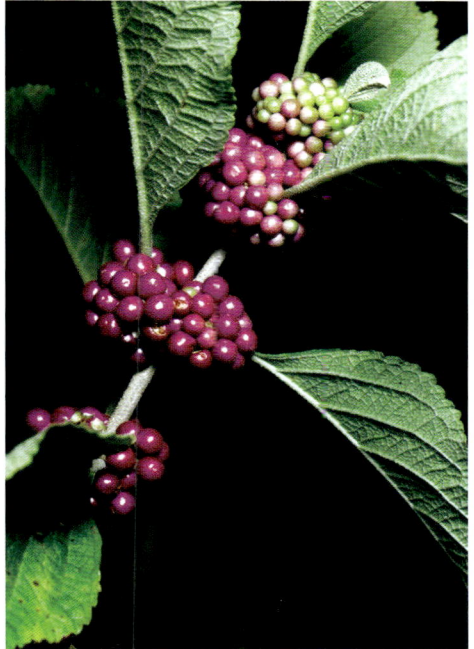

Pink

Calopogon barbatus (Walter) Ames
Bearded Grass-pink
Orchidaceae, Orchid Family

Habit: Herbaceous perennial, to 20 cm tall.

Leaves: Alternate; stalkless; blades 1 or 2, emerging near base of stem, to 20 cm long and 1 cm wide.

Inflorescences: 3-7 flowers at tip of stem, open simultaneously.

Flowers: Bright pink, 2-2.5 cm long, 5 petals, 1 paddle-shaped or hourglass-shaped petal above, and one scoop-shaped petal protruding from center of flower, petals wider below the middle.

Fruit: An erect, cylindrical capsule.

Habitat and Distribution: Frequent; savannahs, wet meadows, pinelands and acidic roadside ditches; from central peninsula Florida, west to Louisiana, and north to North Carolina.

Comment: Very showy flowers in March and April. Identified by the petals being wider below the middle.

Calopogon pallidus Chapm.
Pale Grass-pink
Orchidaceae, Orchid Family

Habit: Herbaceous perennial, to 50 cm tall.

Leaves: Alternate; stalkless; blades 1 or 2, ribbed, to 20 cm long and to 5 mm wide.

Inflorescences: 5-12 flowers at tip of stem, opening successively up the stem.

Flowers: Whitish pink or rarely deep magenta, 1 sword-shaped petal opening downwards, 1 wedge- or paddle-shaped petal upwards, and a scoop-shaped petal protruding from center of flower, crescent-shaped petals on each side.

Fruit: A cylindrical capsule.

Habitat and Distribution: Occasional; marshy prairies, wet meadows, bogs, wet pinelands and acidic roadside ditches; throughout Florida, west along the coast to Louisiana, and north to Virginia.

Comment: The showy flowers can be found from March through July. Easily identified by the crescent-shaped petals on each side.

Calopogon barbatus
(Walter) Ames

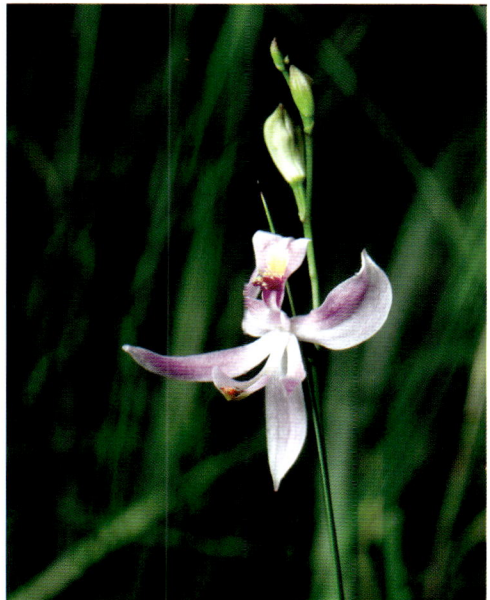

Calopogon pallidus Chapm.

Pink

Calopogon tuberosus (L.) Britton
Common Grass-pink
Orchidaceae, Orchid Family

Habit: Herbaceous perennial, to 75 cm tall.

Leaves: Alternate; stalkless; blades 1 or 2, up to 50 cm long and to 4 cm wide.

Inflorescences: 3-17 flowers at tip of stem, opening successively up the stem.

Flowers: Deep to pale pink, 1 paddle-shaped upper petal, 1 scoop-shaped petal emerging from the center of the flower.

Fruit: Capsule cylindrical.

Habitat and Distribution: Frequent; wet pine savannahs, prairies, bogs, marshes and swamps; throughout Florida, west to Texas, and north to Minnesota and Newfoundland.

Comment: Common Grass-pink blooms from March into August and is quite showy.

Canavalia rosea (Swartz) DC.
Beach Bean
Fabaceae (Leguminosae), Pea or Bean Family

Habit: Prostrate, trailing, twining woody vine, to 10 m long. Stems hairy, upper stems herbaceous.

Leaves: Alternate; stalked; blades divided into 3 leaflets, leaflets fleshy and leathery, broadly oval or almost round, often folded and with a cleft at tip, 4-10 cm long.

Inflorescences: Flowering along stem on long, several-flowered stalks from leaf axils.

Flowers: Pink to purple, typically shaped pea flowers, 2.5-3 cm long.

Fruit: Pods broad, flat, and thick, 10-15 cm long, with raised ridges along both sides of the suture.

Habitat and Distribution: Frequent; coastal beaches and dunes; south and central peninsula Florida.

Comment: Flowers can be seen in all seasons. This vine climbs over most beach vegetation and is a very attractive wildflower. It can be trained as an ornamental on fences and trellises.

Calopogon tuberosus (L.) Britton

Canavalia rosea (Swartz) DC.

Pink

Carphephorus corymbosus (Nutt.) Torr. & A. Gray
Lavender Paintbrush
Asteraceae (Compositae), Aster or Sunflower or Daisy Family

Habit: Herbaceous perennial, 0.3 to 1.2 m tall. Stem solitary, hairy, leafy, emerging from basal rosette of leaves.

Leaves: Alternate; basal leaves stalked, stem leaves stalkless; blades smooth, glandular punctate, basal lance-shaped, widest near the rounded tip, 5-20 cm long, 0.8-2.5 cm broad near tip, stem somewhat oblong, smaller.

Inflorescences: A terminal, disc- or dome-shaped cluster of several to numerous heads, heads 0.7-1 cm long, with 12-20 flowers per head.

Flowers: Pink to purple.

Fruit: Narrow, pointed, 2-4 mm long, 10-ribbed, with tuft of barbed bristles at top, about 3 times as long as nutlet.

Habitat and Distribution: Common; dry pinelands and sandhills; throughout peninsula into the central panhandle of Florida, and north into south Georgia.

Comment: Blooming from summer into fall, the pinelands can be full of Lavender Paintbrush flowers. It is a gorgeous, showy wildflower.

Carphephorus odoratissimus (J.F. Gmel.) H. Hebert
Deer's-tongue or Vanilla Plant
Asteraceae (Compositae), Aster or Sunflower or Daisy Family

Habit: Herbaceous perennial, to 2.1 m tall. Stem solitary, smooth, leafy, emerging from basal rosette of leaves.

Leaves: Alternate; stalkless; blades smooth, glandular punctate, margins toothed; basal leaves narrowly oval, tapering toward base, to 50 cm long and 10 cm wide; stem lance-shaped, smaller than basal.

Inflorescences: Flat-topped, terminal cluster with numerous heads, heads 3.5-5 mm long, with 4-10 flowers per head.

Flowers: Pink to purple.

Fruit: Narrow, pointed, 2-3.5 mm long, with tuft of barbed bristles at top, slightly longer than nutlet.

Habitat and Distribution: Frequent to common; moist to wet savannahs, pine flatwoods, prairies and poorly drained areas; throughout Florida, west to Louisiana, and north to North Carolina.

Comment: Fall into winter flowering, this wildflower can be spectacular, especially after the woods have been burned the previous year. The plant has the odor of coumarin, similar to vanilla, giving the plant one of its common names, Vanilla Plant. The plant is harvested, dried and sold to cigarette manufacturers as a component of cigarette tobacco.

Carphephorus corymbosus
(Nutt.) Torr. & A. Gray

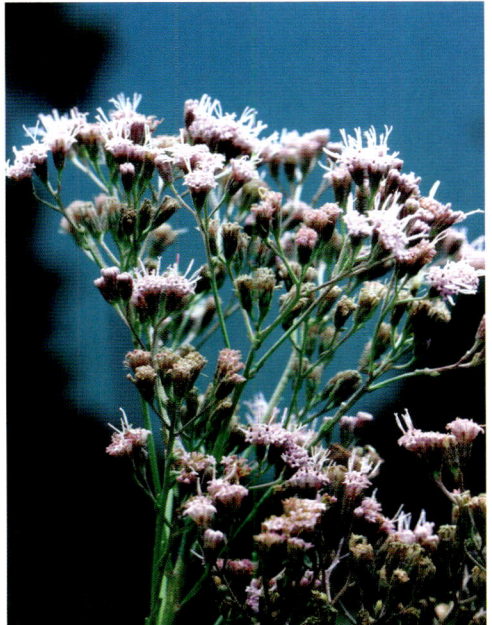

Carphephorus odoratissimus
(J.F. Gmel.) H. Hebert

Pink

Carphephorus paniculatus (J.F. Gmel.) H. Hebert
Hairy Trilisa
Asteraceae (Compositae), Aster or Sunflower or Daisy Family

Habit: Herbaceous perennial, to 1.3 m tall. Stems solitary, hairy, leafy, emerging from basal rosettes.

Leaves: Alternate; stalkless; blades smooth, glandular punctate; basal narrowly oval, tapering toward base, 5-35 cm long, to 4 cm wide; stem smaller than basal.

Inflorescences: Dense, long, cylindrical arrangements of numerous heads at tip of stem, heads 4-6 mm long, with 4-10 flowers per head.

Flowers: Pink to purple.

Fruit: Narrow, pointed, 2-3 mm long, with tuft of barbed bristles at top, twice as long as nutlet.

Habitat and Distribution: Frequent; moist to wet pine savannahs, flatwoods, bogs, and other wet woods margins; throughout Florida, west to southeast Alabama, and north to North Carolina.

Comment: Summer and fall flowering, the elongated arrangement of flowers provides quite a show.

Carphephorus pseudoliatris Cass.
Blazing star Chaffhead
Asteraceae (Compositae), Aster or Sunflower or Daisy Family

Habit: Herbaceous perennial, to 1 m tall. Stems solitary, hairy, leafy, emerging from basal rosettes.

Leaves: Alternate; stalkless; blades smooth, glandular punctate; basal needle-like, 12-40 cm long and 1-2 mm wide; stem progressively smaller.

Inflorescences: Small, dense, flat-topped clusters at tip of stems, heads 6-9 mm long, with 35 or more flowers per head.

Flowers: Pink to purple.

Fruit: Narrow, pointed, about 3 mm long, with tuft of bristles at top, about twice the length of nutlet.

Habitat and Distribution: Occasional; moist to dry, sandy pinelands and flatwoods, and bogs; from panhandle Florida, west to Louisiana, and north to southwest Georgia.

Comment: Flowering is in fall. The numerous small flowers provide a great show.

Carphephorus paniculatus
(J.F. Gmel.) H. Herbert

Carphephorus pseudoliatris Cass.

Pink

Catharanthus roseus (L.) G. Don
Madagascar Periwinkle
[Vinca rosea L.]
Apocynaceae, Dogbane Family

Habit: Erect herbaceous perennial, 20-70 cm tall.

Leaves: Opposite; stalked; blades dark green, glossy, elliptic-oblong, 4-8 cm long.

Inflorescences: Solitary or paired flowers in upper leaf axils.

Flowers: Pink, rose purple, or white, with a dark maroon eye, tubular, 2-3 cm long, with 5 broad lobes, each to 3 cm long.

Fruit: 2 dry capsule-like follicles, slender, 1.5-3 cm long.

Habitat and Distribution: Frequent; disturbed places, scrub land, and open habitats; south and central peninsula Florida, and rarely north to North Carolina; a native of Madagascar, now pantropical.

Comment: Flowering from summer until frost, Madagascar Periwinkle is a tough plant with very showy blooms. It is a popular plant in cultivation and persists and spreads in dry, open sites. The plant has a number of medicinal uses, but can be TOXIC if ingested.

Cercis canadensis L.
Redbud Tree
Fabaceae (Leguminosae), Pea or Bean Family

Habit: Small, deciduous tree, usually 5-10 m tall. Trunk often curved, bark fairly smooth.

Leaves: Alternate; long-stalked; blades distinctly heart-shaped, 7-12 cm long, appearing after flowering, margins smooth.

Inflorescences: Tight clusters of flowers on short spurs along the stems appear before the leaves.

Flowers: Light pink to magenta, 0.8-1.3 cm long, bilaterally symmetrical, with 5 petals, the upper 3 fanned out, and the lower two clasped together.

Fruit: Pods oblong, flat, 4-10 cm long, 0.8-1.8 cm broad, reddish purple turning brown.

Habitat and Distribution: Frequent; rich woodlands; from central peninsula Florida to northern Mexico, Texas and Oklahoma, north to Nebraska, Iowa, Michigan and New York.

Comment: As a spring flowering ornamental and wildflower Redbud is popular with landscapers and homeowners. This small tree can be spectacular in flower. Flowering with the earliest species, the dense clusters cover the stems before leaves appear. The curved trunk can also be an attractive addition to landscapes.

Catharanthus roseus (L.) G. Don

Cercis canadensis L.

Pink

Cirsium horridulum Michx.
Horrible Thistle
Asteraceae (Compositae), Aster or Sunflower or Daisy Family

Habit: Herbaceous biennial, 0.2-1.5 m tall. Stem usually solitary, from a basal rosette, woolly, leafy, branching below flowers.

Leaves: Alternate; stalkless; blades very spiny, largest blades near base, often dissected, 10-30 cm long, 3-10 cm wide, young blades hairy becoming smooth with age.

Inflorescences: Solitary to several, large, dense heads at stem and branch tips, heads 5-6 cm in diameter, each surrounded by small, spiny, leaf-like bracts, only disc flowers present.

Flowers: Pink, purple, yellow, or rarely white.

Fruit: Nutlet, smooth, 4-6 mm long, with bristles 3-4 cm long at top.

Habitat and Distribution: Common; open woods, fields, and disturbed sites; throughout Florida, west to Texas, and north to Maine; Mexico.

Comment: Flowering can be seen in all warm months southward and spring and summer northward. While the large heads of flowers are showy, Horrible Thistle lives up to its name when viewing the multitude of spiny leaves. Northward a yellow-flowered variant is sometimes seen and is called Yellow Thistle.

Cirsium lecontei Torr. & A. Gray
LeConte's Thistle
Asteraceae (Compositae), Aster or Sunflower or Daisy Family

Habit: Herbaceous perennial, 0.4-1.1 m tall, from one to several thick roots. Stem usually solitary, hairy, occasionally with few branches, from basal rosette.

Leaves: Alternate; stalkless; blades spiny, often lobed or simply spiny-toothed, largest blades near base, most blades on lower stem, 10-30 cm long, 1-4 cm wide, thinly hairy underneath.

Inflorescences: Solitary, stalked heads at ends of stems or branches, each 1-3 cm in diameter, all of the narrow bracts surrounding the head are long-tapering and sharply pointed, some ending in a single spine at tip, only disc flowers present.

Flowers: Rosy-purple.

Fruit: Nutlet, smooth, 5 mm long, with bristles 3 cm long at top.

Habitat and Distribution: Occasional; low moist to wet habitats, pinelands and savannas; from central and western panhandle Florida, west to Louisiana, and north to North Carolina.

Comment: This slender summer and fall blooming plant has flowers perched at the tip of the stem not unlike a shaving brush.

Cirsium horridulum Michx.

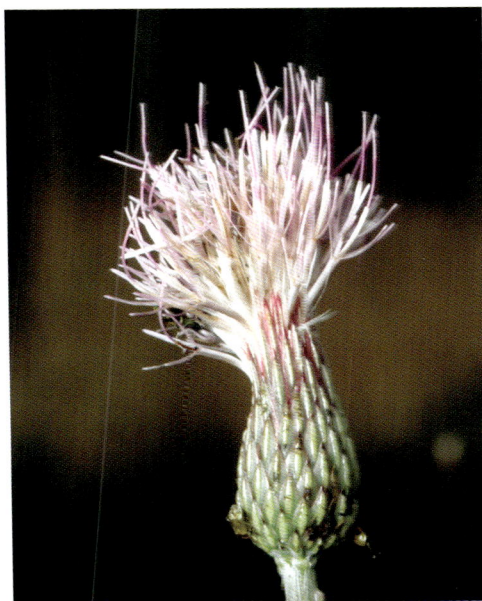

Cirsium lecontei Torr. & A. Gray

Pink

Cirsium nuttallii DC.
Nuttall's Thistle
Asteraceae (Compositae), Aster or Sunflower Family

Habit: Herbaceous biennial, 1-3.5 m tall. Stems branching, hairless or with spreading hairs, winged toward base.

Leaves: Alternate; lacking a stalk; blades spiny, deeply lobed and the lobes toothed again, sheathing stem at base, larger leaves 25-60 cm long.

Inflorescences: Solitary terminal heads on almost naked branches, outer bracts of the head tipped with an outward pointing spine, heads domed and puff-like above the cup-shaped base, only disc flowers present.

Flowers: Pink to pale purple or rarely white.

Fruit: Nutlets 3-4 mm long, smooth, with a tuft of white bristles on top.

Habitat and Distribution: Common; wet to dry sandy soils, in disturbed habitats, hammocks, fields, and along roadsides; peninsula Florida, west to Louisiana and north to South Carolina, disjunct in Virginia, mostly on the coastal plain.

Comment: This tall thistle flowers from the spring into the fall. While the flowers are not very showy the simple stem and roughly lobed leaves are interesting.

Corallorhiza wisteriana Conrad
Spring Coralroot
Orchidaceae, Orchid Family

Habit: Herbaceous perennial, 10-40 cm tall, from an underground runner (rhizome). Stems succulent, unbranched, brownish yellow to reddish brown.

Leaves: Absent.

Inflorescences: Long, terminal, spike-like.

Flowers: Greenish yellow to white with purplish markings, sometimes lip pure white or flowers marked with red, 5-7 mm long, lower lip prominent.

Fruit: An elliptic capsule, 9-11 mm long.

Habitat and Distribution: Frequent; hammocks, pine flatwoods, rich deciduous woods and occasionally in turf; from central peninsula Florida, north to New Jersey and Montana, west to Arizona and Mexico.

Comment: A very early flowering orchid, December and January into April, Spring Coralroot sometimes fades into the background as a single plant, but when seen in a mass of several hundreds or thousands of stems it can be spectacular.

Cirsium nuttallii DC.

Corallorhiza wisteriana Conrad

Pink

Coreopsis nudata Nutt.
Swamp Tickseed or Georgia Tickseed
Asteraceae (Compositae), Aster or Sunflower or Daisy Family

Habit: Herbaceous perennial, 0.4-1.2 m tall, from short underground runners or a thick crown. Stems smooth, gray-waxy.

Leaves: Alternate; stalkless; blades linear, rounded, 10-40 cm long, 1-2 mm thick.

Inflorescences: Stalked, terminal, with 1-10 heads.

Flowers: Pink or lavender rays and yellow disc, rays 1.5-3 cm long, with toothed ends, and disc 1-1.5 cm wide.

Fruit: Elliptic, about 3 mm long, winged, scales 1-1.5 mm long at top.

Habitat and Distribution: Frequent; wet pinelands, cypress ponds, and bogs; from north peninsula and panhandle Florida, north to coastal plain Georgia, and west to southeast Louisiana.

Comment: The large showy heads can be seen flowering in spring in wet habitats.

Cuthbertia graminea Small
Grassleaf Roseling or Pink Spiderwort
[*Tradescantia rosea* Vent. var. *graminea* (Small) E.W. Anderson & Woodson; *Callisia graminea* (Small) G.C. Tucker]
Commelinaceae, Spiderwort Family

Habit: Herbaceous perennial, 5-40 cm tall. Stem thin, smooth, usually branched.

Leaves: Alternate; stalkless; blades erect, linear, grass-like, to 30 cm long, 1-3 mm wide.

Inflorescences: Terminal clusters.

Flowers: Pink, 2-2.5 cm broad, with 3 equal petals, each about 1 cm long.

Fruit: A spherical capsule, 4 mm diameter.

Habitat and Distribution: Occasional; sandhills and dry pine-oak woods; from central peninsula to central panhandle Florida, and north to Virginia.

Comment: A terrific little grass-like wildflower noticed mostly while flowering in spring and summer. It is limited to very dry habitats. Grassleaf Roseling is offered by nurseries.

Coreopsis nudata Nutt.

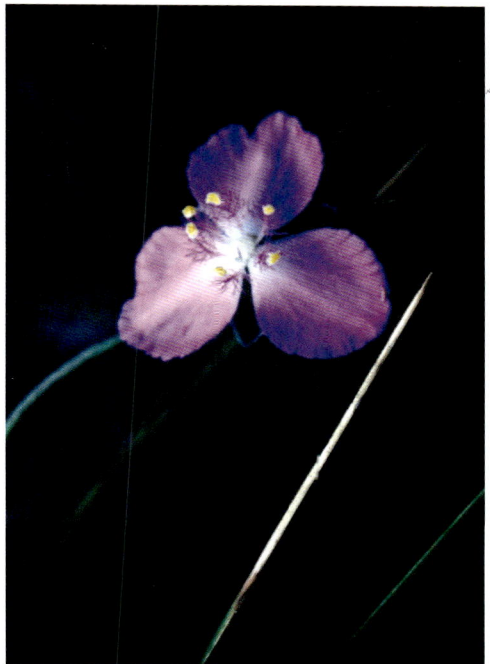

Cuthbertia graminea Small

Pink

Cuthbertia rosea (Vent.) Small
Wideleaf Roseling
[*Tradescantia rosea* Vent. var. *rosea; Callisia rosea* (Vent.) D.R. Hunt]
Commelinaceae, Spiderwort Family

Habit: Herbaceous perennial, to 40 cm tall. Stem erect, smooth, simple or few from a clump.

Leaves: Alternate; stalkless; blades lax, narrow, 20-30 cm long, 3-12 mm wide.

Inflorescences: Terminal clusters.

Flowers: Pink or rose, 2-2.5 cm broad, with 3 equal petals.

Fruit: A spherical capsule, about 4 mm diameter.

Habitat and Distribution: Infrequent to rare; sandy woods; from central peninsula Florida, north to North Carolina.

Comment: Wideleaf Roseling is much less common than Grassleaf Roseling in nature, but is also offered by nurseries. It flowers in spring and summer and does best in dry sites.

Dalea feayi (Chapm.) Barneby
Globe-headed Prairie-clover
Fabaceae (Leguminosae), Pea or Bean Family

Habit: Herbaceous perennial, 30-80 cm tall, in clumps. Stems branching, erect, smooth, slightly woody, and leafy.

Leaves: Alternate; stalked; blades once-divided, with 3-9 narrow pointed leaflets, leaflets 5-12 mm long.

Inflorescences: Terminal, globe-like heads with numerous leaf-like bracts.

Flowers: Light pink or rose, 7-9 mm long, with 5 petals, 4 stamen-like, 5 stamens.

Fruit: Pods straight, 3 mm long.

Habitat and Distribution: Frequent; dry pinelands, sandhills and scrub; throughout peninsula Florida into central panhandle Florida and southern Georgia.

Comment: Normally summer and fall flowering, it can bloom all year in the southern part of its range. Frequently mistaken for a member of the Aster Family due to the tight heads of flowers. The 5 stamens and round heads can be used to separate Globe-headed Prairie-clover from other Prairie-clovers.

Cuthbertia rosea (Vent.) Small

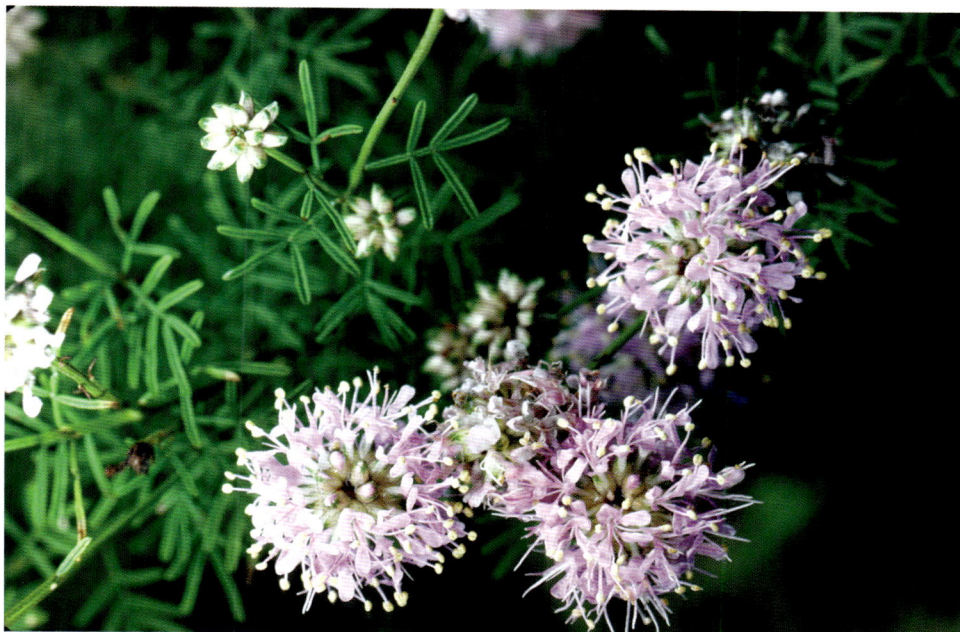

Dalea feayi (Chapm.) Barneby

Pink

Decodon verticillatus (L.) Elliott
Swamp Loosestrife
Lythraceae, Loosestrife Family

Habit: Perennial shrub, 0.6-2.5 m long. Branches arching, often rooting at tips.

Leaves: Opposite or in whorls; stalked; blades lance-shaped, to 15 cm long and 4 cm wide.

Inflorescences: Flower clusters in leaf axils.

Flowers: Petals pink, narrow, 8-12 mm long.

Fruit: A rounded capsule, 4-7 mm diameter; seeds angular, reddish, 1.5-2 mm long.

Habitat and Distribution: Frequent; swamps, marshes and ponds; from central peninsula Florida, west to Louisiana, and north to Indiana, Missouri, Minnesota, Quebec, and Nova Scotia.

Comment: This shrub could be more frequently cultivated for areas where water stands. The flowers are attractive and seen in summer.

Desmodium incanum DC.
Creeping Beggarweed
[Desmodium canum Schinz & Thell.]
Fabaceae (Leguminosae), Pea or Bean Family

Habit: Herbaceous perennial, to 1 m tall, from above or below ground runners. Stem ascending to erect, hairy.

Leaves: Alternate; stalked; blades with 3 elliptic to oval leaflets, leaflets 2-6 cm long, hairy.

Inflorescences: Terminal and axillary spikes.

Flowers: Rose to purple, typically shaped pea flowers, 5-8 mm long.

Fruit: Pod hairy, to 3 cm long, flat on top and constricted between seeds; seeds 3-8, flat on top, lower part rounded, with hooked hairs.

Habitat and Distribution: Frequent to common; hammocks, woods, pinelands, turf and other disturbed sites; throughout Florida, south Georgia, south Alabama, and south Texas; West Indies, Central, and South America; a native of the tropics, now pantropical.

Comment: Flowering in warm months northward and all year southward. Creeping Beggarweed is an extremely common turf weed, responding aggressively to irrigation. Normally to about 60 cm tall it will easily reach 1 m tall in wet disturbed habitats. The fruits attach easily to clothing or fur by the hooked hairs and are widely distributed. This weed should extend its range quickly.

Decodon verticillatus (L.) Elliott

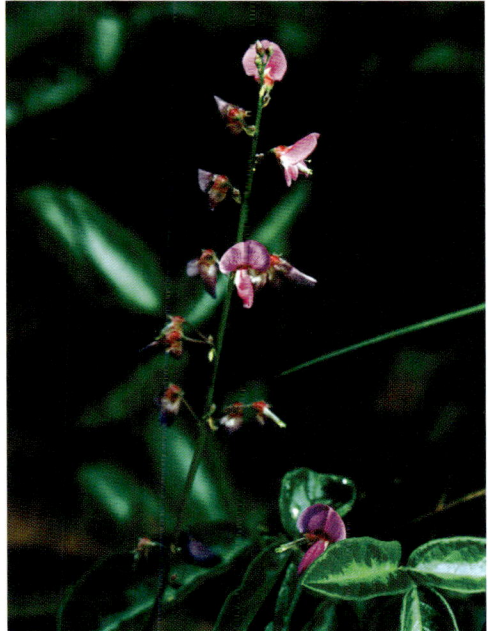

Desmodium incanum DC.

Pink

Drosera capillaris Poir.
Pink Sundew
Droseraceae, Sundew Family

Habit: Herbaceous perennial, 0.5-4 cm tall. Stem erect, smooth.

Leaves: Alternate, in a basal rosette; stalked; blades 3-8 cm long, reddish, spoon-shaped, covered with glandular hairs.

Inflorescences: Spike-like terminal arrangement, along one side of stem.

Flowers: Pink, 2-20, 5 petals, each 6-7 mm long.

Fruit: Capsule rounded, 3-4 mm long, partially enclosed by sepals; seeds brown, 0.5 mm long, ridged.

Habitat and Distribution: Frequent; wet, sandy, acid soils such as bogs, ditches and flatwoods; throughout Florida, north to Virginia, west to Texas, and Mexico south into northern South America; West Indies.

Comment: This diminutive, interesting plant, common in wet acid soils, flowers in spring. A sticky exudation at the tip of the glandular hairs on the leaves cause tiny insects to stick and be adsorbed.

Drosera filiformis Raf.
Dew-thread or Thread-leaf Sundew
Droseraceae, Sundew Family

Habit: Herbaceous perennial, to 24 cm tall. Stem erect, smooth.

Leaves: Alternate, in a basal rosette; stalked; blades 8-24 cm long, to 1 mm wide, blades indistinct from stalk, stalk and blade covered with purplish, glandular hairs.

Inflorescences: Spike-like terminal arrangement, along one side of stem.

Flowers: Pink to purple, 4-6, 5 petals, each 7-15 mm long.

Fruit: Capsule rounded, about 4 mm long, partially enclosed by sepals; seeds black, elliptical, about 0.5 mm long, ridged.

Habitat and Distribution: Very infrequent to rare; moist to wet sandy soils of pond and lake bottoms and margins, roadsides; Washington and Bay Counties in panhandle Florida, southeast North Carolina to Massachusetts.

Comment: Flowering occurs in spring southward and into September in the northern part of the range. The pink flowers are certainly attractive, but the eye-catchers are the long thin leaves covered with purplish glandular hairs. The hairs sparkle in the sun. A sticky exudation at the tip of the glandular hairs causes tiny insects to stick and be adsorbed. Dew-thread is listed as Endangered by the State of Florida.

Drosera capillaris Poir.

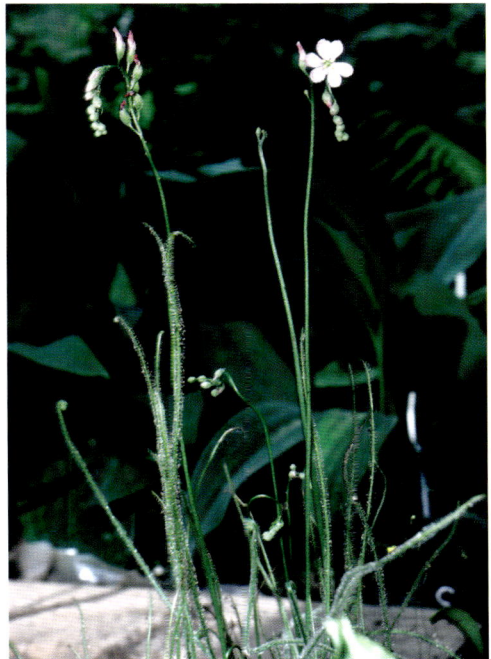

Drosera filiformis Raf.

Pink

Elephantopus carolinianus Raeusch.
Leafy Elephant's-foot
Asteraceae (Compositae), Aster or Sunflower or Daisy Family

Habit: Herbaceous perennial, to 1.0 m tall, from a thick taproot. Stems 1-few, leafy, hairy, branched upwards on stem.

Leaves: Alternate; few or no basal blades, stalkless; blades broadly elliptic to elliptic, slightly hairy on top, densely hairy underneath, 6-25 cm long and 4-11 cm wide, margins with blunt teeth.

Inflorescences: Heads contained in terminal glomerules surrounded by 3 leaflike bracts, 4-flowered, only disc flowers present.

Flowers: Pink to pale purple or white, each 7-8 mm long, with 5 lobes.

Fruit: Nutlets oblong, ridged, hairy, 3.5-4 mm long, with a tuft of bristles 4-5 mm long at the tip.

Habitat and Distribution: Frequent; moist shaded habitats with sandy soils, forests; central peninsula Florida, north to New Jersey, Ohio, Illinois and Kansas, and west to Texas.

Comment: Flowering occurs in summer and fall. The small flowers are difficult to see due to the surrounding leaflike structures. Leafy Elephant's-foot looks more like a weed than a wildflower. This species of Elephant's-foot looks less like an Elephant's-foot than others due to the leafy stem and the few basal blades. Other species (such as the following) have a dense circle of basal leaves.

Elephantopus elatus Bertol.
Florida Elephant's-foot
Asteraceae (Compositae), Aster or Sunflower or Daisy Family

Habit: Herbaceous perennial, 0.7-1.2 m tall, from a thick taproot. Stems 1-several, hairy, with no or few leaves.

Leaves: Alternate, in a flat basal rosette; stalkless; blades densely hairy, especially on lower surface, broadly elliptic often broader at the tip, 8-25 cm long and 2-7 cm wide, with scalloped edges.

Inflorescences: Heads contained in terminal glomerules surrounded by 3 leaflike bracts, 4-flowered, only disc flowers present.

Flowers: Pink to pale purple, each 7-9 mm long, with 5 long slender lobes.

Fruit: Nutlets oblong, ridged, hairy, 4-5 mm long, with a tuft of bristles 5-5.5 mm long on the top.

Habitat and Distribution: Common; flatwoods, deciduous woodlands, margins of swamps, sandhills, and disturbed sandy habitats; throughout Florida, west to Louisiana and north to South Carolina along the coastal plain.

Comment: Florida Elephant's-foot flowers in the summer and fall, but the blooms are scarcely noticed because of their small size and the surrounding leaflike structures. The plant gets its common name from the resemblance of the arrangement of the basal leaves fancifully to an elephant's foot. The leaves are in a large circle around the rootstock.

Elephantopus carolinianus
Raeusch.

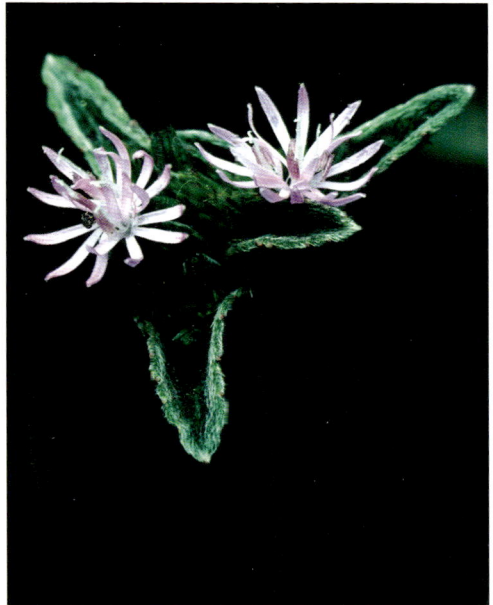

Elephantopus elatus Bertol.

Pink

Emilia fosbergii Nicolson
Cupid's Shavingbrush or Florida Tassel-flower
Asteraceae (Compositae), Aster or Sunflower or Daisy Family

Habit: Herbaceous annual, 10-80 cm tall, from a taproot. Stems erect or ascending, simple or with a few branches.

Leaves: Alternate; stalkless; blades variable in size and shape, usually at least somewhat clasping the stem, lowest leaves largest, margins toothed.

Inflorescences: Cylindrical heads, solitary or in a flat-topped arrangement, terminal, disc flowers only.

Flowers: Pink to coral to red disc flowers spreading out about 2-5 mm above the green bracts.

Fruit: Nutlets narrow, hairs on the 5 angles.

Habitat and Distribution: Frequent; disturbed sites; peninsula Florida; a native of the Old World tropics, now pantropical.

Comment: The bright pink flowers poking out of the head of this wildflower, not unlike a shaving brush, often provoke comments. The flowers bloom all warm months. This is a very weedy species.

Eupatorium coelestinum L.
Mist Flower or Ageratum
[*Conoclinium coelestinum* (L.) DC.]
Asteraceae (Compositae), Aster or Sunflower or Daisy Family

Habit: Herbaceous perennial, to 1 m tall, from underground runners (rhizomes).

Leaves: Opposite; stalked; blades hairy, oval to triangular, with three main veins, margins scalloped to toothed. Stems and leaves may have purple-red pigmentation.

Inflorescences: Heads in clusters at tips of branches.

Flowers: Violet, purple or showy bright blue tubular flowers, 35-70 per head, narrow, 2-3 mm long and with protruding slender stigmas giving appearance of a spray.

Fruit: Nutlets narrow, pointed, 1-2 mm long, with bristles about 3 mm long at top.

Habitat and Distribution: Frequent to common; wet habitats such as floodplains, ditches, and river and stream banks; throughout Florida, west to Texas, north to Kansas, Illinois and Ohio, and east to New Jersey; West Indies.

Comment: Common in cultivation and weedy in nurseries, gardens, and along roadsides. This dense bushy native blooms from summer into fall. A hardy staple of southern gardens, it freezes to the ground in winter. With mulch, Mist Flower can be grown in most dry habitats, but does best in moist areas or with irrigation. It is easily grown from seeds or by transplanting the runners.

Emilia fosbergii Nicolson

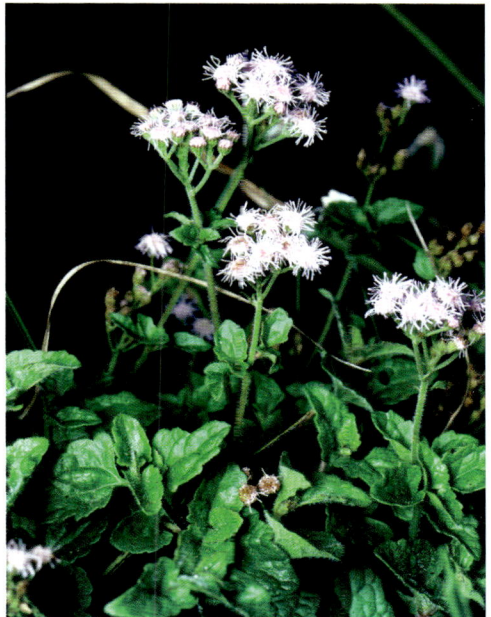

Eupatorium coelestinum L.

Pink

Eupatorium fistulosum Barratt
Joe-pye Weed, Hollow-stemmed Joe-pye Weed, Queen-of-the-meadow
[*Eupatoriadelphus fistulosus* (Barratt) R.M. King & H. Robins.]
Asteraceae or Compositae, Aster or Sunflower or Daisy Family

Habit: Herbaceous, erect, robust, non-clumping perennial, to 3 m tall. Stems purplish, waxy, hollow.

Leaves: In whorls of 3 to 7; short-stalked; blades elliptic to lance-shaped, often narrow, margin toothed.

Inflorescences: Round-topped terminal masses of small slender heads, only disc flowers present.

Flowers: Pink-purple, tubular, 3 to 8, very small, 3-5 mm long.

Fruit: Nutlets slender, 5-ribbed, bristles 6 mm long at top.

Habitat and Distribution: Occasional; stream banks, waste areas, bottomlands, moist woods and meadows; from central peninsula Florida, west to Oklahoma and Texas, and north to Iowa and Maine.

Comment: Blooms mid-summer until frost. The hollow stems sometimes used as a drinking straw when needed outdoors.

Eupatorium incarnatum Walter
Ageratum
[*Fleischmannia incarnata* (Walter) R.M. King & H. Rob.]
Asteraceae (Compositae), Aster or Sunflower or Daisy Family

Habit: Herbaceous perennial, 0.3-1.2 m tall. Stems sprawling or leaning through vegetation, branching.

Leaves: Opposite; stalked; blades triangular, 2-7 cm long and 2-5 cm wide, margins toothed.

Inflorescences: Loose, rounded, terminal clusters, only disc flowers present.

Flowers: Pink to white, 12-30 per head, tubular, about 3 mm long, with protruding slender stigmas and sweet vanilla like fragrance.

Fruit: Nutlets, narrow, 5-ribbed, bristles 3-4 mm long at top.

Habitat and Distribution: Occasional to infrequent; rich woods, swamps and moist disturbed habitats; from central peninsula Florida, west to Texas and Mexico, and north to Arkansas, Missouri, Kentucky, West Virginia and Virginia.

Comment: Flowering in summer and fall.

Eupatorium fistulosum Barratt

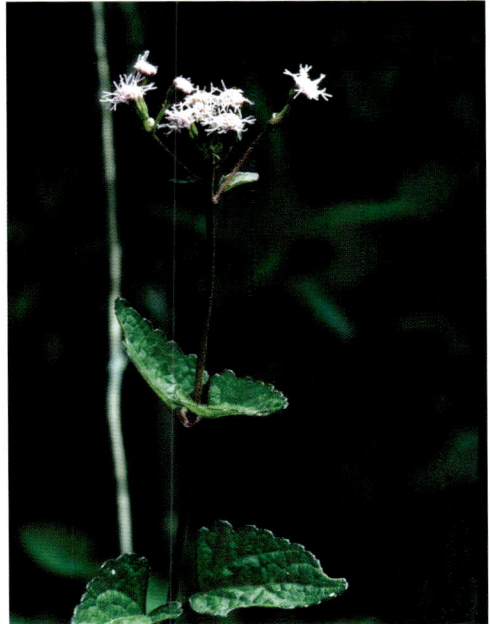

Eupatorium incarnatum Walter

Pink

Fumaria officinalis L.
Fumitory or Earthsmoke
Fumariaceae, Fumitory Family

Habit: Herbaceous annual, 0.2-0.8 m tall. Stems smooth and branching.

Leaves: Alternate; stalked; blades deeply divided into narrow, feathery segments.

Inflorescences: Elongated masses along stem tips.

Flowers: Pink or purplish, more or less tubular, 6-9 mm long, with 4 petals, upper petal sac-shaped at base.

Fruit: An almost spherical capsule, 2-3 mm diameter.

Habitat and Distribution: Occasional to locally common; fields, roadsides, and disturbed sites; throughout Florida, and scattered throughout most of the U.S. Native to Eurasia.

Comment: Winter and spring flowering, this little annual can produce a spectacular mass of color.

Galactia erecta (Walter) Vail
Erect Milkpea
Fabaceae (Leguminosae), Pea Family

Habit: Herbaceous perennial, 20-40 cm tall, from an underground runner. Stems several, erect, zigzag, with a few scattered hairs.

Leaves: Alternate; stalked; blades divided with 3 oblong, narrowly elliptic to linear leaflets, each 1.5-4 cm long, hairless.

Inflorescences: Axillary clusters of 1-6 flowers, clusters 1-2 cm long.

Flowers: Pale purple to white, 6-9 mm long, typically pea-shaped.

Fruit: A pod, 2-4 cm long, 5-8 mm broad, with 6-10 seeds.

Habitat and Distribution: Infrequent; dry to moist sandy soils of pinelands and mixed or deciduous woods; throughout panhandle Florida, west to east Texas, and north to North Carolina.

Comment: The flowers of Erect Milkpea can be found from spring through summer. While not especially showy, the habitats in which they occur are not usually filled with wildflowers, so these flowers often catch the eye.

Fumaria officinalis L.

Galactia erecta (Walter) Vail

Pink

Galactia minor Duncan
Small-leaf Prostrate Milkpea
[*Galactia floridana* Torr. & Gray var. *microphylla* Chapm.; *Galactia microphylla* (Chapm.) Rogers ex Isley]
Fabaceae (Leguminosae), Pea or Bean Family

Habit: Perennial, trailing vine, to 60 cm long. Stems with short fine hairs, short internodes, 1-2 cm long.

Leaves: Alternate; stalked; blades once-divided, with 3 narrow oval leaflets, leaflets each about 1.5 cm long.

Inflorescences: Flowers stalked, solitary or in clusters from leaf axils.

Flowers: Pink to pink-purple, typically pea-shaped, 1.2-1.4 cm long, with flaring upper petals.

Fruit: Pods firm, 2-3 cm long, 4-5 mm wide.

Habitat and Distribution: Infrequent; sandy pinelands, oak woods, sandhills and occasionally in sandy disturbed sites; northern Florida, north to South Carolina, and west to Mississippi.

Comment: This colorful wildflower is a smaller version of Prostrate Milkpea. Small-leaf Prostrate Milkpea blooms in the warm months.

Galactia regularis (L.) BSP.
Twining Milkpea
[*Galactia volubilis* (L.) Britton, misapplied]
Fabaceae (Leguminosae), Pea or Bean Family

Habit: Perennial, twining, climbing vine, 1-2.5 m long. Stems hairy.

Leaves: Alternate; stalked; blades once-divided, with 3 oval to broadly elliptic leaflets, leaflets 1.5-4.5 cm long.

Inflorescences: Flowers stalked, solitary or in clusters from leaf axils.

Flowers: Pink-purple, typically pea-shaped, 7-12 mm long, with flaring upper petals.

Fruit: Pods hairy, 2-6 cm long, 3-5 mm wide.

Habitat and Distribution: Common; pinelands, open woods, borders, sandhills, old fields, marshes, and disturbed areas; throughout Florida, west to east Texas, and north to Missouri, Pennsylvania and New York.

Comment: Twining Milkpea is a common garden weed that, if left, can become an exciting wildflower. The pink blooms with their flaring upper petal provide a delicate but noticeable color among foliage in summer and fall.

Galactia minor Duncan

Galactia regularis (L.) BSP.

Pink

Galactia volubilis (L.) Britton
Prostrate Milkpea
[*Galactia macreei* M.A. Curtis; *Galactia regularis* (L.) BSP., misapplied]
Fabaceae (Leguminosae), Pea or Bean Family

Habit: Perennial, trailing or low, loosely climbing vine, to 1.5 m long. Stems with short fine hairs. Internodes 1.5-3 cm long.

Leaves: Alternate; stalked; blades once-divided, with 3 oval leaflets, leaflets 2-5 cm long.

Inflorescences: Flowers stalked, solitary or in clusters from leaf axils.

Flowers: Pink-purple to lavender, typically pea-shaped, 1.1-1.5 cm long, with flaring upper petals.

Fruit: Pods hairy, 2-5 cm long, 4-5.5 mm wide.

Habitat and Distribution: Frequent; pinelands, hammocks, scrub, sandhills and other dry sites; throughout Florida, west to Louisiana and east Texas, and north to Pennsylvania.

Comment: Prostrate Milk-pea flowers in warm months. Large populations provide a show of large pealike flowers along the ground in dry sandy habitats. This wildflower could certainly be better utilized as a ground cover for difficult open sandy areas.

Garberia heterophylla (W. Bartram) Merr. & F. Harper
Garberia
[*Garberia fruticosa* (Nutt.) A. Gray]
Asteraceae (Compositae), Aster or Sunflower or Daisy Family

Habit: Perennial shrub, 1-2.5 m tall. Stems woody, hairy, bushy branched.

Leaves: Alternate, evergreen; short-stalked; blades rounded, 1.5-3.5 cm long, glandular dotted.

Inflorescences: Terminal flat-topped clusters of heads, only disc flowers present.

Flowers: Pink or purple, usually with 3-5 flowers per head, all tubular, about 1 cm long.

Fruit: Nutlets slender, ribbed, 7-8 mm long, hairy, with bristles at top.

Habitat and Distribution: Frequent to common; sand pine and oak scrub, and sandhills; from central into north peninsula Florida

Comment: Primarily flowering in fall, Garberia can be a wonderful shrub in dry sandy areas. Used as a border or specimen plant, the gray evergreen foliage and showy heads of flowers in fall are quite colorful.

Galactia volubilis (L.) Britton

Garberia heterophylla
(W. Bartram) Merr. & F. Harper

101

Pink

Geranium carolinianum L.
Carolina Geranium or Crane's-bill
Geraniaceae, Geranium Family

Habit: Herbaceous winter annual, 20-50 cm tall, from a taproot. Stems branched, with dense short hairs.

Leaves: Opposite; very long-stalked; blades rounded, deeply divided into 5-9 radiating segments, each segment further divided.

Inflorescences: Flowers paired, in a loose leafy arrangement at stem tips.

Flowers: Pale pink to pink violet, petals 5, equal, each about 7 mm long, somewhat rectangular, notched at end.

Fruit: Dry, 2-5 cm long, 5-chambered, with a prominent beak, 5-seeded.

Habitat and Distribution: Common; fields, gardens, pastures, and disturbed sites; throughout Florida, north to Maine and British Columbia, and west to Texas and California.

Comment: Carolina Geranium blooms in spring. It is a common weed of gardens, old fields and row crops. The prominent beak is distinctive.

Geranium maculatum L.
Wild Geranium
Geraniaceae, Geranium Family

Habit: Herbaceous perennial, 20-70 cm tall, from a thick underground runner. Stems branched, hairy.

Leaves: Opposite; very long-stalked; blades rounded, deeply divided into 5-7 radiating segments, each segment further divided.

Inflorescences: Flowers paired, in a loose leafy arrangement at stem tips.

Flowers: Pink to purplish-pink (rarely white), petals 5, equal, each about 1.2-2.5 cm long.

Fruit: Dry, 2.2-3.0 cm long, 5-chambered, with a prominent beak, 5-seeded.

Habitat and Distribution: Frequent to common; forested habitats; mostly on the piedmont and northward, central Georgia, west to Oklahoma and Nebraska, north to Manitoba and Maine.

Comment: The large pink flowers of Wild Geranium can easily be seen in wooded habitats in late spring from April into June. Wild Geranium can be found for sale in some nurseries. A white cultivated form is also sometimes available.

Geranium carolinanum L.

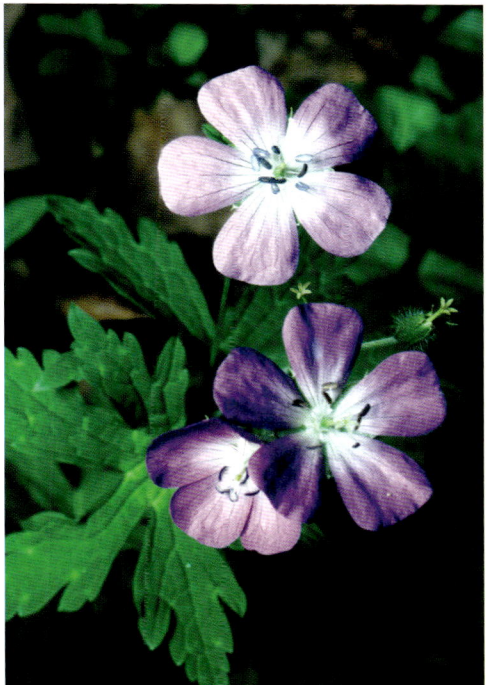

Geranium maculatum L.

Pink

Glandularia pulchella (Sweet) Tronc.
Moss Verbena
[*Verbena tenuisecta* Briquet]
Verbenaceae, Vervain Family

Habit: Prostrate, herbaceous perennial, 10-30 cm tall. Stems spreading, square in cross section, hairy.

Leaves: Opposite; stalked; blades hairy, 1-4 cm long, deeply divided, divisions narrow.

Inflorescences: Terminal, flat-topped or round-topped spikes.

Flowers: Pink, purple, or white, over 1 cm wide, with narrow tube 1 cm long, and 5 petals with notched tips.

Fruit: 4 nutlets, each 3 mm long.

Habitat and Distribution: Frequent to locally common; fields, pastures, roadsides, and other disturbed sites; throughout Florida, west to Texas, and north to North Carolina, small populations have been found in Illinois, Kentucky, Missouri, Arizona and California; a native of South America.

Comment: A common border plant with very colorful flowers in summer and fall. Moss Verbena has escaped into grassy habitats where these low growing plants can compete for light. It makes a dramatic wildflower when in bloom.

Hedeoma graveolens Chapm. ex A. Gray
Mock-pennyroyal
[*Stachydeoma graveolens* (Chapm. ex A. Gray) Small]
Lamiaceae (Labiatae), Mint Family

Habit: Herbaceous perennial or low shrub, 20-60 cm tall. Stems erect, hairy, older stems woody, younger stems greenish brown, usually branching to form a compact, roundish plant.

Leaves: Opposite; stalkless; blades thick textured, glandular dotted, rounded to oblong to spade-shaped, 1-1.5 cm long, margins rolled downward.

Inflorescences: Flowers solitary in leaf axils at stem tips.

Flowers: Pale lilac to rose-purple, tubular, 1 cm long, with 2 lips, upper lip notched, lower lip with 3 lobes, center lobe notched and with a yellowish purple-dotted band.

Fruit: Consists of 4 round nutlets, each 1 mm diameter.

Habitat and Distribution: Occasional; sandy elevated locations or ridges in pinelands and flatwoods; endemic in central panhandle Florida.

Comment: Mock-pennyroyal flowers during the warm months, from spring into fall. The blooms are quite attractive.

Pink

Glandularia pulchella
(Sweet) Tronc.

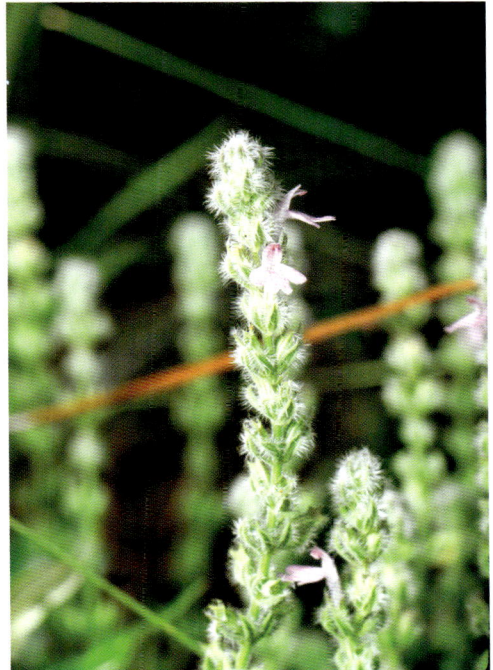

Hedeoma graveolens
Chapm. ex A. Gray

Pink

Heliotropium amplexicaule Vahl
Wild Heliotrope
Boraginaceae, Borage Family

Habit: Herbaceous perennial, 10-50 cm tall, from a deep woody root. Stems hairy, several from the root, sometimes slightly woody.

Leaves: Alternate; stalkless; blades narrow, oblong, margins toothed, 2-8 cm long.

Inflorescences: Flowers arranged along one side of the curled, terminal, flowering stalk.

Flowers: Violet to rose purple to blue to white, with yellow eye, bell-shaped, about 5 mm long, with 5 petals.

Fruit: 2 nutlets, 2-3 mm long.

Habitat and Distribution: Occasional; dry sites, fields, roadsides and other disturbed places; from central peninsula Florida, west to Texas, and north to Missouri and Virginia; a native of Argentina and Uruguay.

Comment: Flowering in spring and summer, Wild Heliotrope is a good border or bedding plant. It is weedy, but is a very appealing wildflower.

Hibiscus grandiflorus Michx.
Swamp Hibiscus
Malvaceae, Mallow Family

Habit: Perennial herb or shrub, to 3 m tall. Stems green or woody, green stems covered with velvety hair.

Leaves: Alternate; long-stalked; blades velvety, 3-lobed, sometimes arrowhead-shaped, to 18 cm long, and with toothed edges.

Inflorescences: Flowers long-stalked, solitary in leaf axils at stem tips.

Flowers: Pink to white with a deep purple central eye, somewhat nodding, 20 to 30 cm wide, 5 petals rounded and wider toward ends, each 12-14 cm long.

Fruit: A velvety, oval to cup-shaped capsule, 5-chambered, and 2-3 cm long; seeds smooth, dark, 3 mm long, with 2 or more per chamber.

Habitat and Distribution: Occasional; marshes, swamps, and other wet areas; throughout Florida, west to southeast Louisiana, and north to southeast Georgia.

Comment: Swamp Hibiscus has several color combinations of flowers. Some colonies have pink or purple flowers and some have a red central eye. These summer flowering plants prefer rich moist soil and at least some period during the day when they have full sun. This species freezes back and recovers from the rootstock.

Heliotropum amplexicaule Vahl

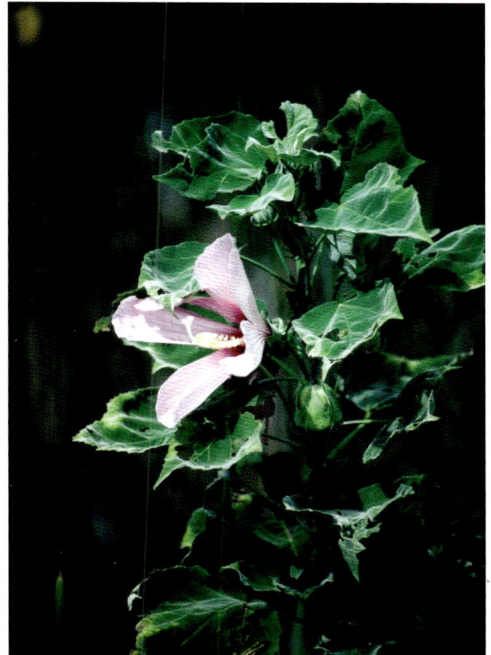

Hibiscus grandiflorus Michx.

Pink

Ipomoea cordatotriloba Dennst.
Sharp-pod Morning-glory
[*Ipomoea trichocarpa* Elliott]
Convolvulaceae, Morning-glory Family

Habit: Perennial twining vine, low climbing. Stems twining, brown, slightly hairy.

Leaves: Alternate; stalked; blades heart-shaped to deeply 3-lobed. Sepals unequal in length, linear, hairy.

Inflorescences: Clusters of 1-5 flowers from leaf axils.

Flowers: Pink to purple or rarely white, funnel-shaped, 3-5 cm diameter, 2.8-4.5 cm long, with radiating banded markings on petals.

Fruit: Capsule almost spherical, 6-9 mm long, hairy on top, 2-chambered, with 2 smooth seeds per chamber.

Habitat and Distribution: Frequent; sandy disturbed sites; throughout Florida, west to Texas, and north to South Carolina.

Comment: The large showy flowers are seen in the warm months. Sharp-pod Morning-glory is a nasty crop weed, but can be enjoyed as a cultivated plant in a contained area.

Ipomoea pes-caprae (L.) R. Br.
Railroad Vine
Convolvulaceae, Morning-glory Family

Habit: Perennial, prostrate, trailing vine, stems to 80 m long. Stems fleshy, green, rooting at joints.

Leaves: Alternate; long-stalked; blades leathery, round to somewhat oblong, notched at tip, often folded, 4-10 cm long.

Inflorescences: Flowers solitary or in clusters from leaf axils.

Flowers: Pink to lavender to purple, funnel-shaped, 4-6 cm long, with radiating banded markings on petals and some notches along outward edge of petals. Sepals unequal in length, somewhat rounded.

Fruit: Capsule almost spherical, about 1.5 cm diameter, with hairy seeds.

Habitat and Distribution: Frequent; sand dunes, beaches, strand, coastal flats and disturbed coastal sites; throughout coastal Florida, west along the Gulf Coast to Texas, and north along the coast into Georgia.

Comment: Trailing over bare sand along the coasts, Railroad Vine makes long strings fancifully resembling a railroad. Flowering can occur in any month. This species should be used more frequently as an ornamental on bare soils.

Ipomoea cordatotriloba Dennst.

Ipomoea pes-caprae (L.) R. Br.

Pink

Ipomoea sagittata Poir.
Glades Morning-glory
Convolvulaceae, Morning-glory Family

Habit: Perennial, twining vine, low-climbing, from creeping rootstock. Stems smooth.

Leaves: Alternate; stalked; blades narrowly to broadly lance-shaped, base arrow-shaped with two elongated lobes, 4-10 cm long. Sepals unequal in length, apex rounded with an abrupt tip.

Inflorescences: Solitary flowers in leaf axils.

Flowers: Pink to rose-purple to purple, funnel-shaped, 6-9 cm long.

Fruit: Capsule rounded, about 1 cm in diameter; seeds with long hairs on the angles.

Habitat and Distribution: Frequent; fresh and brackish marshes, dunes, beaches, and in mangroves; throughout Florida, west to Texas and Mexico; West Indies; Guatemala; western Mediterranean.

Comment: Glades Morning-glory is a promising wildflower. The large showy blooms occur in summer and fall. The stems are low-growing so that the flowers are easily seen. The stems also twine tightly so they do not appear to be in a dense mass. The unusual narrow arrowhead-shaped leaf blades often provoke comment.

Justicia crassifolia (Chapm.) Chapm. ex Small
Large-flowered Water-willow or Thick-leaved Water-willow
Acanthaceae, Acanthus Family

Habit: Herbaceous perennial, 20-40 cm tall. Stems smooth, grooved on upper parts.

Leaves: Opposite; stalkless; blades fleshy, narrow, smaller toward base of stem, 3-10 cm long.

Inflorescences: Long-stalked, terminal or axillary spikes.

Flowers: Rose-lavender to purple, tubular, 2-3 cm long, with 2 lips, upper lip smaller, 2 lobed, lower lip larger, 2 cm long, 3-lobed.

Fruit: Capsule about 2 cm long; seeds with pebbly surface.

Habitat and Distribution: Rare; swamps, bogs, wet flatwoods, and other low, wet sites; Franklin and Gulf Counties in panhandle Florida.

Comment: Large-flowered Water-willow is an endemic that is listed as Endangered by the State of Florida. The flowers are showy and seen in spring and summer.

Pink

Ipomoea sagittata Poir.

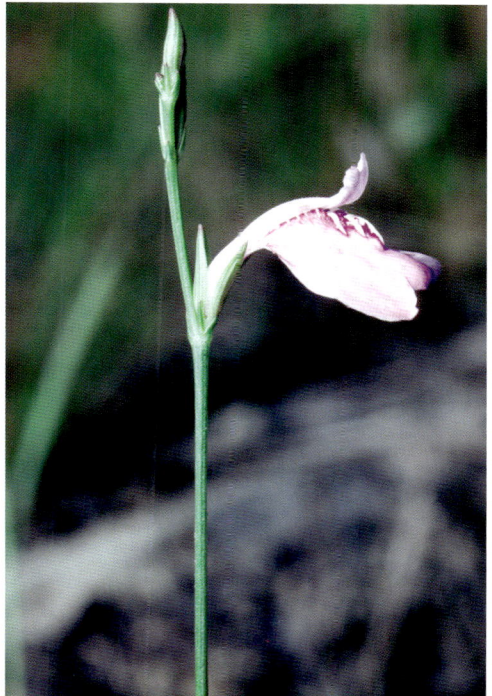

Justicia crassifolia
(Chapm.) Chapm. ex Small

Pink

Kalanchoe daigremontiana Raym.-Hamet & H. Perrier
Devil's-backbone
[*Bryophyllum daigremontiana* (Raym.-Hamet & H. Perrier) A. Berger]
Crassulaceae, Stonecrop or Orpine Family

Habit: Herbaceous, succulent perennial, to 1 m tall. Stems smooth.

Leaves: Opposite; stalkless; blades fleshy, lance-shaped, 5-15 cm long, lower surface green, mottled with purple or reddish-brown, margin with teeth, margin produces plantlets.

Inflorescences: Opposite, branched clusters at tips of stems.

Flowers: Pinkish to dusty purple or lavender, tubular, about 2.5 cm long, drooping, 4 petals.

Fruit: A dry capsule, splitting along one side to release seeds.

Habitat and Distribution: Rare to occasional; disturbed habitats, thickets; south peninsula Florida, West Indies; a native of Madigascar, now widespread in tropical climates.

Comment: Flowers can frequently be seen from spring thorough summer, but blooming can occur all year. The flowers are certainly attractive, but the main attraction of the plant is the mottled coloring and the curiosity of the many plantlets growing from the leaf margins.

Kalanchoe tubiflora (Harv.) Raym.-Hamet
Chandelier Flower
[*Bryophyllum tubiflorum* Harv.]
Crassulaceae, Stonecrop or Orpine Family

Habit: Herbaceous, succulent perennial, to 1 m tall. Stems grayish, erect.

Leaves: Opposite; stalkless; blades fleshy, slender, cylindrical, mottled with purple, 3-15 cm long, tips with teeth, plantlets produced from between teeth.

Inflorescences: Opposite, branched clusters at tips of stems.

Flowers: Pink, salmon, scarlet or orange, tubular to bell-shaped, 2-3 cm long, drooping, 4 petals.

Fruit: A dry capsule, splitting along one side to release seeds.

Habitat and Distribution: Occasional; wooded or open disturbed sites; southern and central peninsula Florida; a native of South Africa and Madagascar.

Comment: This commonly cultivated plant frequently escapes into the wild. The peculiar leaves and unusual form of Chandelier Flower provoke many comments. Even if leaves are separated from the plant they will produce plantlets at their tips. Flowers appear in winter.

Kalanchoe daigremontiana
Raym.-Hamet & H. Perrier

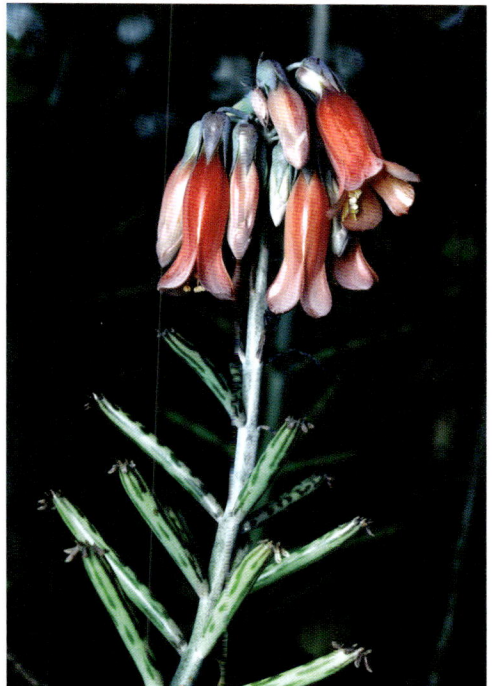

Kalanchoe tubiflora
(Harv.) Raym.-Hamet

Pink

Kalmia hirsuta Walter
Hairy Wicky
Ericaceae, Blueberry or Heath Family

Habit: Small perennial, evergreen shrub, 20-60 cm tall, from a hard underground base. Stems several to numerous, twigs hairy.

Leaves: Alternate; stalkless; blades variable, broadly or narrowly oval, 5-15 mm long and 2-8 mm wide.

Inflorescences: 1-4 flowers clustered at base of new leaves.

Flowers: Pale to deep pink, broadly bell shaped, 1-1.5 cm diameter, with 5 short, pointed lobes, and ring of darker pink spots inside.

Fruit: Capsule round, about 3 mm long, with tiny seeds.

Habitat and Distribution: Frequent; flatwoods, pine savannahs, scrub and borders of bogs; from central peninsula Florida, west to southeast Louisiana, and north to southeast South Carolina.

Comment: A very attractive, spring and summer flowering, delicate shrub. The flowers are very pretty, but the color is delicate and not showy at distance. Hairy Wicky should be better utilized as a small specimen or border plant.

Kosteletzkya virginica (L.) C. Presl ex A. Gray
Marsh-mallow, Seashore-mallow, Saltmarsh-mallow
Malvaceae, Mallow Family

Habit: Herbaceous perennial, somewhat shrubby, 0.5-1.7 m tall. Stems hairy, lower portion sometimes partly woody.

Leaves: Alternate; stalked; blades triangular, sometimes 3-lobed, 6-14 cm long, hairy, margins toothed, sometimes with two basal lobes.

Inflorescences: Terminal, leafy, branched clusters.

Flowers: Pink or whitish, nodding, hibiscus-like, 5-8 cm wide, with 5 overlapping petals, each 3-4.5 cm long.

Fruit: A 5-angled, hairy capsule, 6-8 mm long. Seeds smooth, dark brown, 3-5 mm long, 1 seed/chamber.

Habitat and Distribution: Common; swamps, fresh and brackish marshes, shores and other wet areas; throughout Florida, west to Texas, and north to Virginia and Delaware; Cuba.

Comment: Marsh-mallow flowers in spring and summer. Although a perennial, sometimes the plant functions as an annual, returning from seeds each year. Resembling a small Hibiscus flower this can be quite showy, especially in masses in marshes.

Kalmia hirsuta Walter

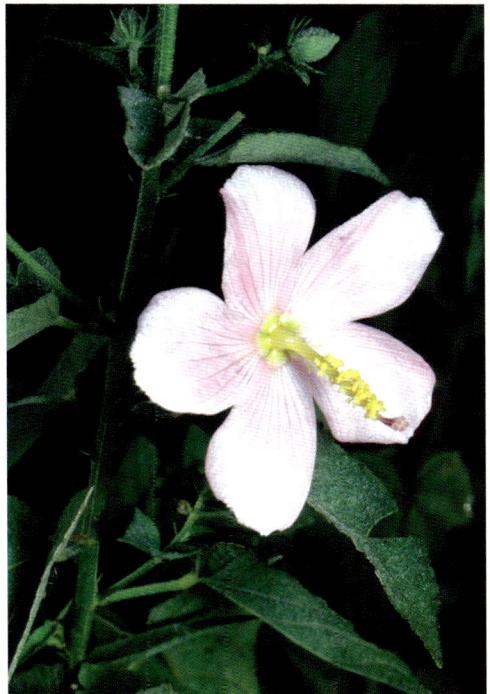

Kosteletzkya virginica
(L.) C. Presl ex A. Gray

Pink

Lamium amplexicaule L.
Henbit
Lamiaceae (Labiatae), Mint Family

Habit: Herbaceous annual, to 35 cm tall, from a taproot. Stems square in cross section, hairy, branching at base then arching upward, often rooting at lower joints.

Leaves: Opposite; lower leaves stalked, upper leaves stalkless; blades round to kidney shaped, sometimes somewhat 3-lobed, 1-3.5 cm broad, margins toothed.

Inflorescences: Clustered in leaf axils at stem tips.

Flowers: Purple, about 1.2-1.8 cm long, with slender tube and 2 lips, upper lip arched over mouth of tube, lower lip 3-lobed, with darker spots on central lobe.

Fruit: 4 nutlets, 2-2.4 mm long.

Habitat and Distribution: Occasional or locally common; open sites, fields, pastures, roadsides, lawns, gardens and other disturbed and waste places; from central peninsula Florida, north throughout the U.S. into Canada; a native of Eurasia and north Africa.

Comment: This small weed has attractive flowers in winter and spring.

Lespedeza bicolor Turcz.
Shrub Bush-clover
Fabaceae (Leguminosae), Pea or Bean Family

Habit: Perennial shrub, 1-3 m tall. Stems 0.5-1.5 cm diameter, smooth to hairy.

Leaves: Alternate; stalked; blades once-divided, with 3 leaflets, leaflets are oval, variable in shape, 2-5 cm long, hairy below or on both surfaces.

Inflorescences: Dense flowers along, 1-40 cm long, spike-like branches of a terminal panicle.

Flowers: Pinkish purple, typically pea-shaped, 0.8-1.2 cm long.

Fruit: Pods broad, 6-8 mm long, with tapered ends, hairy.

Habitat and Distribution: Rare in Florida, infrequent elsewhere; fields, roadsides, disturbed areas, woodlands and borders, mountain slopes, flatwoods, and banks; from central peninsula and panhandle Florida, west to Louisiana, and north to Arkansas, Kentucky and Virginia; a native of Japan.

Comment: Flowering in warm months from the spring through the fall in the southern range and summer and fall in the northern range. The multitude of flowers makes this an interesting colorful shrub for border plantings. Originally Shrub Bush-clover was cultivated for wildlife food and soil improvement.

Lamium amplexicaule L.

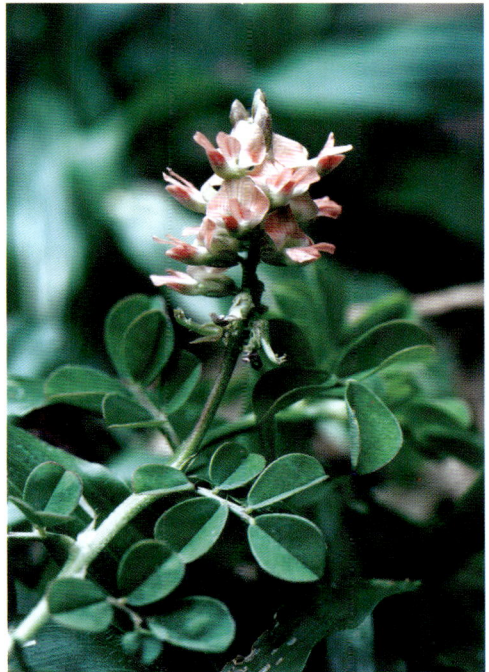

Lespedeza bicolor Turcz.

Pink

Liatris gracilis Pursh
Common Blazing-star
Asteraceae (Compositae), Aster or Sunflower or Daisy Family

Habit: Herbaceous perennial, 0.4-1 m tall, from a thickened rootstock. Stems hairy to nearly smooth, leafy, rarely branched.

Leaves: Alternate; stalkless; blades narrow and grass-like, to 35 cm long and 1 cm wide, largest leaves at base, become progressively smaller upwards, with a few long scattered stiff hairs on margin.

Inflorescences: Stalked heads forming long cylindrical masses along upper portion of stem, only disc flowers present.

Flowers: Pink or pink-purple, 3 to 7 per head, tubular, 6-8 mm long, with 5 pointed lobes.

Fruit: Nutlets ribbed, 2.5-4 mm long, with bristles at top.

Habitat and Distribution: Common; sandy woods, pinelands and flatwoods; throughout Florida, west to southwest Alabama, and north to South Carolina.

Comment: In many pinelands this summer- and fall-flowering species is the most common blazing-star seen. When planting Common Blazing-star provide a bit of additional room as it tends to bend and loop over.

Liatris graminifolia (Walter) Willd.
Grassleaf Blazing-star
Asteraceae (Compositae), Aster or Sunflower or Daisy Family

Habit: Herbaceous perennial, to 1.5 m tall, from a thickened rootstock. Stems hairy to nearly smooth, leafy.

Leaves: Alternate; stalkless; blades very slender, 5-30 cm long and 2-7 mm wide, largest leaves at base, with a few long stiff hairs at base on margin.

Inflorescences: Short-stalked or stalkless heads forming long cylindrical masses along upper portion of stem, only disc flowers present.

Flowers: Pink or pink-purple, 7-14 tubular flowers per head, 11-16 mm long, with 5 pointed lobes.

Fruit: Nutlets ribbed, 2.5-3 mm long, with bristles at top.

Habitat and Distribution: Occasional; old fields, flatwoods, and open, oak woods; from central peninsula Florida, to Alabama, and north to New Jersey.

Comment: Tolerating somewhat drier conditions, Grassleaf Blazing-star blooms in summer and is suitable for most Florida gardens.

Liatris gracilis Pursh

Liatris graminifolia (Walter) Willd.

Pink

Liatris spicata (L.) Willd.
Dense Blazing-star
Asteraceae (Compositae), Aster or Sunflower or Daisy Family

Habit: Herbaceous perennial, 0.6-2.5 m tall, from a thickened rootstock. Stems smooth or sometimes hairy, leafy, often retain fibers of old basal leaves.

Leaves: Alternate; stalkless; blades slender and grass like, 10-40 cm long and 0.5-2.0 cm broad, largest leaves at base, leaves smaller upwards.

Inflorescences: Stalkless heads, forming long cylindrical masses along upper portion of stem, only disc flowers present.

Flowers: Pink or pink-purple, 5-18 per head, tubular, 7-11 mm long, with 5 pointed lobes.

Fruit: Nutlets ribbed, hairy on ribs, 3-8 mm long, with bristles at top.

Habitat and Distribution: Occasional; low and moist sites, bogs, savannas, flatwoods, rocky woods and sandhills; throughout Florida, north to New York, Michigan and Missouri, and west to southeast Louisiana, Wyoming and New Mexico.

Comment: This blazing-star, blooming in summer and fall, like others is very attractive. It requires more moisture than other species.

Liatris tenuifolia Nutt.
Fineleaf Blazing-star
Asteraceae (Compositae), Aster or Sunflower or Daisy Family

Habit: Herbaceous perennial, 0.6-2.0 m tall, from a thickened rootstock. Stems smooth or nearly so, majority of leaves basal, fibers of old basal leaves retained.

Leaves: Alternate; stalkless; blades needle-like, to 40 cm long and 6 mm broad, upper stem with very few smaller leaves.

Inflorescences: Stalked heads, forming long cylindrical masses along upper portion of stem, only disc flowers present.

Flowers: Pink or pink-purple, 3-6 per head, tubular, 6-8 mm long, and with 5 pointed lobes.

Fruit: Nutlets ribbed, hairy, 3-3.5 mm long, with bristles at top.

Habitat and Distribution: Frequent to common; open oak or pine woods, sandhills, and clearings; throughout Florida, west to southern Alabama, and north on the coastal plain to South Carolina.

Comment: Like other blazing-stars the spectacular purple spikes of flowers in summer and fall make Fineleaf Blazing-star a great addition to any garden. An added bonus is that it will grow in the drier conditions found in much of Florida.

Liatris spicata (L.) Willd.

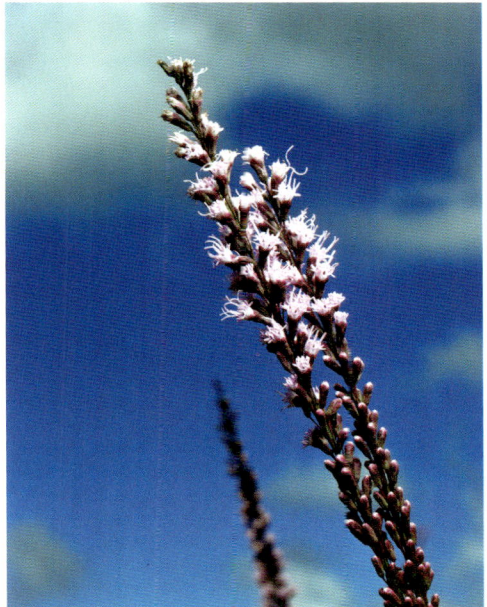

Liatris tenuifolia Nutt.

Pink

Lindernia anagallidea (Michx.) Pennell
Many-flowered False Pimpernel
[*Lindernia dubia* (L.) Pennell var. *anagallidea* (Michx.) Cooperr.]Scrophulariaceae,
Figwort or Snapdragon Family

Habit: Herbaceous annual, 5-20 cm tall. Stems smooth, slender, much branched and spreading, 4-angled in cross section.

Leaves: Opposite; stalkless; blades rounded to somewhat lance-shaped, to 2.0 cm long.

Inflorescences: Long-stalked, solitary flowers in leaf axils.

Flowers: Pale lavender to white, 6-9 mm long, tubular, upper lip tiny, 2-lobed, lower lip larger, with 3 spreading lobes.

Fruit: A broadly elliptic capsule, 2.5-3.5 mm long.

Habitat and Distribution: Occasional; sandy shores and margins of ponds, streams and other water bodies, as well as wet flatwoods; throughout Florida, west to Texas, and north to North Dakota, Quebec and New Hampshire.

Comment: The pale flowers on this small weak plant are scarcely noticed in summer and fall.

Lupinus villosus Willd.
Lady Lupine
Fabaceae (Leguminosae), Pea or Bean Family

Habit: Herbaceous annual or biennial, to 60 cm tall. Stems densely covered with silvery shaggy hair, somewhat woody, spreading, forming dense clump.

Leaves: Alternate; stalked, stipules at base of stalk conspicuous; blades broadly elliptic, hairy, 4-15 cm long.

Inflorescences: Long, terminal, dense spike.

Flowers: Pink to lavender or purple, with deep red purple spot on upper petal, typically pea-shaped.

Fruit: Pods flattish, elliptic, 2.5-4 cm long, covered with long shaggy hairs.

Habitat and Distribution: Frequent; dry woods, sandhills, scrub, sandy pastures and roadsides; from central peninsula Florida west to Louisiana, and north to North Carolina.

Comment: A striking spring-flowering wildflower that is no less noticeable in other seasons due to the mass of silvery hairs covering the plant. Can be a great wildflower for difficult dry, open, sandy habitats.

Lindernia anagallidea
(Michx.) Pennell

Lupinus villosus Willd.

Pink

Lycium carolinianum Walter
Christmas Berry or Carolina Wolfberry
Solanaceae, Nightshade or Potato Family

Habit: Perennial shrub, 0.3-3.0 m tall. Stems long, curving, with many short, thorny branches.

Leaves: Alternate; stalkless; blades succulent, narrow and broader at tip, to 2.5 cm long, often clustered on spur shoots.

Inflorescences: Solitary flowers on long stalks in leaf axils.

Flowers: Lavender, blue or white, about 1 cm diameter, with 4 or 5 spreading petals.

Fruit: Berry bright red, oval, about 1 cm long.

Habitat and Distribution: Frequent; coastal dunes, marshes, mangrove swamps and shell mounds; throughout Florida, west to Texas and northeast Mexico, and north to Georgia; West Indies.

Comment: Christmas Berry is a bushy-branched, thorny shrub. It flowers in summer and fall. The flowers are attractive, but the showy red fruit can hang on for months. The plant prefers habitats that are frequently inundated, but can occur in upland areas that receive salt spray. Christmas Berry can be cultivated in wet and irrrigated soils lacking salt.

Lygodesmia aphylla (Nutt.) DC.
Roserush, Skeleton-weed, Rushweed, Flowering-straw
Asteraceae (Compositae), Aster or Sunflower Family

Habit: Herbaceous perennial, 0.3-0.8 cm tall, from deep creeping roots. Stems slender, usually without branches, nearly leafless, rushlike, with milky sap.

Leaves: Alternate; stalkless; blades few, mainly basal, narrow and grasslike, to 30 cm long and 1-3 mm wide. Flower heads usually solitary at ends of stems, heads with only ray flowers.

Inflorescences: Heads usually solitary at ends of stems, only ray flowers present.

Flowers: Pink, pale purple, or rarely white, 2-3 cm diameter, with about 10 rays with fringed ends.

Fruit: Nutlets 10-12 mm long, with an equally long tuft of bristles at the top.

Habitat and Distribution: Common; dry pinelands and sandy scrub; throughout the peninsula to central panhandle Florida and north to south Georgia.

Comment: Skeleton-Weed flowers in the summer northward and all year southward. The large heads are showy. This tall skinny wildflower deserves its common name. The slim almost unnoticed stems seem to be ill-suited to support the large heads.

Lycium carolinianum Walter

Lygodesmia aphylla (Nutt.) DC.

Pink

Lyonia lucida (Lam.) K. Koch
Shiny Lyonia or Fetterbush
Ericaceae, Heath or Blueberry Family

Habit: Shrub or small tree, to 4 m tall. Stems rigid, erect, with sharply angled twigs.

Leaves: Alternate; stalked; blades evergreen, glossy, leathery, broadly elliptic, 2-9 cm long, glandular dotted underneath, translucent veins along the smooth, rolled margin.

Inflorescences: Flower clusters along the previous year's branches at the bases of leaves.

Flowers: Pink to rose or white, nodding, somewhat urn-shaped, about 1 cm long.

Fruit: A rounded, smooth, 5-chambered, woody capsule, 3-5 mm long.

Habitat and Distribution: Common; wet pinelands, cypress ponds, swamps and other wet woods; throughout Florida, north to southeast Virginia, and west to Louisiana; Cuba.

Comment: Flowering is in spring. The common name, Shiny Lyonia, reflects the shiny nature of the leaves. It can be cultivated as a specimen plant, in masses, or as a foundation plant. It grows best in acidic soils. The shiny, evergreen foliage and the showy spring flowers make this a good ornamental. It is thought to be quite POISONOUS when ingested by animals.

Lythrum alatum Pursh
Winged Loosestrife
Lythraceae, Loosestrife Family

Habit: Herbaceous perennial, to 1.3 m tall. Stems 4-angled, erect, branched, smooth, and very leafy.

Leaves: Opposite on lower stem, alternate on upper stem; stalkless; blades elliptic to lance-shaped, to 7 cm long.

Inflorescences: Solitary flowers in upper leaf axils.

Flowers: Pinkish-purple to bluish, small, tubular, with 6 petals, each about 6 mm long.

Fruit: A capsule, to 5 mm long.

Habitat and Distribution: Infrequent to locally common; open wet habitats, woods, pinelands, marshes and prairies; Maine to North Dakota, south to Wyoming, Texas and south peninsula Florida.

Comment: This summer-flowering species frequently evokes a second look because of the tiny pinkish flowers and the spare, wandlike, branching pattern.

Lyonia lucida (Lam.) K. Koch

Lythrum alatum Pursh

Pink

Macroptilium lathyroides (L.) Urb.
Phasey Bean
[*Phaseolus lathyroides* L.]
Fabaceae (Leguminosae), Pea or Bean Family

Habit: Herbaceous annual, to 1 m tall. Stems branching, smooth or hairy.

Leaves: Alternate; stalked; blades once-divided, with 3 broadly elliptic to linear-oblong leaflets, leaflets 1-7 cm long.

Inflorescences: Long-stalked spikes from leaf axils.

Flowers: Deep pink purple to vivid red-purple to maroon to almost black, typically pea-shaped, showy, 2-2.4 cm long.

Fruit: Pods very narrow, 8-10 cm long, hairy, beaked at tip.

Habitat and Distribution: Frequent; pastures, along roadsides and in other disturbed sites; throughout Florida, north into southeast Georgia, Louisiana; West Indies, Mexico, south into South America; a native of tropical America.

Comment: Spring and summer flowering northward and all year southward. Phasey Bean flowers are very showy and extremely colorful, always provoking comment. Originally introduced for forage in Florida and the West Indies, it has escaped and become a common weed along roadsides and in old fields.

Malus angustifolia (Aiton) Michx.
Flowering Crabapple or Southern Crabapple
Rosaceae, Rose Family

Habit: Deciduous shrub or small tree, to 8-10 m tall, often sprouting from roots. Outer bark gray, inner bark reddish brown. Branches spreading to form a rounded crown, twigs stiff, reddish brown, forming short, spur shoots that are thorn-like.

Leaves: Alternate; short-stalked; blades elliptic to ovate, 2.5-6 cm long, firm, margins toothed.

Inflorescences: Clusters of 3-5 flowers at tips of the short spur shoots.

Flowers: Pink or white, fragrant, 2-3 cm diameter, with 5 rounded, spreading petals, and numerous stamens in center of flower.

Fruit: A yellowish green, round, smooth, 2-3 cm diameter small apple.

Habitat and Distribution: Occasional; hammocks, thickets, woodland edges, hedge rows, and old fields; from panhandle Florida, west to Louisiana, and north to Missouri, Kentucky and Virginia.

Comment: Flowering spectacularly in spring. The pink flowers fade to white. These colorful plants can be used as specimen trees in the landscape or combined for a mass effect. The fruit is edible.

Macroptilium lathyroides (L.)
Urb.

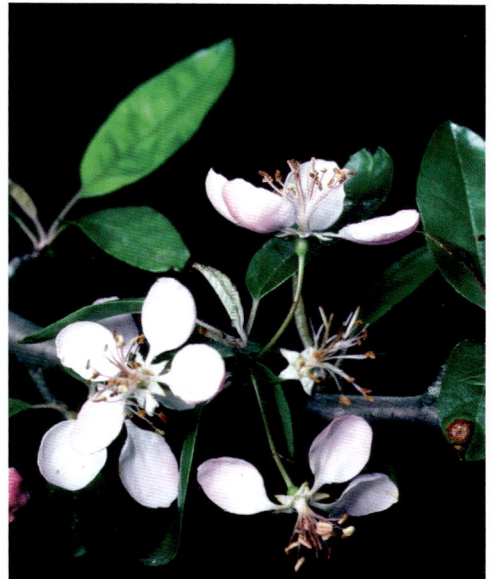

Malus angustifolia (Aiton) Michx.

Pink

Melochia spicata (L.) Fryxell
Hairy Melochia or Bretonica Peluda
[*Melochia hirsuta* Cav.; *Melochia villosa* (Mill.) Fawc. & Rendle]
Sterculiaceae, Chocolate Family

Habit: Herbaceous or somewhat woody annual or short-lived perennial, to 1.5 m tall. Stems erect, branches lying flat to ascending, covered with long hairs.

Leaves: Alternate; stalkless or short-stalked; blades ovate to lance-shaped, sometimes somewhat 4-sided, to 7 cm long, hairy on both surfaces, margins toothed.

Inflorescences: Terminal or axillary, head-like, stalkless clusters near stem tips.

Flowers: Pink, purple, or violet, petals to 1 cm long.

Fruit: A capsule, to 3 mm in diameter.

Habitat and Distribution: Occasional; disturbed sites and pinelands; central and southern peninsula Florida; south through the West Indies and Mexico into Central and South America.

Comment: Flowering can be any time during the year. The flowers are not really showy but attractive enough to provoke comments. Hairy Melochia has a weedy nature.

Mimosa quadrivalvis L.
Sensitive Brier
[*Schrankia microphylla* (Dryand. ex Sm.) J.F. Macbr.]
Fabaceae (Leguminosae), Pea or Bean Family

Habit: Herbaceous sprawling perennial, to 2 m long. Stems prostrate and spreading, angled, and covered with curved prickles.

Leaves: Alternate; long-stalked; blades twice-divided, the 1-5 primary paired divisions each with 4-17 pairs of leaflets, each 2-8 mm long, leaflets fold together when touched.

Inflorescences: Dense terminal heads from leaf axils.

Flowers: Pink, small.

Fruit: Pods 4-sided, prickly, 5-12 cm long and 3-5 mm wide, not jointed.

Habitat and Distribution: Frequent; roadsides, clearings, open woods, pinelands, scrub, sandhills; throughout Florida, north to Virginia, Kentucky, Iowa and South Dakota, and west to Texas.

Comment: Blooming in summer and fall, the heads of Sensitive Brier can scarcely be separated from Mimosa-vine. The stem is usually arching over and into vegetation versus the prostrate stem of Mimosa-vine. Of course, the curved prickles on the stem and fruit of Sensitive Brier absolutely enable a quick and definitive determination.

Melochia spicata (L.) Fryxell

Mimosa quadrivalvis L.

Pink

Mimosa strigillosa Torr. & A. Gray
Mimosa-vine, Creeping Mimosa, Powderpuff, Baby's-bath-brush, Vergonzosa
Fabaceae (Leguminosae), Pea or Bean Family

Habit: Herbaceous perennial, to 2 m long. Stems prostrate, hairy, sometimes woody at base, often forming annual, above ground runners.

Leaves: Alternate; long-stalked; blades twice-divided, the 4-6 primary paired divisions each with 6-15 pairs of leaflets, each 2-4 mm long, leaflets fold together when touched.

Inflorescences: Dense terminal heads from leaf axils.

Flowers: Rose pink, 2-2.5 mm long.

Fruit: Pods 1-2.5 cm long and 5-10 mm broad, hairy, jointed.

Habitat and Distribution: Occasional; moist grasslands, openings, turf, and other open disturbed sites; from peninsula Florida north into southeast Georgia, also in eastern Texas, southern Arkansas and eastern Mississippi; South America; Pacific Islands.

Comment: A spring and early summer flowering wildflower. Mimosa-vine can be planted into turf, mowed and fertilized with the turf, and, ceasing mowing while blooming, enjoyed as a wildflower. The colorful heads seem to pop-up out of the grass.

Oenothera speciosa Nutt.
Showy Evening-primrose
Onagraceae, Evening-primrose Family

Habit: Herbaceous perennial, to 50 cm tall, from underground runners. Stems ascending, branching, hairy.

Leaves: Alternate; stalked; blades broadly elliptic, to 8 cm long, sometimes with toothed or lobed edges.

Inflorescences: Flowers solitary in axils of upper leaves, from nodding buds.

Flowers: Pinkish, with 4 large, broad, delicately-veined petals, each 2-4 cm long.

Fruit: A club-shaped, angled capsule, 1-1.4 cm long.

Habitat and Distribution: Infrequent to rare, widely scattered; disturbed sites, roadsides, open woods, prairies, fields; from peninsula Florida, west to Texas and north to North Carolina; a native from Texas to Kansas and Missouri and south into northeast Mexico.

Comment: The large pink flowers occur in late spring and are very showy in populations from the thin underground runners. Showy Evening-primrose is available from some native nurseries and can be used as a low bedding plant. A form with white flowers also occurs.

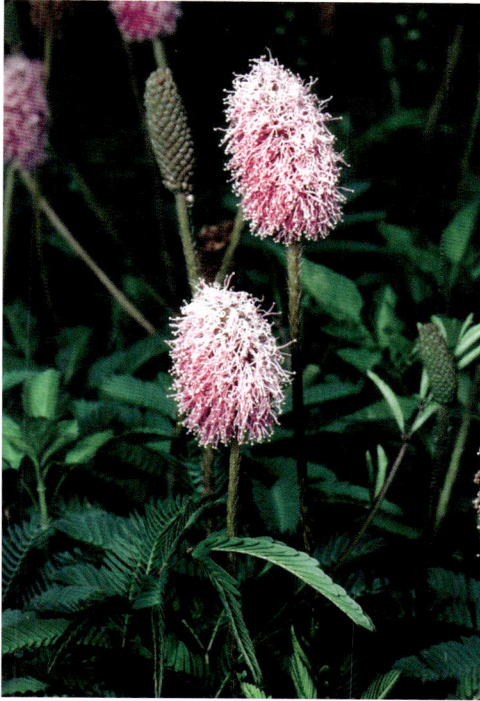

Mimosa strigillosa Torr. & A. Gray

Oenothera speciosa Nutt.

Pink

Orbexilum pedunculatum (Mill.) Rydb.
Sampson's Snakeroot
[*Psoralea psoralioides* (Walter) Cory]
Fabaceae (Leguminosae), Pea or Bean Family

Habit: Herbaceous perennial, 30-90 cm tall, from tuberous roots. Stems with tiny hairs, clustered, few branches.

Leaves: Alternate; stalked; blades once-divided, with 3 leaflets, elliptic to lance-shaped, leaflets 3-7 cm long, with glandular dots on surface.

Inflorescences: Long-stalked, spike-like, from leaf axils.

Flowers: Lavender-lilac to purplish, narrow, typically shaped pea flowers, 5-7 mm long.

Fruit: Pod disc-shaped, wrinkly, 4-5 mm long, hairless.

Habitat and Distribution: Infrequent; dry open pinelands and woods; from northern peninsula Florida, west to Texas, and north to Illinois and Virginia.

Comment: The flower spike occurs in spring and summer. The small flowers and sparse branching make this plant hard to spot.

Oxalis debilis Kunth
Pink Wood Sorrel
[*Oxalis corymbosa* DC.]
Oxalidaceae, Woodsorrel Family

Habit: Herbaceous perennial, 8-30 cm tall, from bulbils. No aerial stem.

Leaves: Alternate; very long-stalked; blades with 3 hairy, heart-shaped leaflets, like a shamrock, to 3.5 cm long and 5 cm wide.

Inflorescences: Long, hairy stalks from the plant base, support terminal clusters of stalked flowers.

Flowers: Pink to rose purple fading to blue, rarely white, with 5 petals, often lightly striped, arching outward, 1-1.5 cm long.

Fruit: A narrow, elliptic, capsule, about 1 cm long.

Habitat and Distribution: Occasional to locally common; moist, disturbed sites; throughout Florida and west to Texas; West Indies; Old World tropics; a native of tropical America.

Comment: Flowering in any warm month, most of these showy flowers are seen in the spring and fall. The shamrock-like leaves are comment provoking. It is sold as a pot plant. Older plants are frequently discarded into fence rows and adjacent vacant lots.

Orbexilum pedunculatum
(Mill.) Rydb.

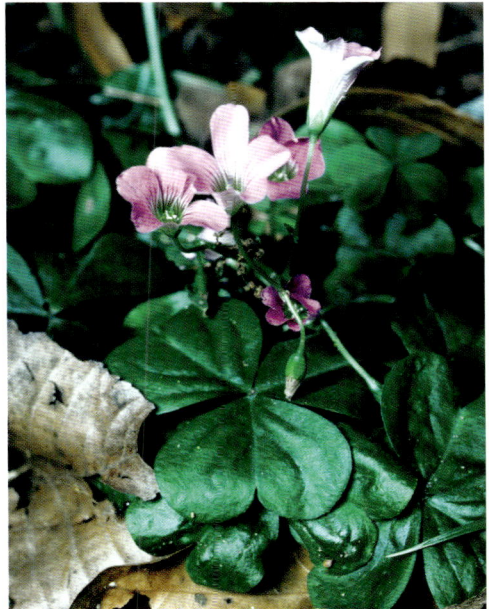

Oxalis debilis Kunth

Pink

Pedicularis canadensis L.
Lousewort
Scrophulariaceae, Figwort or Snapdragon Family

Habit: Herbaceous perennial, 10-40 cm tall, forming large clumps from short underground runners. Stems erect, hairy.

Leaves: Alternate; long-stalked; blades elliptic, 5-15 cm long, 2-5 cm wide, with many narrow lobes, margins toothed.

Inflorescences: Terminal, in a hairy, round, or elongate cluster.

Flowers: Various combinations of red, purple, yellow, and rust, 1.8-2.5 cm long, with large, downward arching upper lip, and smaller, 2-crested lower lip, upper and lower lips can be same or different colors.

Fruit: An oblong, flattened capsule, about 1.5 cm long.

Habitat and Distribution: Occasional; open upland woodlands, meadows, prairies, seepage slopes, and other moist sites; from north peninsula and panhandle Florida, west to Texas, northern Mexico and Colorado, and north to Manitoba, Quebec and Maine.

Comment: Lousewort flowers in spring. It is used as a border plant. The common name is derived from a European belief that cattle grazing on this genus would be come infested with lice. The lice infestations probably had no relation to ingestion of Lousewort.

Penstemon australis Small
Slender Beard-tongue or Southeastern Beard-tongue
Scrophulariaceae, Figwort or Snapdragon Family

Habit: Herbaceous perennial, 20-80 cm tall. Stems finely hairy, unbranched upward.

Leaves: Opposite, basal rosette; basal stalked, stem stalkless; basal blades broader at the tip, stem blades lance-shaped, 5-14 cm long, 2-3 cm broad, margins smooth to toothed.

Inflorescences: Flowers loose, erect, on a branched stalk.

Flowers: Pink to violet-purple, tubular, 1.5-2.5 cm long and 6-8 mm wide at mouth with a 2-lobed upper lip curling upward and a 3-lobed lower lip projecting outward, white with pale violet stripes inside.

Fruit: A cone-shaped capsule, 6-10 mm long.

Habitat and Distribution: Frequent; pinelands, sandhills and other dry sandy sites; from central peninsula Florida, west to Alabama and north to Virginia.

Comment: This small member of the Snapdragon Family blooms in the spring. For best visual effect Slender Beardtongue should be planted in groups of at least five to seven in full sun.

Pedicularis canadensis L.

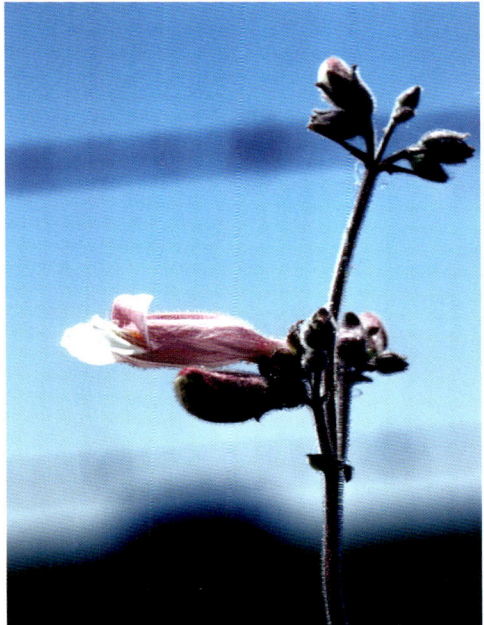

Penstemon australis Small

Pink

Phlox drummondii Hook.
Annual Garden Phlox
Polemoniaceae, Phlox Family

Habit: Herbaceous annual, 10-70 cm tall. Stem erect, slightly hairy, usually branched.

Leaves: Alternate on upper stem and opposite on lower stem; essentially stalkless; blades oblong to elliptic, to 8 cm long, hairy.

Inflorescences: Terminal clusters.

Flowers: Pink, lavender, magenta, white or variegated, 1.6-2.5 cm broad, with 5 spreading petals, and often with a different colored eye.

Fruit: A papery capsule.

Habitat and Distribution: Frequent and locally common; open grasslands, woodlands, and dry, sandy, and disturbed sites; throughout Florida, west to Texas, and north to Virginia; a native of Texas.

Comment: This spring-blooming wildflower is spectacular in masses. It has been widely planted throughout the world and frequently escapes. There are many named color variations.

Phlox floridana Benth.
Florida Phlox
Polemoniaceae, Phlox Family

Habit: Herbaceous perennial, to 80 cm tall. Lower stems smooth.

Leaves: Opposite; stalkless; blades linear to lance-shaped, to 10 cm long, hairless.

Inflorescences: Terminal, glandular hairy cluster.

Flowers: Pink to purple, often with a pale eye, 1-2.5 cm wide, with 5 spreading petals.

Fruit: A papery capsule.

Habitat and Distribution: Occasional; sandy soils of sandhills, flatwoods, hammocks; from central peninsula Florida to adjacent counties of Alabama and Georgia.

Comment: Florida Phlox blooms in late spring to early summer. It is commercially available, used primarily as a bedding plant.

Phlox drummondii Hook.

Phlox floridana Benth.

Pink

Phyla nodiflora (L.) Greene
Match-Head, Frog-fruit, Cape-weed, Turkey-tangle, Mat-grass, Hierba de la Virgen Maria
[*Lippia nodiflora* L.]
Verbenaceae, Verbena or Vervain Family

Habit: Herbaceous perennial, 0.2-1.3 m long, to 10 cm tall. Stems creeping and ascending, hairy.

Leaves: Opposite; stalked; blades elliptic, usually broadest at the rounded tip, 1-7 cm long, teeth at tip, tiny hairs on both surfaces.

Inflorescences: Rounded or cylindrical heads terminal on long stalks from leaf axils.

Flowers: Pink to rose-purple or white, small, 5-lobed, 2-2.5 mm long.

Fruit: Two nutlets, to 1 mm long.

Habitat and Distribution: Common; moist, sandy soils of virtually all but the wettest habitats, especially in disturbed sites; throughout Florida, west to California, and north to Oklahoma, Arkansas, Kentucky and Virginia.

Comment: This little wildflower can bloom in any warm month. Match-head is frequently called Mat-grass or Carpet-grass in California where it is used as a ground cover. In more humid climates along the Gulf Coast, it will not form a solid ground cover. The stalks and small heads resemble a wooden match in both structure and color of the head.

Physostegia godfreyi Cantino
Godfrey's False Dragon-head
Lamiaceae (Labiatae), Mint Family

Habit: Herbaceous perennial, to 1 m tall, from underground runner. Stems smooth, usually single, 4-angled.

Leaves: Opposite; stalkless at the top of the stem and long-stalked on lower stem; blades narrowly elliptic, smooth, margins smooth or wavy or with a few blunt teeth.

Inflorescences: Opposite flowers in a spike-like terminal arrangement.

Flowers: Pale rose with purple interior venation, somewhat tubular, 1-2 cm long, upper and lower lips, lower lip 3-lobed and splotched with purple.

Fruit: 4 warty nutlets, about 2 mm long.

Habitat and Distribution: Occasional; in flatwoods, bogs, swamps and ditches; endemic in central panhandle Florida in Liberty, Franklin, Calhoun, Gulf, and Bay Counties.

Comment: Godfrey's False Dragon-head flowers in summer. The flowers are large and attractive, but the plant is not frequently seen.

Phyla nodiflora (L.) Greene

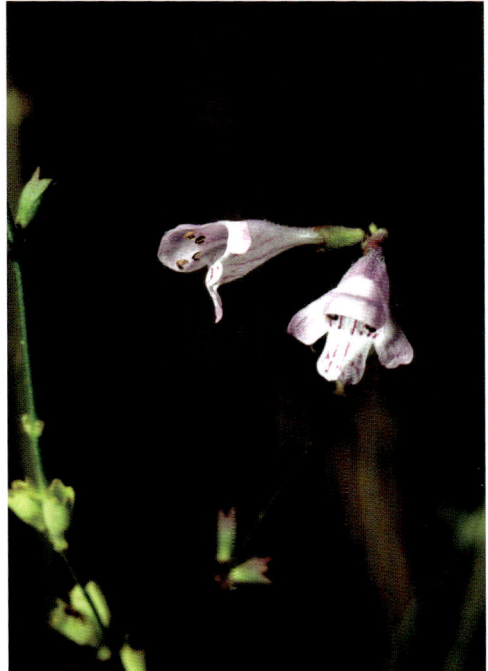

Physostegia godfreyi Cantino

Pink

Phytolacca americana L.
Pokeweed
Phytolaccaceae, Pokeweed Family

Habit: Herbaceous perennial, 1-3 m tall, from a huge taproot. Stems smooth, branched, becoming reddish in older plants.

Leaves: Alternate; stalked; blades elliptic to somewhat oval to lance-shaped, to 30 cm long.

Inflorescences: Cylindrical, drooping or erect spikes from upper joints.

Flowers: Pinkish white to green, minute.

Fruit: A purple to black, somewhat flattened berry, 7-12 mm diameter.

Habitat and Distribution: Common; margins and primarily in disturbed sites; throughout Florida, west to Texas, and north to Quebec, Ontario, and Maine.

Comment: Pokeweed flowers in all warm months. The flowers, although small, are attractive massed in spikes. The reddish to purplish stems are also attractive, but very fleshy and easily broken. The fruits are very juicy and popular with birds who spread the seeds far and wide. Kids often eat the fruit which can cause an upset stomach if too many are ingested. All parts of the plant are POISONOUS if eaten. The younger leaves are often eaten as a potherb, but must be properly prepared. The lower parts of the plant, especially the enlarged roots, are EXTREMELY POISONOUS if ingested. An exceptionally weedy species.

Pluchea rosea R.K. Godfrey
Godfrey's Fleabane
Asteraceae (Compositae), Aster or Sunflower or Daisy Family

Habit: Herbaceous perennial, 0.4-1.1 m tall. Stems single or many, covered with fine, sticky hairs.

Leaves: Alternate; stalkless; blades elliptic to oblong to rounded, 2-10 cm long, with toothed margins.

Inflorescences: Terminal clusters of heads.

Flowers: Pink to purple, tubular.

Fruit: Black, slender, densely hairy, to 1 mm long, white bristles at top.

Habitat and Distribution: Common; areas where water tables fluctuate in wet flatwoods, savannas, marshes, low woods and disturbed sites; throughout Florida, north to North Carolina, and west to Texas and Mexico; West Indies.

Comment: Godfrey's Fleabane flowers throughout spring and summer. The round clusters of pink blooms are frequently noticed as is the distinct odor when the plant is handled.

Phytolacca americana L.

Pluchea rosea R.K. Godfrey

Pink

Pogonia ophioglossoides (L.) Ker-Gawl.
Rose Pogonia
Orchidaceae, Orchid Family

Habit: Herbaceous perennial, to 40 cm tall, from a short underground runner (rhizome). Stem smooth, single.

Leaves: Leaf solitary, rarely 2; stalkless; blades smooth, broadly elliptic, 2-10 cm long, 1-2 cm wide, sheathing stem.

Inflorescences: Flowers 1 to 4, terminal.

Flowers: Rose-pink to lavender or rarely white, fragrant, about 4 cm long, lower lip with purple fringed margin and yellow center.

Fruit: An ellipsoid capsule, 2-3 cm long.

Habitat and Distribution: Frequent; bogs, marshes, prairies, seeps and wet pinelands; from central peninsula Florida, north to Newfoundland and Manitoba, and west to Texas.

Comment: A delightful little wildflower of wet acidic soils blooming in March and May. The pink flowers seen either singly or in clusters always stand out among other plants.

Polygala chapmanii Torr. & A. Gray
Chapman's Milkwort
Polygalaceae, Milkwort Family

Habit: Herbaceous annual, to 60 cm tall. Stems smooth, single, often with a few branches.

Leaves: Alternate; stalkless; blades linear, to 2 cm long.

Inflorescences: Spike-like terminal clusters.

Flowers: Pinkish to lavender, about 3 mm long.

Fruit: A rounded capsule, about 1.5 mm long.

Habitat and Distribution: Frequent; wet pinelands, seepage areas and bogs; from central panhandle Florida to Mississippi.

Comment: The small flowers are delicately attractive in spring and early summer. These plants are often overlooked as wildflowers.

Pogonia ophioglossoides
(L.) Ker-Gawl.

Polygala chapmanii
Torr. & A. Gray

Pink

Polygala crenata C.W. James
Scalloped Milkwort or Crenate Milkwort
Polygalaceae, Milkwort Family

Habit: Herbaceous perennial, to 30 cm tall. Stems several to numerous, simple and unbranched.

Leaves: Alternate; stalkless; blades somewhat oval and usually broader toward the tip, 0.3-1.5 cm long, margins smooth.

Inflorescences: Short terminal spikes.

Flowers: Pinkish to rose-purplish to purple, about 4-5 mm long, margins fringed.

Fruit: An oblong capsule, about 3 mm long.

Habitat and Distribution: Occasional; wet pinelands, swamps, bogs, and depressions; from panhandle Florida to southeast Louisiana.

Comment: Spring and summer blooming, the clusters of stems with small pyramidal or rounded spikes of flowers are a delight to find.

Polygala cruciata L.
Drumheads
Polygalaceae, Milkwort Family

Habit: Herbaceous annual, 5-50 cm tall. Stems smooth, unbranched or branched.

Leaves: Mostly in whorls of 3 or 4 around stem or alternate upwards; stalkless; blades linear to linear-elliptic, dotted.

Inflorescences: Terminal, cylindrical heads, each to 3 cm long.

Flowers: Rose purple to purplish green, small, 1.5-3 mm long, fringed.

Fruit: A capsule, to 2 mm long.

Habitat and Distribution: Frequent; wet pinelands, seepage and bogs; throughout Florida, west to east Texas, and north to Kentucky, Ohio, Michigan, Illinois, Minnesota, and Virginia.

Comment: Drumheads flower from summer into fall. The colorful heads clustered on short stems are fun to see down among the taller ground cover.

Polygala crenata C.W. James

Polygala cruciata L.

Pink

Polygala grandiflora Walter
Large-flowered Milkwort
Polygalaceae, Milkwort Family

Habit: Herbaceous perennial, 20-60 cm tall. Stems hairy, with erect branching.

Leaves: Alternate; short-stalked; blades elliptic to lance-shaped to linear, 1.5-4 cm long, margins smooth.

Inflorescences: Loose clusters at joints along the upper stem.

Flowers: Pink with purple and green, 5-7 mm long, with two larger oval 'wings' clasping smaller inward curling petals.

Fruit: A capsule, to 5 mm long.

Habitat and Distribution: Common; pinelands, sandhills, dunes, and other habitats with sandy soils; throughout Florida, west to Mississippi, and north to North Carolina along the coastal plain.

Comment: Flowering in warm months, even though the flowers are not especially large, Large-flowered Milkwort is an easily noticed wildflower.

Polygala hookeri Torr. & A. Gray
Hooker's Milkwort
Polygalaceae, Milkwort Family

Habit: Herbaceous annual, to 30 cm tall. Stems smooth, branching.

Leaves: In whorls of 3s and 4s; stalkless; blades linear, 0.5-1.5 cm long, dotted, margins smooth and rolled, distance between whorls larger than length of leaves.

Inflorescences: Loose, terminal, spike-like clusters.

Flowers: Pink or reddish, tinged with green, 3-3.5 mm long.

Fruit: A rounded capsule, about 2 mm long.

Habitat and Distribution: Occasional; wet pinelands and bogs; from central panhandle Florida, along the coastal plain west to Texas, and north to North Carolina.

Comment: Hooker's Milkwort is a thin little summer-blooming wildflower.

Polygala grandiflora Walter

Polygala hookeri Torr. & A. Gray

149

Pink

Polygala incarnata L.
Procession Flower or Pink Milkwort
Polygalaceae, Milkwort Family

Habit: Herbaceous annual, 20-70 cm tall. Stems smooth, covered with white film, branching infrequently.

Leaves: Alternate; stalkless; blades fleshy, very slender, 0.5-1.7 cm long, margins smooth.

Inflorescences: Thin, terminal, spikes.

Flowers: Pink, 5-7 mm long, fringed.

Fruit: A rounded capsule, to 2.4 mm long.

Habitat and Distribution: Frequent; sandhills, flatwoods, old fields, woods, prairies, bogs and woodland borders; throughout Florida, west to Texas and Mexico, and north to Nebraska, Iowa, Wisconsin, Michigan Ontario, Long Island, New Jersey, and Pennsylvania.

Comment: Flowering in all warm months, Procession Flower is a skinny often unnoticed tiny bouquet that surprises those who look closely.

Polygonella robusta (Small) G.L. Nesom & V.M. Bates
Large-flowered Sandhill Wireweed
Polygonaceae, Buckwheat Family

Habit: Perennial shrub, 60-90 cm tall. Stems smooth, woody, brittle, often branching.

Leaves: Alternate; stalkless; blades very narrow, 2-6 cm long, sheathing at base, sheath fringed with bristles.

Inflorescences: Spike-like displays from upper leaf axils.

Flowers: Pink or white, tiny, fringed.

Fruit: Nutlets approximately 1 mm wide, hard, 3-angled.

Habitat and Distribution: Common; scrub and sandhills; from central peninsula to central panhandle Florida.

Comment: Blooming in summer and fall this small shrub is very colorful. The flowers are small but the spikes with numerous flowers are conspicuous. Large-flowered Sandhill Wireweed can become a ground cover in the sparsely vegetated sandy habitats where it occurs.

Pink

Polygala incarnata L.

Polygonella robusta
(Small) G.L. Nesom & V.M. Bates

Pink

Polygonum hydropiperoides Michx.
Mild Water-pepper
Polygonaceae, Buckwheat Family

Habit: Herbaceous annual or perennial, to 1 m tall, commonly forming mats. Stems erect or ascending, smooth or hairy, frequently rooting at joints.

Leaves: Alternate; short-stalked; blades narrow, to 17 cm long, sheathing at base, sheath fringed with bristles.

Inflorescences: Loose spikes from joints near tip of stem.

Flowers: Pink, white, or green, small, approximately 3 mm long.

Fruit: Nutlets brown, lustrous, triangular, to 2.3 mm long.

Habitat and Distribution: Common; virtually all wet native or disturbed habitats; throughout Florida, New Brunswick to British Columbia, south through Mexico.

Comment: Mild Water-pepper flowers from spring into fall. Flowering occurs later in summer and into early fall in colder climates. Freezing to the ground, it can persist from joints that have been covered by dense vegetation, sediments, or standing water. At sites where massive mats occur the plants can provide an impressive display.

Portulaca amilis Speg.
Broadleaf Pink Purslane
Portulacaceae, Purslane Family

Habit: Herbaceous annual, to 10 cm tall, mostly prostrate or reclining. Stems sprawling, sometimes with ascending tips, succulent.

Leaves: Alternate; stalkless; blades flat, to 2 cm long and 7 mm wide, broader at tip, succulent, with conspicuous axillary hairs.

Inflorescences: Terminal heads, surrounded by several leaves.

Flowers: Pink to pink-purple, to 2 cm wide, with 5 petals.

Fruit: A capsule that breaks open around the middle; seeds numerous, small, shiny, black to dark brown.

Habitat and Distribution: Frequent; sandy soils of disturbed sites; from central peninsula Florida, west to western panhandle Florida, north to North Carolina; a native of South America.

Comment: Flowering from spring into fall, Broadleaf Pink Purslane has showy flowers. The broad leaves, pink flowers and conspicuous hairs in the leaf axils make this plant easy to identify.

Pink

Polygonum hydropiperoides
Michx.

Portulaca amilis **Speg.**

Pink

Portulaca grandiflora Hook.
Rose-moss
Portulacaceae, Purslane Family

Habit: Herbaceous annual, to 25 cm long, prostrate. Stems succulent, sprawling to ascending.

Leaves: Alternate; stalkless; blades round, linear, 2.5 cm long and 2 mm wide, succulent, with axillary hairs.

Inflorescences: Terminal, solitary or in heads, surrounded by several leaves.

Flowers: Pink, maroon, or yellow, to 5.5 cm wide, with 4-5 or more petals.

Fruit: A capsule that breaks open around the middle; seeds silvery gray, small.

Habitat and Distribution: Rare to occasional; disturbed sites; from central peninsula Florida into northeast U.S. Native to South America.

Comment: Flowering from spring into fall southward and in summer northward. The large very showy flowers of this infrequent escape from cultivation are easily noticed. Rose-moss is extremely common in cultivation as a potted plant.

Raphanus raphanistrum L.
Wild Radish or Jointed Charlock
Brassicaceae (Cruciferae), Mustard Family

Habit: Herbaceous winter annual, 30-80 cm tall, from a taproot. Stems covered with stiff hairs.

Leaves: Alternate; stalked; blades larger on lower parts of plant, smaller and narrower on upper parts, lower leaves often dissected, upper leaves lobed, margins of all blades toothed.

Inflorescences: Stalked, in an open, elongated, terminal assortment, flowers opening progressively upward.

Flowers: Pink, purple, or more commonly yellow or white, 2-3 cm wide, with 4 petals, petals with apparent veins.

Fruit: Pods narrow, 2-6 cm long and 3-5 mm diameter, with constrictions between seeds.

Habitat and Distribution: Frequent, locally common; disturbed sites, particularly in cultivated fields; throughout Florida and throughout the temperate regions of the world; a native of the Mediterranean region of Europe.

Comment: Flowering in spring, Wild Radish is familiar to many as a weed of roadsides and gardens. In mass it makes a showy wildflower.

Portulaca grandiflora Hook.

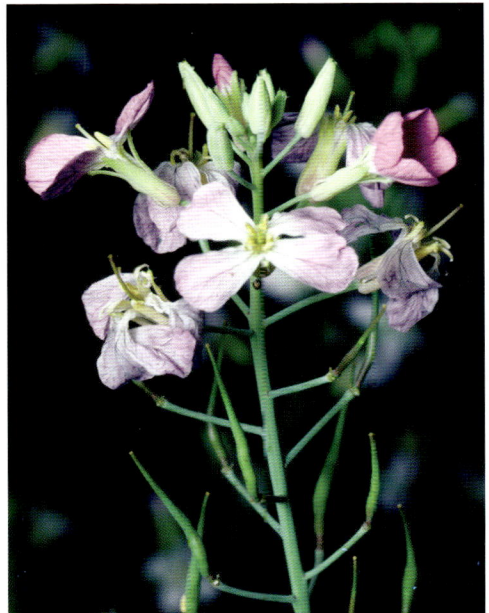

Raphanus raphanistrum L.

Pink

Rhexia alifanus Walter
Smooth Meadow-beauty or Tall Meadow-beauty
Melastomataceae, Meadow-beauty Family

Habit: Herbaceous perennial, 0.3-1 m tall. Stems smooth, erect, wand like.

Leaves: Opposite; stalked; blades smooth, narrow, elliptic to oval, 3-8 cm long, to 1 cm wide, with 3 major veins, margins smooth.

Inflorescences: Terminal, loosely branched, spreading clusters.

Flowers: Pink-purple to rose, to 5 cm wide, with 4 petals, each 2-2.5 cm long, tube with glandular hairs, anthers obviously curved.

Fruit: Capsule urn-shaped, 6-12 mm long, 5-7 mm wide, with glandular hairs.

Habitat and Distribution: Occasional; moist pinelands, flatwoods, and ditches; from central peninsula Florida, north to North Carolina, and west to Texas.

Comment: Smooth Meadow-beauty is an attractive wildflower, blooming in spring and summer. Best cultivated in a natural setting with wet soils, the pretty flowers are large enough to be noticed individually or in groups.

Rhexia cubensis Griseb.
Cuban Meadow-beauty
Melastomataceae, Meadow-beauty Family

Habit: Herbaceous perennial, to 60 cm tall. Stems have hairs with glandular tips.

Leaves: Opposite; stalked; blades linear to ovate, with 3 major veins, glandular hairs on both surfaces, margins toothed.

Inflorescences: Terminal, loosely branched, spreading clusters.

Flowers: Pink or sometimes white, to 4 cm wide, petals 4, each 1.5-2.0 cm long, anthers obviously curved.

Fruit: Capsule urn-shaped, 10 to 14 mm long, 5-7 mm wide, with glandular hairs.

Habitat and Distribution: Frequent; low, open, wet sites, wet pinelands, and ditches; throughout Florida, west to Mississippi, and north to North Carolina.

Comment: Flowers are extremely attractive and showy in spring and summer. The large, yellow anthers and large, showy, pink petals can easily be seen from a distance. Petals fall very easily when brushed. It is sometimes offered by native nurseries.

Rhexia alifanus Walter

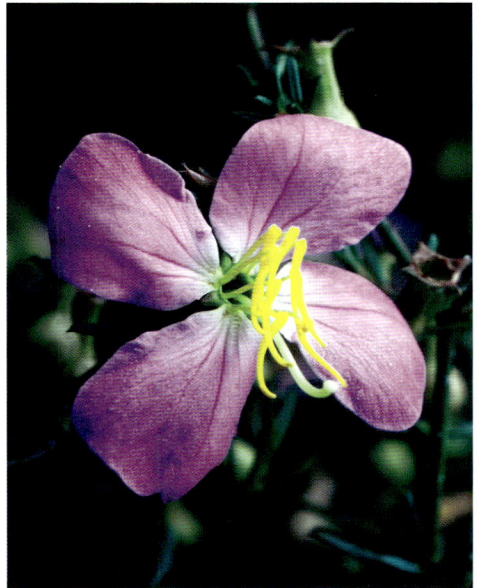

Rhexia cubensis Griseb.

Pink

Rhexia mariana L.
Pale Meadow-beauty
Melastomataceae, Meadow-beauty Family

Habit: Herbaceous perennial, 20-80 cm tall, from underground runners. Stems glandular hairy, branched or unbranched.

Leaves: Opposite; short-stalked; blades linear to oval, with 3 major veins, to 6 cm long, 2 cm wide, glandular hairy, margins toothed.

Inflorescences: Terminal, loosely branched, spreading clusters.

Flowers: Pink to pale lavender or white, 3-5 cm wide, petals 4, each 10-2.5 cm long, anthers obviously curved.

Fruit: Capsule urn-shaped, to 1 cm long, glandular hairy.

Habitat and Distribution: Common; low sites and wet areas, marshes, wet pinelands, ditches; throughout Florida, west to Texas, and north to Oklahoma, Missouri, Illinois, Indiana, and Massachusetts.

Comment: The large flowers can be seen from spring until frost. The masses of plants produced from the underground runners produce drifts of pale pink flowers on open shores and other areas with little other vegetation. The fragile petals can be easily detached.

Rhexia petiolata Walter
Coastal Plain Meadow-beauty
Melastomataceae, Meadow-beauty Family

Habit: Herbaceous perennial, 20-60 cm tall. Stems smooth, sometimes branched from base.

Leaves: Opposite; stalkless; blades broadly elliptic to oval, with 3 major veins, 1-2 cm long, to 1.5 cm wide, smooth or sparsely glandular hairy, margins toothed and hairy.

Inflorescences: Terminal, loosely branched, spreading clusters.

Flowers: Pink to violet, 2.8-3.4 cm wide, petals 4, each 1.4-1.7 cm long, anthers small and straight.

Fruit: Capsule urn-shaped, 1.2 cm long, smooth.

Habitat and Distribution: Frequent; moist to wet sites, wet pinelands, marshes, wet swales, and bogs; from central peninsula Florida, west to Texas, and north to Oklahoma, Missouri, Illinois, Indiana, and Virginia.

Comment: Coastal Plain Meadow-beauty blooms in spring and summer. It is a handsome, showy wildflower.

Rhexia mariana L.

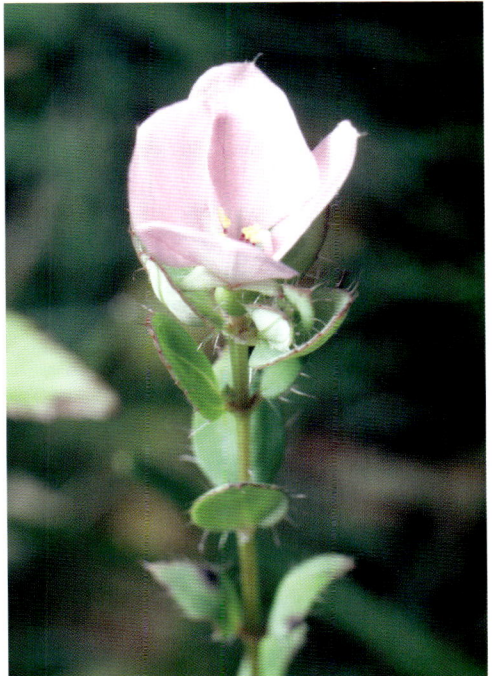

Rhexia petiolata Walter

Pink

Rhexia virginica L.
Common Meadow-beauty
Melastomataceae, Meadow-beauty Family

Habit: Herbaceous perennial, to 1 m tall, roots sometimes tuberously thickened. Stems glandular hairy, usually branched.

Leaves: Opposite; stalkless or nearly so; blades elliptic to ovate, with 3 major veins, to 7 cm long, to 3 cm wide, margins toothed and hairy.

Inflorescences: Terminal, loosely branched, spreading clusters.

Flowers: Pink-purple to rose, 2-5 cm wide, petals 4, each to 2.5 cm long, anthers obviously curved.

Fruit: Capsule urn-shaped, 8-10 mm long, glandular hairy.

Habitat and Distribution: Frequent; low, wet sites, wet pinelands, pond margins, seepage, and ditches; from north peninsula Florida, west to Texas, and north to Missouri, Kansas, Indiana, Wisconsin, and Nova Scotia.

Comment: Summer and fall blooming, Common Meadow-beauty is a noted wildflower through most of the Gulf Coastal Plain and the eastern U.S.

Rhododendron canescens (Michx.) Sweet
Wild Azalea, Piedmont Azalea,
Ericaceae, Blueberry Family

Habit: Perennial shrub, to 3-5 m tall. Stems woody, new twigs densely hairy.

Leaves: Alternate; deciduous; stalked; blades hairy, elliptic, 3-9 cm long.

Inflorescences: Terminal clusters.

Flowers: Pink to white, 1.5-2.5 cm long, 2-3 cm across, with narrow tube, 5 unequal flared petals at mouth, 5 elongated stamens protruding from mouth.

Fruit: A woody capsule, hairy, about 1.5 cm long.

Habitat and Distribution: Frequent; moist woods and along streams, wet flatwoods, hydric hammocks and seepage areas; from Marion County in central peninsula Florida, west to Texas, and north to Ohio, Maryland and Delaware.

Comment: Flowering in early spring just as or before the leaves appear. The very sweet fragrance and large showy flowers are spectacular, especially in mass. The young light green leaves are also attractive. It is underutilized as a landscape plant, but is available from most nurseries in its range.

Rhexia virginica L.

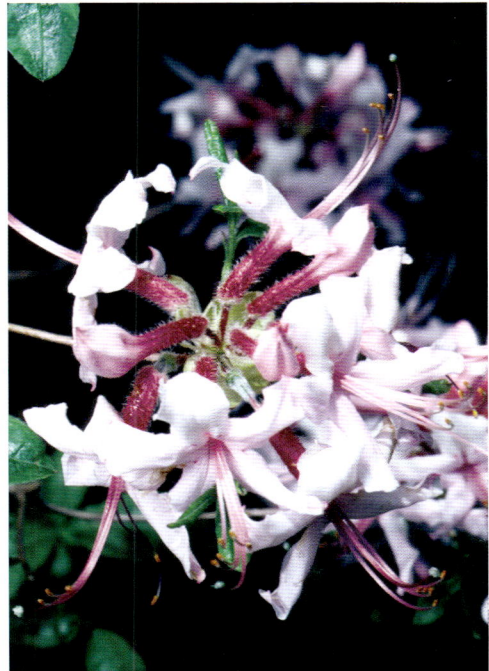

Rhododendron canescens
(Michx.) Sweet

Pink

Rhododendron chapmanii A. Gray
Chapman's Azalea
[*Rhododendron minus* Mich. var. *chapmanii* (A. Gray) W.H. Duncan & Pullen]
Ericaceae, Blueberry Family

Habit: Perennial shrub, to 2 m tall. Stems spreading, branches ascending and rigid.

Leaves: Alternate, evergreen; stalked; blades 3-5 cm long, with yellow dots, especially on lower surfaces, surfaces generally curved downward.

Inflorescences: Large radial or dome-like clusters.

Flowers: Pink, bell-shaped, about 3 cm long, 3-4 cm wide, with 5 unequal petals, stamens about as long as petals.

Fruit: A woody capsule, scaly, about 1 cm long.

Habitat and Distribution: Rare; pine flatwoods and shrub bog and bay borders; in only 6 northern Florida counties - Clay County in the north peninsula and Franklin, Gadsden, Gulf, Leon and Liberty Counties in the Florida panhandle.

Comment: Chapman's Azalea flowers in March and April. The large clusters of showy pink flowers are especially attractive. It is listed by the U.S. and Florida as Endangered. In recent years, this plant has been legally propagated and is available as a landscape plant. The leaves resemble Sand Live Oak leaves. It works well as a border shrub in broken shade.

Rhodomyrtus tomentosa (Aiton) Hassk.
Downy Myrtle or Rose Myrtle
Myrtaceae, Myrtle Family

Habit: Perennial shrub, to 2.5 m tall. Stems hairy.

Leaves: Opposite, evergreen; stalked; blades broadly elliptic to ovate, three-veined, to 7 cm long, hairy underneath, margins smooth.

Inflorescences: Flowers solitary or in threes from leaf axils.

Flowers: Pink to rose, to 2.5 cm wide, 5-lobed.

Fruit: A berry to 1.5 cm long, hairy, crowned with a persistent part of flower.

Habitat and Distribution: Occasional; locally common; flatwoods, central and southern peninsula Florida; a native of Australia and India through China into the Philippine Islands.

Comment: Flowering in spring, Downy Myrtle is quite showy. The fruits turn purple when ripe and are edible. Propagation is by seeds. It is very invasive in native habitats.

Rhododendron chapmanii A. Gray

Rhodomyrtus tomentosa
(Aiton) Hassk.

Pink

Richardia grandiflora (Cham. & Schltdl.) Schult. & Schult. f.
Large-flowered Pusley
Rubiaceae, Madder Family

Habit: Herbaceous perennial, to 8 cm tall, from taproot. Stem prostrate, branching, hairy.

Leaves: Opposite; stalked; blades elliptic, broader at tip, to 4 cm long, hairy.

Inflorescences: Terminal or axillary heads.

Flowers: Pink, violet, blue, or white, to 2 cm long, tubular.

Fruit: A small, elliptic capsule.

Habitat and Distribution: Occasional to locally common; dry sandy disturbed sites, turf, open fields and roadsides; central and south peninsula Florida; a native of South America.

Comment: This summer and fall flowering wildflower is quite showy as a ground cover. It also is very weedy in turf. Large-flowered Pusley is moving rapidly northward in Florida's peninsula due to intentional and unintentional human transport of this wildflower.

Rosa palustris Marshall
Swamp Rose
Rosaceae, Rose Family

Habit: Perennial shrub, 0.3-2 m tall. Stems smooth, often reddish, with downwardly curved prickles.

Leaves: Alternate; stalked; blades divided once, with 2-9 leaflets, leaflets elliptical, often broader at tip, 2-6 cm long, hairy underneath, with toothed edges.

Inflorescences: Flowers stalked, terminal, solitary or clustered.

Flowers: Pale pink, 5-8 cm diameter, with 5 broad, rounded petals.

Fruit: A hip red, oval, about 1 cm diameter and 1.5 cm long, glandular hairy.

Habitat and Distribution: Frequent; swamps, marshes, stream banks, floodplains, and other wet habitats; central peninsular Florida, west to Mississippi, and north to Minnesota and Nova Scotia.

Comment: Becoming more popular in cultivation, Swamp Rose flowers from spring into summer, and irregularly in other warm months.

Richardia grandiflora (Cham. &
Schltdl.) Schult. & Schult. f.

Rosa palustris Marshall

Pink

Rubus argutus Link
Highbush Blackberry or Southern Blackberry
Rosaceae, Rose Family

Habit: Perennial shrub, to 3 m tall, with underground runners. Stems erect or arching, covered with straight or curved prickles.

Leaves: Alternate; stalked; blades divided into 3 or 5 leaflets, leaflets 3-10 cm long, elliptic, smooth, or slightly hairy underneath, with doubly toothed edges, stalks, and midveins with curved prickles.

Inflorescences: Flowers stalked, terminal, solitary, or clustered.

Flowers: Pink, rose, or white, to 5 cm wide, with 5 spreading petals.

Fruit: Aggregate, round, or cylindrical, 1-2.5 cm long, becoming purple or black when ripe.

Habitat and Distribution: Common; meadows, pastures, fence rows, woodland margins, and swamps; from central peninsula Florida, north occasionally as far as Massachusetts, Maryland, Missouri, and Kentucky, west to Louisiana.

Comment: Highbush Blackberry can be a spectacular wildflower when masses bloom in spring. The younger canes usually over grow the older stems from the previous year. The dense mass of older and current stems are almost impenetrable, especially when growing into other vegetation. The fruits, ripening in early summer, are edible, juicy, and sweet, and highly desirable. It also can become a very troublesome weed.

Sabatia angularis (L.) Pursh
Rose-pink Sabatia, Common Marsh-pink, Bitter-bloom
Gentianaceae, Gentian Family

Habit: Herbaceous annual or biennial, 20-80 cm tall. Stems 4-sided, angles sharp, freely branching.

Leaves: Opposite; stalkless; blades rounded, 1.5-5 cm long, 3- to 7-nerved.

Inflorescences: Terminal, oppositely-branched clusters.

Flowers: Pink with a green eye, with 5 petals, each 1.5-2 cm long and half as wide.

Fruit: An ellipsoid capsule, 6-10 mm long.

Habitat and Distribution: Occasional to frequent; moist meadows, marshes, fields, hammocks, mixed woodlands, and margins; from panhandle Florida, west to Texas, and north to Kansas, Michigan, and Connecticut.

Comment: Flowering in summer, Rose-pink Sabatia makes a splash of color. It is almost weedy northward and its frequency contributes to its value as a wildflower.

Rubus argutus Link

Sabatia angularis (L.) Pursh

Pink

Sabatia bartramii Wilbur
Ten-petaled Sabatia or Bartram's Rosegentian
[*Sabatia decandra* (Walter) R.M. Harper, misapplied; *Sabatia dodecandra* (L.) BSP. var. *coriacea* (Elliott) Ahles]
Gentianaceae, Gentian Family

Habit: Herbaceous perennial, to 1 m tall, from underground runners. Stem erect, nearly round or slightly 4-sided.

Leaves: Opposite; stalkless; blades to 8 cm long, succulent, basal leaves elliptic often broader at tip, stem leaves narrower than stem.

Inflorescences: Terminal, alternately-branched clusters, usually with only one flower at the tip of each branch.

Flowers: Pink with yellow eye, with 7-13 petals, each 2-3.5 cm long.

Fruit: An ellipsoid capsule, 10 mm long.

Habitat and Distribution: Frequent; wet pinelands, ditches, pond margins; throughout Florida, west to southeastern Mississippi, and north to southwestern South Carolina.

Comment: The large, many-petaled flowers are seen in summer and fall. It is a great wildflower that always provokes comment.

Sabatia grandiflora (A. Gray) Small
Large-flowered Sabatia or Large-flowered Marsh-pink
Gentianaceae, Gentian Family

Habit: Herbaceous annual, to 1.2 m tall. Stems erect, rounded.

Leaves: Opposite; stalkless; blades to 4 cm long, succulent, basal leaves elliptic often broader at tip, stem leaves linear.

Inflorescences: Terminal, alternately-branched clusters.

Flowers: Pink to rose with yellow eye or star pattern in center, with 5 petals, each 1.8-2.5 cm long, about 1 cm wide.

Fruit: An ellipsoid capsule, 8-10 mm long.

Habitat and Distribution: Common; wet pinelands, marshes, shores, ditches, and other low sites; throughout Florida, and west to southeast Alabama; Cuba.

Comment: Flowering in all warm months, the large petals are quite showy. It could be used more frequently in damp soils as a colorful wildflower.

Sabatia bartramii Wilbur

Sabatia grandiflora
(A. Gray) Small

Pink

Sabatia stellaris Pursh
Star Sabatia or Rose-of-Plymouth
Gentianaceae, Gentian Family

Habit: Herbaceous annual, to 50 cm tall. Stems erect, rounded, or angled.

Leaves: Opposite; stalkless; blades to 4 cm long, thin, with prominent midvein, basal leaves usually narrowly elliptic, stem leaves linear at least as wide as stem or more so.

Inflorescences: Terminal, alternately-branched clusters.

Flowers: Pink or pale rose with yellow red-bordered star-shaped pattern at center, with 5 petals, each 1-1.5 cm long, and half as wide.

Fruit: An ellipsoid capsule, about 1 cm long.

Habitat and Distribution: Frequent; bogs, fresh to salt marshes, wet marl prairies, and coastal swales; throughout Florida, west to Louisiana, and north to Massachusetts; Bahama Islands, Cuba, central Plateau of Mexico.

Comment: A summer blooming wildflower usually seen in coastal areas.

Scutellaria parvula Michx.
Small Skullcap or Little Skullcap
Lamiaceae (Labiatae), Mint Family

Habit: Herbaceous perennial, 1-3 m tall, from underground runners and tubers. Stems erect, 4-angled, glandular hairy, tiny hairs on angles, branched at base.

Leaves: Opposite; stalked on lower stem, stalkless on upper stem; blades small, hairy, broadly oval, 6-15 mm long, 7- to 11-veined, sometimes with slightly toothed margins.

Inflorescences: Terminal, spike-like arrangements.

Flowers: Pink to violet to blue to white, 6-8 mm long, tubular, widening at mouth, with hood-shaped, 3-lobed upper lip and unlobed lower lip at mouth.

Fruit: 4 nutlets, dull brown, bumpy, rounded, flattened on one side, about 1 mm long.

Habitat and Distribution: Rare to infrequent; dry soils of low and upland woods, banks, ledges, flats, and fields; from panhandle Florida, west to Texas, and north to Minnesota, Ontario, Quebec, Long Island, and West Virginia.

Comment: Small Skullcap flowers in spring. The small flowers of this rare wildflower, even in mass, are not very showy. Skullcaps are so named because of a helmet-like crest on the upper part of the green tube around the base of flower.

Sabatia stellaris Pursh

Scutellaria parvula Michx.

Pink

Sesuvium portulacastrum (L.) L.
Sea-purslane or Cenicilla
Aizoaceae, Mesembryanthemum Family

Habit: Herbaceous perennial, prostrate, to 2 m long. Stems succulent, prostrate, or trailing, rooting at nodes.

Leaves: Opposite; stalkless; blades succulent, narrow, 1.5-6 cm long.

Inflorescences: Flowers solitary on 3 mm or more long stalks in leaf axils.

Flowers: Deep pink or pinkish purple, but green underneath, with numerous stamens, star-shaped, 1.4-2 cm across, with short protrusion from center.

Fruit: Capsules about 1 cm long, opening by breaking apart around base, containing black seeds.

Habitat and Distribution: Frequent; seashores, dunes, brackish marshes, and salt flats; throughout coastal Florida, north to North Carolina, west to Texas; south into South America; Old World Tropics.

Comment: Flowering in all warm months, this salt loving succulent forms mats of pink wildflowers along the coast.

Sida ciliaris L.
Pink Sida
Malvaceae, Mallow Family

Habit: Herbaceous or semi-woody perennial, to 30 cm tall. Stems spreading to prostrate, branched from base, hairy with star-shaped hairs.

Leaves: Alternate; stalked; blades linear to rounded, to 2.5 cm long, hairy with star-shaped hairs, margins toothed.

Inflorescences: Stalked, few-flowered clusters from leaf axils.

Flowers: Pink to rose-purple or salmon or white or yellow, 1.2-3 cm wide.

Fruit: Composed of 5-8 hard, rough, bumpy, 1- or 2-toothed segments.

Habitat and Distribution: Rare; dry open habitats, pastures, hammocks, scrub oaks, mesquite thickets, roadsides, and clay flats; Broward, Dade, and Monroe Counties in southeast peninsula Florida, south Texas; Bahama Islands, West Indies south into South America; Old World.

Comment: The unusual color of Pink Sida flowers can be seen in most warm months. In subtropical and tropical areas this wildflower could be more frequently used.

Sesuvium portulacastrum (L.) L.

Sida ciliaris L.

Pink

Silene polypetala (Walter) Fernald & B.G. Schub.
Fringed Catchfly, Pink Catchfly, Fringed Pink
Caryophyllaceae, Pink Family

Habit: Herbaceous perennial, to 6 cm tall, from underground runners or offshoots. Stems form rosettes, hairy.

Leaves: Opposite; stalkless; blades usually broader at tip, 3-9 cm long.

Inflorescences: Stalked, terminal clusters of 3-5.

Flowers: Pink, 5 petals, fringed at tips, 4-5 cm wide.

Fruit: An oval capsule, 7-9 mm long.

Habitat and Distribution: Rare; bluff forests; along the Apalachicola River in Gadsden and Jackson Counties in panhandle Florida into southwest Georgia.

Comment: This ground-hugging wildflower blooms from March into May. The delicate, fringed pink blooms are large but frequently almost covered by leaves and other vegetation. Fringed Catchfly is listed by the U.S. and Florida as Endangered.

Stachys floridana Shuttlew. ex Benth.
Florida Betony
Lamiaceae (Labiatae), Mint Family

Habit: Herbaceous perennial, 10-40 cm tall, from underground runners and tubers. Stems hairy, 4-angled, delicate, freely branching, smooth.

Leaves: Opposite; stalked; blades shovel- or lance-shaped, 1-7 cm long, with toothed edges.

Inflorescences: Terminal whorled clusters.

Flowers: Pink or pale purple with dark dots, tubular, 1-1.5 cm long, 2-lipped, upper lip smaller and hoodlike, lower lip larger and 3-lobed.

Fruit: 4 rounded nutlets, about 1.5 mm long, dark brown, smooth.

Habitat and Distribution: Frequent to common; open woods, roadsides, gardens, turf, and disturbed sites; throughout Florida, west to Texas, and north to Virginia.

Comment: Flowering from spring into fall. This species was probably native to the central Florida peninsula. It has spread, primarily as a weed, by fruits, underground runners, and tubers with ornamental plants. The white, long, thin underground runners are easily broken and very difficult to remove from soil. Tubers are extremely hard to kill with systemic herbicides due to the challenge of its transportation through the long runners. The white segmented tubers resemble a rattlesnake's rattle. The tubers are edible raw or cooked. The crunchy raw tubers are often sliced into salads and have a slightly spicy flavor.

Silene polypetala
(Walter) Fernald & B.G. Schub.

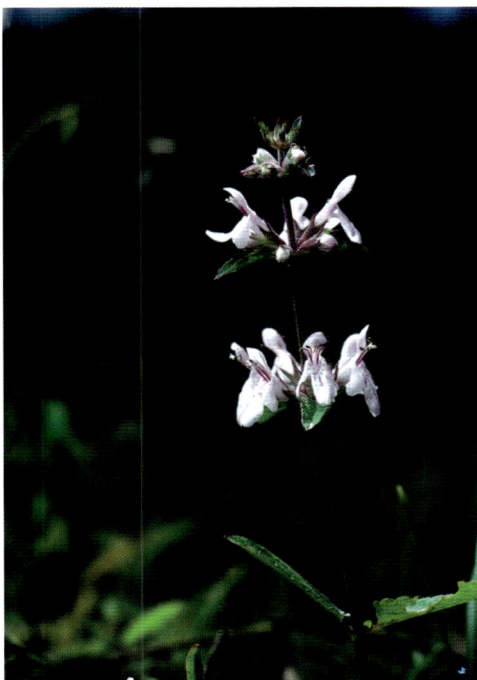

Stachys floridana
Shuttlew. ex Benth.

Pink

Stokesia laevis (Hill) Greene
Stokes' Aster
Asteraceae (Compositae), Aster or Sunflower or Daisy Family

Habit: Herbaceous perennial, 25-50 cm tall. Stems hairy, especially upwards, die back each year to basal rosette.

Leaves: Alternate; short-stalked on lower part of plant and stalkless upward; blades glandular dotted, elliptic, to 30 cm long and smaller upward, upper leaf bases with toothed margins.

Inflorescences: 1-7, stalked, terminal heads, heads surrounded by a few bracts, each with a spiny tip.

Flowers: Pinkish to violet blue to blue to white, ray flowers 2-3 cm long.

Fruit: Short, fat, shiny nutlet.

Habitat and Distribution: Occasional; wet flatwoods, savannas, and bogs; from north peninsula Florida, north to South Carolina, and west to Louisiana.

Comment: Stokes' Aster can be purchased at many nurseries. Flowering in summer, it has a number of color variations. Most of the year the plant appears as a mound of leaves.

Tephrosia virginiana (L.) Pers.
Goat's Rue or Devil's Shoestring
Fabaceae (Leguminosae), Pea or Bean Family

Habit: Herbaceous perennial, 20-70 cm tall, from a woody crown. Stems covered with gray hairs, few or no branches.

Leaves: Alternate; very short stalk; blades once-divided, with 7-35 leaflets, each 1-3 cm long, with gray hairs on both surfaces or only on lower surface.

Inflorescences: Terminal clusters.

Flowers: Bicolored - pink, yellow-pink striped, becoming yellow, typically pea-shaped, petals 1.5-2 cm long.

Fruit: Pods oblong, flat, 2.5-5.5 cm long, 3-5.5 mm wide, hairy.

Habitat and Distribution: Frequent to common; open woods, old fields, dunes, clearings, and roadsides; from central peninsula Florida, west to Texas, and north to Iowa, Minnesota, and New Hampshire.

Comment: Mostly spring flowering, Goat's Rue has uniquely colored flowers, sometimes seeming more yellow than pink and vice versa. Thought to be TOXIC if ingested.

Pink

Stokesia laevis (Hill) Greene

Tephrosia virginiana (L.) Pers.

Pink

Teucrium canadense L.
Wood Sage or American Germander
Lamiaceae (Labiatae), Mint Family

Habit: Herbaceous perennial, 0.3-1.2 m tall, from underground runners. Stems gray hairy, 4-angled, leafy.

Leaves: Opposite; stalked; blades elliptic to lance-shaped, 5-12 cm long, hairy, lower surface silvery, margins toothed.

Inflorescences: Terminal spikes.

Flowers: Pink to blue, tubular, 1.1-1.8 cm long, with 5 lobes, lower lip largest and scoop-shaped.

Fruit: 4 oval, yellowish brown, 2 mm long nutlets.

Habitat and Distribution: Frequent to common; low, wet thickets, marshes, swamps, and hammocks; from Quebec to British Columbia, south throughout the U.S. into Mexico and Cuba.

Comment: Used as an ornamental, sometimes as an edge plant, the showy spikes of blooms can be seen in the summer. Occasionally flowers are a very light pink, almost white.

Tradescantia roseolens Small
Scrub Spiderwort
Commelinaceae, Spiderwort Family

Habit: Herbaceous perennial, 30-60 cm tall. Stems erect, solitary, or branched from the base, slightly hairy.

Leaves: Alternate; stalkless; blades grass-like, to 30 cm long, base sheathing stem.

Inflorescences: Terminal stalked clusters, sometimes drooping.

Flowers: Violet to blue, petals 3, each 1-2 cm long, sepals with only glandular hairs.

Fruit: An oblong capsule, 5-7 mm long.

Habitat and Distribution: Infrequent; scrub and dry woods; from central peninsula Florida north to South Carolina.

Comment: Scrub Spiderwort flowers in spring. The flowers are showy, but the plant has a scattered distribution and is seldom seen. It is a great wildflower for dry open sites.

Teucrium canadense L.

Tradescantia roseolens Small

Pink

Trifolium carolinianum Michx.
Carolina Clover
Fabaceae (Leguminosae), Pea or Bean Family

Habit: Herbaceous annual or weak perennial, 5-30 cm tall. Stems ascending or reclining, sometimes forming mats, somewhat hairy.

Leaves: Alternate; stalked; blades once-divided with 3 leaflets, leaflets oval and broader at tip, each 4-15 mm long, margins with minute teeth toward tip.

Inflorescences: Long-stalked, terminal and axillary, round heads.

Flowers: Violet to purple, loosely clustered, 4-7 mm long.

Fruit: Pods oblong, 2-4 cm long, with 2-4 seeds.

Habitat and Distribution: Occasional; open areas in woods, prairies, sandy fields, and roadsides; from central peninsula Florida, north to North Carolina, west to east Texas.

Comment: The pink-tinged flowers occur in spring. The small flowers, even in heads, of these low plants are often difficult to see among other vegetation.

Urena lobata L.
Caesar Weed or Bur Mallow
Malvaceae, Mallow Family

Habit: Herbaceous perennial or partly woody shrub, 1-3 m tall. Stems erect, branching, often red, hairy with star-shaped hairs.

Leaves: Alternate; stalked; blades rounded, often somewhat 4-sided, 5-10 cm long, usually with pointed shallow lobes and toothed edges, densely star-shaped hairy on lower surface, midvein on underside with a slit-like gland.

Inflorescences: Solitary flowers or in few-flowered clusters in leaf axils.

Flowers: Pink, darker toward center, 2-3.5 cm wide, with 5 petals united at the bases.

Fruit: Capsules rounded, 5-chambered, covered with barbed spines.

Habitat and Distribution: Common; disturbed sites and waste areas, invading native habitats; peninsula Florida; Bahamas; West Indies; generally widespread throughout the subtropics and tropics; a native of the Old World.

Comment: Flowers all year. The barbed spines on the fruit act as an effective dispersal mechanism. The spines attach the fruit to almost anything that brushes it. The blooms are attractive, but the plant is extremely weedy.

Trifolium carolinianum Michx.

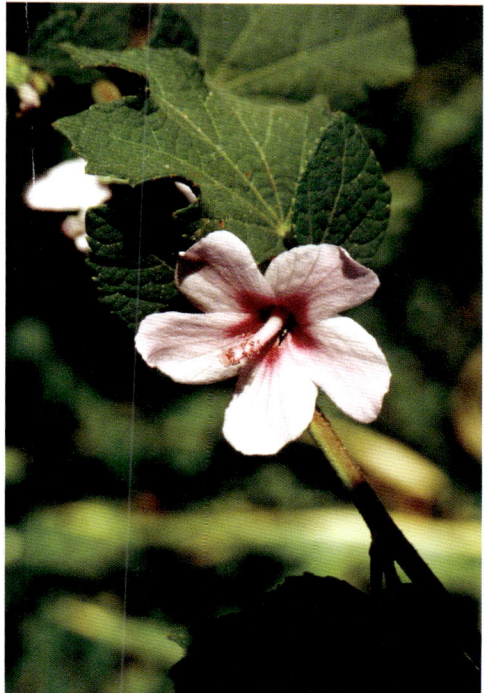

Urena lobata L.

Pink

Utricularia resupinata B.D. Greene ex Bigelow
Small Purple Bladderwort
Lentibulariaceae, Bladderwort Family

Habit: Herbaceous annual or perennial, to 10 cm tall. Stems leaf-like, mostly subterranean, short, root-like, branches resemble grass-blades. Floating plants with densely tangled masses of narrow grass-blade-like branchlets. Branches and branchlets above and below ground with a few urn-like bladders.

Leaves: Alternate; stalked; blades dissected into filiform segments bearing bladder-like traps.

Inflorescences: 1-3 flowers, terminal on slender, wiry, erect stalks with a miniature bract 1-1.5 cm below the flower.

Flowers: Purple, about 1 cm long, facing upward, with a spur.

Fruit: A small, 2-parted capsule; seeds angled.

Habitat and Distribution: Frequent; muddy soils, shallow margins of ponds, and wet flatwoods; throughout Florida, west to south Alabama, north to Indiana, Wisconsin, Quebec, Nova Scotia, New England, and New York.

Comment: Shallow ponds can be dotted with purple flowers at any time from spring into fall when Small Purple Bladderwort is blooming.

Vaccinium darrowii Camp
Darrow's Blueberry or Glaucous Blueberry
Ericaceae, Blueberry Family

Habit: Perennial shrub, to 0.5 to 1 m, rarely to 4 m tall. Stems twiggy, young twigs hairy.

Leaves: Alternate, evergreen; stalkless; blades light waxy blue-green, stiff, elliptic to somewhat rounded, 0.5-1.5 cm long, with no stalked glands on lower surface.

Inflorescences: Axillary clusters of 2-8 flowers.

Flowers: Pink to red or white, 5-8 mm long, urn-shaped with tiny lobes at mouth, drooping.

Fruit: Fleshy, round, 4-6 mm diameter, blue when ripe.

Habitat and Distribution: Frequent to common; dry pine flatwoods, sandhills, creek banks; throughout Florida, west to southeast Texas, and north to south Georgia.

Comment: Darrow's Blueberry flowers in spring. The flowers and foliage are attractive, especially so since the plants are usually found in extensive colonies. The fruit is edible, but is only of fair quality. It should be used more often as an ornamental.

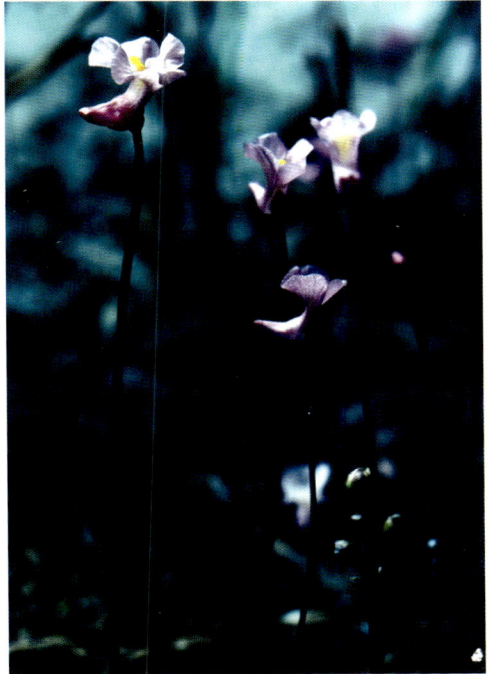

Utricularia resupinata
B.D. Greene ex Bigelow

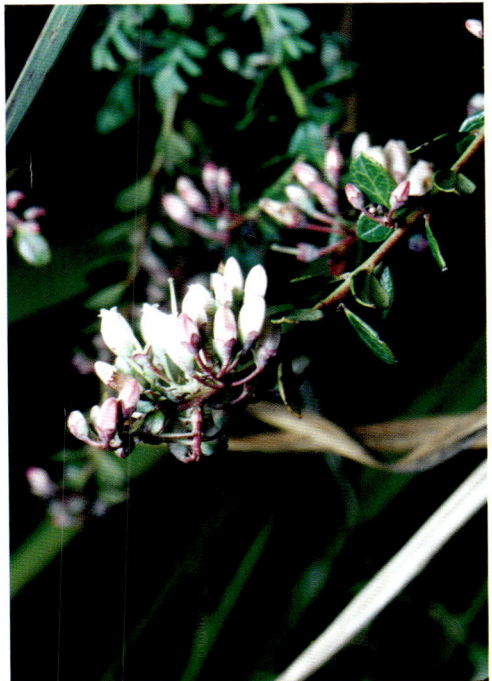

Vaccinium darrowii Camp

Pink

Vernonia angustifolia Michx.
Narrow-leaved Ironweed
Asteraceae (Compositae), Aster or Sunflower or Daisy Family

Habit: Herbaceous perennial, 0.6-1.2 m tall. Stems leafy, branching, finely hairy.

Leaves: Alternate; stalkless; blades very narrow to needle-like, 5-11 cm long and 2-6 mm wide, no basal rosette.

Inflorescences: Numerous stalked heads in a loose, terminal, branched cluster, heads about 1 cm in diameter, only disc flowers present.

Flowers: Pink, 8-30 flowers, flowers 11-13 mm long.

Fruit: Nutlets about 3 mm long, ribbed, ribs hairy, tuft of bristles at top.

Habitat and Distribution: Frequent; dry pinelands, dry woods, and sandy old fields; from central peninsula Florida, west to Mississippi, and north to North Carolina.

Comment: A nice wildflower usually noticed during the summer into fall flowering season. The thin stems and leaves are difficult to see among other vegetation.

Vernonia gigantea (Walter) Trel.
Giant Ironweed or Tall Ironweed
Asteraceae (Compositae), Aster or Sunflower or Daisy Family

Habit: Herbaceous perennial, 1-2 m tall. Stems leafy, finely hairy, branching at top.

Leaves: Alternate; stalked; blades elliptic to lance-shaped, sometimes narrowly so, margins toothed, 6-30 cm long, smaller on upper stems.

Inflorescences: Numerous stalked heads in a loose, terminal, branched cluster, only disc flowers present.

Flowers: Reddish-purple, 9-30 flowers, flowers 9-11 mm long.

Fruit: Nutlets about 3.5 mm long, ribbed, ribs hairy, tuft of bristles at top.

Habitat and Distribution: Common; hammocks, floodplains, pastures, and wet margins; central peninsula Florida, north to Delaware, Ohio, Illinois, and Missouri, west to Oklahoma and Texas.

Comment: This is the tallest of our Ironweeds, sometimes reaching to 3 m or more. This large herbaceous plant flowers in summer and fall and makes a very showy display. In the late fall or at frost, the stems die back to a rosette. It can be planted by seeds on moist soil and largely left on its own.

Vernonia angustifolia Michx.

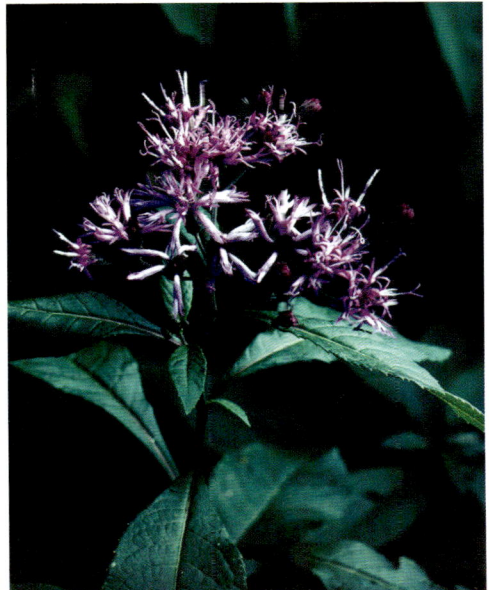

Vernonia gigantea (Walter) Trel.

Pink

Vicia sativa (L.) subsp. *nigra* (L.) Ehrh.
Narrow-leaved Vetch
Fabaceae (Leguminosae), Pea or Bean Family

Habit: Herbaceous annual, to 1 m tall. Stems erect, sprawling or slightly climbing, smooth or lightly hairy.

Leaves: Alternate; stalkless; blades once-divided, with 8-14 oblong or linear leaflets, each 1-4 cm long, leaf terminates in a branched tendril.

Inflorescences: Flowers in axillary pairs.

Flowers: Pink-purple, rose, or white, typically pea-shaped, 1- 2.5 cm long.

Fruit: Pods oblong, 2.5-8 cm long, cylindrical or flattened.

Habitat and Distribution: Occasional; roadsides, fields, borders, clearings, and other disturbed sites; from northern peninsula Florida throughout the U.S., virtually cosmopolitan in temperate climates; a native of Europe.

Comment: Flowers in the early spring. The large flowers and peculiar tendrils at leaf tips make this weedy looking plant an attractive wildflower. Cultivated for forage and green manure.

Warea amplexifolia (Nutt.) Nutt.
Wide-leaf Warea or Clasping Warea
Brassicaceae (Cruciferae), Mustard Family

Habit: Herbaceous annual, 30-70 cm tall. Stems smooth.

Leaves: Alternate; clasping; blades 1-3 cm long, lance shaped to oval.

Inflorescences: Terminal, domed clusters.

Flowers: Purplish to white, with 4 petals, each to 8-9 mm long.

Fruit: Capsules slender, 4-6 mm long, on long drooping stalk.

Habitat and Distribution: Rare; dry soils of sandhills and scrub; Lake, Orange, Osceola, and Polk Counties in central peninsula Florida.

Comment: Flowers in spring and summer are showy. Listed by the U.S. and Florida as Endangered.

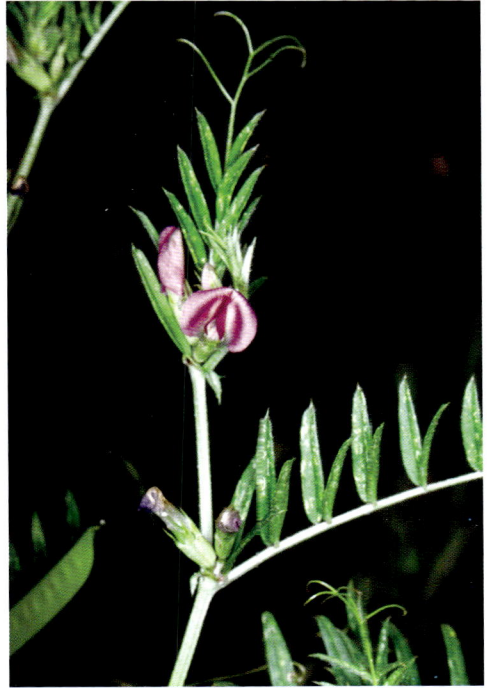

Vicia sativa (L.) subsp. *nigra* (L.), Ehrh.

Warea amplexifolia (Nutt.) Nutt.

Pink

Zephyranthes grandiflora Lindl.
Rosepink Zephyr-lily
Amaryllidaceae, Amaryllis Family

Habit: Herbaceous perennial, to 30 cm tall, from a bulb.

Leaves: Alternate, basal; stalkless; blades slender, grass-like, to 30 cm long, flat.

Inflorescences: Flowers solitary, terminal on a long scape.

Flowers: Whitish-rose to pink, about 10 cm wide, 6-parted.

Fruit: A capsule, rounded, 3-parted, with many flat, shiny seeds.

Habitat and Distribution: Infrequent to rare; disturbed areas; central peninsula and panhandle Florida; a native of Mexico and the West Indies.

Comment: The large pink flowers bloom in spring on thin stalks. Rosepink Zephyr-lily is usually found in areas where it had been cultivated.

Zephyranthes grandiflora Lindl.

Blue

Amorpha fruticosa L.
False Indigo or Bastard Indigo
Fabaceae (Leguminosae), Pea or Bean Family

Habit: Perennial shrub, 1.5-4 m tall, with underground runners (rhizomes). Stems smooth to densely hairy.

Leaves: Alternate; stalked; blades once-divided,10-30 cm long, with 9-31 leaflets, each 1-5 cm long, hairy, and gland-dotted.

Inflorescences: Dense, solitary to several, tapering, spike-like clusters, 5-20 cm long.

Flowers: Red-purple or blue, 5-6 mm long.

Fruit: Pods 5-9 mm long, about 2 mm wide, slightly curved, surface smooth and covered with glandular dots, contains 1 or 2 seeds.

Habitat and Distribution: Frequent; hammocks and along stream and riverbanks, and in open, wet woods; throughout Florida north to New Hampshire, Manitoba, southern Saskatchewan, and North Dakota, south to Texas and adjacent Mexico, west through New Mexico and Arizona to southern California; escaped from cultivation into Oregon, Washington, and Idaho.

Comment: This summer flowering shrub is often not noticed. The stems are thin and the foliage can be sparse. The flowers are such a dark blue that they are overlooked. The flower color is interesting and the shrub can compliment a landscape when planted against a larger tree. Several plants will grow from the same rhizome. Each stem will tend to be of a different length making an attractive cluster. The characteristic that most people remember about False Indigo is the unpleasant smell produced by the glands on the foliage and fruit when either are crushed.

Amsonia ciliata Walter
Blue Dogbane or Fringed Bluestar
Apocynaceae, Dogbane Family

Habit: Herbaceous perennial, 0.6-1.3 m tall, from woody above ground runners. Stems hairy, often branching towards the top.

Leaves: Alternate; stalked; blades very narrow, 3-8 cm long and 1-5 mm broad.

Inflorescences: Clusters at ends of stems.

Flowers: Blue, 1-2 cm across, with slender tube and 5 lobes.

Fruit: Capsules erect, paired, slender, 8-16 cm long and 3-4 mm broad.

Habitat and Distribution: Infrequent; dry pinelands, sandy woodlands, sandhills, limestone hills, grasslands, prairies, cedar brakes, and along railroad rights-of-way; central peninsula Florida, west to Texas, and north to North Carolina.

Comment: The delightful azure flowers can be found in the spring and summer. The narrow leaves and drift of blue flowers combine to form a delicate haze of color.

Amorpha fruticosa L.

Amsonia ciliata Walter

Blue

Amsonia tabernaemontana Walter
Bluestar, Texas-star, Eastern Bluestar
Apocynaceae, Dogbane Family

Habit: Herbaceous perennial, 0.3-1 m tall, from a woody rootstock. Stems usually several from base, smooth, with no to few branches.

Leaves: Alternate; stalked; blades narrowly oval, to 16 cm long and 1-6 cm broad.

Inflorescences: Clusters at ends of stems.

Flowers: Blue to purplish blue, 1-2 cm across, with slender tube and 5 lobes.

Fruit: Capsules erect, paired, 5-12 cm long and 2-3 mm diameter.

Habitat and Distribution: Infrequent; low wooded sites and deciduous upland woods; north peninsula Florida, west to Texas, and north to Kansas, Indiana, Illinois, and New Jersey.

Comment: The attractive Bluestar flowers are seen in the spring.

Asclepias cinerea Walter
Carolina Milkweed
Asclepiadaceae, Milkweed Family (Apocynaceae, Dogbane Family)

Habit: Herbaceous perennial, 30-70 cm tall. Stems slender, solitary, smooth, with milky sap.

Leaves: Opposite; stalkless; blades linear, 3-9 cm long and 1-2 mm broad, smooth.

Inflorescences: Round-topped clusters of 3 to 7 flowers, in leaf axils at stem tips.

Flowers: Lavender, 5 petals, each 5-7 mm long, spreading downward, crown 4-5 mm wide.

Fruit: A capsule, erect, 8-12 cm long and 5-10 mm broad, widest at base, tapering to tip.

Habitat and Distribution: Frequent; usually in wet soils, pinelands, savannahs, or bogs, occasionally in sandhills; central peninsula and panhandle Florida, north to coastal plain South Carolina.

Comment: Carolina Milkweed blooms in late spring and early summer. The lavender flowers in clusters are attractive, but not particularly showy.

Amsonia tabernaemontana Walter

Asclepias cinerea Walter

195

Blue

Aster adnatus Nutt.
Scaleleaf Aster or Tight leaved Aster
[*Symphyotrichum adnatum* (Nutt.) G.L. Nesom]
Asteraceae (Compositae), Aster or Sunflower Family

Habit: Herbaceous perennial, 20-80 cm tall, with tuberous rootstocks. Stems hairy.

Leaves: Alternate; lacking stalks; basal leaves to 2.5 cm long; stem leaves small, scale-like, very close together, ascending and somewhat appressed, the basal part of the leaf adnate to the stem, giving the stem a naked look.

Inflorescences: In heads at tips of stems and branches.

Flowers: Heads to 3 cm across with light blue to lilac rays and yellow disc flowers.

Fruit: Nutlets to 3 mm long, smooth.

Habitat and Distribution: Frequent; dry open pinelands and sandhills on the coastal plain; throughout Florida, west to Mississippi, north to Georgia, and south into the Bahamas.

Comment: Scaleleaf Aster is almost unnoticed except when blooming in the fall and winter. The large blue flowers stand out against the usually dry sparse background. This plant would make a nice addition to a wildflower garden mix.

Aster concolor L.
Silvery Aster
[*Symphyotrichum concolor* (L.) G.L. Nesom]
Asteraceae (Compositae), Aster or Sunflower Family

Habit: Herbaceous perennial, 0.3-1 m tall, from a short crown and often with underground runners (rhizomes). Stems silky with hairs, simple or with a few branches.

Leaves: Alternate; lacking a stalk; blades narrowly elliptic to lance-shaped, 1-4 cm long and 0.3- 1.5 cm wide, grayish and covered with silky hairs.

Inflorescences: Ascending clusters of heads at ends of stems.

Flowers: Heads 2-3.5 cm across, with 8-16 blue rays and yellow disc flowers.

Fruit: Nutlets hairy, to 2.5 mm long, with a white to tan tuft of 5-6 mm long bristles on top.

Habitat and Distribution: Infrequent; dry sandy pinelands; throughout Florida, west to Louisiana and north to Tennessee, Kentucky, and Massachusetts.

Comment: The fall flowering Silvery Aster is an appealing, showy wildflower as both the flowers and the silvery foliage are attractive.

Aster adnatus Nutt

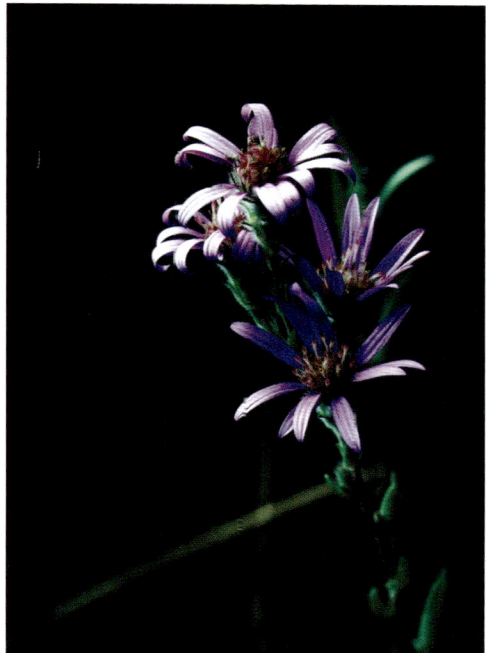

Aster concolor L.

Blue

Aster dumosus L.
Bushy Aster
[*Symphyotrichum dumosum* (L.) G.L. Nesom]
Asteraceae (Compositae), Aster or Sunflower Family

Habit: Herbaceous perennial, 0.3-1.5 m tall, from underground runners (rhizomes). Stems smooth, 1 to many clumped together.

Leaves: Alternate; lacking a stalk; basal leaves lance-shaped and broader at the tip, margins with irregular teeth, usually absent on fall flowering plants and present on spring flowering plants; stem leaves very narrow, 2-12 cm long and 2-8 mm broad, rigid.

Inflorescences: Few to numerous heads in spreading and branching clusters at ends of stems.

Flowers: Heads 1.5-2.8 cm across, with 13-30 very narrow blue, lavender, or rarely whitish rays and yellowish to reddish disc flowers.

Fruit: Nutlets 1.2-2.3 mm long, slightly hairy, with a whitish tuft of bristles 3-4 mm long on top.

Habitat and Distribution: Common; woodlands, meadows, marshes, ditches, turf, and old fields; throughout Florida, west to east Texas, and north to Arkansas, Michigan, and Maine.

Comment: As a wildflower, Bushy Aster is an unusually variable plant mostly blooming in the fall but also flowering in the spring. Spring blooming plants are often small having grown after the winter freeze sometimes being only 10 cm tall. Larger fall blooming plants can be covered with blue heads of flowers. Bushy Aster can be transplanted by means of the runners or started from seeds. The species often is an unnoticed weed in turf. The basal leaves are so different from the stem leaves that few recognize the plant. It also can coexist with the turf even at low mowing heights. If planted on the edge of turf, mowing can be stopped in the spring and fall to allow flowering.

Aster linariifolius L.
Stiff-leaved Aster or Flaxleaf Aster
[*Ionactis linariifolia* (L.) Greene]
Asteraceae (Compositae), Aster or Sunflower Family

Habit: Herbaceous perennial, 0.2-0.7 m tall, from a short rootstock, occasionally with underground runners (rhizomes). Stems erect, thin, few to many forming a clump.

Leaves: Alternate; lacking a stalk; blades very narrow, 1.5-4 cm long and 1-4 mm wide, sandpapery.

Inflorescences: A somewhat flat-topped cluster of a few to many heads at the tips of branches.

Flowers: Heads 1.6-2.8 cm across, with 7-20 blue to violet, rarely pink or white rays and yellow to red disc flowers.

Fruit: Nutlets black, silvery hairy, 2.5-2.8 mm long, with a tan tuft of bristles 4-6 mm long at the top.

Habitat and Distribution: Occasional; dry, open, usually sandy soils of woodlands, woodland borders, and pinelands in extreme north peninsula and panhandle Florida, west to east Texas, and north to Arkansas, Missouri, Wisconsin, Quebec, and Maine.

Comment: These small plants have been overlooked as an attractive fall blooming wildflower. The strict growth and large heads make a nice bedding plant.

Blue

Aster dumosus L.

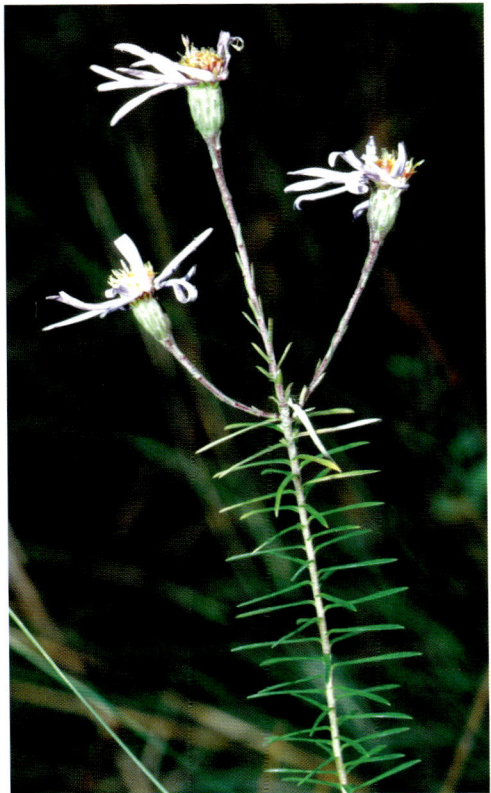

Aster linariifolius L.

Blue

Aster simmondsii Small
Large-headed Bushy Aster
[*Symphyotrichum simmondsii* (Small) G.L. Nesom]
Asteraceae (Compositae), Aster or Sunflower Family

Habit: Herbaceous perennial, 0.2-1.2 m tall. Stems erect, branches erect or spreading, minutely hairy.

Leaves: Alternate; lacking a stalk; blades linear to elliptic, to 12 cm long, smooth, margins toothed on lower leaves, becoming less toothed and smooth upwards on the stem.

Inflorescences: Solitary or branched clusters of heads at the tips of stems.

Flowers: Heads about 2-3 cm across, with lavender to pink-purple or occasionally white rays and yellow disc flowers.

Fruit: Nutlets to 2 mm long, hairy, with a cream tuft of bristles on top.

Habitat and Distribution: Infrequent; moist to dry soils, flatwoods and other pinelands; from north central into south peninsula Florida.

Comment: The few large heads usually found on this fall and winter flowering species are immediately noticeable. Large-headed Bushy Aster is very variable and thought to be a hybrid with several possible parents. The Seminoles reportedly used this species medicinally to bathe victims of sunstroke.

Aster subulatus Michx.
Annual Marsh Aster or Hierba del Marrano
[*Symphyotrichum subulatum* (Michx.) G.L. Nesom]
Asteraceae (Compositae), Aster or Sunflower Family

Habit: Herbaceous annual, 0.3-1.5 m tall, from a taproot. Stems solitary, smooth, often fleshy, freely branched.

Leaves: Alternate; stalkless; blades fleshy, narrow, linear to narrowly elliptic, to 20 cm long and 2-10 mm wide.

Inflorescences: Solitary to many heads in an open, diffuse cluster at stem tips.

Flowers: Heads solitary to many, about 1-1.5 cm wide with less than 20 (rarely to 50) blue to violet or sometimes white rays, 0.1-1 cm long, with yellow disc flowers.

Fruit: Nutlets to 2.5 mm long, hairy, with a brownish tuft of bristles on the top.

Habitat and Distribution: Frequent; brackish and fresh marshes, ponds, depressions, palmetto flats, and other wet habitats, often a weed in irrigated crops and ditches; throughout Florida, north to Maine and New Brunswick, west to Missouri, Kansas, New Mexico, Arizona, and California, south through the West Indies, and Central America into South America.

Comment: Often very common in marshy areas, Annual Marsh Aster can, through sheer numbers, be an impressive wildflower. When growing alone the smaller flower heads of this plant can easily be overlooked.

Aster simmondsii Small

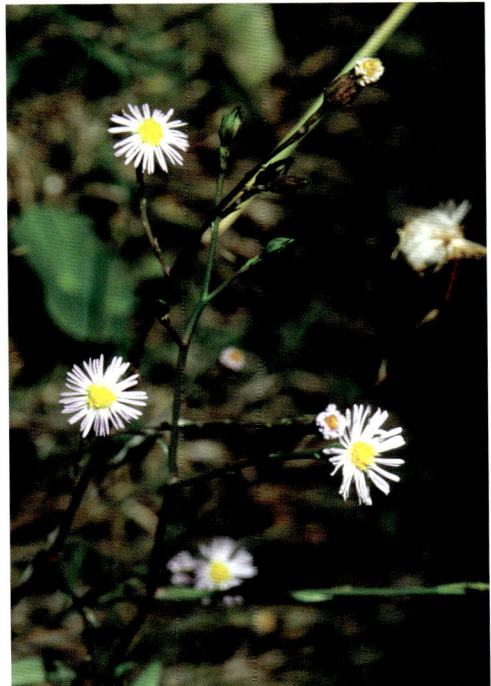

Aster subulatus Michx.

Blue

Bacopa caroliniana (Walter) B.L. Rob.
Blue-hyssop or Lemon Bacopa
Scrophulariaceae, Figwort Family

Habit: Herbaceous perennial, to 1 m long, rooted. Stems succulent, hairy, creeping, or floating.

Leaves: Opposite; bases clasping stems; blades oval to spade-shaped, thick, 0.5-2 cm long, covered with punctate dots.

Inflorescences: Solitary in leaf axils, usually one flower per pair of leaves.

Flowers: Blue, tubular, with 4 or 5 lobes, about 1 cm long.

Fruit: Capsules oval, 4-5 mm long.

Habitat and Distribution: Common; wet sites, ponds, lakes, ditches, marshes, and swamps; throughout Florida, west to Texas, and north to southeast Virginia, disjunct in Kentucky.

Comment: Blue-hyssop flowers in the warm months, all year in south peninsula Florida and during the summer northward. The tips of stems of this plant often form a mat at the surface of water, the blue flowers and foliage appear to be a blue haze. Even more appealing than the flower color is the lemony odor when the leaves are bent or crushed, whence comes the common name Lemon Bacopa. Sometimes Blue-hyssop is cultivated in water gardens.

Barleria cristata L.
Philippine Violet
Acanthaceae, Acanthus Family

Habit: Herbaceous or shrubby perennial, to 1 m tall. Stems hairy.

Leaves: Opposite; short-stalked; blades lance-shaped to elliptic to oval, 5-12 cm long.

Inflorescences: Small clusters or solitary in leaf axils.

Flowers: Blue, violet, pink, or white, bell shaped, 6-7 cm long, with 4 or 5 lobes.

Fruit: An ellipsoid capsule, to 1.5 cm long, 4-seeded, seeds hairy.

Habitat and Distribution: Rare; disturbed sites from southern to northern Florida, in southern peninsula Florida it is becoming more frequent as an escape. Native of India.

Comment: Blooming in all warm months Philippine Violet is frequently planted as an ornamental. The showy flowers and shrubby form fit into many landscape schemes although it can become leggy. Subject to freezing temperatures, it is being used as an annual northwards.

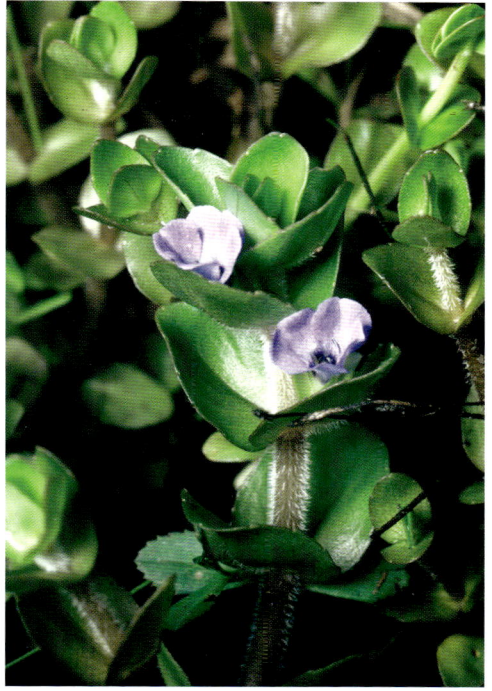

Bacopa caroliniana
(Walter) B.L. Rob.

Barleria cristata L.

Blue

Buchnera americana L.
Blueheart
[*Buchnera floridana* Gand.]
Scrophulariaceae, Figwort Family

Habit: Herbaceous perennial, 30-80 cm tall. Stems rough.

Leaves: Opposite; lacking a stalk; blades oval, larger leaves 2-8 cm long, becoming smaller on upper stems, rough hairy, with irregularly and sharply toothed margins.

Inflorescences: Loose clusters at ends of stems.

Flowers: Purple or white, about 1 cm long, tubular, with 5 lobes to 9 mm long.

Fruit: Capsules oval or pear shaped, 5-6 mm long.

Habitat and Distribution: Common; sandy or gravely soils, pinelands, flatwoods, prairies, meadows, and disturbed habitats; throughout Florida, west to Texas, and north to Kansas, Illinois, Michigan, southern Ontario, and New York.

Comment: The blue flowers found in the spring and fall are eye-catching, but not large. The white variation is even more striking.

Campanula floridana S. Wats. ex A. Gray
Florida Bellflower or Blue Bell
Campanulaceae, Bellflower Family

Habit: Herbaceous perennial, often spreading, 20-40 cm long or tall. Stems branching, weak, slender.

Leaves: Alternate; lacking a stalk and clasping the stems; blades narrowly elliptical to linear, margins with tiny teeth, 1 to 4 cm long.

Inflorescences: Solitary flowers in upper leaf axils.

Flowers: Blue or white, star-shaped, with 5 narrow, pointed petals, to about 1 cm long.

Fruit: Capsules rounded, to 4 mm long.

Habitat and Distribution: Endemic and locally common; low, moist to wet sites; southern peninsula to eastern panhandle Florida.

Comment: This small, low wildflower is quite attractive when blooming. Flowering occurs in spring. It can make a wonderful addition to any low damp meadow. When not in flower, Florida Bellflower blends into the other low vegetation.

Blue

Buchnera americana L.

Campanula floridana
S. Wats. ex A. Gray

Blue

Centrosema virginianum (L.) Benth.
Climbing Butterfly-pea
Fabaceae (Leguminosae), Pea or Bean Family

Habit: Perennial trailing or climbing vine, to 1.5 m long. Stems branched, climbing by twining, smooth or hairy.

Leaves: Alternate; stalked; blades trifoliate with 3 linear to ovate leaflets, each 2-7 cm long and 1.5-2.5 cm wide.

Inflorescences: One to 4, stalked flowers from leaf axils.

Flowers: Pale bluish to pink with white center, upper petal large, rounded, spreading, 2.5-4 cm long.

Fruit: Pods very narrow, 7-12 cm long and 3-4 mm wide.

Habitat and Distribution: Frequent; usually in open areas, pinelands, hammocks, swales, dry woods, old fields, borders of wet woods, and roadsides, throughout Florida, north to New Jersey and Oklahoma, west to Texas, and south through Bermuda, the West Indies, into Uruguay and Argentina; also introduced into India.

Comment: The large flower of Climbing Butterfly-pea seen in the summer and fall is quite showy. The plant usually is trailing over low vegetation along margins. The foliage is not usually noticed until closer inspection and the flowers appear like large colorful spots provoking many comments. In cultivation the vine can be allowed to climb over shrubs that do not flower in the summer providing an additional season of color. It also can be planted to grow on fences. Climbing Butterfly-pea is frequently confused with Butterfly-pea, *Clitoria mariana*, but can be separated because the sepal tube is shorter than the lobes and bracts.

Cichorium intybus L.
Chicory or Blue-sailors
Asteraceae (Compositae), Aster or Sunflower Family

Habit: Herbaceous perennial, to 1 m tall, from long, fleshy, branched taproot. Stems with milky sap, few, branching towards top.

Leaves: Alternate; stalked at base becoming stalkless upwards; blades lance-shaped and broadest towards tip, to 35 cm long, margins smooth, toothed, and/or lobed.

Inflorescences: One to 3 heads in the axils of the upper leaves.

Flowers: Heads only with blue (rarely pink or white) rays, rays to 2.5 cm long and with 5-toothed tips.

Fruit: An angled nutlet, 2-2.6 mm long, with a crown of papery scales.

Habitat and Distribution: Infrequent to frequent; roadsides, fields, fence rows, and other disturbed open habitats; throughout cool temperate North America; native to Eurasia and North Africa.

Comment: The striking azure-blue flowers can be found in the summer and fall. At flowering the lower leaves are usually withered. The roots can be ground, roasted and mixed with coffee or even used as a coffee substitute.

Centrosema virginianum
(L.) Benth.

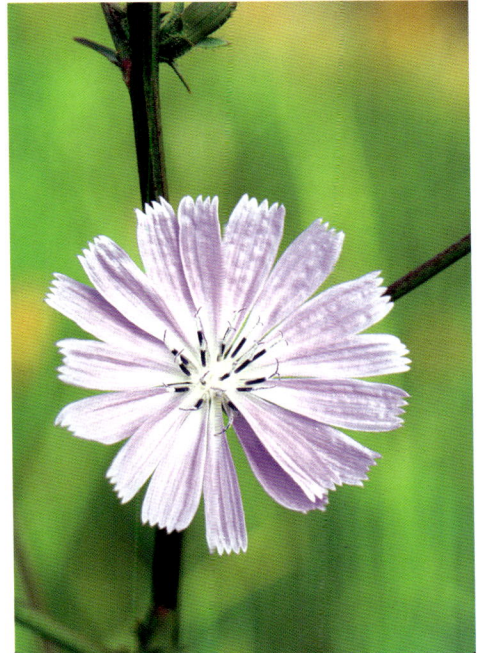

Cichorium intybus L.

Blue

Clematis baldwinii Torr. & A. Gray
Pine-hyacinth
Ranunculaceae, Buttercup Family

Habit: Herbaceous perennial, from a hard, branched rootstock, 20-60 cm tall. Stems erect, hairy, usually without branches.

Leaves: Opposite; with short or no leafstalk; blades simple or 3- to 5-lobed or deeply divided to resemble leaflets, margins smooth, 2-10 cm long.

Inflorescences: One to several flowers on long stalks from leaf axils.

Flowers: Nodding, bell-shaped, with tips curved back, petals absent, sepals bluish, purple, or pinkish.

Fruit: Nutlets to 5 mm long, with a plume-like tip 0.5-1 cm long, covered with silky hairs.

Habitat and Distribution: Frequent; moist and wet pinelands and wet margins; southern and central Florida.

Comment: Flowering can be all warm months, but mostly starts in late spring. As a wildflower the unusual habit, flower, and fruit are all attractive.

Clematis reticulata Walter
Vase Vine
Ranunculaceae, Buttercup Family

Habit: Herbaceous, perennial vine. Stems weakly climbing or spreading, slightly hairy to smooth, many-angled.

Leaves: Opposite; stalked; blades variable, usually divided, with 3-9 oval, often lobed, leaflets, leathery, with raised veins.

Inflorescences: Solitary flowers on long stalks from leaf axils.

Flowers: Nodding, bell-shaped or vase-shaped, with tips curved back, petals absent, sepals bluish.

Fruit: Nutlets rounded, silky, 4-6 mm wide, with an extended plume on end.

Habitat and Distribution: Frequent; pinelands and dry woody sites; central Florida west to Texas and north to South Carolina.

Comment: A spring and summer bloomer, Vase Vine has attractive foliage and quite unusual flowers and fruits. The clustered hairy fruits are quite showy. It can be trained on fences. The vine frequently spreads into other adjacent vegetation and is almost unseen until the fruit matures.

Clematis baldwinii
Torr. & A. Gray

Clematis reticulata Walter

Blue

Clitoria mariana L.
Butterfly-pea
Fabaceae (Leguminosae), Pea or Bean Family

Habit: Herbaceous perennial, 0.3-1.2 m long. Stems ascending, trailing, or twining, smooth or slightly hairy.

Leaves: Alternate; stalked; blades trifoliate with 3 ovate to lance-shaped leaflets, each 2-7 cm long and 1-4 cm wide.

Inflorescences: One to three stalked flowers from leaf axils.

Flowers: Light blue to purplish, with a large spreading oval upper petal, 4-5.5 cm long and 3-4 cm wide, lined with purple.

Fruit: Pod narrow, flattened, 3-5 cm long and 5-7 mm wide.

Habitat and Distribution: Frequent; sandy soils in open woods, old fields, along roadsides, dry flatwoods, sandhills, scrubs, hammocks, and clearings; throughout Florida, north to New York, New Jersey, Wisconsin, Minnesota, Ohio, Indiana, Missouri, and Oklahoma, west to Texas and disjunct in Arizona; also in Asia.

Comment: Butterfly-pea seen blooming in the spring and summer is very showy and possibly the largest flower of herbaceous bean species in the United States. The plant usually is ascending in open areas or trailing over low vegetation along margins. The foliage is relatively unnoticed, but the flowers are outstandingly colorful. In cultivation the vine can be allowed to lean into shrubs that do not flower in the summer providing an additional season of color. It also can be planted to grow along fences. Butterfly-pea is frequently confused with Climbing Butterfly-pea, *Centrosema virginianum*, but can be separated because the sepal tube is longer than the lobes and bracts.

Commelina diffusa Burm. f.
Spreading Dayflower
Commelinaceae, Spiderwort Family

Habit: Herbaceous annual, 0.2-1 m long. Stems smooth, creeping, forming mats, and rooting at stem nodes.

Leaves: Alternate; base sheathing the stem; blades lance-shaped, 2.5-8 cm long and 0.5-2 cm broad, slightly fleshy.

Inflorescences: A boat-shaped, leaf-like spathe with free margins from leaf axils.

Flowers: Blue, with 3 fragile petals, all nearly equal in size.

Fruit: Capsule with 3 to 5 seeds; seeds 2.5 mm long, with a pitted surface.

Habitat and Distribution: Common; moist to wet sandy soils, natural, cultivated, and other disturbed sites; throughout Florida to Texas and Delaware.

Comment: An attractive warm weather blooming fleshy plant. The flowers of this species are very fragile. They wither in the late morning as the day becomes hotter. When picked the petals wilt almost instantly. The stems with their shiny leaves and small blue flowers can rapidly cover a large area and thus become weedy. In a confined setting or if left in native habitats it is a good wildflower.

Clitoria mariana L.

Commelina diffusa Burm. f.

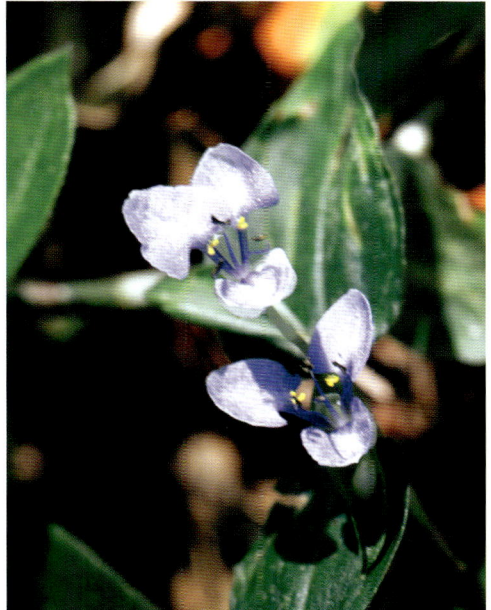

Blue

Commelina erecta L.
Erect Dayflower
Commelinaceae, Spiderwort Family

Habit: Herbaceous perennial, 20-90 cm tall or long, stems knotty lying flat or erect, from a cluster of thick black roots.

Leaves: Alternate; base sheathing the stem and flared at top; blades usually lance-shaped or occasionally oval, 5-15 cm long and 0.5-4 cm broad, fleshy.

Inflorescences: A boat-shaped, leaf-like spathe with margins fused at the base, from leaf axils.

Flowers: Petals 3, fragile, the 2 blue petals much larger than the third white petal.

Fruit: Capsules with 3 to 5 seeds; seeds hairy or granular, 3-3.5 mm long.

Habitat and Distribution: Common; dry woods and dry cultivated sites; throughout Florida, west to Arizona and Wyoming and north to New York and Wisconsin.

Comment: Flowering occurs from spring into fall. The flowers of this species, as with Spreading Dayflower, are very fragile. They wither in the late morning as the day becomes hotter. When picked the petals wilt almost instantly. Showy in groups they are to be seen as planted and not picked flowers. Several birds are fond of the seeds. It is rare to find these offered by native nurseries.

Conradina canescens A. Gray
Scrub Mint or Scrub Conradina
Lamiaceae (Labiatae), Mint Family

Habit: Perennial shrub, to 0.5 m tall. Stems woody, bushy branched, erect, silvery hairy.

Leaves: Opposite, evergreen; stalkless; blades aromatic, needle-like, clustered, 0.7 to 2 cm long, usually about 1 cm long, densely gray hairy on both surfaces.

Inflorescences: Small clusters of 1-3 in leaf axils.

Flowers: Bluish to white to lavender, upper lip erect, lower lip 3-lobed and spotted.

Fruit: A tiny nutlet.

Habitat and Distribution: Locally common; open pinelands, scrub, and clearings; panhandle Florida into adjacent Alabama and southeastern Mississippi.

Comment: This spring blooming shrub with its clustered foliage and numerous bluish flowers is showy. Scrub Mint is a great specimen plant for dry soils and can be utilized in larger clusters. The foliage is attractive all year. Scrub Mint is very similar to *Conradina brevifolia* Shinners, Short-leaved Rosemary, which has shorter leaves (6-8 mm long) and usually more flowers per cluster (1-6). Short-leaved Rosemary, listed by the U.S. and Florida as Endangered, occurs in the sand pine scrub of the Lake Wales Ridge in Highlands and Polk Counties of central peninsula Florida.

Commelina erecta L.

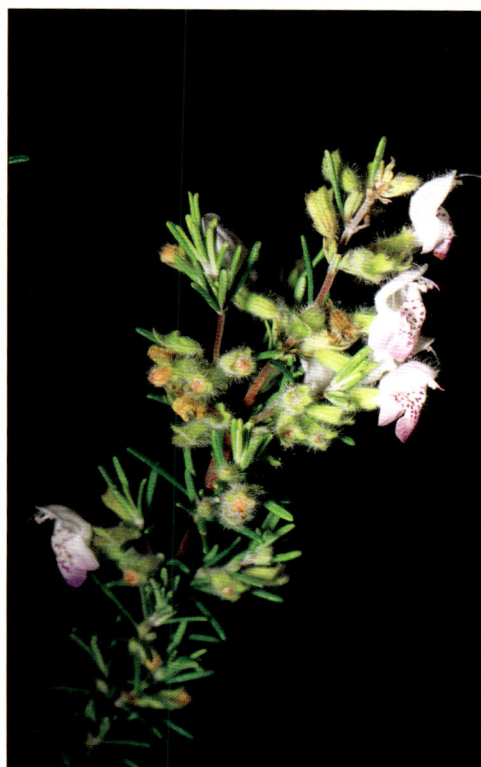

Conradina canescens A. Gray

Blue

Conradina glabra Shinners
Apalachicola Rosemary
Lamiaceae (Labiatae), Mint Family

Habit: Perennial shrub, to 0.8 m tall. Stems woody, bushy branched, erect.

Leaves: Opposite; stalkless; blades evergreen, aromatic, needle-like, clustered, to 1.5 cm long, densely gray hairy underneath.

Inflorescences: Small cluster of 1-4, usually 3-4, in leaf axils.

Flowers: Bluish to lavender to white, upper lip erect, lower lip 3-lobed and spotted.

Fruit: A tiny nutlet.

Habitat and Distribution: Endemic and rare; sandhills; Liberty and Santa Rosa Counties in panhandle Florida.

Comment: Flowering in March and June, this shrub is attractive and deserves to be cultivated. Plants that will grow in extremely dry soils are not common. More and more development is occurring on high dry sites. This species is listed by the U.S. and Florida as Endangered.

Dyschoriste oblongifolia (Michx.) Kuntze
Twinflower or Blue Twinflower
Acanthaceae, Acanthus Family

Habit: Herbaceous perennial, 10-50 cm tall. Stems hairy, branching in pairs.

Leaves: Opposite; lacking stalks; blades elliptic to rounded and broader at the tip, 2.5-4.5 cm long.

Inflorescences: Flowers solitary or clustered in leaf axils.

Flowers: Blue or purple, tubular, 1.5-3 cm long, with 5 rounded lobes, 1-2 cm across at mouth.

Fruit: Capsules brown, oblong, 10-14 mm long and about 3 mm diameter.

Habitat and Distribution: Common; pinelands, sandhills, and savannahs; throughout Florida and north to South Carolina.

Comment: Blooming during the warm months this little wildflower is easily noticed while flowering, but blends in with other small plants and usually is not seen when not in flower. It can make a nice addition to a partially sunny dry site.

Conradina glabra Shinners

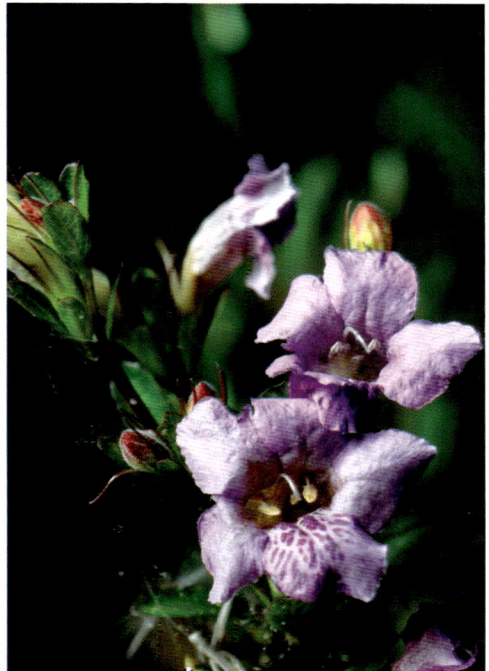

Dyschoriste oblongifolia
(Michx.) Kuntze

Blue

Eichhornia crassipes (Mart.) Solms
Water-hyacinth
Pontederiaceae, Pickerel Weed Family

Habit: Herbaceous, aquatic perennial, 0.4-1 m tall, stems free-floating or rooted in mud.

Leaves: Alternate; stalked, stalks usually spongy and inflated, allowing plants to float, often with a constriction at the top; blades oval, 2-15 cm long.

Inflorescences: Stalked, showy, loose spike, slightly taller than the leaves.

Flowers: Blue to light lavender with yellow streaks, 3-5 cm across, with 6 lobes, upper lobe with yellow patch.

Fruit: Capsule 3-lobed, elliptical, about 1 cm long.

Habitat and Distribution: Common; waterways, marshes, and lakes; throughout Florida, north to Virginia, west to Texas and Missouri; California; West Indies; native of South America.

Comment: Water-hyacinth blooms in all warm months. The large spikes of flowers are very showy. This species is an aggressive, invasive exotic in aquatic habitats and must be properly contained if cultivated. The plant is often found in loose unconsolidated sediments, but the long dense feathery root system prefers spreading into open water.

Erigeron quercifolius Lam.
Southern Fleabane or Oakleaf Fleabane
Asteraceae (Compositae), Aster or Sunflower Family

Habit: Herbaceous biennial or short-lived perennial, 10-80 cm tall. Stems erect, hairy, freely branching.

Leaves: Alternate, mostly in a basal rosette; stalkless or stalked; blades lance-shaped to ovate and broader at the tip, 3-14 cm long and 1-4 cm broad with lobed edges and with teeth; stem leaves 2-10, clasping, smaller.

Inflorescences: Few to many heads in a loose terminal arrangement.

Flowers: Heads to about 1.5 cm across, with 100-200 fine blue or light blue-lavender to white rays and yellow disc flowers.

Fruit: Nutlets 0.6-0.7 mm long with a tuft of bristles on the top.

Habitat and Distribution: Common; moist open sandy habitats, fields, pinelands, hammocks, roadsides, and disturbed areas; throughout Florida, west to Louisiana and north to Maryland on the coastal plain.

Comment: Frequently noticed as it flowers in the spring and summer for the usually profuse numbers of the small blue flowers. Southern Fleabane is a common roadside wildflower whose large numbers of plants and flowers can provide a show even seen at considerable speed. This species also becomes weedy in gardens and poor turf.

Eichhornia crassipes (Mart.) Solms

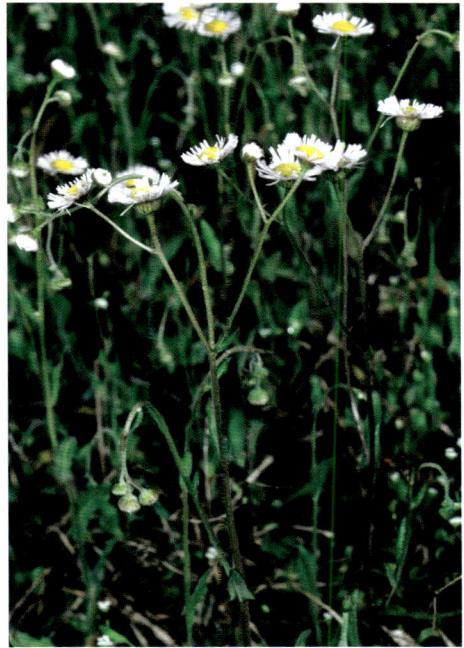

Erigeron quercifolius Lam.

Blue

Gentiana catesbaei Walter
Catesby's Gentian
Gentianaceae, Gentian Family

Habit: Herbaceous perennial, 20-80 cm tall. Stems hairy, rough, simple, or branched at top.

Leaves: Opposite; short-stalked to stalkless; blades narrow, elliptic or lance-shaped, to 8 cm long and 3 cm wide.

Inflorescences: Compact, erect, terminal clusters of 1-9 flowers.

Flowers: Blue, violet, or whitish, broadly tubular or bell-shaped, 3-5 cm long.

Fruit: An elliptic capsule, about 3 cm long; seeds elliptic, winged, brown, about 2 mm long.

Habitat and Distribution: Infrequent; bogs and seepages southward and sandy pinelands and savannahs northward; northern peninsula to central panhandle Florida and north to Virginia.

Comment: This fall flowering wildflower is definitely an unusual element in the widely scattered areas in which it is found.

Glandularia maritima (Small) Small
Seaside Verbena
[*Verbena maritima* Small]
Verbenaceae, Verbena Family

Habit: Herbaceous perennial, to 20 cm long. Stems spreading, creeping.

Leaves: Opposite; base narrowed to stem; blades somewhat oval, to 4 cm long, lobed and/or large-toothed.

Inflorescences: Domed terminal spikes.

Flowers: Rose-purple to purple, 2 cm long, sepal lobes awl-like, to 1 mm long.

Fruit: 4 slender nutlets, about 4 mm long.

Habitat and Distribution: Endemic and infrequent; dunes and pinelands and hammocks; central and southern eastern coastal peninsula Florida.

Comment: Seaside Verbena flowers all year, mainly in spring. Quite showy, this groundcover is infrequently used as an ornamental. It is listed by Florida as Endangered.

Gentiana catesbaei Walter

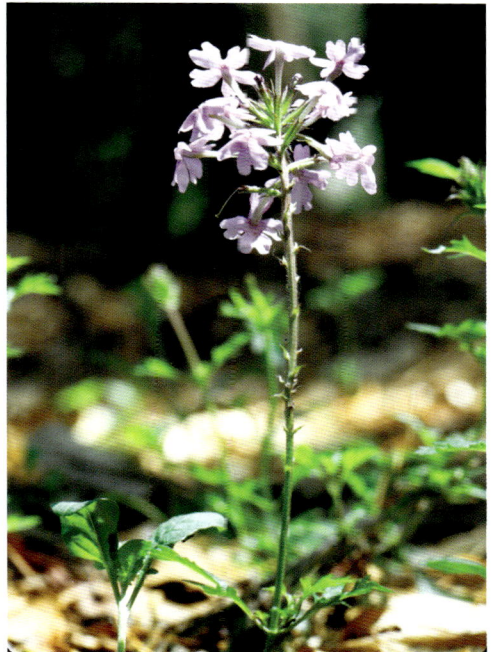

Glandularia maritima
(Small) Small

Blue

Hedyotis crassifolia Raf.
Tiny Bluet
Rubiaceae, Madder Family

Habit: Herbaceous annual, to 10 cm tall. Stems thin, erect, smooth.

Leaves: Opposite; stalked; blades joined by a flange, somewhat rounded, 3-8 mm long.

Inflorescences: Flowers solitary in upper leaf axils.

Flowers: Deep blue with a dark yellow or red eye, 4-parted, 4-9 mm long, 6-8 mm wide.

Fruit: A capsule 4-5 mm wide.

Habitat and Distribution: Infrequent; sandy soils of deciduous woods, fields, meadows, and disturbed sites; panhandle Florida north to Virginia and Illinois, and west to Arkansas and Texas.

Comment: This inconspicuous plant flowers in winter and spring. It is hard to spot when blooming and almost impossible to see at other times.

Hydrolea corymbosa J. Macbr. ex Elliott
Skyflower
Hydrophyllaceae, Waterleaf Family

Habit: Herbaceous perennial, 20-70 cm tall. Stems slender, hairy on upper stem, smooth on lower stem, unarmed or weakly spiny.

Leaves: Alternate; stalkless to short-stalked; blades elliptic to lanceolate, 2-6 cm long.

Inflorescences: Small clusters at ends of stems.

Flowers: Clear blue, about 2 cm across, star-shaped with 5 petals.

Fruit: A capsule, about 5 mm long.

Habitat and Distribution: Frequent; wet ditches, swamps, marshes, and shallow ponds; throughout Florida and north to south Georgia.

Comment: The sky-blue flowers in summer and fall are spectacular. The foliage is shiny. The plants are usually found in shallow standing water or in areas where water frequently stands.

Hedyotis crassifolia Raf.

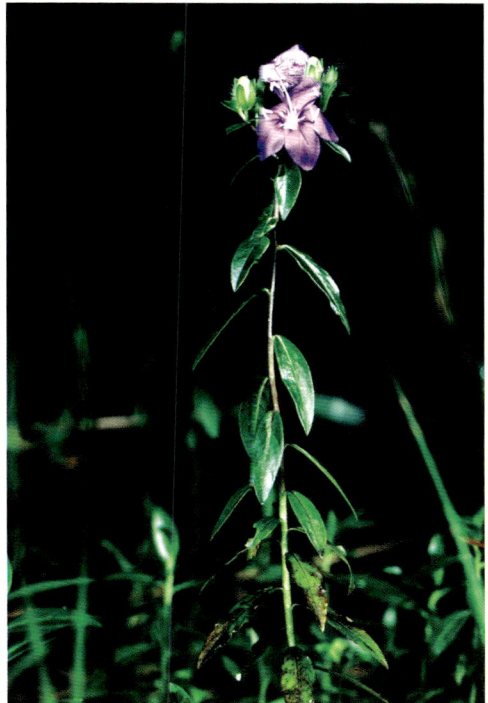

Hydrolea corymbosa
J. Macbr. ex Elliott

221

Blue

Hyptis mutabilis (Rich.) Briq.
Bitter Mint
Lamiaceae (Labiatae), Mint Family

Habit: Herbaceous perennial, to 2.5 m tall. Stems square, rough hairy, somewhat woody, frequently many branched.

Leaves: Opposite; long-stalked; blades egg-shaped to diamond-shaped, with toothed margins, 2-8 cm long.

Inflorescences: Small, separate clusters, forming spikes along the tips of stems.

Flowers: Bluish violet to pale lavender, 3-6 mm long, tubular, with 5 small lobes at mouth.

Fruit: A rounded nutlet, about 1.5 mm long.

Habitat and Distribution: Common; moist, open areas and disturbed sites; throughout Florida north into Louisiana and south Georgia; native to tropical America.

Comment: Flowering all year, the inconspicuous flowers are not very noticeable. Bitter Mint is a very invasive weed that can be very dense in open disturbed sites.

Ipomoea hederacea Jacq. var. *integriuscula* A. Gray
Entire-leaf Morning-glory
Convolvulaceae, Morning-glory Family

Habit: Herbaceous, annual vine, to 3 m long. Stems twining, covered with dense hairs.

Leaves: Alternate; long-stalked; blades heart-shaped to oval to somewhat triangular, not divided or lobed, tips pointed, 5-12 cm long and wide.

Inflorescences: Solitary flowers or few-flowered clusters in leaf axils.

Flowers: Light blue later changing to rose purple, bell-shaped, 3-5 cm long, and 3-5 cm wide.

Fruit: A roundish and flattened capsule, about 1 cm diameter.

Habitat and Distribution: Infrequent; fields, roadsides, and waste places; throughout Florida, north to Maryland and Kansas, and west to Texas; native to tropical America.

Comment: Flowering all year in warmer months. This vine, cultivated for its showy flowers, is best placed on a fence or trellis. It can become extremely weedy and is a troublesome weed in row crops.

Hyptis mutabilis (Rich.) Briq.

Ipomoea hederacea Jacq. var. *integriuscula* A. Gray

Blue

Ipomoea indica (Burm. f.) Merr.
Blue Morning-glory
Convolvulaceae, Morning-glory Family

Habit: Herbaceous vine, perennial in the south and annual northward. Stems twining, hairy.

Leaves: Alternate; long-stalked; blades hairy, 4-9 cm long, heart-shaped to rounded, either without lobes or with 3 lobes.

Inflorescences: Solitary flowers or few-flowered clusters in leaf axils.

Flowers: One to several in hairy clusters, blue, purple, or white, bell-shaped, 5-7 cm long and 5-8 cm wide.

Fruit: A rounded, 3-chambered capsule somewhat flattened at the top, containing 4 seeds.

Habitat and Distribution: Very common; hammocks, thickets, and disturbed sites; throughout Florida; pantropical.

Comment: The showy blue flowers can be found in all warm months. Blue Morning-glory is frequently cultivated, usually on a fence or trellis. It is a very weedy species and is actively spreading northward.

Ipomoea purpurea (L.) Roth
Tall Morning-glory
Convolvulaceae, Morning-glory Family

Habit: Herbaceous, perennial vine. Stems climbing or trailing, hairy.

Leaves: Alternate; long-stalked; blades heart shaped, 4-16 cm long and wide.

Inflorescences: Solitary flowers or few-flowered clusters in leaf axils.

Flowers: Blue, purple, pink, or white, bell-shaped, 4-7 cm long and wide; stigma 3-lobed.

Fruit: A round capsule, about 1 cm in diameter.

Habitat and Distribution: Rare; disturbed sites and thickets; central peninsula into panhandle Florida, more common north throughout most of the northeastern U.S. into Canada, west to Texas; native of tropical America.

Comment: The large, showy flowers in summer and fall make this species popular in cultivation. A common weed in row crops and other disturbed sites.

Ipomoea indica (Burm. f.) Merr.

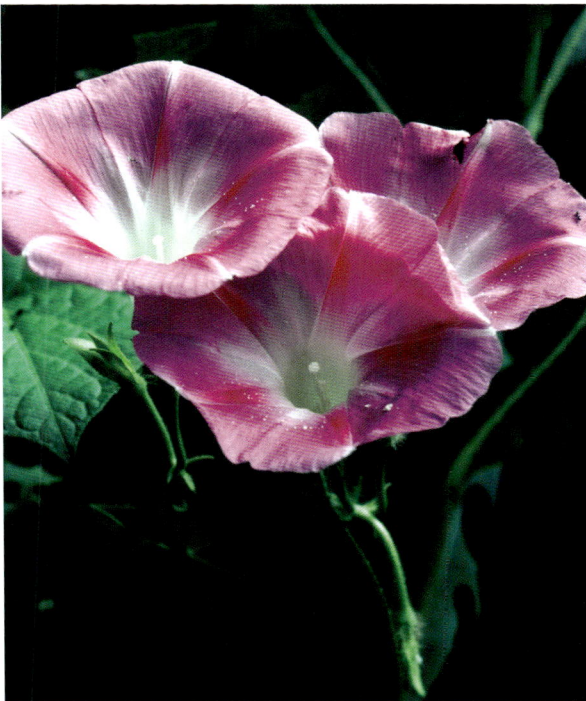

Ipomoea purpurea (L.) Roth

Blue

Iris hexagona Walter
Angle-pod Blue Flag or Dixie Iris
Iridaceae, Iris Family

Habit: Herbaceous perennial from an underground runner, 0.5-1.2 m tall. Stem leafy, zig-zagging.

Leaves: Alternate; sheathing at base; blades 2-ranked (flattened in one plane), yellow-green, sword-shaped, to 1 m long.

Inflorescences: Terminal groups of 1 or more.

Flowers: Blue, purple, or white with white and green variegation, about 10 cm long, with 3 erect, narrow petals and 3 larger, drooping colored sepals.

Fruit: An oblong capsule, to 8 cm long.

Habitat and Distribution: Common; ditches, swamps, and marshes; throughout Florida, west to Texas and north to South Carolina.

Comment: Angle-pod Blue Flag blooms in spring. It is frequently cultivated for the showy flowers and blooms. It does best if planted in standing water or in a wet spot. It can easily be propagated by seeds or portions of the underground runner.

Lactuca graminifolia Michx.
Common Wild Lettuce
Asteraceae (Compositae), Aster Family

Habit: Herbaceous biennial, 0.6-1.5 m tall, from a taproot. Stems smooth, gray to green to reddish, with milky sap.

Leaves: Alternate, mostly in a basal rosette, less frequent on the upper stem; lacking a stalk; blades to 30 cm long and 1-5 cm wide, mostly linear with scattered long, irregular, smooth lobes.

Inflorescences: Terminal, open, spreading, branched clusters of urn-shaped heads.

Flowers: Heads 1.4-1.8 cm long; only ray flowers present, blue or rarely white, 18-24, of different lengths in the same head.

Fruit: Nutlets oblong, reddish brown, about 5 mm long and 2 mm diameter, with a tuft of white bristles 6-9 mm long on the top.

Habitat and Distribution: Common; sandy, open habitats in fields, woods, disturbed sites and along roadsides; throughout Florida, west to Texas, Arizona, and northern Mexico, and north to Georgia and North Carolina.

Comment: Spring and summer flowering, this weedy Wild Lettuce has pretty flowers on an unusual looking plant.

Iris hexagona Walter

Lactuca graminifolia Michx.

Blue

Lantana montevidensis (Spreng.) Briq.
Trailing Lantana or Polecat-geranium
Verbenaceae, Vervain Family

Habit: Perennial shrub, to 1 m long, from a woody base. Stems trailing, covered with hairs, lying flat, often forming mats, rooting at the nodes.

Leaves: Opposite; short-stalked; blades ovate to elliptic, 1-3 cm long, rough and sandpapery, glandular, margins with coarse teeth.

Inflorescences: Long-stalked terminal heads.

Flowers: Lilac to red with a white eye, tubular, 0.8-1.8 cm long, with 5 lobes, heads and flowers glandular.

Fruit: A drupe with a hard stone, to 5 mm diameter, almost black when ripe.

Habitat and Distribution: Infrequent; woods, fence rows, roadsides, and other disturbed sites; throughout Florida; Georgia, Alabama, California, Texas; West Indies; Costa Rica; native of northern South America.

Comment: Showy flowers occur in all warm months on this popular bedding plant. As a ground cover it can become somewhat straggly and needs to have older stems removed or be replanted.

Linaria canadensis (L.) Chaz.
Old Field Toadflax or Blue Toadflax
[*Nuttallanthus canadensis* (L.) D.A. Sutton]
Scrophulariaceae, Figwort Family

Habit: Herbaceous winter annual or biennial, 15-70 cm long. Stems freely branching from base.

Leaves: Alternate on upper stems, opposite or in 3s at base; stalkless; blades needle-like, to 2 cm long.

Inflorescences: Spike-like, at the ends of nearly leafless, smooth stems.

Flowers: Blue to violet, tubular, 5-10 mm long, basal spur 5-9 mm long, with 2-lipped mouth, upper lip narrow, two lobed, lower lip wide, 3-lobed.

Fruit: A roundish capsule, 2-3 mm in diameter.

Habitat and Distribution: Common; dry, disturbed areas; throughout Florida, north to Massachusetts, Quebec, and Minnesota, west to Texas, south into Mexico; Pacific coast.

Comment: A very common weedy wildflower seen blooming in sandy, grassy fields and along roadsides in spring. If left unmowed the masses of flowering plants can make a good show.

Lantana montevidensis
(Spreng.) Briq.

Linaria canadensis (L.) Chaz.

Blue

Linaria floridana Chapm.
Florida Toadflax
Scrophulariaceae, Figwort Family

Habit: Herbaceous annual or biennial, 10-40 cm tall.

Leaves: Alternate on upper stems, whorled in 3s at base; stalkless; blades narrow, to 2.5 cm long.

Inflorescences: Spike-like at stem and branch tips.

Flowers: On glandular stalks, blue to violet, 5-6 mm long, basal spur about 1/2 mm long, 2-lipped.

Fruit: A capsule, about 2 mm long.

Habitat and Distribution: Frequent; dry sandy sites, often in sandhill and scrub habitats; throughout Florida, north into south Georgia, and west into Mississippi.

Comment: This spring blooming wildflower is showy in masses, but only found in very dry habitats.

Linaria texana Scheele
Texas Toadflax
Scrophulariaceae, Figwort Family

Habit: Herbaceous annual or biennial, to 70 cm tall. Stems branching freely from the base.

Leaves: Alternate on upper stems, opposite or in 3s at base; stalkless; blades narrow, to 3 cm long.

Inflorescences: Spike-like, along nearly smooth, leafless ends of stems.

Flowers: Tubular, violet-blue, 10-14 mm long, basal spur 5-9 mm long.

Fruit: A capsule, 2.5-3.5 mm long.

Habitat and Distribution: Occasional; sandy soils in open woods, fields, pastures, and roadsides; northern peninsula Florida, north to Virginia, Tennessee, and British Columbia, west to Texas and California, south into Mexico.

Comment: Texas Toadflax, also spring-flowering, has the largest flower and is the showiest of the blue flowered toadflaxes. It is frequently confused with Old Field Toadflax which has a flower less than 1 cm long.

Linaria floridana Chapm.

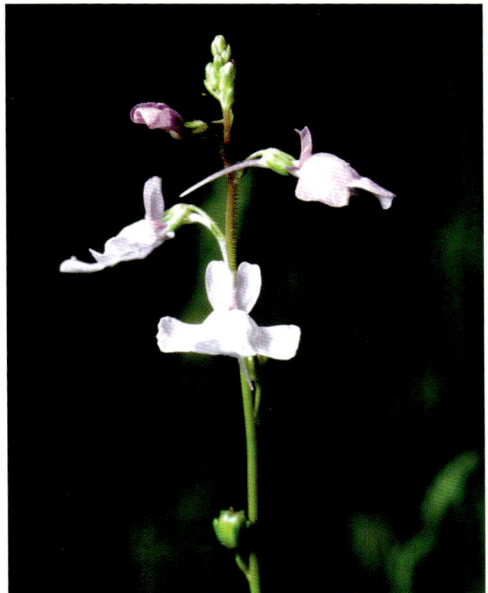

Linaria texana Scheele

Blue

Lindernia grandiflora Nutt.
Round-leaved False Pimpernel
Scrophulariaceae, Figwort Family

Habit: Herbaceous annual, to 40 cm long. Stems lying down, creeping, slender, smooth, forming mats.

Leaves: Opposite; nearly stalkless; blades oval or rounded, to 1 cm long, glandular, with tiny glandular dots.

Inflorescences: Solitary flowers on long stalks from leaf axils.

Flowers: Blue to purple, mottled with white, tubular, 8-10 mm long, with 2 yellow ridges in throat, 2-lipped, lower lip with 3 spreading, mottled lobes, upper lip notched.

Fruit: A capsule, 4-6 mm long, seeds winged.

Habitat and Distribution: Common; moist, sandy soils of wetland margins, fields, and roadsides; throughout peninsula into central panhandle Florida and into south Georgia.

Comment: Flowering in most warm months Round-leaved False Pimpernel is a very colorful ground-hugging wildflower. Most ditches and wetland edges contain this frequently unnoticed plant nestled into the groundcover.

Lobelia feayana A. Gray
Bay Lobelia
Campanulaceae, Bellflower Family

Habit: Herbaceous annual, to 30 cm tall or long. Stems smooth, mostly creeping.

Leaves: Alternate; stalked; blades rounded, sometimes toothed, to 1.5 cm long.

Inflorescences: Terminal, in loose spikes.

Flowers: Stalked, blue to purple with white centers, about 1 cm long, with 2 lips, lower lip with 3 lobes, upper lip 2-lobed.

Fruit: A capsule, to 4 mm wide.

Habitat and Distribution: Common; open, moist, and wet sites; throughout peninsula into eastern panhandle Florida, endemic.

Comment: Primarily flowering in the spring and fall, flowers are seen in most warm months. A delightful wildflower that could be seeded into wet spots.

Lindernia grandiflora Nutt.

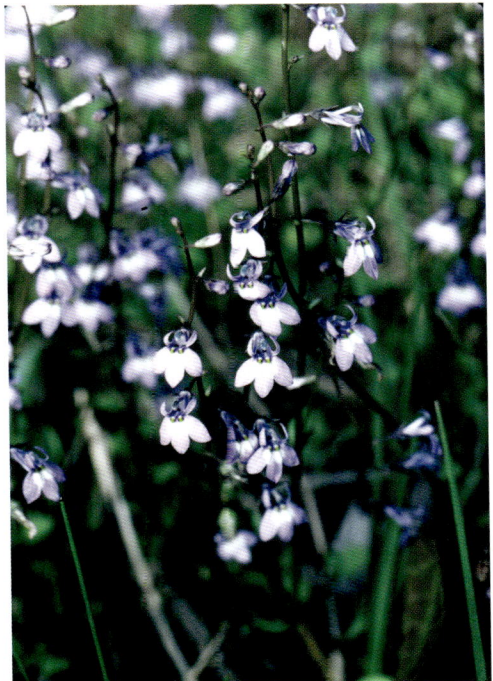

Lobelia feayana A. Gray

Blue

Lobelia glandulosa Walter
Coastal Plain Lobelia or Glade Lobelia
Campanulaceae, Bellflower Family

Habit: Herbaceous perennial, to 1.5 m tall. Stems erect, smooth, usually simple.

Leaves: Alternate; stalked; blades narrow, 2-15 cm long, often with toothed edges, sometimes teeth on upper leaves have glands.

Inflorescences: Terminal, spike-like.

Flowers: Short or no stalks, blue to purple, with hairy throat, to 1.5 cm long, 2-lipped, upper lip with 2 small lobes, lower lip with 3 lobes, white center, sepals often with glandular teeth.

Fruit: A capsule, 5-8 mm wide.

Habitat and Distribution: Common; swamps, wet pinelands, marshes and along ditches; throughout Florida, west to Alabama, and north to Virginia.

Comment: Blooming in warm months, Coastal Plain Lobelia is a wildflower that catches the eye. The spike of blue flowers at the tip of erect stems often pokes out of the surrounding groundcover.

Lobelia homophylla E. Wimm.
Little White Lobelia
Campanulaceae, Bellflower Family

Habit: Herbaceous annual, 20-60 cm tall. Stems smooth.

Leaves: Opposite; stalkless; blades somewhat rounded, to 4 cm long, margins with few teeth.

Inflorescences: Terminal, spike-like.

Flowers: Stalked, blue with white center, to 11 mm long.

Fruit: A capsule, about 6 mm long and 3 mm wide.

Habitat and Distribution: Endemic and frequent; low, wet sites, pinelands, roadsides, and open wooded habitats; peninsula Florida.

Comment: The light blue flowers of Little White Lobelia can be found from spring into fall. Not very showy, this small wildflower is hard to spot.

Lobelia glandulosa Walter

Lobelia homophylla E. Wimm.

Blue

Lobelia puberula Michx.
Downy Lobelia
Campanulaceae, Bellflower Family

Habit: Herbaceous perennial, to 1.5 m tall. Stems covered with short, stiff hairs.

Leaves: Alternate; stalked; blades elliptic to lance-shaped, up to 12 cm long, with toothed edges, sometimes teeth on upper leaves with glands.

Inflorescences: Terminal, spike-like.

Flowers: Short or no stalks, blue to purple, with white centers, rarely entirely white, tubular, to 2 cm long, 2-lipped, upper lip 2-lobed, lower lip 3-lobed, sometimes sepals with glandular teeth.

Fruit: A capsule, 6-8 mm wide.

Habitat and Distribution: Occasional; moist wet soils of woods, swamps, prairies, and meadows; central peninsula Florida, west to Texas and Oklahoma, and north to New Jersey.

Comment: Like most Lobelias, Downy Lobelia blooms in warm months with spikes of showy flowers. Seeded into wet sites it can be a wonderful surprise when it blooms.

Lupinus diffusus Nutt.
Sky-blue Lupine
Fabaceae (Leguminosae), Pea or Bean Family

Habit: Herbaceous annual or biennial, to 0.8 m tall, from a taproot. Stems sprawling or ascending, densely hairy, and woody at base.

Leaves: Alternate, mostly basal; stalked; blades elliptic, 4-12 cm long and 2-5 cm broad, silky-hairy.

Inflorescences: Dense spike-like clusters at stem tips above the leaves.

Flowers: Typically pea-shaped, varying shades of blue with a white eye-spot on the upper petal, 1.2-1.5 cm long.

Fruit: A pod, hairy, 3-5 cm long, less than 1 cm broad.

Habitat and Distribution: Frequent; open sandy areas, scrub, sandhills, and dry hammocks; throughout Florida, west to Mississippi, and north to North Carolina.

Comment: Sky-blue Lupine is a showy wildflower blooming mostly in the spring, or all year in south peninsula Florida. The plant is quite attractive and is as distinct as the flowers.

Lobelia puberula Michx.

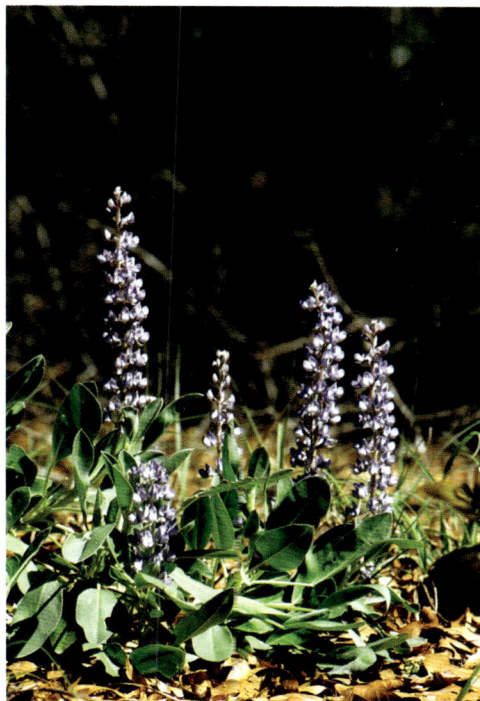

Lupinus diffusus Nutt.

Blue

Lythrum alatum Pursh var. *lanceolatum* (Elliott) Torr. & A. Gray ex Rothr.
Winged Loosestrife
[*Lythrum lanceolatum* Elliott]
Lythraceae, Loosestrife Family

Habit: Herbaceous perennial, to 1.2 m tall. Stem much branched, usually with many stems.

Leaves: Mostly alternate, some opposite; essentially stalkless; blades elliptic to lanceolate, to 7 cm long and 1 cm wide.

Inflorescences: Solitary flowers in most of the upper leaf axils.

Flowers: Blue, purple, light violet, 4 to 6 petals, each 3-6 mm long.

Fruit: A cylindrical capsule, about 5 mm long; numerous seeds.

Habitat and Distribution: Frequent; wet flatwoods, marshes, swamps, wet margins, meadows, wet depressions, and ditches; throughout Florida, north to Virginia and Tennessee, and west to Oklahoma and Texas.

Comment: The small blue flowers in great profusion on the many-branched plants are often found in large populations blooming in summer. In combination the conglomeration of flowers are quite attractive.

Mazus pumilus (Burm. f.) Steenis
Asian Mazus
[*Mazus japonicus* (Thunb.) Kuntze]
Scrophulariaceae, Figwort Family

Habit: Herbaceous annual, to 15 cm tall, from a basal rosette. Stems one to several, hairy.

Leaves: Opposite; short-stalked; blades broadest at tip, to 4 cm long, irregular teeth on margin.

Inflorescences: Solitary flowers, alternate on a terminal stalk.

Flowers: Blue to pale lavender, marked with yellow and white, to 10 mm long, 2-lipped.

Fruit: A capsule.

Habitat and Distribution: Very infrequent to occasional to locally common; moist to wet habitats, lawns, ditches, open bottomlands, and disturbed areas; throughout Florida, widely scattered northward through North Carolina into Pennsylvania and Missouri, west into eastern Texas; Pacific slope; West Indies; native to east Asia.

Comment: The flowers of this little weed occur in warm months. The flowers are attractive, although small. The plants are often massed and usually found on open ground in poor habitats.

Lythrum alatum Pursh var. *lanceolatum* (Elliott) Torr. & A. Gray ex Rothr.

Mazus pumilus (Burm. f.) Steenis

239

Blue

Melia azedarach L.
Chinaberry Tree
Meliaceae, Mahogany Family

Habit: Small to medium tree, to 15 m tall, with a compact, rounded crown.

Leaves: Alternate; long-stalked; large, twice to three times divided, leaflets pointed, oval, 2-8 cm long, with large teeth on margins.

Inflorescences: Large clusters in leaf axils along branches.

Flowers: Blue to lilac, small, cylindrical, tubular, with 5 petals, each 6-8 mm long.

Fruit: A round, yellow berry, 1-1.5 cm diameter, containing 1 ridged bony stone with a seed.

Habitat and Distribution: Frequent; naturalized in thickets, old fields, and disturbed areas; throughout Florida, north to Virginia, and west to Oklahoma and eastern Texas; a horticultural plant from Asia.

Comment: A distinctive spring bloomer. The large clusters of flowers are showy. It was originally planted as a shade tree and for its large rounded crown. Chinaberry is rapidly invading native habitats. The fruits are soft and yellow when ripe. Birds ingesting ripe fruits can appear disoriented. Fruits are POISONOUS when ingested by humans and livestock.

Passiflora incarnata L.
Maypop or Passion-flower

Passifloraceae, Passion-flower Family

Habit: Perennial vine, to 8 m or more long, producing underground runners. Stems prostrate or climbing with tendrils, smooth or slightly hairy.

Leaves: Alternate; long-stalked, with two round glands at top of stalk; blades 5-15 cm long, deeply 3-lobed, with finely toothed edges. Simple tendrils from the leaf axils enable the vine to climb.

Inflorescences: Solitary or occasionally two, long-stalked flowers in leaf axils.

Flowers: Bluish, lavender, or whitish, 4-7 cm diameter, with an intricate, fringed crown with colored bands.

Fruit: An oval or roundish berry, 4-7 cm long, green turning yellowish when ripe, filled with seedy, edible pulp.

Habitat and Distribution: Sporadic to locally common; open, dry woods, old fields, fence rows, row crops and disturbed sites; throughout Florida, north to Virginia and Missouri, west to Texas; Bermuda.

Comment: The spectacular flowers occur in spring and summer. The underground runners and massive production of seeds at any disturbed site, including row crops, can result in a substantial tangle of vines. The flowers are unusual and showy, making Maypop popular. The vine could be cultivated on a trellis from a container in an area where the fruits can be collected, to avoid spreading the seeds.

240

Melia azedarach L.

Passiflora incarnata L.

Blue

Pediomelum canescens (Michx.) Rydb.
Buckroot or Three-leaved Prairie-turnip
[*Psoralea canescens* Michx.]
Fabaceae (Leguminosae), Pea or Bean Family

Habit: Herbaceous perennial, 0.3-0.9 m tall, from a thick rounded or elongated root. Stems solitary, branched near the top, covered with gray hairs.

Leaves: Alternate; short-stalked; palmately trifoliate with 3 leaflets or some leaves towards the top of the stem with 2 leaflets or 1 leaflet, elliptic to somewhat rounded, covered with gray hairs and glandular dots, 2-6 cm long and 1-3 cm broad.

Inflorescences: Loose, few-flowered clusters in the axils of the leaves.

Flowers: Dark blue to violet with yellow-green markings, typically pea-shaped, 8-12 mm long.

Fruit: Pod oval, about 1 cm long, covered with glandular dots.

Habitat and Distribution: Frequent; dry open habitats, sandhills, flatwoods, woods, and old fields; central peninsula Florida, west to southern Alabama, and north to Virginia.

Comment: Flowering in the spring and summer, Buckroot is not showy, but the manner of growth is unusual.

Phlox divaricata L.
Blue Phlox
Polemoniaceae, Phlox Family

Habit: Herbaceous perennial, to 50 cm tall. Stems usually leaning down, forming open mat.

Leaves: Opposite, evergreen; nearly stalkless; blades elliptic to lance-shaped, to 5 cm long.

Inflorescences: Dense disc-like or hemispheric clusters on ends of erect flowering shoots.

Flowers: Blue to blue-violet, rarely white, with slender tube 1-1.5 cm long and 5 spreading lobes, each 1-1.5 cm long, mouth of tube darker than petals.

Fruit: A capsule, 4-6 mm long.

Habitat and Distribution: Rare; rich, moist woods; panhandle Florida, north to Vermont, Quebec, Minnesota, and South Dakota, and west to Texas.

Comment: Spring blooming and very showy flowers, make this a very popular cultivated native. A number of named variations are sold.

Pediomelum canescens
(Michx.) Rydb.

Phlox divaricata L.

243

Blue

Piloblephis rigida (W. Bartram ex Benth.) Raf.
Pennyroyal
Lamiaceae (Labiatae), Mint Family

Habit: Shrubby perennial, 10-70 cm tall. Stems diffusely branched, spreading, or erect, very leafy, hairy.

Leaves: Opposite; stalkless; blades very narrow and needle-like, to 1.2 cm long, densely covering stems.

Inflorescences: Dense, spike-like clusters, to 6 cm long, at ends of stems.

Flowers: Pale bluish-purple, fragrant, tubular, 7-8 mm long, 2-lipped, upper lip smaller 2-lobed, lower lip 3-lobed, speckled.

Fruit: Composed of 4 nutlets.

Habitat and Distribution: Common and endemic; pinelands and sandhills; peninsula Florida.

Comment: Flowering can occur all year. This small, dense shrub is characteristic in Florida pinelands. The foliage is fragrant with a distinct aroma. Pennyroyal is a desirable wildflower for dry sites.

Pinguicula planifolia Chapm.
Chapman's Butterwort
Lentibulariaceae, Bladderwort Family

Habit: Herbaceous perennial, to 35 cm tall. Stems/flower stalks scapose from a basal rosette, glandular hairy.

Leaves: Basal rosette; stalkless; oblong to elliptic, to 8 cm long, very thin, nearly transparent, flat, purplish-red, short glandular hairs on top.

Inflorescences: Solitary flower at stem tip.

Flowers: Violet to purplish-red, throat darker violet, about 3 cm long and wide, with 5 deeply notched lobes, spur 3-4 mm long.

Fruit: A capsule, about 5 mm wide.

Habitat and Distribution: Occasional; bogs, swamps, ditches, and sites where seepage occurs; Leon County westward in the Florida panhandle, west to southeast Mississippi.

Comment: Blooming from February into April, Chapman's Butterwort flowers are easily seen but not easily located. This plant is listed as Threatened by Florida.

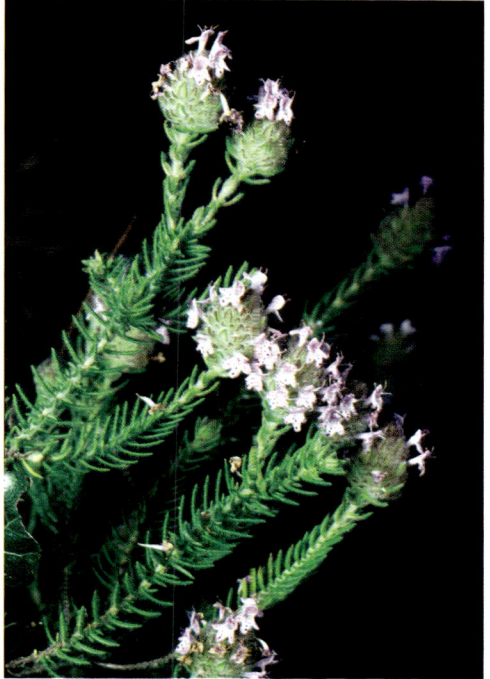

Piloblephis rigida
(W. Bartram ex Benth.) Raf.

Pinguicula planifolia Chapm.

Blue

Pinguicula pumila Michx.
Small Butterwort
Lentibulariaceae, Bladderwort Family

Habit: Herbaceous perennial, 5-15 cm tall. Stems/flower stalks 1-10, scapose from a basal rosette.

Leaves: Basal rosette; stalkless; blades elliptic to somewhat rounded, 1-3 cm long, greasy to touch, succulent.

Inflorescences: Solitary flower at tip of stem.

Flowers: Blue, sometimes marked with purple and yellow, bell-shaped, less than 2 cm long, with 5 shallowly notched lobes, spur 3-4.5 mm long.

Fruit: A capsule, 2-3 mm in diameter.

Habitat and Distribution: Frequent; moist, acidic pinelands; throughout Florida, west to Texas and north to North Carolina.

Comment: Flowering from winter into late spring some seepage areas are blanketed with a carpet of Small Butterworts. The sticky, light green basal leaves complement the distinct flowers.

Pontederia cordata L.
Pickerelweed
Pontederiaceae, Pickerelweed Family

Habit: Herbaceous perennial, 0.4-2.0 m tall, from thick underground runners. Stem soft, smooth.

Leaves: Alternate, basal; long-stalked; blades erect, lance-shaped to rounded.

Inflorescences: Spike-like clusters at stem tips.

Flowers: Blue or rarely white, marked with yellow, 12-20 mm long, tubular, 2-lipped, upper lip with 3 fused lobes and yellow markings, lower lip with 3 separate lobes.

Fruit: Ellipsoid, one-seeded, ridged, with rough teeth on ridges, 4-8 mm long.

Habitat and Distribution: Common; marshes, ditches, ponds, streams, lakes, and other shallow, wet areas; throughout Florida, north to Prince Edward Island, Nova Scotia, and Ontario, west to Missouri, Oklahoma and Texas; south through Central America into South America; West Indies.

Comment: Flowering from spring into fall. Frequently planted in standing water for the showy flowers and shiny green foliage.

Pinguicula pumila Michx.

Pontederia cordata L.

Blue

Prunella vulgaris L.
Heal-all or Selfheal
Lamiaceae (Labiatae), Mint Family

Habit: Herbaceous perennial, 5-80 cm tall. Stems ascending or spreading, 4-angled, hairy, leafy.

Leaves: Opposite; stalked; blades elliptic to broadly lance-shaped, 3-9 cm long, margins smooth or toothed.

Inflorescences: Small, dense terminal spikes.

Flowers: Blue-violet or sometimes white, 2-lipped, upper lip hood-like, lower lip 3-lobed with frayed edge on center lobe.

Fruit: 4 shiny, ribbed, rounded nutlets, about 2.5 mm long.

Habitat and Distribution: Infrequent to occasional; moist disturbed areas; north peninsula Florida; north throughout North America; native of Eurasia, now widely naturalized and nearly cosmopolitan.

Comment: Heal-all blooms from spring until frost. It is a good border plant for shady sites. As the common name suggests this plant was once thought to be a cure for almost anything. The foliage is edible cooked or raw.

Ruellia brittoniana Leonard ex Fernald
Mexican Ruellia
Acanthaceae, Acanthus Family

Habit: Herbaceous perennial, to 1 m tall. Stems erect, become smooth with age.

Leaves: Opposite; stalked; blades, narrow and somewhat lance-shaped, to 30 cm long.

Inflorescences: Long-stalked, branched clusters from leaf axils.

Flowers: Blue-violet to purple, petunia-like, funnel-shaped, 2-4.5 cm long, with 5 lobes, each about 2 cm long.

Fruit: An elliptical capsule, about 2 cm long.

Habitat and Distribution: Frequent; waste and disturbed sites; throughout Florida, north into South Carolina, west into Texas; native of Mexico.

Comment: Mexican Ruellia has dense evergreen foliage and will become bushy if not frozen back during the winter. The large, attractive light lavender flowers, occurring from spring into fall, together with the green foliage make this an attractive plant for landscaping. It will grow almost everywhere. Seeds disperse easily and widely from explosive capsules which makes placing it near native habitats risky in terms of escape.

Prunella vulgaris L.

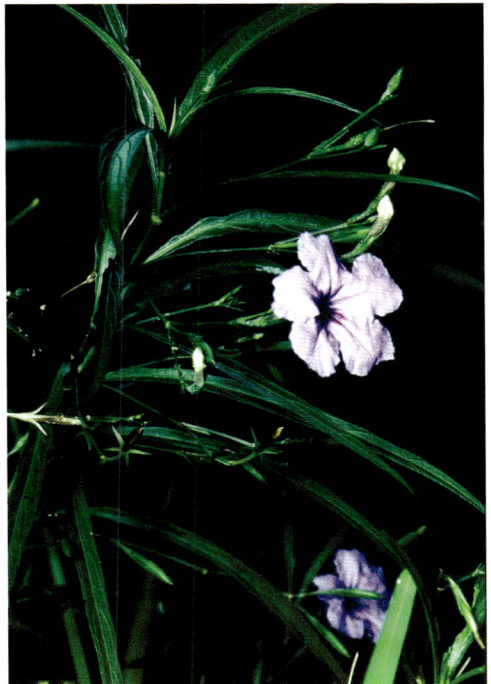

Ruellia brittoniana
Leonard ex Fernald

Blue

Ruellia caroliniensis (J.F. Gmel.) Steud.
Wild-petunia
Acanthaceae, Acanthus Family

Habit: Herbaceous perennial, to 90 cm tall. Stems erect or rarely flat, hairy.

Leaves: Opposite; stalked; blades elliptic to ovate, 10 cm long, hairy.

Inflorescences: Stalkless clusters from leaf axils.

Flowers: Blue to pale purple, or rarely white, funnel-shaped, 2.5-5 cm long, with 5 spreading lobes, each 1-1.5 cm long.

Fruit: An elliptic capsule, about 1.7 cm long.

Habitat and Distribution: Common; dry woods, pinelands, prairies, sandhills, flatwoods, and clearings; throughout Florida, north to New Jersey, Ontario, and Indiana, west to Texas.

Comment: Wild-petunia is extremely adaptable in terms of growing conditions. It is one of the few wildflowers that will grow equally well in full sun or very shady conditions. Growing best in better drained, loamy soils, it can be found growing and doing well from the open bright sun of sandhills to wet shady hammocks. Blooming from early spring into the fall or early winter in more southerly climates, it will attract several pollinators, especially butterflies. The plants become dormant during the winter. Underground suckers and seeds serve to spread this plant. Seeds spread rapidly and suckers more slowly. Capsules pop open and spread seeds to some distance. As a wildflower this species is terrific although the foliage is not notable. In cultivation, stems should be trimmed in late fall.

Salvia lyrata L.
Lyre-leaved Sage
Lamiaceae (Labiatae), Mint Family

Habit: Herbaceous annual, 10-60 cm tall, from a basal rosette. Stems hairy, 4-angled, with few or no leaves.

Leaves: Opposite; stalkless; blades broadest at tip, lyre-shaped, lobed, unlobed, or dissected, 5-17 cm long, margins toothed.

Inflorescences: Hairy, spike-like clusters at ends of stems.

Flowers: Blue to violet, tubular, 1.5-3 cm long, 2-lipped, upper lip smaller, lower lip larger.

Fruit: Dark brown, rounded nutlets, about 2 mm long.

Habitat and Distribution: Common; open woods, thickets, turf, and other disturbed sites; throughout Florida, north to Connecticut, Ontario, Illinois, and Missouri, south to Texas.

Comment: Lyre-leaved Sage flowers from spring into frost southward and in spring northward. The large flowers are showy and the plant is offered by some nurseries. It is regarded as a turf weed along the coastal plain, and is especially visible as a roadside wildflower.

Ruellia caroliniensis (J.F. Gmel.) Steud.

Salvia lyrata L.

Blue

Salvia riparia Kunth
Southern Sage
[*Salvia privoides* Benth.]
Lamiaceae (Labiatae), Mint Family

Habit: Herbaceous annual, to 1 m tall. Stems hairy, 4-angled.

Leaves: Mostly basal, opposite; stalkless to short-stalked; blades rounded, 1-3 cm long, with toothed edges.

Inflorescences: Widely spaced clusters along a slender spike at stem tips.

Flowers: Blue to purplish, tubular, 5-6 mm long, 2-lipped, glandular hairy.

Fruit: Composed of four nutlets, each to 2 mm long.

Habitat and Distribution: Frequent; sandy soils of open woods and other disturbed, shady, dry sites; central and southern peninsula Florida; West Indies, Mexico, Central America.

Comment: Southern Sage flowers in most of the warm months, but the small flowers are scarcely noticed among the other ground covering vegetation.

Scutellaria floridana Chapm.
Florida Skullcap
Lamiaceae (Labiatae), Mint Family

Habit: Herbaceous perennial, to 60 cm tall. Stems ascending, 4-sided, hairy.

Leaves: Opposite; stalkless; blades linear, margins rolled, dotted with glands, to 3 cm long, margins smooth.

Inflorescences: Widely spaced clusters along a slender spike at tips of stems.

Flowers: Blue to purplish, tubular, to 2.5 cm long, 2-lipped; sepals with a persistent helmet or cap on top.

Fruit: Composed of four nutlets.

Habitat and Distribution: Rare; wet flatwoods and grassy openings; Franklin, Gulf, and Liberty Counties in the Florida panhandle.

Comment: Florida Skullcap has showy flowers that can be seen from April into August. This low plant mixes into other vegetation and can be difficult to spot. It is listed as Endangered by the U.S. and by the State of Florida.

Salvia riparia Kunth

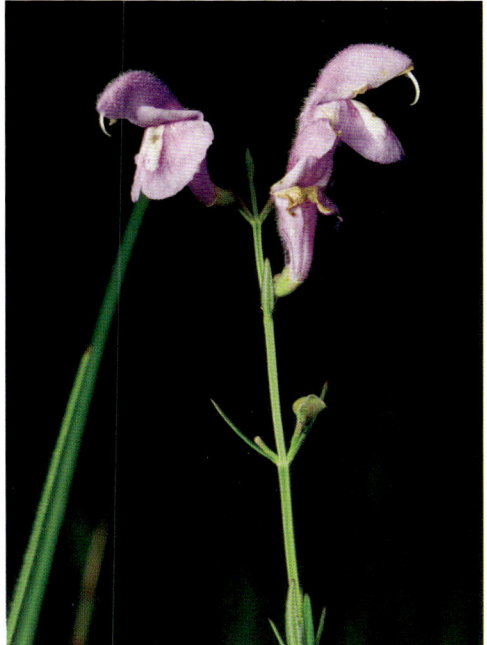

Scutellaria floridana Chapm.

Blue

Scutellaria integrifolia L.
Rough Skullcap
Lamiaceae (Labiatae), Mint Family

Habit: Herbaceous perennial, 20-70 cm tall. Stems erect, branched, 4-angled, hairy.

Leaves: Opposite; stalked to stalkless; lower blades triangular, to 4 cm long, with toothed edges, upper blades slender, 2-6 cm long, with smooth or toothed edges.

Inflorescences: Spikes at stem tips.

Flowers: Blue to pale violet or pink, tubular, 1.5-2.5 cm long, with 2 large lips, upper lip hood-like, lower lip broad; sepals with a persistent helmet or cap on top.

Fruit: Composed of 4 bumpy, dark brown nutlets, about 1 mm long.

Habitat and Distribution: Frequent; dry, open pinelands, sandhills, hammocks, meadows, fields, roadsides, and wetter sites; central peninsula Florida, west to Texas, and north to Ohio, Kentucky, Tennessee, and Massachusetts.

Comment: Rough Skullcap flowers in the summer northward and from spring into fall southward. This wildflower is offered by nurseries. It can be planted in dry to moist soils and makes a wonderful bedding plant.

Sisyrinchium angustifolium Mill.
Wide-winged Blue-eyed-grass
Iridaceae, Iris Family

Habit: Grass-like, herbaceous perennial, 15-50 cm tall. Stems thin, 2.5-4 mm wide, with 2 noticeable wide wings.

Leaves: Alternate; stalkless; blades slender, grass-like, to 30 cm long and 2-4 mm broad.

Inflorescences: Radiating clusters of a few flowers on a stem which is as tall or taller than the leaves, with 2 sheathing leaves occurring below each flower cluster.

Flowers: Blue, violet, or white, 6-parted.

Fruit: A roundish capsule, 4-6 mm long.

Habitat and Distribution: Common; moist meadows, fields, ditches, roadsides, prairies, and open woodlands; throughout Florida, west to Texas, north to Minnesota and Newfoundland.

Comment: Flowering is most frequent in the spring, but can extend through the summer. Locally common in roadside ditches this species makes an excellent show of blue, as a roadside wildflower. Occasionally sold by native nurseries, it is used as a bedding plant.

Scutellaria integrifolia L.

Sisyrinchium angustifolium Mill.

Blue

Sisyrinchium atlanticum E.P. Bicknell
Narrow-winged Blue-eyed-grass
Iridaceae, Iris Family

Habit: Grass-like herbaceous perennial, 10-50 cm tall. Stems thin, 0.5-2 mm side, with 2 very thin wings.

Leaves: Alternate; stalkless; blades slender, grass-like, to 30 cm long and 2-4 mm wide.

Inflorescences: Small radial clusters on flowering stalk, stalk to 40 cm tall, with 2 sheathing leaves below each flower cluster.

Flowers: Blue or sometimes white, flowers 6-parted.

Fruit: A roundish capsule, 3-5 mm long.

Habitat and Distribution: Frequent; wet flatwoods, open woods, meadows, edges of salt marshes, roadsides, and disturbed open sites; throughout Florida, north to Maine, Nova Scotia, Michigan, Indiana, Minnesota, and Missouri, west to southeast Texas.

Comment: Flowering is most frequent in the spring, but can extend through the summer. Locally common in roadside ditches this wildflower makes an excellent show of blue. Occasionally sold by native nurseries, it is used as a bedding plant.

Sisyrinchium rosulatum E.P. Bicknell
Annual Blue-eyed-grass
Iridaceae, Iris Family

Habit: Herbaceous grass-like winter annual, to 35 cm tall. Stems spreading, ascending, branching.

Leaves: Alternate, in a basal rosette, with shorter leaves along the stems; stalkless; blades narrow, grasslike, 2-8 cm long.

Inflorescences: Flowers solitary or a few in a cluster at the ends of flowering stalks.

Flowers: Blue, lavender-rose, or white, with a yellow center rimmed by a rose-purple eye-ring, 6-parted.

Fruit: A capsule, 3-4 mm long.

Habitat and Distribution: Occasional or locally common; moist pinelands, old fields, roadsides, and lawns; throughout Florida, west to Texas, north to Arkansas and North Carolina; native to South America. .

Comment: Annual Blue-eyed-grass is a colorful little weed blooming in the spring and summer. It prefers open moist areas with low vegetation.

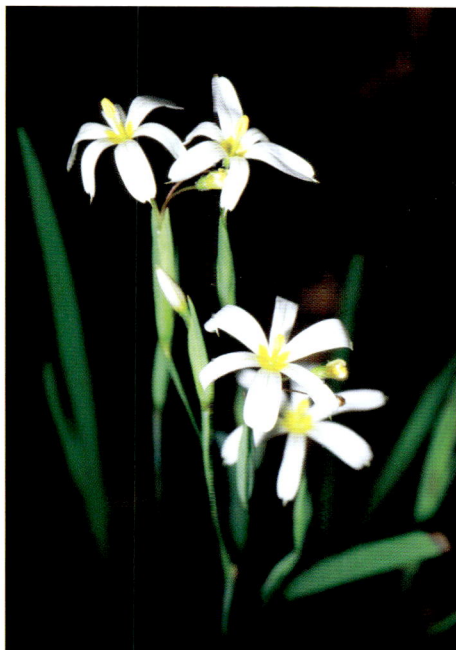

Sisyrinchium atlanticum E.P. Bicknell

Sisyrinchium rosulatum
E.P. Bicknell
(Photographs courtesy of XID Services, Inc.)

Blue

Stachytarpheta jamaicensis (L.) Vahl
Blue Porter Weed
Verbenaceae, Vervain Family

Habit: Herbaceous perennial, to 1 m tall. Stems branched, hairless, somewhat 4-angled.

Leaves: Opposite; stalked; blades broadly elliptic to rounded, 2-10 cm long, with toothed edges.

Inflorescences: Scaly-looking, terminal spikes, 10-15 cm or more long.

Flowers: Blue to violet with white eye, 8-10 mm long, opening a few at a time, tubular, with 5 large, fl aring lobes.

Fruit: Composed of 2 hard, elliptic segments, each about 5 mm long.

Habitat and Distribution: Occasional to frequent; coastal dunes, shell mounds, cleared sites, and other disturbed areas; central and southern peninsula Florida; Alabama; Hawaii; Bermuda, West Indies, Central America, South America; introduced into Old World Tropics.

Comment: Flowers can be produced all year. Blue Porter Weed is becoming a popular landscape plant. It is very tolerant of most soils and moisture conditions.

Stylodon carneum (Medik.) Moldenke
Pygmy Verbena
[*Verbena carnea* Medik.]
Verbenaceae, Vervain Family

Habit: Herbaceous perennial, to 90 cm tall. Stems ascending to erect, 4-angled, simple or few-branched, hairy.

Leaves: Opposite; stalkless; blades elliptic to broader at the tip to rounded to lance-shaped, 2-12 cm long, sandpapery, margins toothed.

Inflorescences: Long, thin, glandular-hairy spikes at branch and stem tips.

Flowers: Blue to pink, somewhat tubular, hairy, 6-9 mm long, 5-lobed, each 4-6 mm long.

Fruit: A 4-lobed nutlet, about 3 mm long.

Habitat and Distribution: Occasional; sandy, dry, open woods, and pinelands; central peninsula Florida, north to North Carolina, west to Texas.

Comment: Pygmy Verbena blooms in spring and summer. The small flowers are frequently overlooked among the other vegetation.

Stachytarpheta jamaicensis
(L.) Vahl

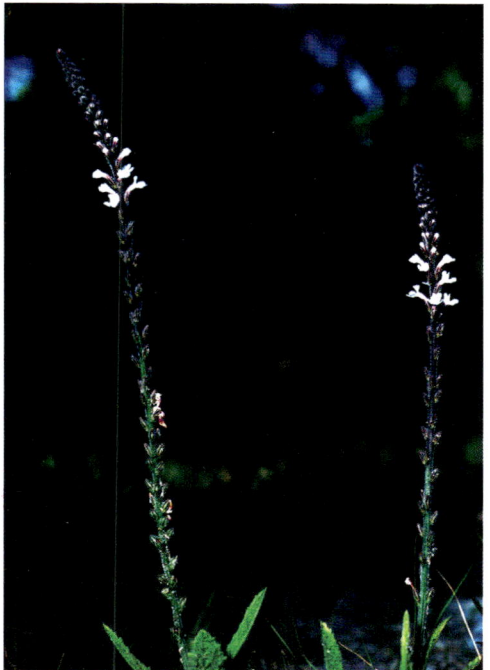

Stylodon carneum
(Medik.) Moldenke

Blue

Tradescantia hirsutiflora Bush
Hairy-flower Spiderwort
[*Tradescantia hirsuticaulis* Small, misapplied]
Commelinaceae, Spiderwort Family

Habit: Herbaceous perennial, 20-50 cm tall, from long fleshy roots. Stems erect or ascending, densely hairy.

Leaves: Alternate; stalkless; blades grass-like, 7-32 cm long, base sheathing stem.

Inflorescences: Terminal or lateral, stalked clusters, sometimes drooping.

Flowers: Sepals with glandular and nonglandular hairs; petals blue to rose, about 2 cm long, with 3 petals.

Fruit: An oblong capsule, 4-7 mm.

Habitat and Distribution: Infrequent; dry woods, scrub, sandhills, and prairies; panhandle Florida, west to Texas, Arkansas, and Oklahoma, north to North Carolina.

Comment: Hairy-flower Spiderwort blooms in spring and certainly deserves its common name. Flowers are showy.

Tradescantia ohiensis Raf.
Common Spiderwort
Commelinaceae, Spiderwort Family

Habit: Herbaceous perennial, 20-80 cm tall, from thin fleshy roots. Stems erect to ascending, smooth, somewhat fleshy.

Leaves: Alternate; stalkless; blades grass-like, 15-45 cm long, base sheathing stem.

Inflorescences: Terminal or lateral, stalked clusters.

Flowers: Sepals smooth or with tuft of hairs at tip; petals blue, 3-petaled, 1-1.8 cm long.

Fruit: An oblong-rounded capsule, 4-6 mm long.

Habitat and Distribution: Common; meadows, prairies, dry woods, sandy roadsides, and other dry disturbed sites; central peninsula Florida, west to Texas, north to Minnesota and Massachusetts.

Comment: This spring-flowering wildflower is becoming more popular in cultivation. The almost fleshy foliage is attractive in the warm months and the flowers are quite showy.

Tradescantia hirsutiflora Bush

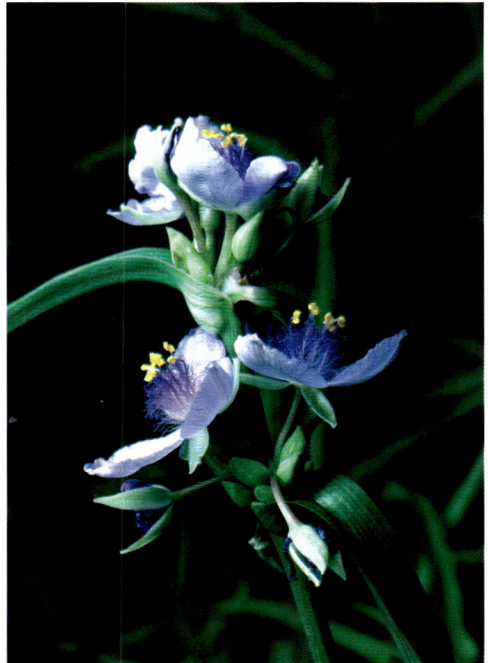

Tradescantia ohiensis Raf.

Blue

Trichostema dichotomum L.
Blue Curls
Lamiaceae (Labiatae), Mint Family

Habit: Herbaceous annual, to 1 m tall. Stems often woody at base, hairy, branching occurs in pairs.

Leaves: Opposite; stalked; blades elliptic to lance-shaped, 2-7 cm long, margins smooth.

Inflorescences: Terminal, openly-branched, branches in pairs with a flower at each pair.

Flowers: Blue, with short tube and 2 lips, upper lip is 4-lobed and the lower lip is narrow with a light blue base with dark spots at base; stamens long and arched.

Fruit: Composed of 4 nutlets, surface textured, 1.5-2.9 mm long.

Habitat and Distribution: Common; dry sandy or rocky woods, sandhills, flatwoods, and disturbed sites; throughout Florida, west to Texas and north Missouri, Michigan, and Maine.

Comment: Blue Curls is late summer into early fall blooming. It is an annual which is useful where an earlier blooming plant has retired for the season. The unusual blue flowers with curving stamens are difficult to accurately describe. It reseeds easily and seldom needs much encouragement.

Triodanis perfoliata (L.) Nieuwl.
Venus' Looking-glass
[*Specularia perfoliata* (L.) DC.]
Campanulaceae, Bellflower Family

Habit: Herbaceous annual, 0.1-1 m tall. Stems branching from base, ascending or erect, hairy.

Leaves: Alternate; stalkless and clasping; blades round, 1-3 cm long, margins toothed.

Inflorescences: One to several stalkless flowers in upper leaf axils.

Flowers: Light blue to purple, lower flowers never open, star-shaped, 4-10 mm long, with 5 petals and a small white center.

Fruit: A 2-chambered, ellipsoid capsule, 9 mm long, opening by pores in the sides about 1-1.5 mm below the top.

Habitat and Distribution: Frequent; open fields, roadsides, gardens, and disturbed sites; central peninsula Florida, north to Maine, Quebec, Montana, and British Columbia, south through Texas and Mexico into tropical America.

Comment: Flowering primarily in spring, the round blades, flowers clustered in the upper leaves, and peculiar holes in the walls of the capsule make this an easy plant to identify. Venus' Looking-glass is quite weedy.

Trichostema dichotomum L.

Triodanis perfoliata (L.) Nieuwl.

Blue

Verbena bonariensis L.
Roadside Verbena
Verbenaceae, Vervain Family

Habit: Herbaceous perennial, 0.6-2 m tall. Stems stiffly erect, rough hairy, angled.

Leaves: Opposite; stalkless and somewhat clasping at base; blades elliptic, 5-15 cm long, margins toothed.

Inflorescences: Compact spikes at stem and branch tips.

Flowers: Blue to violet or purple, tubular, to 4 mm long.

Fruit: Composed of 4 nutlets, each about 1.5 mm long.

Habitat and Distribution: Occasional; old fields and disturbed sites; throughout Florida, west to Texas, north to Tennessee and North Carolina; West Indies; native of Argentina, Brazil, Paraguay, and Uruguay.

Comment: Roadside Verbena blooms most of the spring and summer. This weed is easy to spot by noticing the stiff, erect growth pattern and the opposite branches with compact spikes of flowers at the tips.

Verbena officinalis L.
European Vervain
Verbenaceae, Vervain Family

Habit: Herbaceous annual, 0.2-1 m tall. Stems slender, diffusely branching, smooth.

Leaves: Opposite; stalk-like base; blades 2-7 cm long, once or twice divided, the divisions again dissected into lobes.

Inflorescences: Flowers arranged along loose terminal spikes.

Flowers: Blue, purple, or white, tubular, hairy, 3-5 mm long, with lobes less than 1 mm long.

Fruit: Composed of 4 brown, rough-surfaced nutlets, each about 2 mm long.

Habitat and Distribution: Infrequent; roadsides, fields, and disturbed sites; panhandle Florida, north to Massachusetts, west to Louisiana; native of Europe.

Comment: Blooming in summer and fall. The flowers are attractive in mass, but quite small. The highly divided foliage of this weed is usually the first characteristic that catches attention.

Verbena bonariensis L.

Verbena officinalis L.

Blue

Veronica persica Poir.
Bird's-eye Speedwell or Persian Speedwell
Scrophulariaceae, Figwort Family

Habit: Herbaceous annual, 10-30 cm tall. Stems hairy, low, spreading, with numerous branches forming basal rosette.

Leaves: Opposite; stalked; blades oval, to 3 cm long, margins toothed.

Inflorescences: Flowers stalked and solitary in the axils of upper alternate leaves.

Flowers: Bright blue, with darker lines and pale center, 4-lobed, to 7-12 mm wide.

Fruit: A 2-lobed, flattened capsule, 3-5 mm long 5-9 mm broad.

Habitat and Distribution: Scattered and locally common; gardens, lawns, roadsides, fields and other disturbed sites; northern peninsula Florida, north throughout most of North America; native of southwest Asia.

Comment: Bird's-eye Speedwell is spring flowering and quite weedy.

Vicia acutifolia Elliott
Sand Vetch or Fourleaf Vetch
Fabaceae (Leguminosae), Pea or Bean Family

Habit: Perennial herbaceous vine, 0.5-1.5 m long, from under ground runners (rhizomes). Stems sprawling, climbing by tendrils, or twining, smooth to minutely hairy.

Leaves: Alternate; stalked; blades with 4 (occasionally 2 or 6) leaflets and a simple tendril at the tip, each leaflet 1.5-4 cm long and very narrow.

Inflorescences: Loose clusters of 4-10 flowers in leaf axils.

Flowers: Blue to white and typically pea-shaped, to 8 mm long.

Fruit: Pod narrow, flattened, 2.5-3 cm long, 4- to 8-seeded.

Habitat and Distribution: Common; ditches and other moist sites, flatwoods, wet hammocks, margins of swamps, streams, rivers, and ponds, and disturbed wet habitats; throughout Florida and north to southern Alabama, coastal Georgia, and southeastern South Carolina.

Comment: This somewhat delicate-looking plant blooms in the spring. Sand Vetch is attractive, but not very showy.

Veronica persica Poir.

Vicia acutifolia Elliott

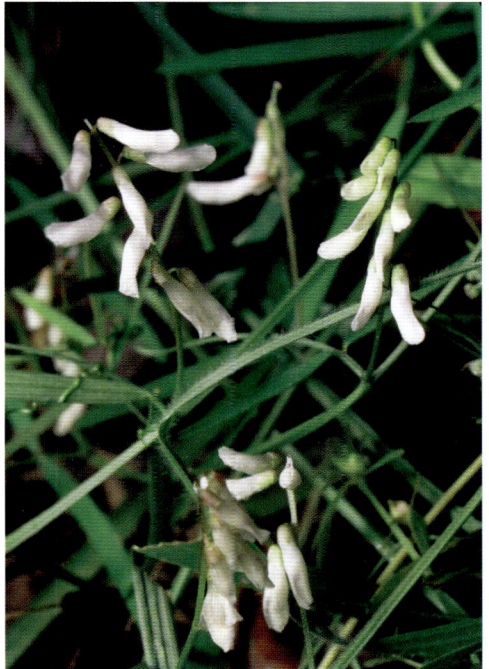

Blue

Vicia floridana S. Wats.
Florida Vetch
Fabaceae (Leguminosae), Pea or Bean Family

Habit: Perennial herbaceous vine, to 80 cm long, from underground runners (rhizomes). Stems sprawling or trailing, nearly smooth.

Leaves: Alternate; stalked; blades with 2-4 (occasionally 6) elliptic leaflets with a simple tendril at the tip, each leaflet 0.5-1.5 cm long.

Inflorescences: Loose clusters of 2-8 flowers in leaf axils.

Flowers: Blue to lavender, typically pea-shaped, to 8 mm long.

Fruit: Pods to 1.5 cm long.

Habitat and Distribution: Frequent; moist and wet sites, ditches, stream, river, and pond margins, and wet hammocks; central peninsula and central panhandle Florida into southeastern Georgia.

Comment: Florida Vetch usually trails and is often overlooked. The small flowers are seen in the spring and summer.

Vicia sativa L.
Common Vetch
Fabaceae (Leguminosae), Pea or Bean Family

Habit: Herbaceous annual vine, to 1 m long. Stems lying flat, ascending, or low climbing, smooth or slightly hairy.

Leaves: Alternate; stalked; blades with 8-14 oblong leaflets with simple or branched tendrils at the tip, leaflets 1-3.5 cm long.

Inflorescences: Paired or solitary flowers in leaf axils.

Flowers: Pink-purple or purple to whitish, 1-2.5 cm long.

Fruit: Pods hairy when young, oblong, 3-8 cm long and 4-8 mm wide.

Habitat and Distribution: Occasional; roadsides, fields, and disturbed habitats; central peninsula Florida north throughout the United States and into Canada; native of Europe.

Comment: The large flowers are found in the spring. Still used for hay and as a cover crop, Common Vetch was originally imported for such use, but now also has become a common agricultural weed.

Vicia floridana S. Wats.

Vicia sativa L.

Blue

Viola bicolor Pursh
Field Pansy
[*Viola rafinesquii* Greene]
Violaceae, Violet Family

Habit: Herbaceous annual, to 40 cm tall. Stems slender, erect or spreading, leafy, branched from the base.

Leaves: Alternate, spreading; long-stalked; blades rounded to lance-shaped with broad tips, margins smooth.

Inflorescences: Solitary flowers on long stalks from leaf axils.

Flowers: Blue, white with blue veins, or all white, about 1 cm across, with 5 petals.

Fruit: An oblong capsule, 5-7 mm long.

Habitat and Distribution: Occasional; lawns, clearings, old fields, and roadsides; panhandle Florida, west to Texas, Colorado, and Nebraska, north to New York.

Comment: Colorful spring-flowering annual. Can be cultivated in most sites. The only native pansy in North America.

Viola palmata L.
Early Blue Violet or Wood Violet
[*Viola esculenta* Elliott ex Greene, *Viola septemloba* LeConte]
Violaceae, Violet Family

Habit: Herbaceous perennial from an underground runner. Stemless.

Leaves: Alternate, spreading; long-stalked; blades often 3-lobed or with 4-6 basal lobes, about as wide as long.

Inflorescences: Solitary flowers on long stalks from leaf axils.

Flowers: Blue to violet, 2-4 cm wide, with 5 petals, streaked with white or rarely white, with a hairy spur.

Fruit: A cylindrical capsule.

Habitat and Distribution: Frequent to common; dry soils of sandhills, flatwoods, woods, and clearings; central peninsula Florida north to Maine and Minnesota, west to Texas.

Comment: A frequent wildflower, usually found in drier soils, blooming in spring.

Viola bicolor Pursh

Viola palmata L.

Blue

Viola sororia Willd.
Florida Violet, Common Blue Violet, Dooryard Violet
[*Viola affinis* LeConte, misapplied]
Violaceae, Violet Family

Habit: Herbaceous perennial from an underground runner. Stemless.

Leaves: Alternate, spreading; long-stalked; blades broadly oval, triangular, or heart-shaped, long hair on both surfaces, margins smooth.

Inflorescences: Flowers solitary on long stalks from leaf axils.

Flowers: Blue or violet, 1.5-3.5 cm across, with 5 petals, light center, and hairy spur.

Fruit: A round capsule, purple or purple-spotted.

Habitat and Distribution: Frequent to common; open, moist, wet woody sites and clearings; throughout Florida, west to Texas and California, north to British Columbia, Quebec, and Newfoundland.

Comment: Flowering in spring, sometimes becoming weedy in lawns. A frequent wildflower.

Viola walteri House
Walter's Violet
Violaceae, Violet Family

Habit: Herbaceous perennial, to 10 cm tall, from underground runner. Stems slender, sometimes several, erect or trailing.

Leaves: Alternate, often forming a basal rosette; long-stalked; blades rounded to heart-shaped, short hairs on upper surface, purple on lower surface, especially on the veins, margins smooth.

Inflorescences: Flowers solitary on long stalks from leaf axils.

Flowers: Blue-violet, 2-2.5 cm across, most petals usually hairy, spur hairy.

Fruit: A round capsule, 4-7 mm long.

Habitat and Distribution: Infrequent; dry, rich woods; central peninsula Florida, west to Texas, north to Ohio and Virginia.

Comment: Spring blooming, this little violet is readily identified with or without flowers by the purple color on the lower leaf surface.

Viola sororia Willd.

Viola walteri House

Blue

Wahlenbergia marginata (Thunb.) A. DC.
Asiatic Bellflower
Campanulaceae, Bellflower Family

Habit: Herbaceous perennial, 10-65 cm tall. Stems slender, often several in a clump, erect, leafy on lower portions.

Leaves: Alternate; stalkless; blades very narrow, 1-4 cm long, margins smooth, toothed and often wavy.

Inflorescences: Long-stalked flowers scattered on upper stems.

Flowers: Blue to violet, bell shaped, 5-8 mm long, with 5 pointed petals.

Fruit: A capsule, cone-shaped or bowl shaped, 3-6 mm long.

Habitat and Distribution: Frequent; sandy fields, roadsides, and other disturbed open sites; central peninsula Florida, north to Alabama and North Carolina; native to Asia.

Comment: Flowering from spring until frost, Asiatic Bellflower is such a thin plant with small blooms that it is usually not noticed. It is considered a weed, but never seems to be problematic.

Wisteria frutescens (L.) Poir.
Wisteria, American Wisteria, Atlantic Wisteria
Fabaceae (Leguminosae), Pea or Bean Family

Habit: Perennial high climbing vine, 2-15 m long. Stems woody, climbing by twining, new branches hairy.

Leaves: Alternate; stalked; blades with 9-15 leaflets, ovate to elliptic, each leaflet tapering, pointed, 2-6 cm long, slightly hairy underneath.

Inflorescences: Dense, hanging clusters, 4-12 cm long, terminal.

Flowers: Typically pea-shaped, petals blue violet to lilac or rarely white, 1.5-2 cm long.

Fruit: Pod narrow and flattened, 4-12 cm long and 1-1.2 cm broad, smooth.

Habitat and Distribution: Occasional; low woodlands and floodplains, often along margins; central peninsula Florida west to east Texas and north to Iowa, Missouri, and Delaware.

Comment: Usually flowering in the spring and early summer, it can be somewhat irregular through the summer with a few scattered blooms. Wisteria behaves as a trailing or climbing vine in nature and can be wonderfully showy along the margins of water courses or wet woods. In cultivation it can be maintained as a shrub or trained over arbors and up trees to produce a great splash of color and fragrance.

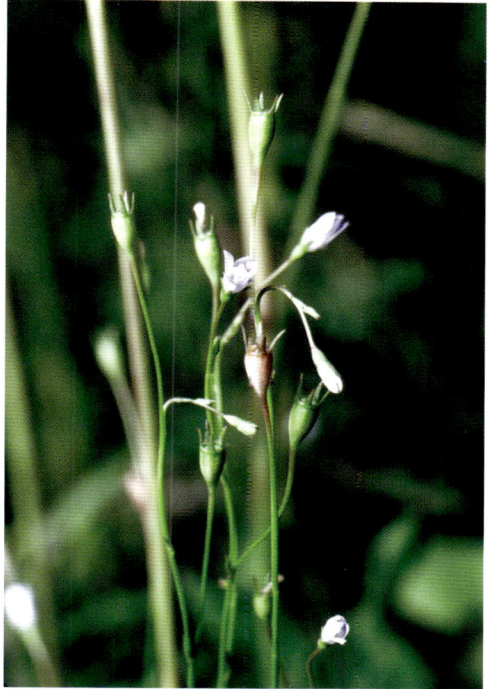

Wahlenbergia marginata
(Thunb.) A. DC.

Wisteria frutescens (L.) Poir.

Orange

Asclepias curassavica L.
Scarlet Milkweed or Veintiunilla
Asclepiadaceae, Milkweed Family

Habit: Herbaceous annual, 0.3-1.5 m tall. Stems slender, smooth at maturity and with milky sap.

Leaves: Opposite; stalked; blades lance-shaped, 5-12 cm long and to 1 cm broad.

Inflorescences: Small terminal clusters.

Flowers: Orange-red, orange, and yellow-orange, with 5 orange-red petals to 1 cm long that are curved down and outward with a yellow to orange, 5-pointed hood in center of flowers.

Fruit: Pods slender, to 10 cm long.

Habitat and Distribution: Infrequent; sandy and disturbed sites, especially in moist to wet soils; throughout Florida, west to southern Texas, south through the West Indies, Mexico south into South America, and in the Old World Tropics.

Comment: This showy ornamental is becoming more frequently used. The flowers occur all the warm seasons of the year. It has a leggy appearance and is best planted behind other shorter plants. However, the blazing orange-red flowers and dark foliage can make a splash of color when planted by themselves or in mass. An equally attractive form of this plant is being sold with pure orange flowers.

Asclepias lanceolata Walt.
Red Milkweed
Asclepiadaceae, Milkweed Family

Habit: Herbaceous perennial 0.4-1.2 m tall. Stem single, with a few pairs of leaves.

Leaves: Opposite; short-stalked; blades linear, 1-25 cm long, 0.5 to 1 cm wide.

Inflorescences: Few, terminal, 3- to 8-flowered clusters, each 2-5 cm wide.

Flowers: Red-orange, petals curved down and outwards, 8-11 mm long, drying purplish; an orange-red or orange hood occurs in the center of flower, with horn-like protrusions.

Fruit: Smooth, erect pods, 7-10 cm long, 1 cm wide.

Habitat and Distribution: Frequent; fresh and brackish marshes, wet pine woods, and floodplain forests; southern Florida west to southeastern Texas, and north to southern New Jersey.

Comment: This summer flowering milkweed is a bright, showy addition to a landscape, especially if other plants are used to fill in the sparse foliage.

Asclepias curassavica L.

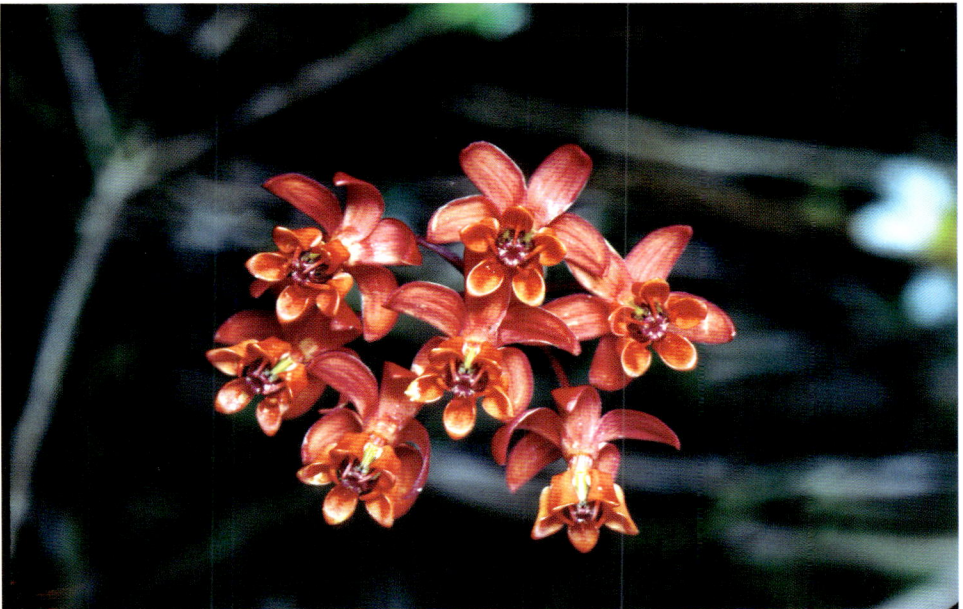

Asclepias lanceolata Walt.

Orange

Asclepias tuberosa L.
Butterfly Weed
Asclepiadaceae, Milkweed Family

Habit: Herbaceous perennial, 20-80 cm tall. Stems erect or spreading, stout, hairy, sap not milky.

Leaves: Alternate, plentiful; short-stalked; blades narrow to oblanceolate, 3-10 cm long, 0.3-2.5 cm broad, hairy, especially beneath.

Inflorescences: Few to many, dense, domed clusters at stem tips.

Flowers: Bright orange, 5 petals , 6-10 mm long, spreading downward, crown 4-6 mm diameter, spreading upward.

Fruit: A capsule 8-12 cm long and 1-1.5 cm broad.

Habitat and Distribution: Frequent to common; dry pinelands, sandhills, and dry fields, dry prairies, and upland woods; throughout Florida, west to Arizona, Colorado, and Nebraska, and north to Michigan, Wisconsin, Minnesota, Vermont, New Hampshire, and southern Ontario.

Comment: Flowering in summer and fall, the bright, showy, striking flowers and compact stature of this plant make it a winner in landscapes. Dry native habitats can be colorfully splotched with this plant, making an interesting milieu of the landscape.

Bignonia capreolata L.
Cross Vine or Quarter Vine
[*Anisostichus capreolata* (L.) Bureau; *Anisostichus crucigera* (L.) Bureau ex Small, misapplied]
Bignoniaceae, Trumpet Creeper Family

Habit: Perennial, woody, high-climbing vine with tendrils.

Leaves: Opposite; stalked; blades divided into 2 oblong to lance-shaped leaflets, each 6-15 cm long, and a terminal tendril.

Inflorescences: Short-stalked, axillary clusters of 2-5 flowers.

Flowers: Red-orange, yellowish inside, bell- or trumpet-shaped, 4-5 cm long, with 5 lobes curling outward.

Fruit: A flattened, slender capsule, to 17 cm long; seeds elliptic, thin, papery, winged.

Habitat and Distribution: Frequent; floodplain forests, thickets, and mixed, deciduous or moist woods; central peninsula Florida, north to Maryland, Virginia, and Illinois, and west to east Texas.

Comment: Cross Vine can climb to 20 m to areas where sunlight is available. When blooming in the tree tops the spring flowers can be hard to spot. Frequently branches of the vine hang down in bright sunny openings and the flower clusters provide an unexpected show in the forest.

Asclepias tuberosa L.

Bignonia capreolata L.

Orange

Campsis radicans (L.) Seemann ex Bureau
Trumpet Vine, Cow-itch Vine, Trumpet-honeysuckle
Bignoniaceae, Trumpet Creeper Family

Habit: Perennial high-climbing vine, to over 20m. Stems woody, trailing, or climbing with aerial roots.

Leaves: Deciduous, opposite; stalked; blades dissected into 7-15 pointed oval leaflets, each 2-8 cm long and 1-4 cm broad, with coarse teeth on the margins.

Inflorescences: Small terminal clusters.

Flowers: Orange to red, joined in a trumpet- shaped tube, 5-9 cm long, with 5 rounded lobes.

Fruit: A long, slender capsule, 10-18 cm long and 2-3 cm diameter, splitting into two halves at maturity. Seeds have papery wings.

Habitat and Distribution: Frequent; woods, fencerows, roadsides, disturbed sites, and floodplain forests; central peninsula Florida, west to Texas, north to Iowa, Illinois, West Virginia, and New Jersey, and occasionally escaped from cultivation as far north as Michigan and Connecticut.

Comment: Spring and summer flowering in the south and summer flowering in the north, Trumpet Vine is a showy wildflower, but often flowers at the top of the canopy and out of sight. If occurring on a fence or planted to grow on a six to ten foot tall post, the stem tips and subsequent flowers can be seen at eye level as stem tips tend to bend downward. The bright showy flowers are often visited by humming birds. In the warmer weather of the southern coastal plain Trumpet Vine can become very weedy. The constant drift of the winged seeds from capsules maturing high in the trees spreads seeds. Long underground roots and runners will frequently sprout to produce additional vines, often providing a dense groundcover in open areas in woods. The ability to climb by aerial roots makes this a good plant to cover bare tree trunks and open walls, however, the root remnants are difficult to remove if you wish to paint or utilize the surface otherwise. In the fall the leaflets turn various shades of orange and red as they prepare to drop for the winter. The common name, Cow-itch Vine, indicates the reputation of the foliage as a cause of contact dermatitis for some individuals.

Canna x generalis L.H. Bailey
Garden Canna
Cannaceae, Canna Family

Habit: Herbaceous perennial, 1-2 m tall. Stems waxy.

Leaves: Large, elliptical, 30-50 cm long and 8-20 cm broad.

Inflorescences: Short, clustered terminal spike.

Flowers: Orange-red often with red markings, to 7 cm long, with 3 erect petals with smooth or undulating edges, frequently with yellowish margins.

Fruit: Capsules 2-3.5 cm long, about as broad. Seeds round, about 5-9 mm in diameter.

Habitat and Distribution: Infrequent; low and wet disturbed sites, including ditches; throughout the coastal plain from North Carolina to Texas.

Comment: Blooming in spring and summer, Garden Canna, frequently escapes from cultivation. The entire plant is quite showy with the large leaves and irregular generous flowers. Being a garden hybrid the color of the flowers and foliage is quite variable. Garden Canna is a frost sensitive warm season perennial, persisting in southern and central Florida.

Campsis radicans (L.) Seemann ex Bureau

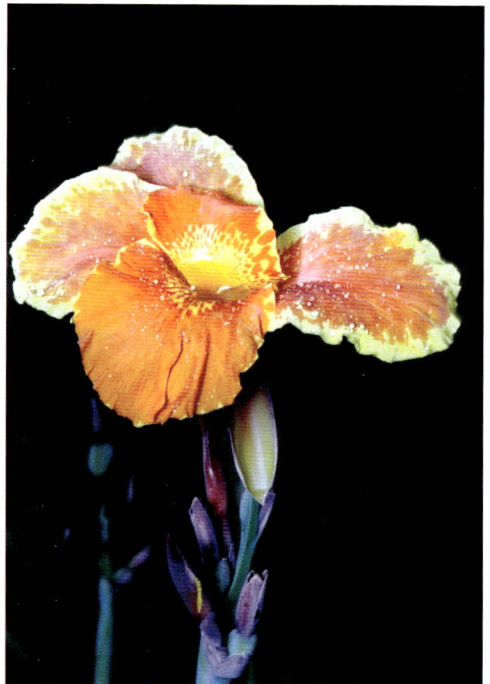

Canna x generalis L.H. Bailey

Orange

Clerodendrum kaempferi (Jacq.) Siebold
Kaempfer's Glorybower
Verbenaceae, Vervain Family

Habit: Perennial evergreen shrub, to 1.8 m tall.

Leaves: Opposite; long-stalked; blades large, oval, heart-shaped at base, scaly underneath, margins smooth or toothed.

Inflorescences: Large, narrow, branched, terminal panicles.

Flowers: Bright orange-red, about 3.5 cm long with a narrow tube.

Fruit: A drupe.

Habitat and Distribution: Rare; disturbed habitats, vacant lots and cleared lands; south peninsula Florida; native of Asia.

Comment: Flowering all year the large eye-catching clusters of orange-red flowers are striking.

Cordia sebestena L.
Geiger Tree
Boraginaceae, Borage Family

Habit: Perennial evergreen shrub or tree, to 10 m tall.

Leaves: Alternate; stalked; blades oval, to 20 cm long, rough-hairy with smooth or wavy margins.

Inflorescences: Large terminal clusters.

Flowers: Orange or scarlet, to 5 cm long.

Fruit: A hard, white drupe, about 2 cm long.

Habitat and Distribution: Rare; sandy or rocky coastal hammocks in south peninsula Florida; native from the West Indies to Venezuela.

Comment: Flowering in the summer and fall, Geiger Tree can be cultivated in many poor soils making this attractive plant with the bright orange flowers a favorite in tropical climates, where it is widely cultivated.

Clerodendron kaempferi
(Jacq.) Siebold

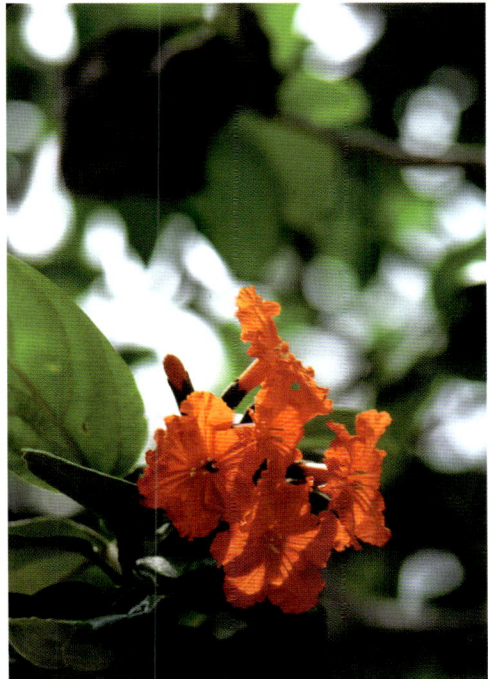

Cordia sebestena L.

Orange

Hamelia patens Jacq.
Fire Bush
Rubiaceae, Madder Family

Habit: Perennial evergreen shrub or small tree, to 8 m tall, but commonly 1-3 m tall. Branches hairy when young.

Leaves: Opposite; stalked; blades large, elliptic to ovate, 8-15 cm long, sometimes folded, hairy on both surfaces with smooth margins.

Inflorescences: Many-flowered, terminal clusters.

Flowers: Orange-red, narrow and tubular in shape, 1-2 cm long, 5-lobed.

Fruit: Berries dark red to black, oval, 6-10 mm long.

Habitat and Distribution: Frequent; coastal sites and edges of hammocks; southern and central peninsula Florida; also in the West Indies, and Mexico south into South America.

Comment: Blooming irregularly throughout the year this tropical shrub is beautiful for foliage and flowers when placed in an open environment. Fire Bush is subject to frost, but will resist slight freezes if protected by overhanging vegetation.

Leonotis nepetifolia (L.) R. Br.
Lion's-ear
Lamiaceae (Labiatae), Mint Family

Habit: Herbaceous annual, 0.3-2 m tall. Stems are erect, simple or branched, square, and softly hairy.

Leaves: Opposite; stalked; oval, wider at base, 5-12 cm long and 4-10 cm broad, with toothed edges.

Inflorescences: Spherical clusters, 3-6 cm in diameter, encircling upper portions of stems.

Flowers: Orange, yellow, or red, tubular, 1.2-2.5 cm long, hairy, with 2-lipped mouth.

Fruit: A capsule, 3-3.3 mm long.

Habitat and Distribution: Infrequent; gardens, roadsides, fencerows, fields, and other disturbed habitats; throughout Florida, west to Texas, and north to Tennessee and North Carolina, native to South Africa.

Comment: Lion's-ear is a summer and fall flowering, striking, tall, slender plant. The flowers are attractive, especially in the large round clusters at the upper stem joints, but the plant looks curious after flowering. When cultivated, best placed against a fence or behind other shorter ornamentals. The common name is derived from the resemblance of the hanging flowers to lion's ears.

Hamelia patens Jacq.

Leonotis nepetifolia (L.) R. Br.

Orange

Lilium catesbaei Walt.
Catesby's Lily, Pine Lily, Leopard Lily
Liliaceae, Lily Family

Habit: Herbaceous perennial, 30-80 cm tall, from scaly bulbs. Stems erect, leafy, unbranched.

Leaves: Alternate; stalkless; blades narrow, linear to linear-lanceolate, grasslike, to 12 cm long and to 1.5 cm wide, shorter on upper stems.

Inflorescences: Flowers solitary or sometimes two, erect, terminal.

Flowers: Orange or red, with 3 petals and 3 sepals, all similar, all 7-11 cm long, pointed, arching outward, base spotted with purple dots.

Fruit: A capsule, 2-5 cm long and 1-1.5 cm diameter.

Habitat and Distribution: Frequent; low moist woods, wet flatwoods, swamps, and savannahs; throughout Florida, west to Louisiana, and north to Virginia.

Comment: Catesby's Lily blooms in the summer and fall with large bright orange-red flowers. It can be a spectacular wildflower, especially when several specimens are clustered. In recent years mowed open wet pinelands, such as those found in transmission line rights-of-way, are the best areas in which to view this plant. In Florida Catesby's Lily is listed as a Threatened species.

Lilium superbum L.
Turk's-cap Lily or Lily-royal
Liliaceae, Lily Family

Habit: Herbaceous perennial, to 3 m tall, from a white bulb and rhizome. Stems erect, unbranched on the lower portion.

Leaves: In whorls of 5-20; stalkless; blades lance-shaped, to 18 cm long and to 3 cm broad.

Inflorescences: Terminal, nodding, 3-25 flowers per plant.

Flowers: Orange to reddish or sometimes orange-yellow, usually spotted, 3 petals and 3 sepals all similar, 5.5-8 cm long and 1-2 cm wide, curling backward.

Fruit: A capsule, 3-5 cm long.

Habitat and Distribution: Rare; moist to wet meadows, swampy woods, mountain coves, and hammocks; Florida panhandle, west to Alabama, and north to southern Indiana, New Hampshire, Massachusetts, and New York.

Comment: This tall erect lily blooms in the summer. It is a handsome wildflower with the clusters of bright, showy, orange flowers at the tops of the stems. Turk's-cap Lily is listed as Endangered in Florida.

Lilium catesbaei Walt.

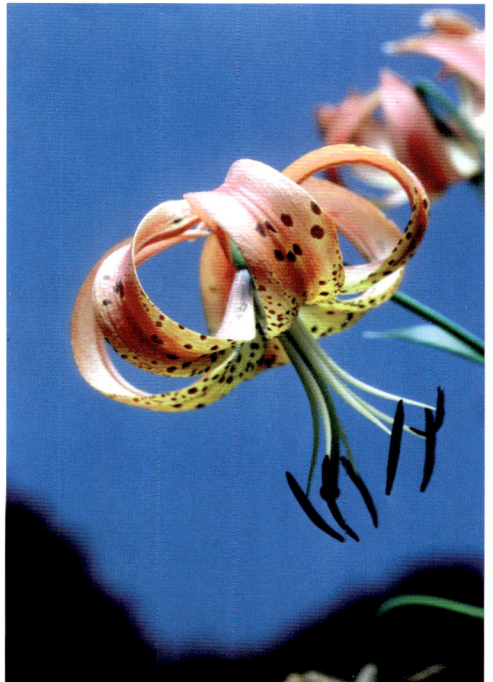

Lilium superbum L.

Orange

Modiola caroliniana (L.) G. Don
Bristly Mallow
Malvaceae, Mallow Family

Habit: Herbaceous perennial, mostly to 0.7 m long. Stems lying flat, spreading, rooting at nodes, older growth woody, newer growth hairy with star-shaped hairs.

Leaves: Alternate; stalked; blades 2-7 cm long and 1.5-3.5 cm broad, 6-7 lobes or deep cuts, with toothed edges, lightly hairy.

Inflorescences: Solitary flowers in leaf axils.

Flowers: Bright orange-red, blackish toward center, to 1 cm across, with 5 petals.

Fruit: A capsule, flat on top with 15-25 partitions, partitions with hair on back and with two awns each, arranged in ring, with 1 seed in each; seeds dark brown, smooth, 1-1.3 mm long.

Habitat and Distribution: Occasional; woods, meadows, pastures, lawns, fields, roadsides, and other disturbed habitats; central peninsula Florida, west to Texas, and north to Virginia or rarely to Massachusetts; south through Mexico into Argentina.

Comment: This often overlooked prostrate wildflower blooms in the warm months from spring into the fall. The orange-red flowers are so delicate and dark in color that they are easily overlooked. With dark green foliage and frequent flowers, Bristly Mallow should be cultivated.

Polygala lutea L.
Candy Weed or Orange Milkwort
Polygalaceae, Milkwort Family

Habit: Herbaceous annual, 5-50 cm tall. Stems low growing, clustered, smooth, often branched.

Leaves: Alternate, with a basal rosette, succulent; stalkless; blades elliptical or broader toward tip, to 4 cm long and 1 cm broad, smaller upwards.

Inflorescences: Dense, terminal, cylindrical heads, each 1-3.5 cm long and 1.2-2 cm in diameter.

Flowers: Orange or golden, slender, to 6 mm long, and drying yellow.

Fruit: A small capsule, about 1.5 mm long, broader at the top; seeds black, 1-1.3 mm long, hairy.

Habitat and Distribution: Common; low, moist woods, wet flatwoods, bogs, swamps and other wet sites, usually with acidic soils; throughout Florida, west to Louisiana, and north to New York, New Jersey, and Pennsylvania.

Comment: As a wildflower blooming in the spring and summer, Candy Weed will always catch your attention. The very bright flowers appear to explode with color. The lower stem and roots have a pleasant smell.

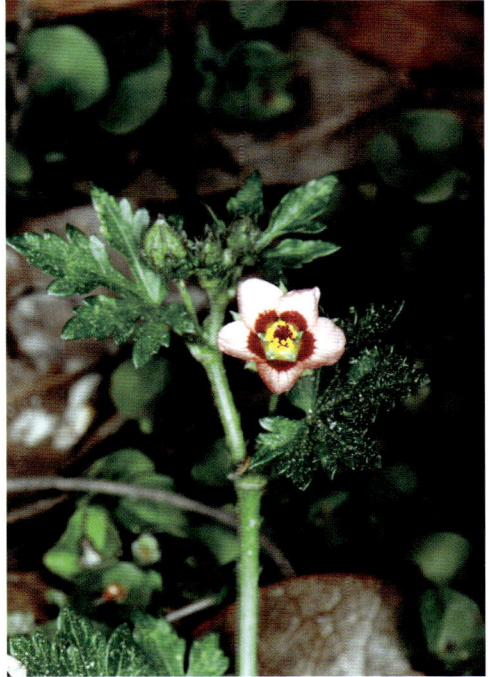

Modiola caroliniana (L.) G. Don

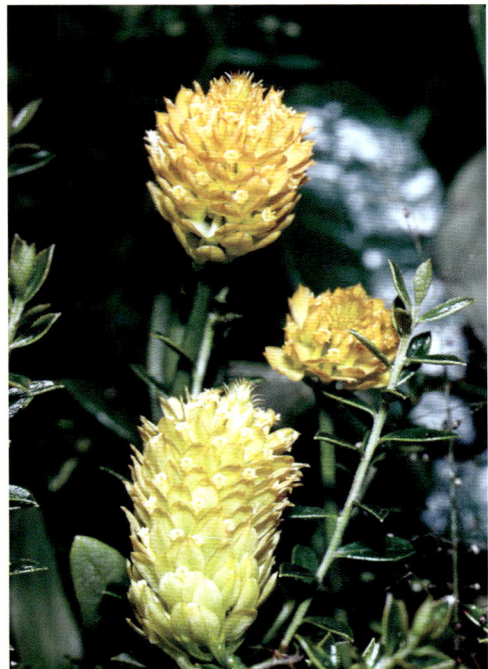

Polygala lutea L.

Orange

Pyrostegia venusta (Ker Gawl.) Miers
Flame Vine
[*Pyrostegia ignea* (Vell.) C. Presl]
Bignoniaceae, Trumpet Creeper Family

Habit: Perennial vine. Stems climbing by thin, three-parted tendrils.

Leaves: Opposite; stalked; blades divided into 2 oval leaflets, wider at base, 4-8 cm long, margins with shallow blunt teeth, and with a tendril emerging between leaflets.

Inflorescences: Hanging, branched, terminal clusters of several flowers.

Flowers: Red-orange, tubular, 5-7 cm long, with 5 recurved lobes.

Fruit: Capsules narrow, to 30 cm long.

Habitat and Distribution: Rare; hammocks and disturbed areas; southeast peninsula Florida; native of South Africa.

Comment: Flowering in spring, Flame Vine makes a brilliant display, particularly if cultivated on an arbor so that the blooming tips can be easily seen. This vine is commonly planted in the tropics.

Sesbania punicea (Cav.) Benth.
Spanish Gold
[*Daubentonia punicea* (Cav.) DC.]
Fabaceae (Leguminosae), Pea or Bean Family

Habit: Perennial shrub or small tree, 1-3 m tall. Stems single or clustered and branched.

Leaves: Alternate; stalked; blades with 12-40 oblong leaflets, each 1-3 cm long.

Inflorescences: 5-10 cm long, drooping clusters from leaf axils.

Flowers: Orange to orange-red, typically pea-shaped, 1.5-2.5 cm long.

Fruit: Pods are 4-8 cm long, 4-winged, and do not split at maturity. Seeds 4-8.

Habitat and Distribution: Frequent; wet disturbed sites such as banks, shores, ditches, roadsides, and swamps; on the coastal plain from central peninsula Florida, west to east Texas, and north to North Carolina; native of South America.

Comment: Spanish Gold blooms from the spring into the fall. The foliage is attractive and the colorful, large, orange flowers always draw a comment. It does best in cultivation with irrigation or damp soils in full sunlight.

Pyrostegia venusta (Ker Gawl.) Miers

Sesbania punicea (Cav.) Benth.

Orange

Zamia floridana A. DC.
Coontie or Florida Arrow-root
[*Zamia pumila* L.]
Zamiaceae, Zamia Family

Habit: Perennial shrub with subterranean stem and erect to ascending leaves, to 1 m high. Stem short, stout, woody, sometimes branched.

Leaves: Alternate; stalked; blades to 1 m long and high, fern- or palm-like with 8-30 divisions, stiff, evergreen, leathery, arching outward from the below ground stem.

Inflorescences: Male and female flowers on separate plants, both produce solitary to numerous stalked cones.

Flowers: Male cones solitary to numerous, stalked, brown, 4-6.5 cm long and 1.2-1.8 cm wide, drooping and falling apart after releasing pollen. Female seed-bearing cones usually solitary, brown and orange, 6-15 cm long and 3-5.5 cm wide, lumpy, fleshy, erect then drooping, borne at end of short stalk.

Fruit: Seeds angular and rounded, shiny orange, often left in a pile at the top of the stem among the leaves after the cone falls apart at maturity.

Habitat and Distribution: Occasional to frequent; pinelands, hammocks, coastal dunes, and flats, sandhills, and shell middens; south Georgia, south through peninsula Florida and the West Indies.

Comment: Flowering all year, but mainly in the fall, Coontie is an extensively cultivated native species. The evergreen foliage is very attractive and the plant is used in a variety of settings. The bright orange seeds are ornamental, but often not seen among the dense leaves. Plants in native settings seldom have the dense numbers of leaves that cultivated specimens have, making the seeds easier to view. The starchy underground stem was the main source of starch for earlier native and non-native settlers in Florida. All parts of this plant are POISONOUS, especially the seeds. Both the seeds and the stem have caused deaths. The starch is specially prepared and the water used for cooking is POISONOUS. Coontie is listed as Commercially Exploited by the State of Florida.

Zamia floridana A. DC.

Yellow

Acacia farnesiana (L.) Willd.
Sweet Acacia, Huisache, Popinac, Opopanax, Cassie, West Indian Blackthorn
Fabaceae (Leguminosae), Pea Family

Habit: Shrub, to 4 m tall. Stems with pairs of spines, spines 1-4 cm long.

Leaves: Alternate, deciduous; stalked; blades twice-divided, 2-5 paired divisions, each division with 10-30 pairs of leaflets; stalked, hairy, a gland located on top of stalk; leaflets linear-oblong, 4.5-6.5 mm long.

Inflorescences: Rounded heads on stalks from leaf axils, heads to 1.3 cm in diameter.

Flowers: Yellow-orange, approx. 2.5 mm long.

Fruit: A pod, 3.5-7 cm long and 1 cm wide, rounded.

Habitat and Distribution: Occasional to infrequent; margins above salt marshes, roadsides, coastal hammocks, pinelands, cactus-mesquite, open disturbed coastal sites; throughout Florida, north to southeastern coastal Georgia, and west into Texas, southern Arizona, California; pantropical; widely planted and escaped from cultivation.

Comment: Sweet Acacia is widely cultivated as an ornamental specimen plant and industrially for the oil from the flowers which is used for making perfume. Most flowering is in the late winter and spring before the new leaves appear. Occasional flowering occurs throughout the rest of the year.

Acacia pinetorum F.J. Herm.
Pineland Acacia or Key Acacia
[*Vachellia insularis* Small; *Vachellia peninsularis* Small]
Fabaceae (Leguminosae), Pea Family

Habit: Shrub, to 4 m tall. Stems slender, with pairs of spines, spines 1-3 cm long.

Leaves: Alternate, deciduous; stalked; blades twice divided, 2-4 paired divisions, each division with 9-15 pairs of leaflets; stalked, a gland located on top of stalk; leaflets oblong, each about 2 mm long.

Inflorescences: Rounded heads on stalks from leaf axils, heads to 9 mm in diameter.

Flowers: Yellow, about 2.5 mm long.

Fruit: A pod, 3-7 cm long, somewhat swollen, constricted between seeds, narrow, curved, woody when mature.

Habitat and Distribution: Occasional; coastal areas, hammocks, shell mounds, and pinelands; central and southern peninsula Florida; south into the West Indies.

Comment: Usually blooming in spring, Pineland Acacia can flower all year. This shrub could be planted as an attractive specimen plant.

Acacia farnesiana **(L.) Willd.**
photograph courtesy of L.B. McCarty

Acacia pinetorum F.J. Herm.

Yellow

Aletris aurea Walter
Golden Colic-root or Yellow Star-grass
Liliaceae, Lily Family (Nartheciaceae, Bog Asphodel Family)

Habit: Herbaceous perennial, 0.3-1.2 m tall.

Leaves: Alternate, mostly basal; stalkless; blades elliptic, 3-12 cm long and 1-2 cm wide.

Inflorescences: Flowers alternate along a terminal stalk.

Flowers: Orange-yellow, stoutly tubular, to 7 mm long, with short, erect, triangular lobes.

Fruit: A capsule, rounded, long-beaked, to 5-8 mm long.

Habitat and Distribution: Occasional; moist meadows, flatwoods, bogs, marshes, and pinelands; from north peninsula Florida, west to east Texas, and north to southern Maryland.

Comment: As a spring and summer blooming wildflower, the thin stems with small flowers are scarcely noticed, except as a curiosity.

Aletris lutea Small
Yellow Colic-root
Liliaceae, Lily Family (Nartheciaceae, Bog Asphodel Family)

Habit: Herbaceous perennial, 0.3-1.1 m tall.

Leaves: Alternate, mostly basal; stalkless; blades linear or linear lance-shaped, to 15 cm long and 2 cm wide.

Inflorescences: Flowers alternate along a terminal stalk.

Flowers: Yellow, slenderly tubular, 6-10 mm long, with rounded, spreading lobes.

Fruit: A capsule, rounded, long-beaked, to 3-5 mm long.

Habitat and Distribution: Common; wet pinelands, meadows, bogs, and cypress depressions; throughout Florida, west to Louisiana, and north to coastal plain Georgia.

Comment: Yellow Colic-root blooms in spring and summer. The long, thin stems with many yellow flowers are attractive, especially when many plants are in close proximity.

300

Aletris aurea Walter

Aletris lutea Small

Yellow

Argemone mexicana L.
Mexican-poppy
Papaveraceae, Poppy Family

Habit: Herbaceous annual, 30-90 cm tall, from a taproot. Stems erect, solitary, spiny, branched, sap yellow.

Leaves: Alternate; stalkless, clasping stem; blades leathery, deeply lobed at base of stem, less so on upper stem, 3-25 cm long, with toothed and spiny margins.

Inflorescences: Flowers solitary, terminal on branch and stem tips.

Flowers: Yellow, 3-7 cm wide, with 4-6 broad crinkled petals and yellow anthers in the center.

Fruit: A spiny, oblong capsule, 2.5-4.5 cm long.

Habitat and Distribution: Frequent, can be locally common; dry soils, rights-of-way and other open disturbed areas; throughout Florida, north to Massachusetts and Illinois, and west to Texas; south into South America.

Comment: Mexican-poppy is a spectacular spring blooming wildflower. The large yellow flowers can easily be seen at some distance. The forbidding spines of this plant make it easily identifiable.

Arnica acaulis (Walter) BSP.
Leopard's-bane or Southeastern Arnica
Asteraceae (Compositae), Aster Family

Habit: Herbaceous perennial, 15-80 cm tall. Stems densely hairy, glandular.

Leaves: Mostly basal; stalkless; basal blades broadly elliptic to rounded, 4-8, 4-12 cm long and 2-8 cm wide, hairy; stem blades stalkless, 1-4 pairs, much smaller than basal.

Inflorescences: Bell-shaped heads, in a nearly flat-topped cluster at the tip of the stem.

Flowers: Yellow rays, each to 3 cm long with 3- to 4-toothed tips.

Fruit: A nutlet, black, 3-5 mm long, with yellowish white bristles, to 7 mm long.

Habitat and Distribution: Infrequent; sandy soils of pinelands, sandhills, and savannahs; panhandle Florida, and north to Georgia, southeast Pennsylvania and Delaware.

Comment: The large flower heads can be seen blooming in late spring. Flowers are showy. As a wildflower the clusters of blooms on each plant, together with several plants growing in the adjacent area, can provide a great panorama.

Argemone mexicana L.

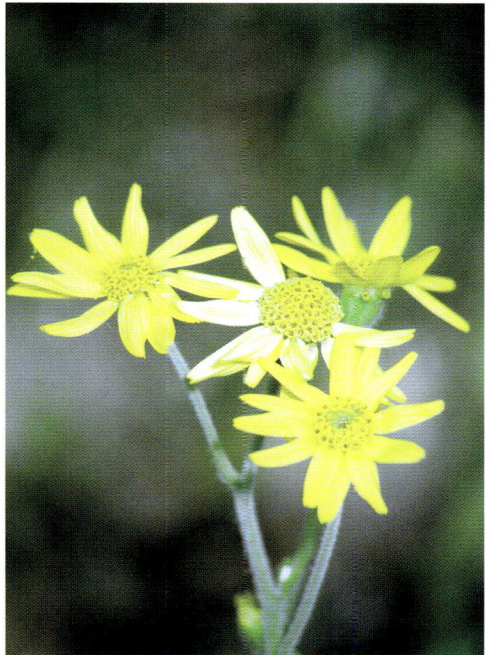

Arnica acaulis (Walter) BSP.

Yellow

Asclepias tuberosa L.
Butterfly-weed, Chigger-flower, Pleurisy-root
Asclepiadaceae, Milkweed Family (Apocynaceae, Dogbane Family)

Habit: Herbaceous perennial, 20-90 cm tall, from a thick crown rootstock. Stems erect or spreading, stout, hairy, sap not milky.

Leaves: Alternate; short-stalked; blades linear to lance-shaped, often broadest near tip, 3-11 cm long, 0.3-3 cm wide, hairy, especially underneath.

Inflorescences: 1 to many, dense domed clusters from leaf axils at stem tips.

Flowers: Bright yellow to orange-red, 5 petals , 6-8 mm long, spreading downward.

Fruit: A capsule, erect, 8-15 cm long and 1-1.5 cm broad, tapering from base to tip.

Habitat and Distribution: Common; dry soils of pinelands, sandhills, pastures, roadsides, prairies, dunes, hillsides, and fields; throughout Florida, west to Arizona and northern Mexico, and north to South Dakota, Minnesota, and New Hampshire.

Comment: Flowering from summer into fall, Butterfly-weed is a common wildflower in dry habitats. The bright yellow to orange-red flowers never fail to provoke a comment. The dried root has been used medicinally.

Astragalus villosus Michx.
Hairy Milkvetch or Bearded Milkvetch
Fabaceae (Leguminosae), Pea or Bean Family

Habit: Herbaceous perennial, 10-20 cm long, from a crown. Stems numerous, spreading, with long soft hairs.

Leaves: Alternate; stalked; blades with 7-17 elliptic to rounded leaflets, each 0.4-1.5 cm long, hairy, especially beneath and along midvein.

Inflorescences: Terminal, dense, spike-like clusters.

Flowers: Yellow to cream to yellowish-green, typically pea-shaped, 0.7-1.2 cm long.

Fruit: A sickle-shaped, long-haired pod, 1.5-2.3 cm long and 3-5 mm wide.

Habitat and Distribution: Occasional; dry woods, sandhills, roadsides, clearings, and disturbed sites; central peninsula Florida, west to Mississippi, and north to Georgia and South Carolina.

Comment: Hairy Milkvetch blooms in spring. The soft yellow clusters of flowers on the downy, spreading plants are attractive and easily seen but not at all showy.

Asclepias tuberosa L.

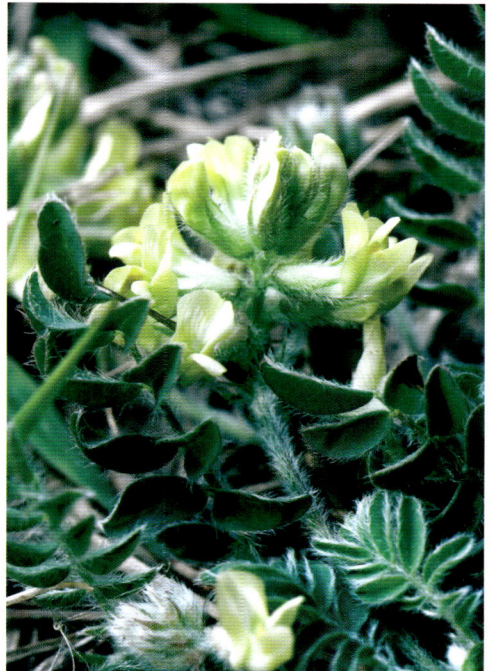

Astragalus villosus Michx.

Yellow

Aureolaria pedicularia (L.) Raf. var. *pectinata* (Nutt.) Gleason
Hairy False Foxglove, Fernleaf Yellow False Foxglove, Annual False Foxglove
[*Aureolaria pectinata* (Nutt.) Pennell]
Scrophulariaceae, Figwort or Snapdragon Family [*Orobanchaceae,* Broomrape Family]

Habit: Herbaceous annual, 0.4-1 m tall. Stems hairy and sticky, many paired branches.

Leaves: Opposite; stalkless; blades to 4 cm long, deeply lobed (like a comb), lance-shaped to somewhat rounded, sticky hairy.

Inflorescences: Flowers solitary, on long stalks from leaf axils.

Flowers: Yellow tinged with brown, bell-shaped, 2.5-4 cm long, with 5 rounded lobes.

Fruit: A capsule, rounded, to 1.2 cm long, sticky hairy.

Habitat and Distribution: Occasional; sandy soils of dry woodlands, especially oak woodlands, sandhills, roadsides, and open forested areas; central peninsula Florida, west to Texas, and north to Missouri and North Carolina.

Comment: Hairy False Foxglove blooms in the summer and fall. It never seems to be common but is easily seen in open sandy areas. The large yellow flowers on hairy plants attract attention, but do not provide a great show. Plants are parasitic on roots of oaks. The plants turn black when dry. Propagated from seeds, it is seldom cultivated.

Balduina angustifolia (Pursh) B.L. Rob.
Yellow Buttons or Coastal Plain Honey-comb Head
Asteraceae (Compositae), Aster or Sunflower Family

Habit: Herbaceous annual, to 1 m tall. Stems slender, 1 to several from base, branched.

Leaves: Alternate; stalkless; blades very narrow, needle-like, to 5 cm long, numerous.

Inflorescences: Terminal stalked heads, heads to 5 cm wide.

Flowers: 5-13 yellow rays, each 1-2 cm long with toothed tips, disc yellow becoming hard and honeycombed.

Fruit: Nutlet, to 2 mm long, with rounded scale-like bristles at top.

Habitat and Distribution: Common; dry, sandy habitats, sandhills, scrub, dunes, and xeric pinelands; throughout Florida, west to southern Mississippi, and north to Georgia, in the coastal plain.

Comment: The bright yellow flower heads can be seen in the summer and fall. Yellow Buttons never ceases to cause comments in the dry sparsely vegetated sands. After flowering, the prominent dry scaly heads on the thin many-branched stems often provoke comment.

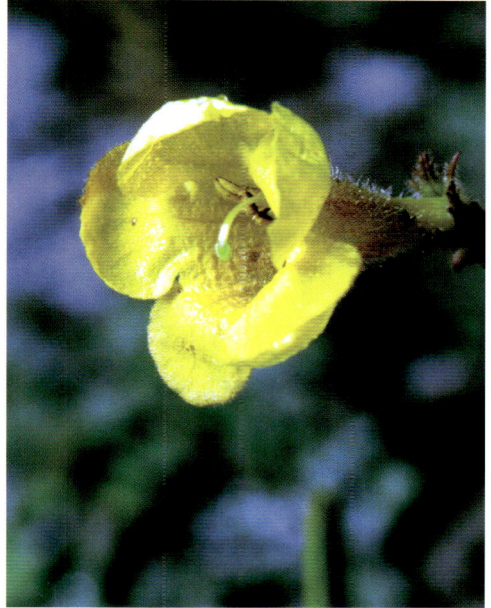

Aureolaria pedicularia (L.) Raf.
var. *pectinata* (Nutt.) Gleason

Balduina angustifolia
(Pursh) B.L. Rob.

Yellow

Balduina uniflora Nutt.
Honey-comb Head
Asteraceae (Compositae), Aster or Sunflower Family

Habit: Herbaceous annual or biennial, to 1 m tall. Stems erect, unbranched or with a few erect, long branches.

Leaves: Alternate; basal stalked, stem stalkless; basal blades elliptic to narrow, stem blades narrow, both to 10 cm long, upper leaves smaller.

Inflorescences: Terminal stalked heads, heads to 7 cm wide.

Flowers: 12-20 yellow rays, each 2-3 cm long with toothed tips, disc yellow becoming hard and honeycombed.

Fruit: Nutlet, 3-4 mm long, with scales at tip.

Habitat and Distribution: Frequent; moist to dry soils of pine lands, savannahs, bays, and bogs; north peninsula and panhandle Florida, west to Louisiana, and north to North Carolina.

Comment: The blooms in summer and fall are showy and easily seen. With only a few flower heads per plant the entire effect is not showy, but very pretty.

Baptisia lanceolata (Walter) Elliott
Pineland Wild-indigo or Gopher Weed
Fabaceae (Leguminosae), Pea or Bean Family

Habit: Herbaceous perennial, to 1 m tall. Stems ascending, often forming a rounded bush.

Leaves: Alternate; stalked; blades with 3 oval leaflets, each 3.5-8 cm long, somewhat leathery.

Inflorescences: Flowers solitary or in clusters of 2-4, terminal or in leaf axils.

Flowers: Yellow, typically pea-shaped, to 2.5 cm long.

Fruit: A pod, 1.5-2.2 cm long and about 1 cm diameter, inflated, hairless at maturity, woody.

Habitat and Distribution: Frequent to locally common; dry soils of woodlands, sandhills, scrub, pastures, and roadsides; central peninsula Florida, west to southern Alabama, and north through south Georgia into southern South Carolina.

Comment: The attractive flowers occur in spring. The flowers appear to be yellow flecks among the dark green leathery leaves. It is a very nice wildflower.

Yellow

Balduina uniflora Nutt.

Baptisia lanceolata (Walter) Elliott

309

Yellow

Baptisia lecontei Torr. & A. Gray
LeConte's Wild-indigo or Pineland Wild-indigo
Fabaceae (Leguminosae), Pea or Bean Family

Habit: Herbaceous perennial, to 1 m tall. Stems erect or ascending, bushy, hairy.

Leaves: Alternate; short-stalked; blades with 3 spatulate leaflets, each 1.5-5 cm long, wider toward tips.

Inflorescences: Terminal clusters of 4-8 flowers.

Flowers: Lemon-yellow, typically pea-shaped, 1.1-1.4 cm long.

Fruit: A pod, to 1 cm long, ellipsoid, leathery, dark brown, finely hairy.

Habitat and Distribution: Frequent; dry woods, flatwoods, scrub, and old fields; central peninsula Florida, west to central panhandle Florida, and north to Georgia.

Comment: The light yellow flowers can be seen in spring and summer. The leaves, lighter than Gopher Weed, do not provide a good contrast for the flowers.

Baptisia simplicifolia Croom
Scare Weed or Coastal Plain Wild-indigo
Fabaceae (Leguminosae), Pea or Bean Family

Habit: Herbaceous perennial, to 1 m tall. Stems erect, with numerous branches.

Leaves: Alternate; stalkless; blades rounded, 5-8 cm long, shiny on upper surface, notched at tip.

Inflorescences: Flowers numerous in terminal clusters.

Flowers: Pale yellow or greenish-yellow, typically pea-shaped, 1.2-1.5 cm long.

Fruit: A pod, to 1 cm long, woody, black, becoming smooth.

Habitat and Distribution: Rare; flatwoods; central panhandle Florida.

Comment: The numerous pale yellow flowers are found in mid- to late summer. The effect of the clusters of flowers and shiny leaves make Scare Weed a very attractive wildflower. Scare Weed is listed by Florida as Threatened.

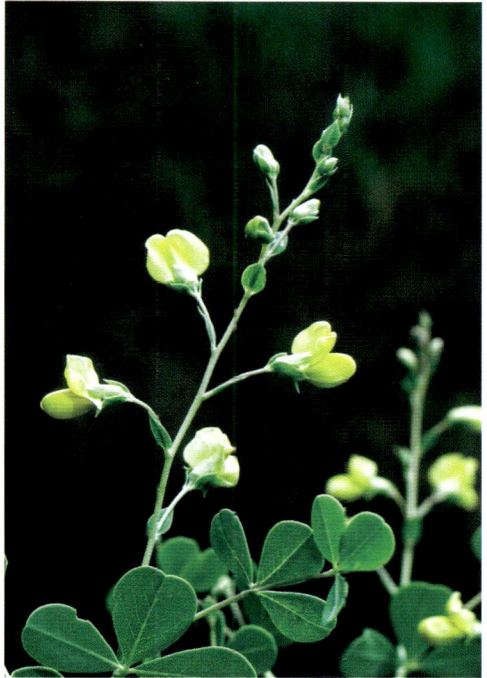

Baptisia lecontei Torr. & A. Gray

Baptisia simplicifolia Croom

Yellow

Berlandiera pumila (Michx.) Nutt.
Red-center Greeneyes or Soft Greeneyes
Asteraceae (Compositae), Aster or Sunflower Family

Habit: Herbaceous perennial, 0.2-0.7 m tall, from a fleshy taproot. Stems woolly, leafy, one or more emerging from basal cluster of leaves.

Leaves: Alternate, stalkless or stalked; blades ovate to lance-ovate, wider at base, 6-13 cm long and 2-7 cm wide, woolly and paler beneath, with blunt-toothed edges.

Inflorescences: Terminal heads, heads to 7 cm wide.

Flowers: Yellow rays, each 1-2 cm long, disc dark reddish.

Fruit: A nutlet, black, to 6 mm long, flattened, 3-4 mm wide.

Habitat and Distribution: Frequent; sandy dry soils of open woods, fields, roadsides, and sandhills; from north central peninsula Florida, west to Texas, and north to South Carolina.

Comment: Blooms can be seen from spring into fall in acid soils. The flowers are showy especially in the dry sandy soils where ground cover is sparse. Red-center Greeneyes can easily be propagated by nutlets.

Berlandiera subacaulis (Nutt.) Nutt.
Florida Greeneyes
Asteraceae (Compositae), Aster or Sunflower Family

Habit: Herbaceous perennial, 0.1-0.4 m tall, from a fleshy taproot. Stems short rough-hairy, with few or no leaves on the stem, one or more emerging from a basal cluster of leaves.

Leaves: Alternate, stalkless or stalked; blades deeply wavy-lobed, 4 to 12 cm long and to 3 cm wide, most occur at the base of the plant.

Inflorescences: Terminal heads, heads to 5 cm wide.

Flowers: Yellow rays, each 1 to 1.5 cm long, disc yellow or green.

Fruit: A nutlet, black, 5 to 6 mm long.

Habitat and Distribution: Common; sandy dry soils of pinelands, sandhills, and disturbed open areas; endemic, peninsula to central panhandle Florida.

Comment: Flowering is summer through fall, all year in the southern peninsula. This plant prefers acid soils. It is offered for sale by only a few native nurseries, but can be easily started from nutlets. The showy flowers make this plant stand out both in nature and in the garden. It can successfully compete with turf making it a successful wildflower when not mowed during the flowering season.

Yellow

Berlandiera pumila (Michx.) Nutt.

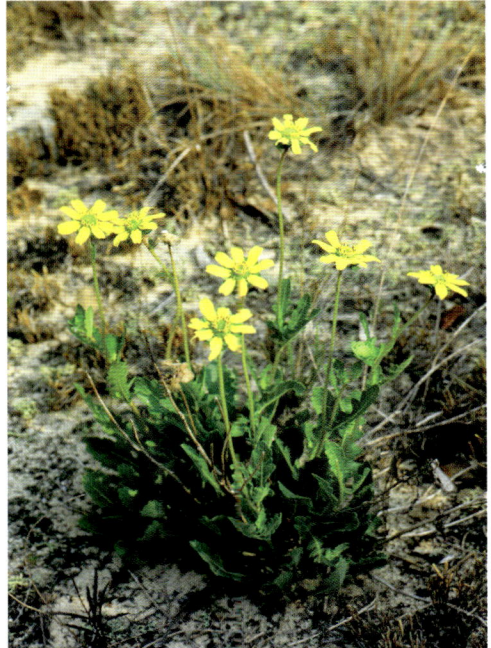

Berlandiera subacaulis (Nutt.) Nutt.

Yellow

Bidens laevis (L.) BSP.
Bur-marigold, Showy Bur-marigold, Wild Golden-glow
Asteraceae (Compositae), Aster or Sunflower Family

Habit: Herbaceous annual or perennial, 0.5 -1.5 m tall. Stems ascending, bending at base, somewhat succulent, branching in pairs.

Leaves: Opposite; stalkless; blades elliptic to lance-elliptic, base sometimes wide and clasping, 4-20 cm long, margins with teeth.

Inflorescences: Terminal heads, heads few.

Flowers: Yellow, heads 3-7 cm wide, with 8 rays, each 1.5-3 cm long, disc golden yellow.

Fruit: A nutlet, 5-8 mm long, flat or 3- or 4-angled, with 2-4 bristles at top, barbs point down.

Habitat and Distribution: Occasional, locally common; wet, open areas, and marshes; throughout Florida and the U.S. except for high mountains; south to South America.

Comment: Bur-marigold is a spectacular fall blooming wildflower. The bright yellow flowers are very showy and when the plants can be seen in mass, as is usually the case in areas with shallow water, the view is amazing.

Bidens mitis (Michx.) Sherff
Marsh Beggar-tick or Coastal Plain Tickseed-sunflower
Asteraceae (Compositae), Aster or Sunflower Family

Habit: Herbaceous annual, 0.3 -1 m tall. Stems erect, branching, smooth or slightly hairy.

Leaves: Opposite; stalked; blades divided into 3-7 slender, pointed lobes, blades 7-12 cm long, margins with teeth.

Inflorescences: Terminal heads.

Flowers: Yellow, heads 2-5 cm wide, with 8 rays, each 1-3 cm long, disc yellow to brownish.

Fruit: A nutlet, 2-5 mm long, flat or somewhat 3-angled, with or without bristles at top, barbs point up.

Habitat and Distribution: Common; moist to wet soils of marshes, cypress domes, swamps, roadsides, disturbed sites; throughout Florida, west to southeast Texas, and north to Maryland.

Comment: Blooming in all warm months, the flowers are not really showy, but provide a wonderful sprinkling of yellow among the low grassy vegetation.

Bidens laevis (L.) BSP.

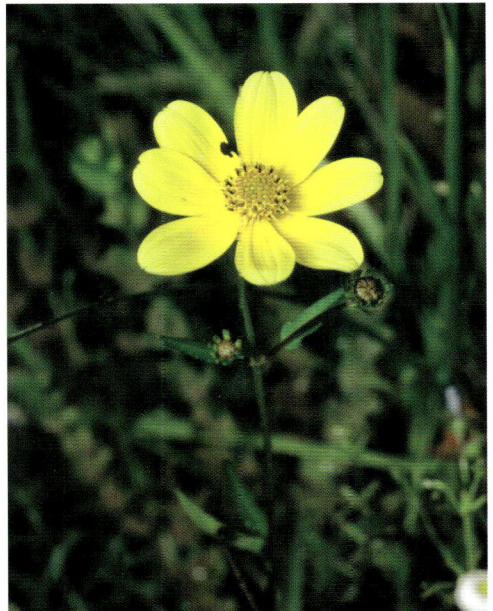

Bidens mitis (Michx.) Sherff

Yellow

Borrichia frutescens (L.) DC.
Sea Oxeye or Sea Oxeye Daisy
Asteraceae (Compositae), Aster or Sunflower Family

Habit: Herbaceous or shrubby perennial, 0.3-1.2 m tall, from underground runners. Stems erect, few or freely branching, leafy.

Leaves: Opposite; stalkless; blades mostly somewhat lance-shaped and wider toward tips, 2-8 cm long, thick, grayish, hairy, margins smooth or sometimes with small teeth towards base.

Inflorescences: Terminal heads.

Flowers: Yellow to orangish rays, 1 cm long, and a yellow or brownish central disc, 1-2 cm diameter.

Fruit: Ray flower nutlets 3-sided, disc flower 4-sided, nutlets both black, 3-4 mm long, with scales on top.

Habitat and Distribution: Common; coastal sandy areas and marshes, locally abundant inland in areas of salt accumulation; throughout Florida, west to Texas, and north to Virginia; Mexico; West Indies; Bermuda.

Comment: Flowering all year, especially southward, large colonies can range over much of the sight distance along the shore and make very colorful displays.

Canna flaccida Salisb.
Golden Canna or Bandana-of-the-everglades
Cannaceae, Canna Family

Habit: Herbaceous perennial, to 1.3 m tall, from a thick underground runner.

Leaves: Alternate; sheathing at base; blades large, elliptic to somewhat rounded, 25-60 cm long.

Inflorescences: Spike-like, terminal clusters.

Flowers: Yellow, to 7 cm long, with 3 large spreading petals.

Fruit: A capsule, 3-lobed, 4-6 cm long and 2-3 cm wide, with a bumpy surface; seeds round.

Habitat and Distribution: Frequent to locally common; swamps, marshes, lakes, spring runs, and wet ditches; throughout Florida, west to Texas, and north to South Carolina.

Comment: Flowers occur in the warm months. Large clusters of blooming plants provide a marvelous show of color. Golden Canna is becoming more frequently available for gardens. Mitigation plantings around retention ponds and other areas of frequent pedestrian and vehicle traffic are providing wildflower displays for the betterment of communities. Propagation can be by dividing the underground runners or from seeds that have been notched or soaked in warm water.

Borrichia frutescens (L.) DC.

Canna flaccida Salisb.

Yellow

Canna x generalis L.H. Bailey
Garden Canna
Cannaceae, Canna Family

Habit: Herbaceous perennial, to 1.8 m tall, from a thick underground runner. Stems waxy.

Leaves: Alternate; sheathing at base; blades large, elliptic to somewhat rounded, 30-50 cm long.

Inflorescences: Spike-like, terminal clusters.

Flowers: Yellow with splotches of orange, orange with red markings, or red, to 7 cm long, with 3 large petals, with smooth or undulating edges.

Fruit: A capsule, 3-lobed, 2-3.5 cm long, about as wide.

Habitat and Distribution: Infrequent, as an escape; disturbed low and wet sites, ditches, abandoned home sites, and vacant lots; throughout Florida, north to North Carolina, and west into Mississippi, to be expected in most southeastern states. A garden hybrid whose parents are native to South America.

Comment: As an escape from cultivation, usually only a few plants of Garden Canna are found at a location. The flowers, seen in warm months, are very attractive. The colors are extremely variable ranging from almost yellow to splotches of yellow among oranges and reds. Propagation can be by dividing the underground runners or from seeds that have been notched or soaked in warm water.

Chamaecrista fasciculata (Michx.) Greene
Partridge Pea
[*Cassia fasciculata* Michx.]
Fabaceae (Leguminosae), Pea or Bean Family

Habit: Herbaceous annual, 0.1-1.0 m tall, from a taproot. Stems erect or spreading, reddish, with fine hairs or usually coarsely hairy.

Leaves: Alternate; stalked, with a stalkless to stalked, saucer-shaped gland below lowest pair of leaflets; blades with 12-36, oblong-linear leaflets, 1-2.5 cm long and 2-6 mm wide.

Inflorescences: Axillary, 2- to 7-flowered clusters, one flower opening at a time.

Flowers: Yellow, 2-4 cm across, with 5 nearly equal petals, the 4 upper petals with a red spot at base.

Fruit: A pod, 3-7 cm long and 4-7 mm side, smooth or hairy.

Habitat and Distribution: Common; pinelands, sandhills, open woodlands, fields, fence rows, roadsides, and other disturbed sites; throughout Florida, west to Texas and South Dakota, and north to Massachusetts and Minnesota.

Comment: The large showy flowers are seen in late spring through the summer. The flowers wilt by the middle of the day. Partridge Pea is cultivated as an ornamental and is extensively planted for cover on open disturbed soils. The roots contain nodules which help provide soil nitrogen. Propagation is by seeds.

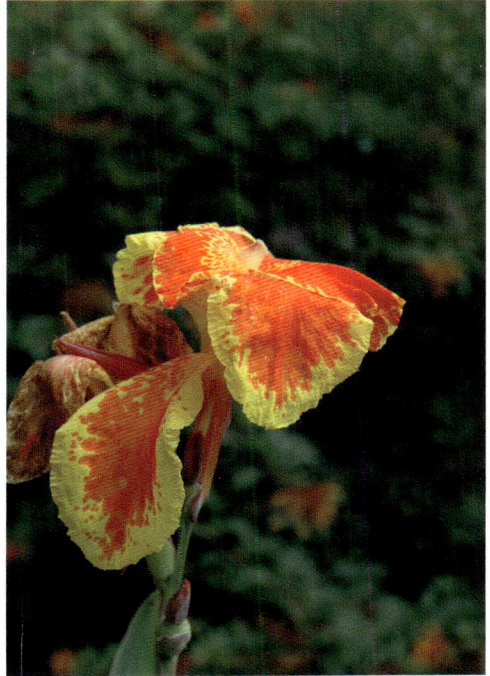

Canna x generalis L.H. Bailey

Chamaecrista fasciculata
(Michx.) Greene

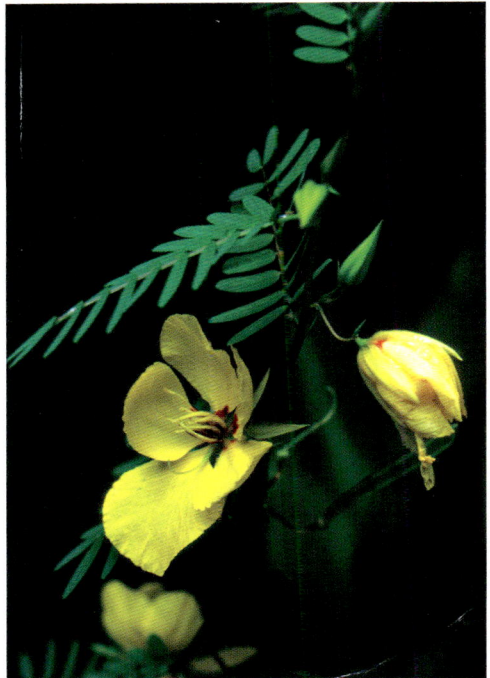

319

Yellow

Chamaecrista nictitans (L.) Moench
Wild Sensitive Plant or Sensitive Pea
[*Cassia nictitans* L.]
Fabaceae (Leguminosae), Pea or Bean Family

Habit: Herbaceous annual, 10-70 cm tall, from a taproot. Stems with fine to long hairs, branched.

Leaves: Alternate, sensitive and folding when touched; stalked, with a thin, stalked gland below the lowest pair of leaflets; blades once-divided, with 14-50, oblong-linear leaflets, 6-15 mm long and 1-3.5 mm wide.

Inflorescences: Axillary, in 1- to 3-flowered clusters.

Flowers: Yellow, to 1 cm wide, with 5 unequal petals, one much larger than others.

Fruit: A pod, 2-5 cm long, 3-5.5 mm wide, usually hairy.

Habitat and Distribution: Frequent to common; pinelands, sandhills, dunes, open woods, fields, roadsides, and other dry disturbed sites; throughout Florida, north to Massachusetts and southern Vermont; west to Ohio, Missouri, Kansas, Texas, and southern Arizona; south into Argentina.

Comment: Flowering occurs from late spring into fall. The flowers are attractive, but not at all showy. Flowers wilt by midday. Propagation is by seeds.

Chapmannia floridana Torr. & A. Gray
Chapman's Pea or Alicia
Fabaceae (Leguminosae), Pea or Bean Family

Habit: Herbaceous perennial, 40-90 cm tall. Stems slender, with sticky hairs.

Leaves: Alternate; stalked; blades once-divided, leaflets usually 5-9, occasionally 3, elliptic to oblong, 0.5-2 mm long.

Inflorescences: Alternate along a terminal, unbranched stalk.

Flowers: Golden yellow, to 1.4 cm long, with 5 petals.

Fruit: An ovoid to oblong pod, nearly round, 1-4 cm long, with 1-4 constrictions between seeds, sticky hairy.

Habitat and Distribution: Frequent; scrub, sandhills, dry pinelands, and dry roadsides; peninsula Florida.

Comment: Flowers can be found from spring into late summer. Chapman's Pea plants are normally taller than the surrounding grasses providing attractive golden highlights dancing above the dull green ground cover.

Chamaecrista nictitans
(L.) Moench

Chapmannia floridana
Torr. & A. Gray

Yellow

Chiococca alba (L.) Hitchc.
Snowberry
Rubiaceae, Madder Family

Habit: Shrub, to 3 m tall. Stems erect, spreading or climbing, smooth.

Leaves: Opposite; stalked; blades ovate to lance-shaped, 1-12 cm long.

Inflorescences: Axillary clusters.

Flowers: Greenish-white to golden yellow, tubular, to 1 cm long, with 5 petals.

Fruit: A round drupe, white, 4-7 mm diameter.

Habitat and Distribution: Frequent; coastal hammocks and margins of thickets; coastal peninsula Florida, Texas; West Indies, Mexico south through Central America.

Comment: Flowering all year, Snowberry is an outstanding wildflower and ornamental. The flowers are very attractive, but the bright white fruits are exceptionally showy. This shrub can be purchased from many nurseries that offer native species.

Chrysogonum virginianum L. var. *australe* (Alex. ex Small) Ahles
Green-and-gold
[*Chrysogonum australe* Alex. ex Small]
Asteraceae (Compositae), Aster or Sunflower Family

Habit: Herbaceous perennial, to 10 cm tall, with stolons. Stems erect, hairy.

Leaves: Opposite; stalked; blades rounded, 2-10 cm long and 1.5-6 cm wide, hairy, with blunt teeth on the margins.

Inflorescences: Solitary heads, terminal or axillary, often long-stalked, heads 2-3.5 cm wide.

Flowers: 5 yellow rays, 0.7-1.5 cm long, and yellow disc flowers.

Fruit: A dark brown to black nutlet, round and flattened, 3-3.5 mm long, 1.5-2 mm wide, hairy, scales attached at base, a tiny cartilaginous crown at tip.

Habitat and Distribution: Infrequent (rare in Fla.); sandy woods; panhandle Florida, north to North Carolina and Tennessee, west to eastern Alabama.

Comment: Green-and-gold can provide numerous blooms from late spring into fall. Easily obtainable from nurseries it is used as a bedding plant. As a wildflower, the showy flowers are sprinkled through the woods. It is propagated by nutlets or from the stolons.

Chiococca alba (L.) Hitchc.

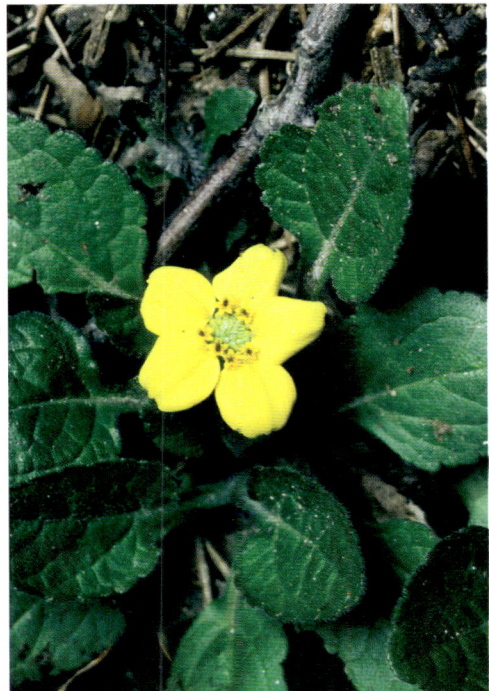

Chrysogonum virginianum L. var. *australe* (Alex. ex Small) Ahles

Yellow

Chrysoma pauciflosculosa (Michx.) Greene
Bush Goldenrod
[*Solidago pauciflosculosa* Michx.]
Asteraceae (Compositae), Aster or Sunflower Family

Habit: Shrub 0.5-1.5 m tall. Stems freely branching from short stocky trunk, leafy, newer twigs sticky.

Leaves: Evergreen, alternate; stalked; blades narrowly oblong to lanceolate with a broader tip, 2-6 cm long and 0.3-1 cm wide, conspicuously pitted-reticulate.

Inflorescences: Terminal, branched clusters of heads.

Flowers: Rays 0-3, yellow, 4-6 mm long, disc flowers 1-5.

Fruit: Nutlet; 2.3-4 mm long, hairy, with tuft of bristles at tip.

Habitat and Distribution: Frequent; sandy open woodlands, scrub, dunes, and sandhills; panhandle Florida, west to Mississippi, and north to North Carolina.

Comment: Late spring through summer blooming, this shrub is quite attractive. The evergreen foliage is distinctive and the flower clusters are eye-catching. Bush Goldenrod is a great wildflower, especially so in the sparsely vegetated habitats in which it grows.

Chrysopsis gossypina (Michx.) Elliott
Hairy Golden Aster or Cottony Golden Aster
[*Heterotheca gossypina* (Michx.) Shinners]
Asteraceae (Compositae), Aster or Sunflower Family

Habit: Herbaceous biennial or very short-lived perennial, 30-80 cm tall. Stems sometimes erect or usually spreading, emerging in tufts, woolly.

Leaves: Alternate; stalkless; blades lance-shaped and broadest at the tip or oblong-elliptic, 3-8 cm long and 1-1.8 cm wide, woolly, upper leaves smaller.

Inflorescences: Terminal, flat-topped, branched, few- to many-flowered clusters.

Flowers: Rays yellow, 8-20, 0.4-1.2 cm long.

Fruit: A cream to brown, linear nutlet, 1.9-2.5 mm long, densely hairy, with tuft of bristles at tip.

Habitat and Distribution: Occasional to frequent; sandy pine or oak woods, sandhills, dunes, and scrub; central peninsula Florida, west to Texas and Oklahoma, and north to Virginia.

Comment: Fall flowering, but noted throughout the summer for the attractive woolly foliage. Closely related varieties occur with smaller heads and different types of hairs.

324

Chrysoma pauciflosculosa
(Michx.) Greene

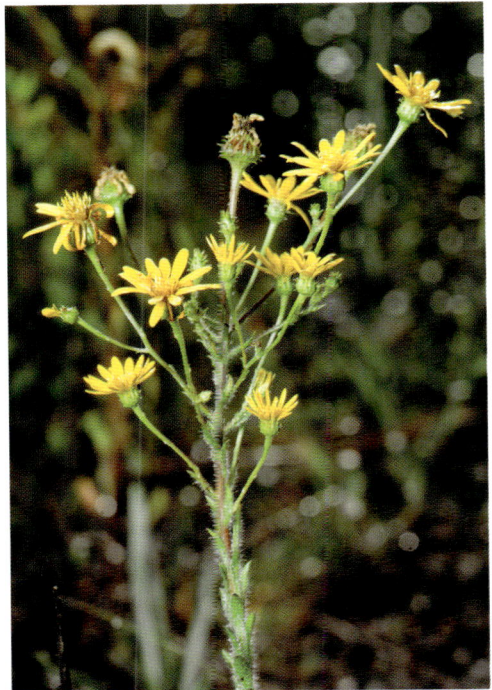

Chrysopsis gossypina
(Michx.) Elliott

Yellow

Chrysopsis mariana (L.) Elliott
Maryland Golden Aster
Asteraceae (Compositae), Aster or Sunflower Family

Habit: Herbaceous perennial, to 1 m tall. Stems erect, shaggy hairy with fine long hairs.

Leaves: Alternate; lower leaves stalked, upper leaves stalkless; blades lanceolate to ovate, broadest at tip, 3-18 cm long and 1-3.5 cm broad, loosely hairy, margins toothed.

Inflorescences: Terminal, flat-topped, branched, few- to many-flowered, glandular clusters of heads.

Flowers: Rays yellow, 13-21, about 1 cm long, narrow.

Fruit: A reddish brown, rounded nutlet, 2-3 mm long, with a tuft of bristles at tip.

Habitat and Distribution: Frequent southward, infrequent elsewhere; dry pinelands, sandhills, and dry woods; central peninsula Florida, west to Texas, and north to Ohio and New York.

Comment: Flowers can be found from summer to fall and occasionally all year at the southernmost locations. Periodically Maryland Golden Aster occurs in enough numbers to make a show, particularly in open habitats.

Chrysopsis scabrella Torr. & A. Gray
Roughleaf Golden Aster or Coastal Plain Golden Aster
[*Chrysopsis floridana* Small; *Chrysopsis lanuginosa* Small; *Heterotheca scabrella* (Torr. & A. Gray) R.W. Long]
Asteraceae (Compositae), Aster or Sunflower Family

Habit: Herbaceous biennial or short-lived perennial, 0.2-1 m tall. Stems erect or spreading, leafy, with stalked, sticky, rough, glandular hairs, and also often loose-woolly hairy.

Leaves: Alternate; stalkless; blades narrowly lanceolate to spatulate, 3-8 cm long, hairy, rough glandular, upper leaves smaller.

Inflorescences: Terminal, flat-topped, branched, few- to many-flowered, glandular clusters of heads.

Flowers: Rays narrow, yellow, about 21, approximately 1 cm long.

Fruit: An obovate nutlet, 2.5-3 mm long, with a tuft of bristles at tip.

Habitat and Distribution: Frequent; pinelands and scrub; throughout Florida and west to Alabama.

Comment: Blooming plants can be found in fall. The rough glandular appearance of this Golden Aster is not very attractive, even when in flower.

Yellow

Chrysopsis mariana (L.) Elliott

Chrysopsis scabrella
Torr. & A. Gray

Yellow

Coreopsis basalis (A. Dietr.) S.F. Blake
Dyeflower or Calliopsis
Asteraceae (Compositae), Aster or Sunflower Family

Habit: Herbaceous annual or biennial, 10-70 cm tall. Stems single and erect or several and laying down at base, freely branching, smooth to hairy.

Leaves: Opposite; stalked; blades undivided and deeply lobed into narrow segments or divided with 1-3 pairs of leaflets, leaflets each 1-5 cm and 0.5-2 cm wide.

Inflorescences: Terminal, loose clusters of heads.

Flowers: Rays yellow, 1-2.5 cm long, 1-2 cm wide, with reddish bases and toothed tips, disc flowers dark red.

Fruit: A dark brown nutlet, 1-2 mm long, almost as wide, without wings or bristles.

Habitat and Distribution: Occasional to common; dry, open, sandy disturbed habitats, roadsides, fields; native to Texas and Southwest Louisiana, now found from central peninsula Florida, west to Texas, and north to North Carolina.

Comment: Dyeflower is one of our most spectacular spring through summer wildflowers, often occurring in masses. Large populations covering acres are eye-popping. Propagation is by nutlets.

Coreopsis floridana E.B. Sm.
Florida Tickseed
Asteraceae (Compositae), Aster or Sunflower Family

Habit: Herbaceous perennial, to 1.2 m tall. Stems relatively thick, smooth, few branches.

Leaves: Alternate; stalked, base clasps stem; blades elliptic to lance-shaped, sometimes narrowly so, to 15 cm long, gradually reduced in size upwards on stems, thick, margins smooth.

Inflorescences: Loose, terminal, few-flowered clusters of large heads.

Flowers: Rays yellow to orange, 1.5-3 cm long, 0.7 to 1.2 cm wide, with toothed tips, disc dark brownish.

Fruit: A grayish brown nutlet, to 5 mm long, warty, with winged margins, and 2 bristles at tip.

Habitat and Distribution: Frequent; endemic throughout Florida, except the extreme western panhandle (to Walton County); wet pinelands, bogs, wet prairies, and cypress depressions.

Comment: Florida Tickseed is a relatively large succulent wildflower. The large flowers are easily noticed from August into November in wet habitats. Propagation is by nutlets.

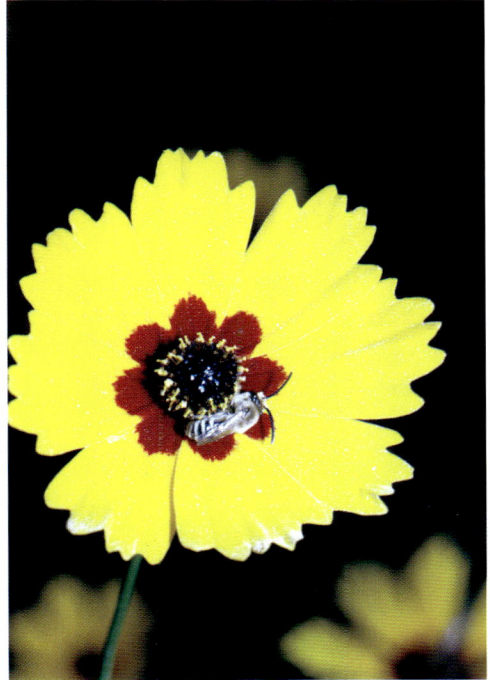

Coreopsis basalis
(A. Dietr.) S.F. Blake

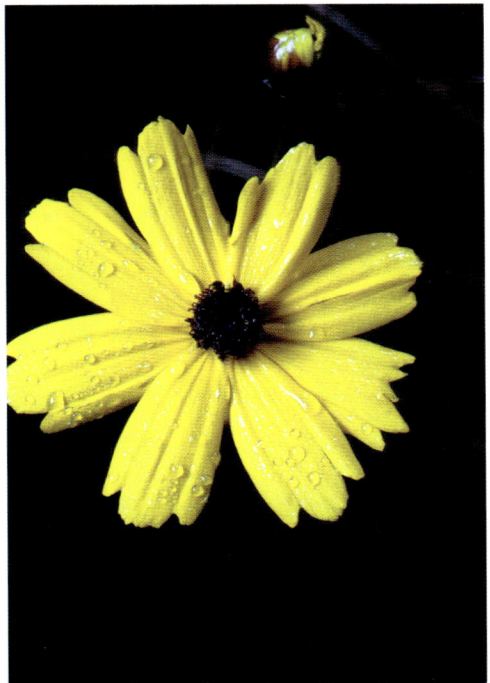

Coreopsis floridana E.B. Sm.

Yellow

Coreopsis lanceolata L.
Lanceleaf Tickseed
Asteraceae (Compositae), Aster or Sunflower Family

Habit: Herbaceous perennial, 20-70 cm tall. Stems erect or spreading, leafy, smooth, or hairy.

Leaves: Opposite; stalked on lower plant, stalkless on upper plant; blades spatulate to narrowly linear-lance shaped, 3-15 cm long, sometimes lobed.

Inflorescences: Heads terminal, solitary or few, on long, thick stalks.

Flowers: Rays yellow, to 3 cm long, disc yellow.

Fruit: A black nutlet, with flattened wings, 2.3-3 mm long, with or without tiny papery teeth at tip.

Habitat and Distribution: Occasional; sandhills, roadsides, sandy woods, and disturbed areas; central peninsula Florida, north to Virginia and southern Ontario, south to Texas and New Mexico.

Comment: Lanceleaf Tickseed blooms in the spring. Probably native to the central and southwestern states, it is now widely cultivated and frequently escapes from cultivation. Propagation is by nutlets and cuttings.

Coreopsis leavenworthii Torr. & A. Gray
Common Tickseed
Asteraceae (Compositae), Aster or Sunflower Family

Habit: Herbaceous annual or short-lived perennial, to 1.5 m tall. Stems usually single, branched.

Leaves: Opposite; stalked on lower plant, stalkless on upper plant; blades to 15 cm long, linear to elliptic, some divided with 2-6 lobes, usually unlobed.

Inflorescences: Heads one to many, terminal.

Flowers: Rays yellow, 1-2 cm long with 3-lobed tips, disc flowers blackish purple, 4-lobed.

Fruit: A brown to purple nutlet, to 3.5 mm long, with broad wings, and 2 bristles at tip.

Habitat and Distribution: Endemic and common; wet soils of pine flatwoods, prairies, and especially ditches and roadsides; throughout peninsular Florida to Walton County in the panhandle.

Comment: Common Tickseed can be found flowering in any warm month. Florida has all too few roadside wildflowers that provide a show. This Tickseed is quite common and can be found in abundance along roads, sometimes providing large long bands of yellow. Propagation is best by nutlets.

Coreopsis lanceolata L.

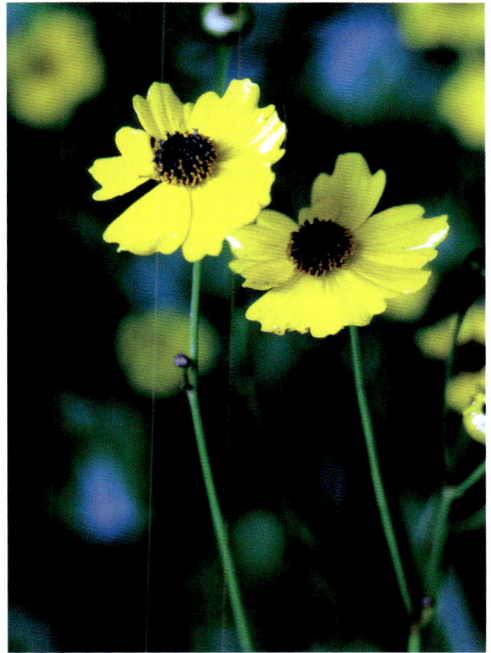

Coreopsis leavenworthii
Torr. & A. Gray

Yellow

Coreopsis linifolia Nutt.
Texas Tickseed
[*Coreopsis gladiata* Walter var. *linifolia* (Nutt.) Cronq.]
Asteraceae (Compositae), Aster or Sunflower Family

Habit: Herbaceous perennial, to 1 m tall, from a rootstock. Stems usually solitary, smooth.

Leaves: Alternate on lower plant, opposite upwards; stalked on lower plant, stalkless on upper plant; blades narrowly elliptical, especially so upwards, to 10 cm long, fresh leaves contain minute dark dots when held to light, margins smooth.

Inflorescences: Terminal heads.

Flowers: Rays yellow, 1-2 cm long, with 3-lobed tips, disc flowers dark purplish red, 4-lobed.

Fruit: A grayish brown nutlet, to 2 mm long, bumpy, with narrow lacerate wings, and two bristles at tip.

Habitat and Distribution: Occasional; wet pine flatwoods, bogs, bays, and wet ditches; northern peninsula Florida, west to Texas, and north to Virginia.

Comment: Texas Tickseed flowers from spring into early summer westward and will continue blooming into fall in most areas. Flower heads tend to be few so that this wildflower, while being quite attractive, does not provide much of a show. Propagation is by nutlets, division, or cuttings.

Corydalis micrantha (Englem. ex A. Gray) A. Gray
Harlequin or Slender Corydalis
Fumariaceae, Fumitory Family

Habit: Herbaceous winter annual, 10-30 or occasionally to 60 cm tall. Stems branching from the base, erect or spreading.

Leaves: Alternate; stalked on lower plant, stalkless on upper plant; blades to 15 cm long, finely divided into linear segments, feathery in appearance.

Inflorescences: Flowers along terminal stalks that extend above the leaves.

Flowers: Bright yellow, tubular, 11-15 mm long, with straight spur at base.

Fruit: A cylindrical capsule, 1.5-3 cm long, erect.

Habitat and Distribution: Frequent; sandy disturbed habitats, bluffs, rocky hills, open woods, and river banks; central peninsula Florida, west to northeast Texas, north to Kansas, Illinois, and Minnesota, east to North Carolina.

Comment: The stalks of bright yellow flowers occur in winter to early spring. Usually a large number of Harlequins are massed in a small area. The resulting display is a show stopper. The display will attract all wildflower watchers in the vicinity to stop, photograph, and exclaim.

Coreopsis linifolia Nutt.

Corydalis micrantha
(Englem. ex A. Gray) A. Gray

Yellow

Croptilon divaricatum (Nutt.) Raf.
Scratch Daisy
[*Haplopappus divaricatus* (Nutt.) A. Gray]
Asteraceae (Compositae), Aster, Sunflower, or Daisy Family

Habit: Herbaceous annual or rarely biennial, 0.3-1.5 m tall, from a taproot. Stems branching, hairy, rough.

Leaves: Alternate; stalkless; blades lance-shaped and broader at the tip, 2.5-10 cm long and 0.3-1.2 cm broad, with spiny tips and toothed edges, smaller on upper stems.

Inflorescences: Widely branched, loose clusters of heads from leaf axils at stem tips.

Flowers: Yellow, 5-7 mm high, with 7-11 narrow rays.

Fruit: A thickened, ellipsoid nutlet, 1.9-2.5 mm long, hairy, with a fan of bristles 4-5 mm long at tip.

Habitat and Distribution: Frequent and locally common; dry pinelands, open fields, and disturbed sites; central peninsula Florida, west to Texas, Arkansas, Kansas, and Oklahoma, and north to Virginia.

Comment: The summer and fall flowers are extremely showy due to the masses of flowers produced, especially from large populations. Sandy clearings can be completely filled with Scratch Daisies.

Crotalaria lanceolata E. Mey.
Lanceleaf Crotalaria
Fabaceae (Leguminosae), Pea Family

Habit: Herbaceous annual, to 1.5 m tall, from a taproot. Stems erect, sparingly branched, with minute hairs.

Leaves: Alternate; stalked; blades with 3 lance-shaped to nearly linear leaflets, each 2-9 cm long and 6-10 mm wide.

Inflorescences: Spike-like, cylindrical clusters at stem tips.

Flowers: Yellow, narrow, pea-shaped, 0.6-0.9 cm long.

Fruit: A narrowly cylindrical pod, 2-4 cm long, 4-7 mm diameter.

Habitat and Distribution: Frequent to locally common; disturbed sites, roadsides, old fields, field margins, and fence rows; Florida north to South Carolina; native to Africa.

Comment: The spike-like clusters of flowers occur from summer into fall. The flowers are not as showy as other Crotalarias. Lanceleaf Crotalaria often occurs as a weed in row crops. TOXIC to all livestock and fowl.

Croptilon divaricatum (Nutt.) Raf.

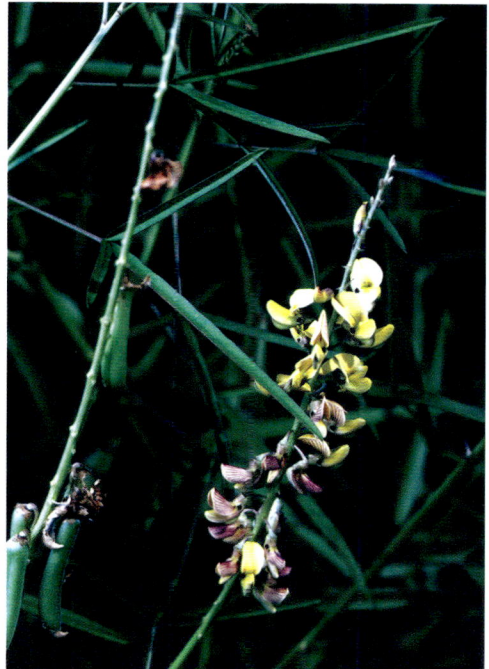

Crotalaria lanceolata E. Mey.

Yellow

Crotalaria ochroleuca G. Don
Slenderleaf Crotalaria
[*Crotalaria brevidens* Benth., misapplied; *Crotalaria intermedia* Kotschy, misapplied]
Fabaceae (Leguminosae), Pea Family

Habit: Herbaceous annual, to 1.5 m tall, from a taproot. Stems erect, with very short hairs.

Leaves: Alternate; stalked; blades with 3 lance-shaped to linear leaflets, each 4-12 cm long and 4-12 mm wide.

Inflorescences: Terminal, spike-like, cylindrical clusters.

Flowers: Yellow, typically pea-shaped, 1.8-2 cm long.

Fruit: An inflated cylindrical pod, 3-5 cm long and about 1.5 cm diameter.

Habitat and Distribution: Occasional to locally common; disturbed areas, fields, roadsides, and vacant lots; central peninsula Florida, north to North Carolina, and west to Louisiana; native of Africa.

Comment: The large flowers in summer make this a showy wildflower. It is easily identified by the large dense spike of flowers and the narrow leaflets. Slenderleaf Crotalaria often occurs as a weed in row crops. This species is TOXIC to all livestock and fowl.

Crotalaria pallida Aiton
Smooth Crotalaria
[*Crotalaria mucronata* Desv.; *Crotalaria striata* DC., misapplied]
Fabaceae (Leguminosae), Pea Family

Habit: Herbaceous annual, 1-2 m tall, from a taproot. Stems erect, hairy.

Leaves: Alternate; stalked; blades with 3 obovate leaflets, 2-7 cm long, smooth above with sparse hairs below.

Inflorescences: Terminal, dense, spike-like clusters.

Flowers: Yellow, pea-shaped, 1.1-1.5 cm long.

Fruit: An upwardly curved, turgid pod, 3-4.5 cm long and 5-6 mm in diameter.

Habitat and Distribution: Frequent; disturbed sites with sandy soils, old fields, roadsides, field margins, and vacant lots; throughout Florida, north to North Carolina, and west to Alabama, Mississippi, and Louisiana; probably a native of Africa, now pantropical.

Comment: Blooming in all warm months, Smooth Crotalaria is a good dependable wildflower in disturbed sandy areas. It often occurs as a weed in row crops. TOXIC to all livestock and fowl.

Crotalaria ochroleuca G. Don

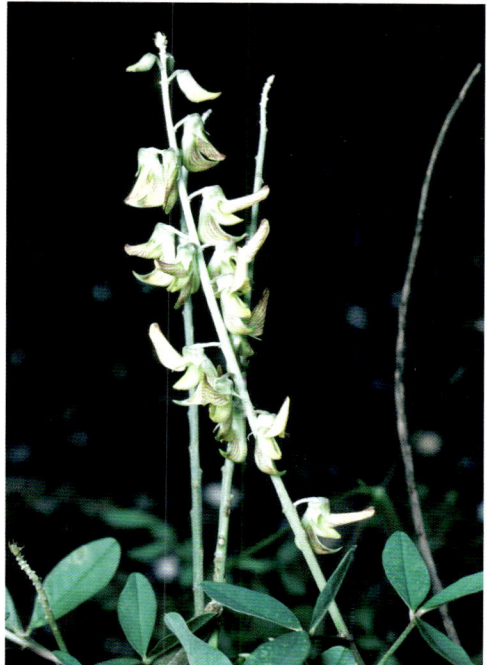

Crotalaria pallida Aiton

Yellow

Crotalaria pumila Ortega
Small Crotalaria
Fabaceae (Leguminosae), Pea Family

Habit: Herbaceous annual or shrubby perennial, to 30 cm tall when low spreading, to 1 m tall when erect. Stems slender, erect or low spreading, solitary or clustered, many branches, sparsely hairy.

Leaves: Alternate; lower leaves stalked; blades with 3 obovate to oblong leaflets; 0.5-3 cm long.

Inflorescences: Short, spike-like clusters of 1-8 flowers, terminal or from leaf axils.

Flowers: Yellow, often orange- or red-tinged, typically pea-shaped, to 1.1 cm long.

Fruit: A cylindrical pod, yellowish, 1-1.5 cm long.

Habitat and Distribution: Frequent; oak woodlands, hammocks, pinelands, grasslands, pinyon-juniper, dunes, and along streams and roadsides; central and southern peninsula Florida, disjunct to western Texas and Arizona; Mexico south into northern South America; Antilles.

Comment: Flowers appear from late spring into fall. Small Crotalaria rarely is a wildflower of notice. The one to few flowers that can be seen at any one time mostly blend into the background among the other vegetation. TOXIC to all livestock and fowl.

Crotalaria purshii DC.
Pursh's Rabbit-bells
Fabaceae (Leguminosae), Pea Family

Habit: Herbaceous perennial, 10-50 cm tall, from a woody taproot. Stems simple or branched, hairy.

Leaves: Alternate; short-stalked, conspicuously winged for most of the length; blades simple, elliptic to oblong, 3-6 cm long and 4-10 mm wide.

Inflorescences: Terminal, spike-like clusters of 2-6 flowers.

Flowers: Yellow, typically pea-shaped, about 1 cm long.

Fruit: A cylindrical pod, 1.5-2.5 cm long and about 1 cm diameter.

Habitat and Distribution: Occasional; dry pinelands and sandhills; central peninsula Florida, west to Louisiana, and north to Virginia.

Comment: Scattered in dry pinewoods habitats, the flowers can be found in spring and summer. Usually, the leaning stems are not hard to spot when in flower as the vegetation in its habitats is quite open. Pursh's Rabbit-bells is TOXIC to all livestock and fowl.

Crotalaria pumila Ortega

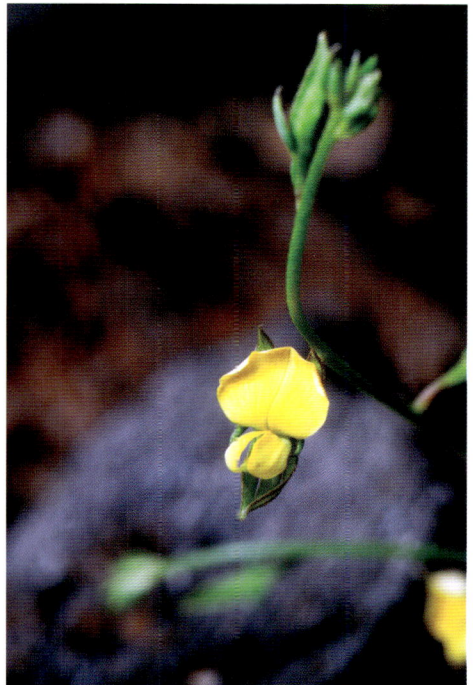

Crotalaria purshii DC.

Yellow

Crotalaria retusa L.
Rattle Weed
Fabaceae (Leguminosae), Pea Family

Habit: Herbaceous annual, 30-90 cm tall, from a woody taproot. Stems erect, with minute hairs, often somewhat silvery.

Leaves: Alternate; short-stalked; blades simple, rounded, wider at tips, 3-8 cm long.

Inflorescences: Terminal, 5- to 20-flowered stalks.

Flowers: Yellow, large, typically pea-shaped, 2-3 cm long.

Fruit: A cylindrical pod, 2.5-4 cm long and 1-1.5 cm in diameter.

Habitat and Distribution: Occasional; old fields, roadsides, and vacant lots; primarily south peninsula Florida, scattered north along the coastal plain into North Carolina, disjunct into east Texas; a native of Asian tropics, now pantropical.

Comment: Cultivated for the large showy flowers, seen from summer into fall, Rattle Weed is mostly an occasional escape. It has become better established in southern Florida. Rattle Weed is TOXIC to all livestock and fowl.

Crotalaria rotundifolia J.F. Gmel.
Rabbit-bells
Fabaceae (Leguminosae), Pea Family

Habit: Herbaceous perennial, to 30, occasionally 70, cm long. Stems lying flat or ascending, branched, hairy.

Leaves: Alternate; nearly stalkless; blades simple, rounded to oblong to sometimes linear, 1-5 cm long, hairy.

Inflorescences: Terminal, loose clusters of 2 to 5 flowers.

Flowers: Yellow, typically pea-shaped, 7-13 mm long.

Fruit: A cylindrical pod, 1.5-2.5 cm long and 7-12 mm in diameter.

Habitat and Distribution: Common; sandy soils of dry woods, fields, beaches, dunes, and lawns; throughout Florida, west to Louisiana, and north to North Carolina.

Comment: Rabbit-bells can be seen flowering during the warm months. As a wildflower it is usually noticed when appearing in turf or along other areas with low vegetation. Rabbit-bells is TOXIC to all livestock and fowl.

Crotalaria retusa L.

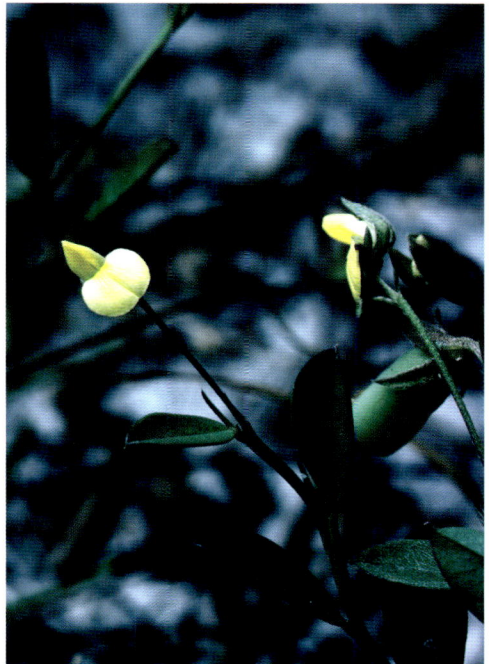

Crotalaria rotundifolia J.F. Gmel.

Yellow

Crotalaria spectabilis Roth
Showy Crotalaria
Fabaceae (Leguminosae), Pea Family

Habit: Herbaceous annual, 0.5-2 m tall. Stems erect, often branched, leafy, with fine hairs.

Leaves: Alternate; short-stalked or stalkless; blades simple, large, rounded, wider toward tips, to 20 cm long, with persistent noticeable rounded flaps beneath each leaf attachment.

Inflorescences: Terminal, numerously-flowered stalks.

Flowers: Yellow, large, typically pea-shaped, 1.7-2.5 cm long.

Fruit: A cylindrical pod, 3-4.5 cm long and 1-2 cm in diameter.

Habitat and Distribution: Common; dry sandy soils of pinelands, fields, roadsides, margins, and open disturbed woodlands; throughout Florida, west to Missouri and Texas, and north to Illinois and Virginia; a native of southern Asia, now pantropical.

Comment: Showy Crotalaria is as good as its name sounds. The large yellow flowers are hard to miss. It has become quite weedy frequently appearing in old fields as a large spectacular mass of yellow. Showy Crotalaria is TOXIC to all livestock and fowl.

Descurainia pinnata (Walter) Britton
Pinnate Tansy-mustard
Brassicaceae (Cruciferae), Mustard Family

Habit: Herbaceous winter annual, 20-80 cm tall. Stems erect, with gray hairs.

Leaves: Alternate; stalked; blades 1-10 cm long and 0.5-2 cm broad, lower blades finely twice-divided and lobed, upper blades once-divided, both feathery in appearance and densely gray hairy.

Inflorescences: Terminal, loose, cylindrical clusters, to 25 cm long.

Flowers: Yellowish, small, 2-4 mm wide, with 4 petals.

Fruit: A dry, 2-parted, club-shaped to oblong pod, 5-20 mm long and 1-2 mm wide.

Habitat and Distribution: Common; open disturbed sites, washes, sandy fields, slopes, and open woods; central peninsula Florida, throughout North America, Canada south into northern Mexico.

Comment: Flowering in spring, Pinnate Tansy-mustard is an attractive small wildflower. The gray foliage is a good contrast for the yellow flowers. When extensively grazed this plant can be TOXIC to cattle.

Crotalaria spectabilis Roth

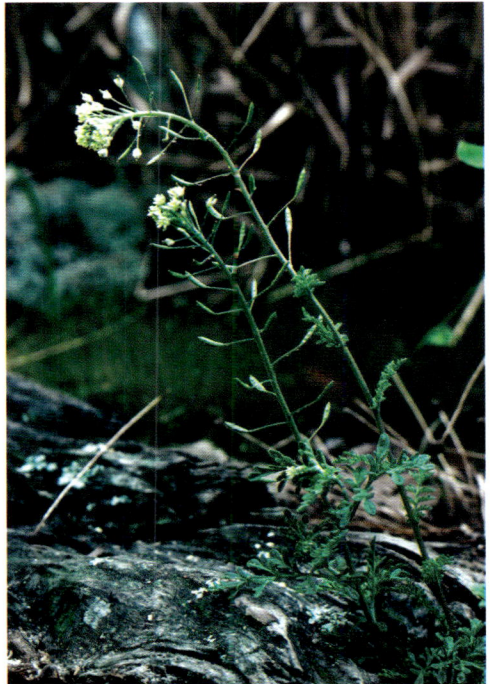

Descurainia pinnata
(Walter) Britton

Yellow

Diospyros virginiana L.
Persimmon
Ebenaceae, Ebony Family

Habit: Deciduous tree, usually to 15 m or occasionally to 30 m tall, often colonial from root shoots. Twigs and leaves are usually soft hairy when young.

Leaves: Alternate; stalked; blades elliptic to ovate, usually 7-15 cm long and 3-7 cm wide.

Inflorescences: Male and female flowers on separate trees in leaf axils, male flowers clustered, female flowers solitary.

Flowers: Inconspicuous, yellowish, vase-shaped, and 4- or 5-lobed.

Fruit: A rounded berry, up to 4 cm in diameter, with 3-8 large seeds.

Habitat and Distribution: Common; dry to wet sandy soils of pinelands, pastures, old fields, deciduous woods, clearings, sandhills, edges of shallow marshes; southern Connecticut and Long Island south to southern Florida, west to Pennsylvania to southeastern Iowa, and eastern Texas.

Comment: The tiny flowers appear in spring, but are seldom noticed. In addition to the attractive orange fruit the leaves turn red purple in the fall. The immature fruit is astringent and puckery, but can become sweet when ripe. Fruit is quite variable in quality and is often not palatable until affected by an early frost. Wild animals are very fond of the fruits, especially raccoon and opossum. The heavy, hard, close grained wood is prized for special uses such as shuttles and mallets.

Duchesnea indica (Andrews) Focke
Mock Strawberry or Indian Strawberry
Rosaceae, Rose Family

Habit: Herbaceous perennial, from a short aboveground runner. Stems creeping, hairy.

Leaves: Alternate; stalked; blades with 3 elliptic to oval leaflets, leaflets wider toward tips, 2-8 cm long, 1-4 cm wide, margins with blunt teeth.

Inflorescences: Solitary flowers in leaf axils.

Flowers: Yellow, 1.5-1,8 cm wide, with 5 petals.

Fruit: An aggregate with seeds on surface (resembling a strawberry), red, fleshy, about 1 cm in diameter.

Habitat and Distribution: Occasional, locally common; lawns, pastures, open woods, seepage areas, marshes, and open disturbed sites; central peninsula Florida, west to eastern Texas and California, and scattered north throughout most of the eastern United States; native to India.

Comment: The showy flowers seen from early spring until frost occur with the bright red fruits during late spring into fall. The combination is very ornamental, provoking frequent comments. The strawberry-like fruits are not poisonous, but do not taste good.

Diospyros virginiana L.

Duchesnea indica (Andrews) Focke

Yellow

Echites umbellata Jacq.
Rubber Vine or Devil's Potato-root
Apocynaceae, Dogbane Family

Habit: Perennial shrubby vine, to 2 m long. Stems twining, semi-woody, with milky sap.

Leaves: Opposite; stalked; blades ovate to somewhat oblong, 4-12 cm long, often slightly curled and/or folded lengthwise, somewhat leathery.

Inflorescences: 3- to 7-flowered axillary clusters.

Flowers: Yellowish to white, 4-6 cm long, with a long tube and 5 spreading petals.

Fruit: Composed of two, dry, pod-like, spreading, narrowly cylindrical follicles, 10-25 cm long; seeds with a tuft of hairs.

Habitat and Distribution: Occasional; pinelands and scrubby thickets; coastal central and throughout southern peninsula Florida; south through the West Indies and Mexico across Central America into northern South America.

Comment: Flowering can occur all year. Rubber Vine twines over low vegetation. The flowers are not particularly showy, but do stand out.

Euthamia caroliniana (L.) Greene ex Porter & Britton
Flat-topped Goldenrod
[*Euthamia minor* (Michx.) Greene; *Euthamia tenuifolia* (Pursh) Greene; *Solidago microcephala* (Greene) Bush]
Asteraceae (Compositae), Aster, Sunflower, or Daisy Family

Habit: Herbaceous, erect, perennial, to 1 m tall, from underground runners (rhizomes). Stems thin, smooth, freely branching.

Leaves: Alternate; stalkless; blades linear, to 8 cm long and 3 mm or less wide, 1- to 3-veined, lax (bending down) and covered with a sticky layer.

Inflorescences: Terminal, flat-topped, large and showy, with many small, sticky heads.

Flowers: Bright yellow, heads with 7-16 ray flowers and 3-7 tiny disc flowers.

Fruit: A nutlet, less than 1 mm long with a tuft of bristles 3-4 mm long at tip.

Habitat and Distribution: Common; sandy soils of wet to dry pinelands, sandhills, old fields, pastures, and often in coastal areas; along the Coastal Plain from Nova Scotia and Massachusetts south into south peninsula Florida, and west into southern Louisiana.

Comment: Blooms July into October. This is a good wildflower for dry areas. It can be planted by rhizomes or seeds and does best in full sun.

Echites umbellata Jacq.

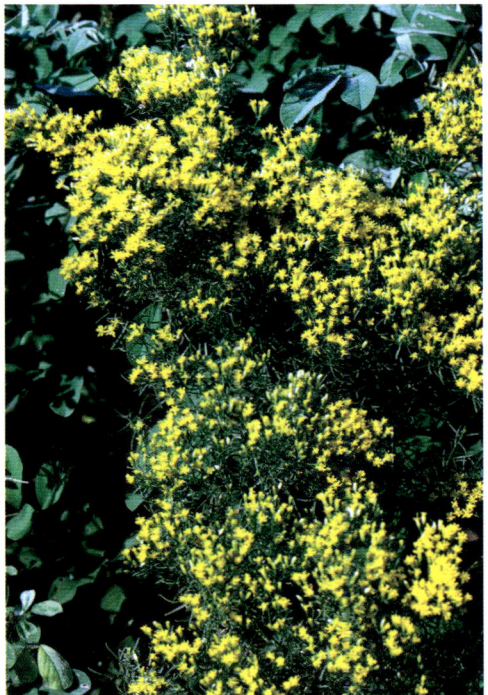

Euthamia caroliniana
(L.) Greene ex Porter & Britton

Yellow

Flaveria floridana J.R. Johnst.
Florida Yellow-top
Asteraceae (Compositae), Aster, Sunflower, or Daisy Family

Habit: Herbaceous perennial, to 1.2 m tall. Stems smooth, branching, semi-shrubby.

Leaves: Opposite; stalkless; blades linear, 2.5-8 cm long and to 1.5 cm wide.

Inflorescences: Dense, terminal, flat-topped clusters of heads.

Flowers: Heads with 10 - 15 yellow flowers, usually only 1 ray flower and the rest disc flowers.

Fruit: A thin nutlet, 1-2 mm long.

Habitat and Distribution: Frequent; sandy soils of beaches, dunes, and other open coastal areas; endemic in central and southern peninsula Florida.

Comment: The large, showy clusters of bright yellow flowers can easily be seen in the summer and fall along coastal areas in central and lower peninsula Florida.

Flaveria linearis Lag.
Narrow-leaf Yellow-top
Asteraceae (Compositae), Aster, Sunflower, or Daisy Family

Habit: Herbaceous perennial, to 0.9 m tall. Stems smooth, branched, sometimes reddish, somewhat woody.

Leaves: Opposite; stalkless; blades linear, to 16 cm long and to 8 mm wide.

Inflorescences: Dense, terminal, flat-topped clusters of heads.

Flowers: Heads with 5-8 yellow flowers, often with only 1 ray flower and the rest disc flowers.

Fruit: A thin nutlet, about 2 mm long.

Habitat and Distribution: Frequent; sandy soils of open coastal areas, flats, marsh margins, thickets, and pinelands; central panhandle south through Florida; West Indies, Yucatan.

Comment: The large collection of yellow clusters at the top of the plant can be seen in summer and fall. Narrow-leaf Yellow-top is most frequently found near the coast, but occasionally is inland.

Flaveria floridana J.R. Johnst.

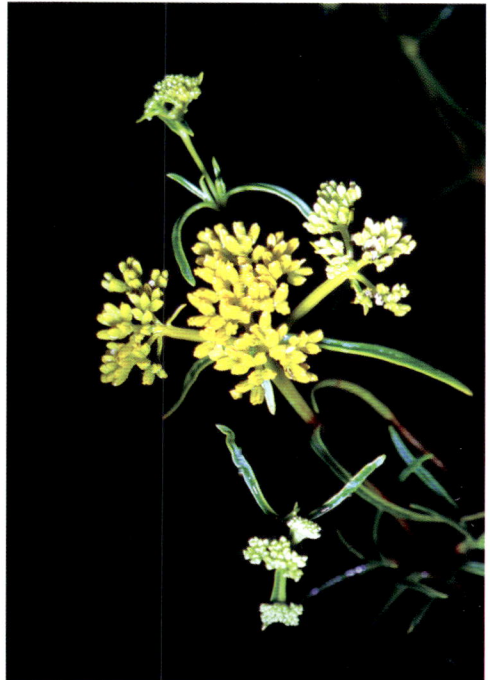

Flaveria linearis Lag.

349

Yellow

Gelsemium rankinii Small
Odorless Jessamine
Loganiaceae, Logania Family

Habit: Perennial vine. Stems woody, trailing, or twining and high climbing.

Leaves: Opposite, evergreen; stalked; blades lance-shaped to rounded, 3-9 cm long and 1-2.5 cm wide.

Inflorescences: Flowers solitary or in small clusters in leaf axils.

Flowers: Dull yellow, to 3 cm long, tubular, flaring, with 5 rounded lobes at mouth, odorless.

Fruit: A two-parted capsule, 1-1.6 cm long and 0.6-0.8 cm wide; seeds flat, 3-4 mm long, not winged.

Habitat and Distribution: Frequent; swamps, bogs, and wet, wooded sites; panhandle Florida, west to Louisiana, and north to North Carolina.

Comment: Flowers appear with the first warm breeze in winter and early spring. The flowers and shiny, evergreen leaves are very attractive.

Gelsemium sempervirens (L.) W.T. Aiton
Yellow Jessamine
Loganiaceae, Logania Family

Habit: Perennial vine. Stems woody, trailing, or twining and high climbing.

Leaves: Opposite, evergreen; stalked; blades lance-shaped to somewhat rounded, 4-9 cm long and to 1.5 cm wide.

Inflorescences: Flowers solitary or in small clusters in leaf axils.

Flowers: Bright yellow, 2-3 cm long, tubular, flaring, with 5 lobes at mouth, very fragrant.

Fruit: A two-parted capsule, 1.5-2.5 cm long and 0.8-1.2 cm wide; seeds flat, to 1.5 mm long, conspicuously thin-winged.

Habitat and Distribution: Common; upland hammocks and other wooded habitats, flatwoods, thickets, and fence rows; central peninsula Florida, west to Texas and Arkansas, and north to Virginia.

Comment: One of the earliest plants to flower at the first suggestion of spring and continuing well into spring. As a wildflower Yellow Jessamine is terrific. The showy flowers and shiny evergreen leaves are very easily noticed growing over shrubby vegetation and fences at the edge of open areas. Many appreciate the appealing fragrance. Yellow Jessamine is becoming more common as a landscape plant and is being used for ground covers. It can be trimmed quite short and trained into beds. The vines are commonly used for wreaths.

Gelsemium rankinii Small

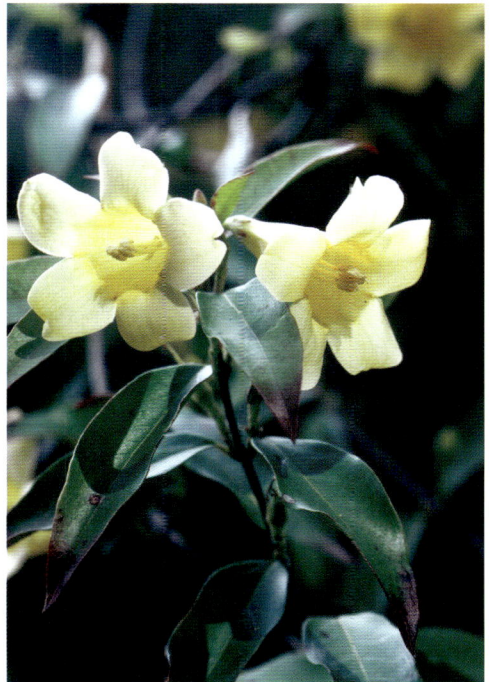

Gelsemium sempervirens
(L.) W.T. Aiton

Yellow

Grevillea robusta A. Cunn.
Silk-oak Tree
Proteaceae, Protea Family

Habit: Tree to 45 m tall, usually to 17 m tall, pyramidal when young, becoming irregular in age.

Leaves: Alternate, evergreen; stalked; blades to 30 cm long, once-divided into many deeply cut leaflets, leaflet margins rolled, silvery underneath.

Inflorescences: One-sided clusters, to 10 cm long, on short leafless branches.

Flowers: Golden yellow, mostly of stamens.

Fruit: A tan, leathery, capsule-like pod.

Habitat and Distribution: Occasional as an escape, frequent as an ornamental; well-drained sandy soils of disturbed areas; south peninsula Florida; native of Australia.

Comment: Blooms occur in April and May. The fern-like foliage on younger trees is very attractive and the flowers are spectacular. When flowering the plant is magnificent. Silk-oak needs plenty of room as it becomes a very large tree. As it ages it becomes very messy and ragged.

Helenium amarum (Raf.) H. Rock
Bitterweed or Spanish Daisy
Asteraceae (Compositae), Aster, Sunflower, or Daisy Family

Habit: Herbaceous annual, 20-50 cm tall, from a taproot. Stems smooth, branched above.

Leaves: Alternate; stalkless; blades 2-7 cm long, narrow, needle-like, lowermost feathery-lobed, quickly deciduous.

Inflorescences: Solitary terminal heads on loose clusters of branches.

Flowers: Heads with 5-10 yellow rays, each 5-12 mm long, wedge-shaped, hairy beneath, with toothed ends, spreading out or downward, disc yellow and domed.

Fruit: A dark brown, hairy nutlet, to 1 mm long, with 6-8 tiny scales at tip.

Habitat and Distribution: Common; sandy soils of fields, open woods, prairies, pastures, roadsides, and disturbed areas; throughout Florida, west to Texas, Arkansas, Missouri, north to Connecticut, Maryland, and Virginia; south into Mexico and the West Indies.

Comment: The yellow flowers are produced from the spring into the fall. Often in mass the effect is very colorful. Milk from cows grazing on Bitterweed has a bad taste. Extremely weedy, overgrazing rapidly spreads the population.

Grevillea robusta A. Cunn.

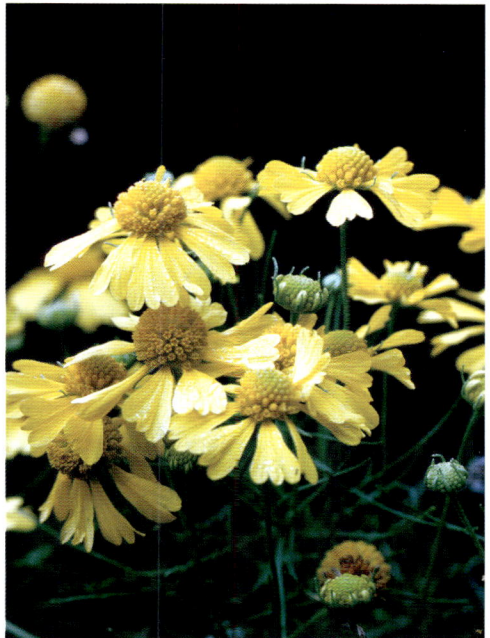

Helenium amarum (Raf.) H. Rock

Yellow

Helenium autumnale L.
Common Sneezeweed or Fall Sneezeweed
Asteraceae (Compositae), Aster, Sunflower, or Daisy Family

Habit: Herbaceous perennial, 0.5-2 m tall. Stems clump forming, winged, 1 to several stems from base.

Leaves: Alternate; stalkless; blades linear-elliptic, 6-15 cm long, margins toothed.

Inflorescences: Solitary terminal heads on loose clusters of branches.

Flowers: Heads with 13-21 yellow rays, each 1.5-2.5 cm long, disc yellow and domed.

Fruit: A light brown or gray nutlet, about 1.5 mm long, ribs hairy, with 5-10 tiny scales at tip.

Habitat and Distribution: Occasional; moist soils of meadows, ditches, flatwoods, floodplains, bogs, and other wet habitats; central peninsula Florida, throughout the United States and adjacent Canada.

Comment: Summer and fall blooming, more frequent in the fall. The large heads with many yellow rays are quite visible as wildflowers.

Helenium pinnatifidum (Nutt.) Rydb.
Swamp Sneezeweed
Asteraceae (Compositae), Aster, Sunflower, or Daisy Family

Habit: Herbaceous perennial, 0.3-1 m tall. Stems usually unbranched, one or more from base, usually hairy.

Leaves: Alternate, persistent; stalkless; blades broadly linear to elliptic or somewhat broader near tip, 3-25 cm long, thick, usually with toothed margins, upper stem leaves smaller, forming short wings to 5 mm long on stems at the point of attachment.

Inflorescences: Solitary terminal heads.

Flowers: Heads with 16-40 yellow fan-shaped rays, each with finely toothed ends, and a yellow domed disc.

Fruit: A brown hairy nutlet, about 1 mm long, with 5-10 tiny scales at tip.

Habitat and Distribution: Common to infrequent; wet pinelands, marshes, swamps, and bogs; throughout Florida, and north to North Carolina.

Comment: Often seen flowering in spring along roadsides and into the edges of wet woods and marshes. Swamp Sneezeweed is very attractive and is often seen growing together with Spring Sneezeweed.

Helenium autumnale L.

Helenium pinnatifidum
(Nutt.) Rydb.

Yellow

Helenium vernale Walter
Spring Sneezeweed
Asteraceae (Compositae), Aster, Sunflower, or Daisy Family

Habit: Herbaceous perennial, 0.3-1 m tall. Stems usually unbranched, one or more from base, smooth.

Leaves: Alternate; stalkless; blades broadly linear to elliptic or somewhat broader towards tip, 5-30 cm long, margins usually smooth, upward leaves smaller, forming wings on stems to 2 cm long.

Inflorescences: Solitary terminal heads.

Flowers: Heads with 16-40 yellow fan-shaped rays, each with finely toothed ends, and a yellow domed disc.

Fruit: A brown, smooth, dotted glandular nutlet, about 1 mm long, with 5-10 tiny scales at tip.

Habitat and Distribution: Infrequent; wet meadows, bogs, marshes, swamps, and wet pinelands; north peninsula Florida, west to Louisiana, and north to North Carolina.

Comment: The very attractive large yellow heads appear in spring. Spring Sneezeweed is often found growing with Swamp Sneezeweed.

Helianthemum carolinianum (Walter) Michx.
Carolina Rockrose
Cistaceae, Rockrose Family

Habit: Herbaceous perennial, 5-38 cm tall. Stems with long soft hairs, arising from basal rosette.

Leaves: Alternate, basal and 2-5 along stems; short-stalked; blades broadly elliptic to somewhat rounded, 2-5 cm long and 0.7-2 cm wide, star-shaped hairs on both surfaces.

Inflorescences: Solitary, opposite a leaf.

Flowers: Yellow, 3-4 cm wide, with 5 petals.

Fruit: A capsule, round, to 1 cm long; seeds reddish black, pimpled.

Habitat and Distribution: Occasional; sandy soils of open woodlands, pinelands, fields, and roadsides; central peninsula Florida, west to east Texas, and north to North Carolina.

Comment: The large flowers with fragile petals can be seen in summer. The star-shaped hairs on the leaves, fuzzy stem, and basal rosette give the identification away. Most frequently large numbers of the plants can be found growing in close proximity.

Helenium vernale Walter

Helianthemum carolinianum
(Walter) Michx.

Yellow

Helianthemum corymbosum Michx.
Pine Barren Rockrose
Cistaceae, Rockrose Family

Habit: Herbaceous perennial; 15 to 35 cm tall. Stems with star-shaped hairs.

Leaves: Alternate; short-stalked; blades elliptic to somewhat rounded, 1.5 to 4 cm long, dark green and star-shaped hairy on the upper surface, and pale green and long hairy on the lower surface.

Inflorescences: Dense terminal clusters.

Flowers: Two kinds of flowers are present - closed flowers with no petals; and open flowers, 16 to 20 cm wide, with 5, showy, yellow petals.

Fruit: A capsule, rounded, hairy, 5-7 mm long; seeds reddish brown.

Habitat and Distribution: Common; dry sandy soils of openings, maritime forests, sandhills, open hammocks, and dunes; throughout Florida, north to North Carolina.

Comment: Blooming from spring through summer and all year in southern peninsula Florida, this showy wildflower has fragile flowers. When flowers are picked or the plant is disturbed the petals readily fall. The plant provides conspicuous clusters of yellow in open, sandy areas.

Helianthemum georgianum Chapm.
Georgia Rockrose
Cistaceae, Rockrose Family

Habit: Herbaceous perennial, 10-40 cm tall. Stems star-shaped hairy.

Leaves: Alternate; short-stalked; blades slenderly elliptic to elliptic, 2-3.5 cm long, star-shaped, hairy on both surfaces.

Inflorescences: Clusters of usually 3 flowers toward the ends of stems.

Flowers: Yellow, with 5 petals, some opening to 2-3 cm wide, others smaller and remaining closed.

Fruit: A capsule, rounded, 4-5.4 mm long; seeds dark red, pitted.

Habitat and Distribution: Occasional; sandhills, maritime woods, open sandy woods, dunes, and dry hammocks; central peninsula Florida, west to Texas and Oklahoma, and north to North Carolina.

Comment: This spring blooming wildflower is quite attractive but not especially showy.

Helianthemum corymbosum
Michx.

Helianthemum georgianum
Chapm.

Yellow

Helianthemum nashii Britton
Scrub Rock rose
Cistaceae, Rockrose Family

Habit: Herbaceous perennial, to 40 cm tall. Stems with gray hairs.

Leaves: Alternate; short-stalked; blades elliptic to narrowly elliptic, 1-3 cm long, pale green on both surfaces, with prominent midvein beneath.

Inflorescences: Terminal, elongate, loose clusters.

Flowers: Yellow, to 2 cm wide, some with 5 petals, others lacking petals.

Fruit: A capsule, rounded, 3-3.5 mm long.

Habitat and Distribution: Frequent; scrub; endemic to peninsula Florida.

Comment: Flowers can be seen in spring and summer. The large flowers are quite easily seen in the sparse groundcover of the scrub habitat.

Helianthus angustifolius L.
Narrowleaf Sunflower
Asteraceae, Aster or Sunflower Family

Habit: Herbaceous perennial, to 2 m tall from a short basal crown. Stems erect, usually 1, simple or branched upwards, hairy, and rough.

Leaves: Opposite on lower stem, alternate on upper stem; stalkless; blades linear or narrowly lance-shaped, to 20 cm long and 1 cm wide, quite rough, with the edges rolled under.

Inflorescences: Large heads terminal on branches from upper leaf axils.

Flowers: Heads with 10-12 yellow rays, each 1-4 cm long, disc reddish brown, to 1.5 cm diameter.

Fruit: A nutlet, 2.5-3 mm long, brown or brown and black, with two short awns at tip.

Habitat and Distribution: Frequent; most to wet pinelands, wet edges, marshes, bogs, swales, ditches, old fields, and wet disturbed areas; central peninsula Florida, west to Texas, and north to Ohio, Missouri, Indiana, and Illinois, and east to Kentucky, New Jersey, Pennsylvania, and Virginia.

Comment: The outer petals are bright yellow making this one of the best fall wildflowers. Narrowleaf Sunflower blooms in the late summer and fall, but fall is the season of maximum flowering. This species frequently hybridizes with Florida Sunflower (*Helianthus floridanus* A. Gray ex Chapm.), a similar species with yellow disc flowers that is usually not more than 3 feet tall.

Helianthemum nashii Britton

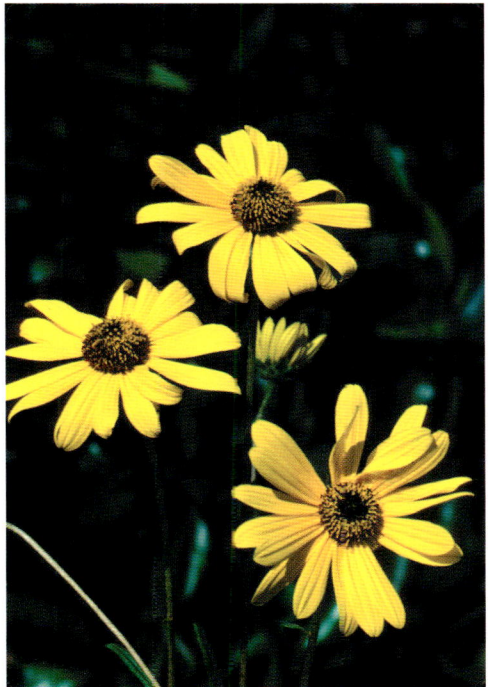

Helianthus angustifolius L.

Yellow

Helianthus debilis Nutt.
Beach Sunflower or Dune Sunflower
Asteraceae, Aster or Sunflower Family

Habit: Herbaceous annual, 0.5-2 m tall, from a taproot. Stems erect or usually lying flat; branches erect or ascending; both can be purple mottled, smooth or rough.

Leaves: Opposite on lower stem, alternate on upper stem; stalked; blades triangular or lance-ovate, wider at base, 5-10 cm long and 2-9 cm wide, rough hairy.

Inflorescences: Terminal heads on branches from upper leaf axils.

Flowers: Heads with 10-20 yellow rays, each 1-3 cm long, disc maroon, 1-2.5 cm diameter.

Fruit: A nutlet, dark gray, about 4 mm long, and very hairy, with two short awns at tip.

Habitat and Distribution: Frequent to occasional; sandy soils of dunes, pinelands, roadsides, fields, and coastal sites; eastern coastal peninsula and western coastal central peninsula Florida, mostly in the Coastal Plain, west to Texas, and north into North Carolina.

Comment: Blooms can occur throughout the year, but are most common during the summer and early fall. This distinctive sunflower is widely planted as an ornamental. The glossy green leaves are rough and triangular in shape providing attractive foliage. Beach Sunflower has two distinctive growth forms, erect or reclining with ascending tips. The erect form, like most coastal plants is not very tall, about 1 meter, and grows outward in a somewhat creeping habit. Several subspecies or varieties have been named.

Heliopsis helianthoides (L.) Sweet
Oxeye Sunflower or Smooth Oxeye
Asteraceae, Aster or Sunflower Family

Habit: Herbaceous perennial, 0.3-1.5 m tall, from short, stout underground runner. Stems erect, smooth.

Leaves: Opposite; stalked; blades ovate or widely lance-shaped, 5-20 cm long and 2.5-12 cm wide, margins with teeth.

Inflorescences: Stalked heads, terminal or from upper leaf axils.

Flowers: Heads with yellow rays, each 3-5 cm long, disc domed, yellow, 1-2.5 cm diameter.

Fruit: A nutlet, brown, oval, 3.5-4.5 mm long, smooth, ray fruit 3-sided, disc fruit 4-sided, tip with tiny scales or smooth.

Habitat and Distribution: Infrequent to rare; thickets, dry to rich open woods, and meadows; British Columbia, the Dakotas, Nebraska, Colorado, and New Mexico, east to Quebec and New York, south to Walton and Jefferson Counties in the Florida panhandle.

Comment: Oxeye Sunflower is a wonderful wildflower, blooming in spring southward and summer northward. It is more common north of the Coastal Plain. Several cultivated forms are offered for sale with brighter and paler yellows, orange-yellows, and double flowers. The plants do best when planted in full sun.

Helianthus debilis Nutt.

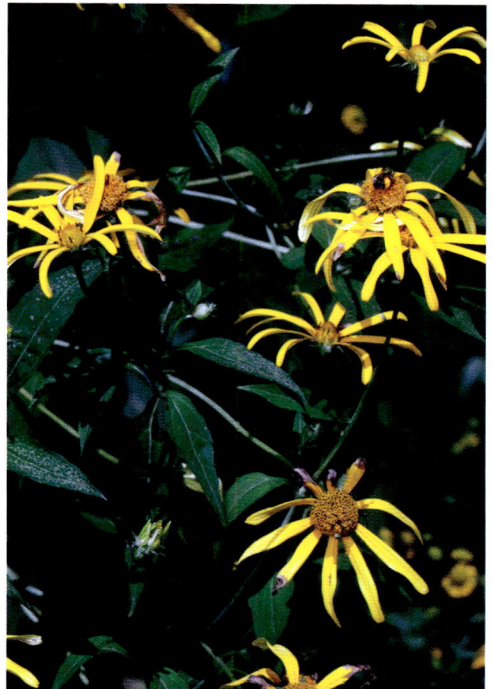

Heliopsis helianthoides
(L.) Sweet

Yellow

Heliotropium polyphyllum Lehm.
Pineland Heliotrope
Boraginaceae, Borage Family

Habit: Herbaceous perennial, 10-60 cm tall or long, from a woody rootstock. Stems prostrate to spreading to erect, hairy.

Leaves: Alternate; stalkless or short-stalked; blades narrowly elliptic, 0.5-2.5 cm long, hairy underneath.

Inflorescences: Terminal, spike-like, flowers arranged along one side of stem.

Flowers: Bright yellow to white, tubular with spreading lobes, 6-8 mm wide, 3-5 mm long, tapering in size toward end of spike.

Fruit: 1.5-3 mm wide, an aggregate of 4 nutlets.

Habitat and Distribution: Frequent; dry to moist soils of pinelands, pond margins, hammocks, and fields; eastern panhandle Florida south through the peninsula; Bahamas; South America.

Comment: As a wildflower, blooming all year, but most frequently in the spring, Pineland Heliotrope is outstanding. The bright yellow or white flowers glitter among the grasses and other low weedy plants that often occur in the habitat.

Heterotheca subaxillaris (Lam.) Britton & Rusby
Camphor Weed
Asteraceae, Aster or Sunflower Family

Habit: Herbaceous annual or biennial, 0.3-1 m tall or occasionally to 1.5 m tall, from a taproot. Stems ascending or erect, hairy.

Leaves: Alternate; basal stalked, upper stalkless, lower deciduous with a winged stalk and flanged at base; blades ovate to elliptic, 2-9 cm long and 1-3 cm wide, sandpapery on both surfaces, margins smooth to toothed.

Inflorescences: Loose terminal clusters of heads.

Flowers: Rays yellow, 15-45, 0.3-1 cm long, disc yellow, 0.7-1.8 cm wide.

Fruit: A nutlet, 2.3-3 mm long, ray fruits 3-angled, smooth, disc fruits somewhat flattened, hairy and with bristles at tip.

Habitat and Distribution: Common; dry sandy soils of sandhills, dunes, roadsides, fields, and other disturbed sites; throughout Florida, north to Delaware and Long Island, west to Arkansas, Texas, and adjacent Mexico.

Comment: Flowers can be seen all year southward and from summer into fall northwards. As a wildflower it often becomes the only easily visible bloom in very disturbed sandy sites. The large numbers of plants with many blooms can be dramatic. Identification is easily made by using the winged and flanged lower leaf stalks. If the lower leaves have fallen, identification is best effected by using the differences between the ray and disc fruits.

Heliotropium polyphyllum Lehm.

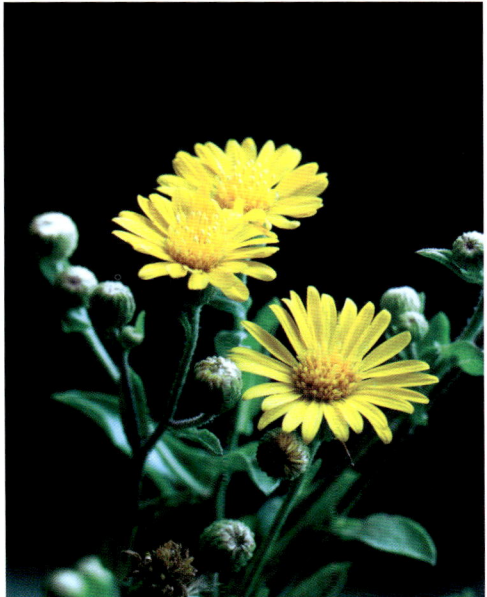

Heterotheca subaxillaris
(Lam.) Britton & Rusby

Yellow

Hibiscus aculeatus Walt.
Comfort-root
Malvaceae, Mallow Family

Habit: Herbaceous perennial, 0.9-1.2 m tall. Stems several, spreading, with rough hairs. Branches spreading or occasionally erect.

Leaves: Alternate; stalked; blades 3 or more, often 5-lobed, 3-12 cm long, rough hairy, with large toothed edges.

Inflorescences: Flowers on short stalks, in a leafy terminal arrangement.

Flowers: Cream to yellow petals fade to pink, with a dark crimson eye, petals 5 to 6 cm long.

Fruit: A hairy, pointed capsule, 1.7-2 cm long; seeds hairy, brown, reticulate, 3.5-4 mm long.

Habitat and Distribution: Frequent; pinelands, occasionally along ponds and streams, disturbed areas; Louisiana, east into coastal plain North Carolina, south into central peninsula Florida.

Comment: Flowering from summer until frost this colorful, usually low growing wildflower is showy and easily cultivated in any rich moist soil with some exposure to full sun. Comfort-root freezes to the ground yearly and regrows from the rootstock.

Hieracium gronovii L.
Gronovius' Hawkweed or Queen-devil
Asteraceae, Aster or Sunflower Family

Habit: Herbaceous perennial, 0.2-1.2 m tall. Stems with spreading hairs, especially on lower stems, sap milky.

Leaves: Alternate, mostly in a basal rosette, much smaller upward; stalkless, much narrowed at base; blades elliptic to lance-shaped and wider towards tip, to 17 cm long, with long simple hairs, star-shaped hairs, and often, but not always, with toothed edges.

Inflorescences: Heads in a narrow, stalked, terminal, loose cluster.

Flowers: Heads with rays to 1 cm long, each narrow, yellow, orange, or whitish with toothed tips, no disc flowers.

Fruit: Nutlets 2.5-4 mm long, with tan tuft of bristles 4-6 mm long at tip.

Habitat and Distribution: Occasional to frequent; dry soils in pinelands, sandhills, open woods, old fields, roadsides, pastures, meadows, and other disturbed sites; central peninsula Florida, west to Texas, and north to Kansas, southern Ontario, and Massachusetts.

Comment: Flowers can be seen in the warm months. The sandy habitats in which it occurs are normally not crowded with low vegetation and the bright yellow to orange heads can easily be seen.

Hibiscus aculeatus Walt.

Hieracium gronovii L.

Yellow

Hieracium megacephalon Nash
Largehead Hawkweed or Coastal Plain Hawkweed
Asteraceae, Aster or Sunflower Family

Habit: Herbaceous perennial, 0.2-0.7 m tall. Stems with spreading hairs, especially on lower stems, sap milky.

Leaves: Alternate, mostly in a basal rosette, much smaller upward; stalkless, much narrowed at base; blades lance-shaped and wider at towards tip, to 17 cm long, with long simple hairs, star-shaped hairs, and often, but not always, with toothed edges.

Inflorescences: Heads in a short, open, terminal cluster.

Flowers: Heads with rays to 1 cm long, each narrow, yellow, orange, or rarely whitish with toothed tips, no disc flowers.

Fruit: Nutlets 3.5-5 mm long, with a tan tuft of bristles 4-6 mm long at tip.

Habitat and Distribution: Common to frequent; dry sandy soils in open pinelands, sandhills, and oak hammocks; throughout the peninsula and into central panhandle Florida, north into south Georgia.

Comment: Very similar to Gronovius' Hawkweed in both form and habitats, Largehead Hawkweed also blooms in the warm months. Largehead Hawkweed has a compact arrangement of flower heads making the flowers somewhat more easily seen.

Hypericum brachyphyllum (Spach) Steud.
Short-leaved Sandweed or Coastal Plain St. John's-wort
Clusiaceae (Guttiferae, Hypericaceae), St. John's-wort or Mangosteen Family

Habit: Shrub, 0.5-1 m tall. Stem erect, branched upward.

Leaves: Opposite, somewhat clustered in axils; stalkless; blades needle-like, 3-10 mm long, covered with punctate dots, margins rolled under.

Inflorescences: Solitary or in small clusters in upper leaf axils.

Flowers: Yellow, 5 petals, to 8 mm long.

Fruit: A narrow, cone-shaped capsule, 4-5 mm long.

Habitat and Distribution: Frequent; low, wet sites, wet flatwoods, bogs, cypress-gum ponds, borrow pits, and wet ditches; throughout Florida, west to southern Mississippi, and north to southern Georgia.

Comment: Flowering in the summer and fall, the profuse branches of Short-leaved Sandweed provide an attractive sight. Individual flowers are not especially showy. The total effect of a population with the dense evergreen leaves and flowers is worth a photograph. Sandweed (Hypericum fasciculatum) is very similar. Sandweed shrubs are taller, 1.5-2 m tall, and the largest leaves are longer, 1.3 to 2.6 cm long. Flowers tend to be in slightly larger clusters of 3 to many giving a bit showier appearance. Sandweed is common and ranges slightly further north into North Carolina.

Yellow

Hieracium megacephalon Nash

Hypericum brachyphyllum
(Spach) Steud.

Yellow

Hypericum chapmanii W.P. Adams
Chapman's St. John's-wort or Apalachicola St. John's-wort
Clusiaceae (Guttiferae, Hypericaceae), St. John's-wort or Mangosteen Family

Habit: Shrub, 2-3 m tall. Stem erect, branched upward.

Leaves: Opposite, somewhat clustered in axils; stalkless; blades needle-like, longest 1.3-2.6 mm long, covered with punctate dots, margins rolled under.

Inflorescences: 1-3 flowers per axil, from upper leaf axils.

Flowers: Yellow, 5 petals, each to 9 mm long.

Fruit: A capsule, about 6 mm long.

Habitat and Distribution: Occasional; wet pine flatwoods depressions and cypress-gum ponds and depressions; coastal areas of the Florida panhandle from Wakulla and Franklin Counties west to Santa Rosa County; endemic.

Comment: Chapman's St. John's-wort flowers in summer. In appearance, it is a larger Sandweed with fewer flowers.

Hypericum cistifolium Lam.
Cluster-leaf St. John's-wort
Clusiaceae (Guttiferae, Hypericaceae), St. John's-wort or Mangosteen Family

Habit: Shrub, 0.3-1 m tall. Stems erect, sparingly branched, somewhat woody, reddish below.

Leaves: Opposite; stalkless; blades lance-shaped or somewhat oblong, 1.5-3 cm long, widest near base, leathery, stiff, covered with punctate dots, with tufts of smaller leaves in axils.

Inflorescences: Many-flowered terminal clusters.

Flowers: Yellow, with 5 petals, each 4-8 mm long.

Fruit: A capsule, 4-5 mm long.

Habitat and Distribution: Common; wet pinelands, swamps, ditches, swales, and pond and marsh margins; throughout Florida, west to Texas, and north to North Carolina.

Comment: Cluster-leaf St. John's-wort is quite common in most damp to wet areas. Identifiable all year it is most noticeable in summer when the many-flowered clusters of yellow flowers appear.

Yellow

Hypericum chapmanii W.P. Adams

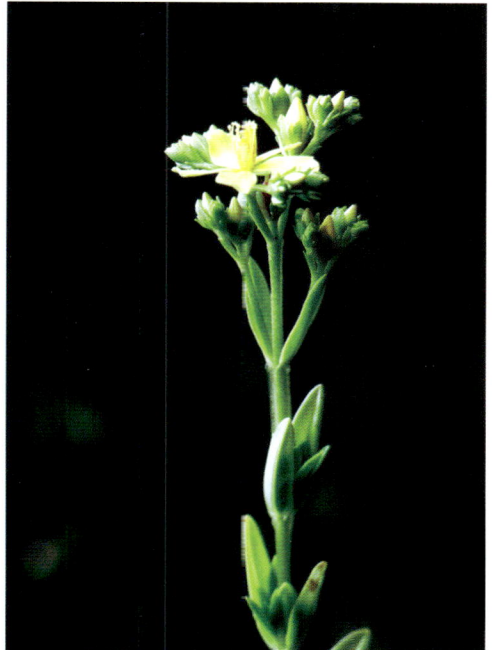

Hypericum cistifolium Lam.

Yellow

Hypericum fasciculatum Lam.
Sandweed
Clusiaceae (Guttiferae, Hypericaceae), St. John's-wort or Mangosteen Family

Habit: Shrub, 1.5-2 m tall. Stem erect, branched upward.

Leaves: Opposite, somewhat clustered in axils; stalkless; blades needle-like, larger blades 13-26 mm long, covered with punctate dots, margins rolled under.

Inflorescences: Clusters of 3 to many flowers in upper leaf axils.

Flowers: Yellow, 5 petals, each to 9 mm long.

Fruit: An oval, cone-shaped, capsule, 3-5 mm long.

Habitat and Distribution: Common; low, wet sites, wet flatwoods, bogs, ditches, cypress-gum ponds, depressions, marshes, and pond margins; throughout Florida, west to southern Mississippi, and north to North Carolina.

Comment: Flowering from spring into fall, the profuse branches of Sandweed provide an attractive sight. The flower clusters are showy. Populations are often found in a band along pond and lake banks and just at the seasonal high water level within marshes and other wet habitats. The differences between the very similar Short-leaved Sandweed and Sandweed are discussed above with Short-leaved Sandweed (*Hypericum brachyphyllum*).

Hypericum hypericoides (L.) Crantz
St. Andrew's-cross
[*Ascyrum hypericoides* L.]
Clusiaceae (Guttiferae, Hypericaceae), St. John's-wort or Mangosteen Family

Habit: Shrub, 0.3-1.5 m tall. Stem single, or several, erect, much branched.

Leaves: Evergreen, opposite; stalkless; blades usually oblong to linear, sometimes wider near tip, to 2.5 cm long, to 8 mm wide, covered with punctate dots.

Inflorescences: Flowers solitary and terminal.

Flowers: Yellow, 4 petals, each 8-10 mm long, sepals green, two quite large and two much smaller, punctate-dotted.

Fruit: An elliptic capsule, 4-9 mm long.

Habitat and Distribution: Common; sandy soils of upland hammocks, pinelands, sandhills, and drier margins of wetlands; throughout Florida, west to Texas and Oklahoma, north to Missouri and Virginia; Bermuda, Bahamas, Cuba, Puerto Rico, Haiti, Dominican Republic, Jamaica, Mexico, Guatemala, and Honduras.

Comment: The peculiar 4-petaled flowers can be found in summer and fall. The flowers are not especially showy and are sometimes hard to see among the plentiful leaves and large sepals. St. Andrew's-cross is the one Hypericum that is common in dry habitats. It is cultivated, but rarely so.

Hypericum fasciculatum Lam.

Hypericum hypericoides
(L.) Crantz

Yellow

Hypericum microsepalum (Torr. & A. Gray) A. Gray ex S. Watson
Flatwoods St. John's-wort
Clusiaceae (Guttiferae, Hypericaceae), St. John's-wort or Mangosteen Family

Habit: Shrub, to 1 m tall. Stems low and bushy-branched, branches leafy.

Leaves: Opposite; stalkless; blades oblong-elliptic, to 1 cm long, with scattered punctate dots, margins rolled under.

Inflorescences: Flowers terminal, solitary.

Flowers: Yellow, with 4 petals, each to 1 cm long.

Fruit: A capsule, 3-4 mm long.

Habitat and Distribution: Common; moist to wet pinelands; Florida panhandle and southwest Georgia.

Comment: This very showy St. John's-wort flowers in spring. The low straggly shrub with many large flowers can scarcely be overlooked.

Hypericum mutilum L.
Dwarf St. John's-wort
Clusiaceae (Guttiferae, Hypericaceae), St. John's-wort or Mangosteen Family

Habit: Herbaceous perennial, 10-80 cm tall. Stems erect, usually loosely branched.

Leaves: Opposite; stalkless; blades oval to lance-shaped, 0.3-3.5 cm long, 5-nerved, with scattered punctate dots, base clasping.

Inflorescences: Terminal clusters.

Flowers: Yellow, with 5 petals, 2-3 mm long.

Fruit: A capsule, 2.5-4 mm long.

Habitat and Distribution: Common; ditches, bogs, wet margins, and other wet areas; throughout Florida, west to Texas, and north to Kansas, Illinois, Manitoba, Ontario, Quebec, and Newfoundland.

Comment: The small yellow flowers appear in the warm months. Flowers are not especially showy. The small fragile-looking plants are usually found along the open sandy edges of ponds and marshes.

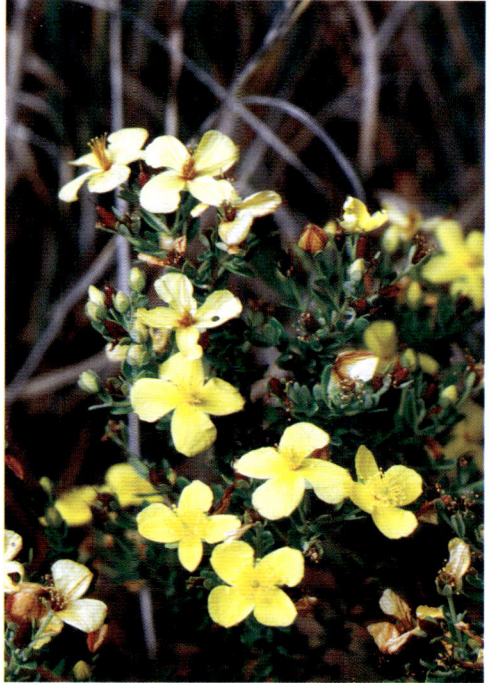

Hypericum microsepalum
(Torr. & A. Gray)
A. Gray ex S. Watson

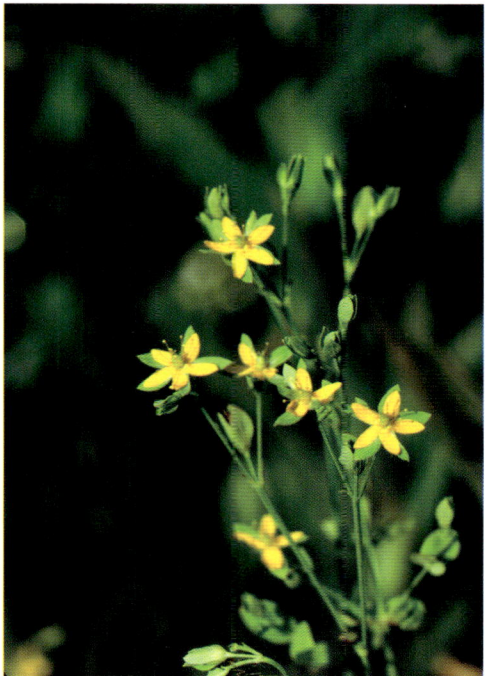

Hypericum mutilum L.

Yellow

Hypericum myrtifolium Lam.
Myrtle-leaved St. John's-wort or Heart-leaved St. John's-wort
Clusiaceae (Guttiferae, Hypericaceae), St. John's-wort or Mangosteen Family

Habit: Shrub, 0.3-1 m tall. Stem erect, grayish, few-branched.

Leaves: Opposite; stalkless; blades oval to somewhat triangular, 1-3 cm long, evergreen, with punctate dots, bases somewhat clasping.

Inflorescences: Terminal clusters.

Flowers: Yellow, with 5 petals, each 1-1.2 cm long.

Fruit: A capsule, 5-8 mm long.

Habitat and Distribution: Frequent; wet pinelands, cypress-gum depressions, bogs, and pond margins; throughout Florida, west to southeastern Mississippi, and north to coastal plain Georgia.

Comment: Flowers, in warm weather, are quite large, but not profuse. Identified by somewhat heart-shaped leaves, it is frequent in wet soils.

Hypericum reductum (Svenson) W.P. Adams
Matted Sandweed or Atlantic St. John's-wort
Clusiaceae (Guttiferae, Hypericaceae), St. John's-wort or Mangosteen Family

Habit: Shrub, 10-50 cm tall. Stems bushy-branched, lying flat, mat forming, often rooting along stems.

Leaves: Opposite, axils clustered; stalkless; small, needle like, usually to 5 mm long, with punctate dots, margins rolled under.

Inflorescences: In leaf axils, usually solitary, occasionally few-flowered.

Flowers: Yellow, with 5 petals, 4-8 mm long.

Fruit: A capsule, 6-8 mm long.

Habitat and Distribution: Frequent; dry, sandy woods and scrub, but occasionally found in small wet depressions and along outer pond margins; throughout Florida, west to southeast Alabama, and north to North Carolina.

Comment: The spring and summer flowers are not easily seen within the tangle of branches of this low bushy shrub.

Hypericum myrtifolium Lam.

Hypericum reductum
(Svenson) W.P. Adams

Yellow

Hypericum setosum L.
Hairy St. John's-wort
Clusiaceae (Guttiferae, Hypericaceae), St. John's-wort or Mangosteen Family

Habit: Herbaceous annual or biennial, 30-75 cm tall. Stems few branched, densely hairy.

Leaves: Opposite; stalkless; blades ovate to lance-shaped, 3-15 mm long, with punctate dots, base somewhat clasping, margins rolled under.

Inflorescences: Flowers in terminal clusters.

Flowers: Yellow, with 5 petals, 4-7 mm long.

Fruit: A capsule, 4-5 mm long.

Habitat and Distribution: Occasional; moist to wet soils of pinewoods, hammocks, ditches, and bogs; central peninsula Florida, west to Texas, and north to Virginia.

Comment: The small flowers, found in summer, are difficult to see among the other low vegetation in which this plant grows.

Hypericum suffruticosum W.P. Adams & N. Robson
Little St. Andrew's-cross or Pineland St. John's-wort
Clusiaceae (Guttiferae, Hypericaceae), St. John's-wort or Mangosteen Family

Habit: Shrub, 7-15 cm tall. Stems thin, branched, spreading, cushion-like, or somewhat matted.

Leaves: Opposite; stalkless; blades elliptic to oblong, 1-8 mm long, with punctate dots outward.

Inflorescences: Solitary flowers in terminal axils.

Flowers: Yellow, with 4 petals, each 5-7 mm long.

Fruit: A capsule, about 4 mm long.

Habitat and Distribution: Occasional; sandy soils in pinelands; north peninsula Florida, west to southeast Louisiana, and north to southeast North Carolina.

Comment: Spring flowering, Little St. Andrew's-cross is a small matted plant with four light yellow petals. Most frequently, it is found in moist to wet open disturbed sandy soils. It can be easily overlooked. St. Andrew's-cross (*Hypericum hypericoides*), discussed above, is a much larger version with stems 1.5 m tall, linear-oblong leaves to 2.5 cm long, and petals to 1 cm long. St. Andrew's-cross, flowering summer through fall, is found in uplands and at the upper driest margins of wetlands. Neither plant has showy flowers.

Hypericum setosum L.

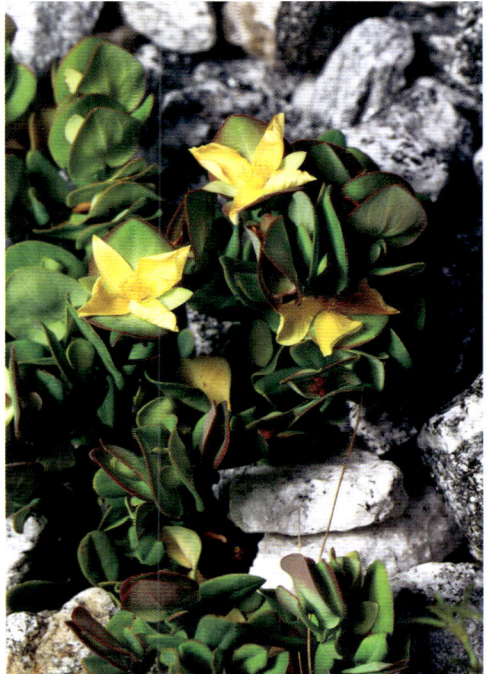

Hypericum suffruticosum
W.P. Adams & N. Robson

Yellow

Hypericum tetrapetalum Lam.
Heart-leaved St. Peter's-wort
Clusiaceae (Guttiferae, Hypericaceae), St. John's-wort or Mangosteen Family

Habit: Shrub, 0.1-1 m tall. Stems becoming woody, reddish.

Leaves: Opposite; stalkless; blades ovate, wider towards base, 1-2.5 cm long, with punctate dots, base clasping.

Inflorescences: Solitary flowers at branch tips.

Flowers: Yellow, with 4 petals, each to 2.1 cm long.

Fruit: A capsule, to 6 mm long.

Habitat and Distribution: Common; low, wet pinewoods and pond margins; throughout the peninsula and west to Okaloosa County in panhandle Florida, and north to southeast Georgia; Cuba.

Comment: The bright yellow flowers can be seen in the warm months. The distinctive leaves and reddish stem are solid identification characteristics.

Hypoxis juncea Sm.
Common Star-grass or Yellow Star-grass
Hypoxidaceae, Yellow Star-grass Family

Habit: Herbaceous perennial, to 25 cm tall, from a vertical rootstock, top of rootstock with fibrous membranous leaf bases.

Leaves: Alternate; stalkless; blades grass- or thread-like, to 30 cm long and to 8 mm wide, hairy.

Inflorescences: Flowers solitary or 2-3 at the tip of a slender leafless stalk, 4-18 cm tall.

Flowers: Yellow, 6-pointed, star-shaped, to 2.5 cm across.

Fruit: A capsule, rounded, 4-8 mm long; seeds black, 1-1.3 mm long, surface pebbled.

Habitat and Distribution: Common; moist to wet soils of pinelands, bogs, banks, and savannahs; throughout Florida, and north on the coastal plain into North Carolina.

Comment: The bright yellow flowers occur in all warm months. The flowers almost always provoke a comment when an observer looks down into the low, frequently sparse vegetation of wet soils.

Hypericum tetrapetalum Lam.

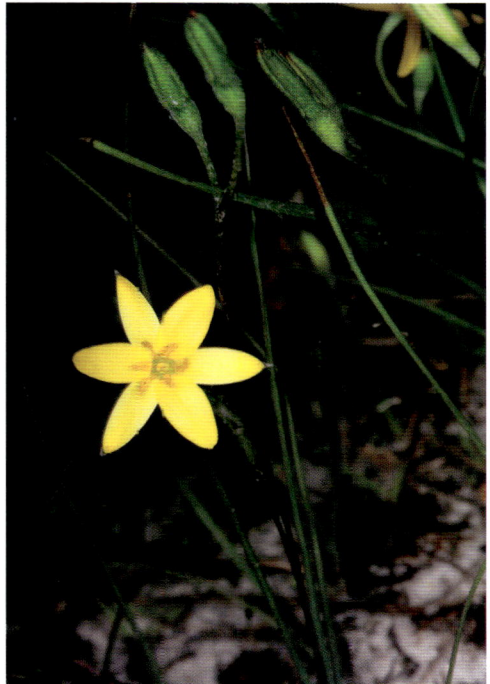

Hypoxis juncea Sm.

Yellow

Hypoxis wrightii (Baker) Brackett
Small-flowered Star-grass, Bristle-seed Star-grass, Bristle-seed Yellow Star-grass
[*Hypoxis micrantha* Pollard]
Hypoxidaceae, Yellow Star-grass Family

Habit: Herbaceous perennial, to 20 cm tall, from a vertical rootstock, top of rootstock with fibrous membranous leaf bases.

Leaves: Alternate; stalkless; blades grass- or thread-like, to 20 cm long, to 2.5 mm wide, hairy.

Inflorescences: Flowers usually solitary at the tip of a slender leafless stalk, 1-10 cm tall.

Flowers: Yellow, 6-pointed, star-shaped, to 1.6 cm across.

Fruit: A capsule, rounded, 5-6 mm long; seeds dark brown, 1-1.5 mm long, surface with sharp-pointed bumps.

Habitat and Distribution: Occasional; moist to wet soils of pinelands, savannas, banks, and ditches; throughout Florida, and along the coastal plain north to Virginia, and west to northeast Texas; West Indies.

Comment: The flowers of Small-flowered Star-grass are found from fall into spring. Blooming during the colder months provides a wildflower when many are not available. The small blooms on the short grass-like plants are scattered among the other ground covers and take extraordinary care to find.

Kallstroemia maxima (L.) Hook. & Arn.
Caltrop
Zygophyllaceae, Caltrop Family

Habit: Herbaceous annual, to about 1 m long. Stems lying flat, radiating from central base.

Leaves: Opposite; stalked; blades with 6-8 oblong leaflets, each to 3 cm long.

Inflorescences: Flowers solitary on long stalks from leaf axils.

Flowers: Yellow, with 5 rounded petals, each 7-8 mm long.

Fruit: A capsule, to 1 cm long, smooth, rounded, and with a beak.

Habitat and Distribution: Occasional; roadsides, open fields, and other disturbed sites; throughout peninsula Florida, coastal plain Alabama, Georgia, South Carolina, and rarely in central Texas; West Indies; Mexico south into South America.

Comment: The large yellow blooms of Caltrop are easily noticed. The plant flowers in all warm months and normally grows in open areas with very low vegetation, such as mowed roadsides and parks. Ultimately, this weedy wildflower will find its way to all such areas with warm temperate to tropical climates.

Hypoxis wrightii (Baker) Brackett

Kallstroemia maxima
(L.) Hook. & Arn.

Yellow

Krigia virginica (L.) Willd.
Dwarf Dandelion
Asteraceae, Aster or Sunflower Family

Habit: Herbaceous annual, 3-40 cm tall, from a taproot. Stems numerous, unbranched, arising from base, hairy, leafless, with milky sap.

Leaves: Alternate, in basal rosette; stalkless; blades linear to lance-shaped and broader towards tip, 2-12 cm long, margins many-lobed or -toothed.

Inflorescences: Small heads, terminal on thin scapes.

Flowers: Heads with yellow-orange rays, each 4-7 mm long.

Fruit: Nutlets reddish brown, to 2 mm long, oblong, with tuft of bristles and scales, 4-5.5 mm long at tip.

Habitat and Distribution: Frequent; sandy soils of woods, pinelands, lawns, roadsides, pastures, and other open disturbed sites; central peninsula Florida, west to Texas, and north to Arkansas, Missouri, Michigan, Vermont, and Maine.

Comment: The very small heads of flowers can be found in spring. The plants really do resemble a very small version of a Dandelion. In protected places they can overwinter.

Lachnanthes caroliniana (Lam.) Dandy
Redroot
Haemodoraceae, Bloodwort Family

Habit: Herbaceous perennial, 0.3-1.2 m tall, from red underground runners, with red roots and red sap. Floral stem erect, smooth below, hairy on upper parts.

Leaves: Alternate, mostly basal; stalkless; blades flat, equitant (enfolding each other, flat fan-like, overlapping), to 30 cm long and 0.5-1.5 cm broad.

Inflorescences: Tight terminal clusters.

Flowers: Yellow, woolly, petals 7-9 mm long, stamens exserted.

Fruit: A capsule, rounded, 2-5 mm in diameter.

Habitat and Distribution: Common; wet flatwoods, bogs, savannahs, wet ditches, and other wet disturbed areas; throughout Florida, west to Louisiana, and north to central Tennessee and North Carolina.

Comment: Flowers on Redroot can be found throughout the summer. Large flowering populations can be spectacular. The bright orange-red runners and roots just under the soil surface are an easy identification feature. Agricultural practices in wetlands and clearing for development frequently move the runners in soil and on equipment to areas in which the plants did not naturally grow. This tough wildflower can appear in some of the strangest places due to agricultural and developmental practices.

Krigia virginica (L.) Willd.

Lachnanthes caroliniana
(Lam.) Dandy

Yellow

Lantana camara L.
Lantana
Verbenaceae, Vervain Family

Habit: Perennial branching shrub, to 2 m tall. Stems 4-angled, hairy, much branched, with a few prickles on stems and branches.

Leaves: Opposite, deciduous; stalked; blades oval, 2-13 cm long, covered with glandular hairs that produce a strong odor, texture is very rough, margins toothed.

Inflorescences: Solitary, many-flowered, stalked, disc- or dome-like heads with concentric rings of variously colored flowers, from leaf axils.

Flowers: Cream, yellow, or pink turning red or orange with a slightly curved, narrow tube.

Fruit: Black drupes, about 7 mm in diameter.

Habitat and Distribution: Common; throughout Florida, east to Georgia and Bermuda, west to Texas; south through the West Indies into Central and South America; Asia; Africa; native to the West Indies.

Comment: Flowering during warm seasons, Lantana's flower colors provide a soft blend of pastel shades. The almost constant flowering provides a blend of colors most frequently recognized as orange. Many cultivated forms are sold which capitalize on various colors. Susceptible to hard frost, mature plants will resist light freezes. This species will hybridize with the native yellow-flowered *Lantana depressa* Small, Florida or Sanibel or Rockland Lantana. All parts of this plant are TOXIC if ingested. Lantana is a primary invader of old citrus groves.

Lantana depressa Small var. *floridana* (Moldenke) R.W. Sanders
Florida Lantana
Verbenaceae, Vervain Family

Habit: Shrub, to 1.1 m tall. Stems not prickly.

Leaves: Opposite; stalked; blades ovate to broadly elliptic, to 7 cm long, margins with teeth.

Inflorescences: Tight, domed clusters on long leaf stalks from the uppermost leaf axils.

Flowers: Yellow, tubular, slightly curved, to 6 mm long, with 4 lobes.

Fruit: A roundish drupe, to 3.5 mm long.

Habitat and Distribution: Occasional; dry soils of hammocks, shell mounds, and pinelands; peninsula Florida from the Dade County Keys north on the east coast to Little Talbot Island north of Jacksonville in Duval County; endemic.

Comment: The bright yellow clusters of flowers in all the warm months are eye-catching. Doubtless Florida Lantana would make a great cultivated plant, but similarly colored cultivars of Common Lantana (*Lantana camera*) have dominated the market. Florida Lantana is listed as Endangered by the State of Florida.

Lantana camara L.

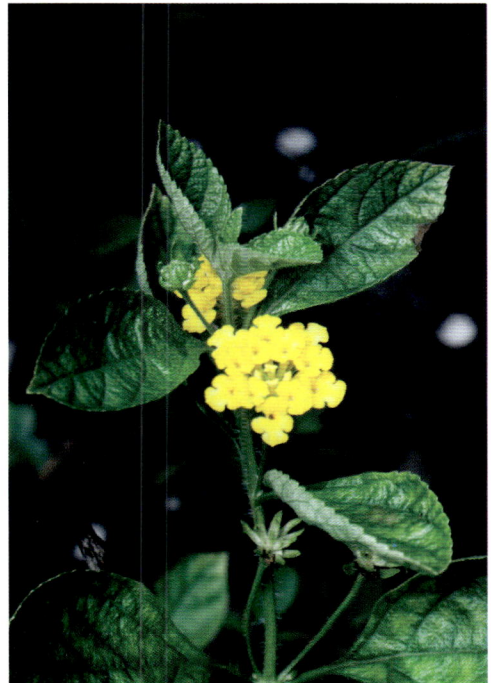

Lantana depressa
Small var. *floridana*
(Moldenke) R.W. Sanders

Yellow

Linum floridanum (Planch.) Trel.
Florida Flax
Linaceae, Flax Family

Habit: Herbaceous perennial, to 1 m tall. Stems smooth, slender, single, or branched.

Leaves: Alternate on upper stem, a few opposite on lower stem; stalkless; blades linear to lance-shaped and broadest near tip, 8-25 mm long, angled upward, sharp-pointed.

Inflorescences: Flowers solitary at the uppermost nodes of stems.

Flowers: Yellow, with 5 broad petals, each 4-8 mm long.

Fruit: A rounded capsule, 2-3.5 mm long.

Habitat and Distribution: Common; moist to wet pinelands, fields, savannahs, bogs, and sandhills; throughout Florida, west to eastern Texas, and north to North Carolina.

Comment: Lemon yellow flowers can be seen in most warm months. The thin spires of branches, each with a few flowers, dance among the surrounding green vegetation.

Linum westii C.M. Rogers
West's Flax
Linaceae, Flax Family

Habit: Herbaceous perennial, to 0.5 m tall. Stems 2-5, smooth, slender.

Leaves: Alternate on the upper stem, opposite on lower stem; stalkless; blades linear-oblong to narrowly elliptic, to 1.5 cm long, tip rounded, uppermost with sharp tips.

Inflorescences: Solitary flowers at uppermost nodes of stems.

Flowers: Yellow, about 6 mm long.

Fruit: A rounded capsule, about 3 mm long.

Habitat and Distribution: Rare; cypress ponds, bogs, boggy pineland depressions, and ditches; central panhandle Florida in Jackson, Calhoun, and Franklin Counties, and in northeast peninsular Florida in Baker County; endemic.

Comment: The pale yellow flowers seen in spring and summer are not easily found, especially so, as West's Flax is quite rare. West's Flax is listed as Endangered by the State of Florida.

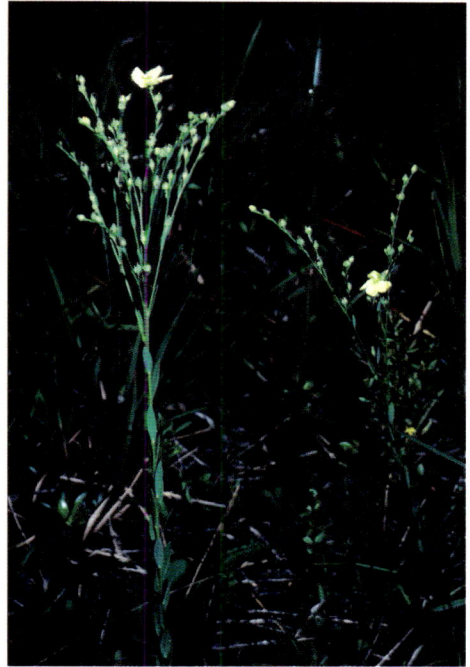

Photos courtesy of Ms. Jessie M. Harris
Linum floridanum (Planch.) Trel.

Linum westii C.M. Rogers

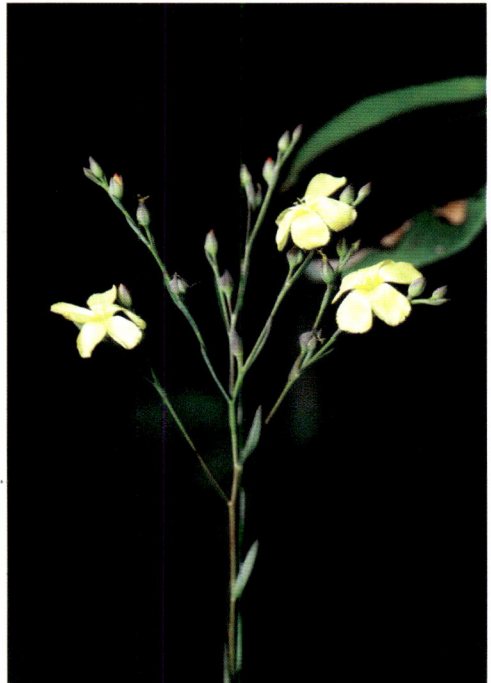

Yellow

Liriodendron tulipfera L.
Tulip-poplar, Tulip Tree, Yellow-poplar
Magnoliaceae, Magnolia Family

Habit: Large tree, to 60 m tall.

Leaves: Alternate, deciduous; long-stalked; blades saddle-shaped, 6-20 cm long and wide, with 4-6 pointed, spreading lobes.

Inflorescences: Solitary flowers, terminal on twigs.

Flowers: Yellow or greenish yellow, cup-shaped, 3-5 cm across, with 6 petals with orange bases.

Fruit: Winged, 2.5-4.5 cm long, 2-seeded; aggregated to form a spirally arranged cone.

Habitat and Distribution: Occasional to common; rich low woods southward, and rich woods of a dryer nature northward; central peninsula Florida, west to Louisiana, and north to southern Michigan and Vermont.

Comment: Tulip-poplar can grow to be an extremely large tree with a trunk diameter of about 3 meters. The spring flowers are quite large, eye-catching, and unusual. The handsome tree is frequently cultivated. The wood is prized by carvers.

Ludwigia arcuata Walter
Piedmont Seedbox or Long-stalked Seedbox
Onagraceae, Evening-primrose Family

Habit: Herbaceous perennial, to 50 cm long. Stems prostrate, leafy, hairy.

Leaves: Opposite; short-stalked to stalkless; blades narrowly to broadly elliptic, sometimes widest toward the tip, 0.8-2 cm long.

Inflorescences: Solitary flowers on stalks longer than the leaves, in leaf axils.

Flowers: Yellow, with 4 petals, 0.8-1 cm long.

Fruit: A 4-sided capsule, 6-8 mm long, curved.

Habitat and Distribution: Common southward, less so northward; wet margins, shores, banks, pools, lakes, marshes, and ditches; central peninsula Florida, west to Alabama, and north to South Carolina.

Comment: Flowering in summer, Piedmont Seedbox is frequently seen matted along the ground in places where water stands. It can be seen on the bottom of pools and ponds in clear water.

Liriodendron tulipfera L.

Ludwigia arcuata Walter

Yellow

Ludwigia decurrens Walter
Winged Water-primrose
Onagraceae, Evening-primrose Family

Habit: Herbaceous annual, 0.2-2.5 m tall. Stems widely branched, 4-angled, 2-winged, hairless.

Leaves: Alternate; stalkless or essentially so; blades lance-shaped, to 18 cm long.

Inflorescences: Flowers short-stalked or stalkless, solitary in leaf axils of branches.

Flowers: Yellow, with 4 petals, 6-12 mm long.

Fruit: A 4-sided capsule, 1-1.8 cm long, broader toward tip.

Habitat and Distribution: Infrequent to occasional; swamps, marshes, shores, wet clearings, and other wet sites; throughout Florida, west to Texas, and north to Missouri, Illinois, Indiana, and Virginia.

Comment: Winged Water-primrose blooms in the warm months. This annual can be identified by the wings along two opposite sides of the four stem angles.

Ludwigia grandiflora (Michx.) Greuter & Burdet
Uruguay Water-primrose or Largeflower Water-primrose
[*Ludwigia uruguayensis* (Camb.) Hara]
Onagraceae, Evening-primrose Family

Habit: Herbaceous perennial. Stems creeping or floating, often forming large mats, rooting at the joints, hairy.

Leaves: Alternate; stalked on lower stem, stalkless on upper; lower blades rounded to spatulate, upper blades lance-shaped to narrowly elliptic, to 15 cm long and 4.5 cm wide.

Inflorescences: Solitary flowers, on 1-2 cm long stalks, in upper leaf axils.

Flowers: Yellow, with 5 or 6 petals, each 1.2-2.2 cm long and to 2 cm wide.

Fruit: A cylindric capsule, 1-2.5 cm long, 3 mm diameter, hairy.

Habitat and Distribution: Rare to infrequent; standing or slow moving water of ditches, ponds, lakes, sloughs, canals, and marshes; north peninsula Florida, west to Texas and Oklahoma, and north to North Carolina; Mexico south into northern Argentina; native of tropical America.

Comment: The large bright yellow flowers can be seen in summer. Uruguay Water-primrose often forms large, dense, floating mats making a good wildflower display.

Ludwigia decurrens Walter

Ludwigia grandiflora
(Michx.) Greuter & Burdet

Yellow

Ludwigia maritima R.M. Harper
Coastal Plain Seedbox
Onagraceae, Evening-primrose Family

Habit: Herbaceous perennial, to 1 m tall. Stems erect, often branched, shaggy hairy.

Leaves: Alternate; stalkless; blades oblong to linear-oblong, to 3 cm long, shaggy hairy, tips rounded.

Inflorescences: Solitary flowers, on stalks to 1 cm long, in upper leaf axils.

Flowers: Yellow, 1.5-3 cm across, with 4 petals, each 7-15 mm long.

Fruit: A cubical, 4-sided capsule, 6-10 mm long, somewhat winged.

Habitat and Distribution: Frequent; wet pinelands, savannahs, and wet disturbed sites, especially after clearing; throughout Florida, west to eastern Louisiana, and north to North Carolina.

Comment: The large flowers of Coastal Plain Seedbox occur in all the warm months. The oblong leaves with rounded tips together with the cubical capsules readily identify this plant.

Ludwigia peruviana (L.) H. Hara
Primrose-willow
Onagraceae, Evening-primrose Family

Habit: Herbaceous perennial or somewhat woody shrub, 1-4 m tall. Stems woody below, branched, hairy.

Leaves: Alternate; short-stalked; blades lance-shaped, 4-15 cm long, shaggy-hairy.

Inflorescences: Solitary flowers, on stalks 1-2 cm long, in upper leaf axils.

Flowers: Yellow, with 4 or 5 petals, each 1-3 cm long.

Fruit: A 4-sided, club-shaped capsule, 1-3 cm long.

Habitat and Distribution: Common; virtually all shallow water, wet margins, ditches, and other wet disturbed sites; throughout Florida, and north to south Georgia; Caribbean, southern Mexico south through South American; Asia; Australia.

Comment: The large showy flowers of Primrose-willow provide a display in any warm month. The plant freezes which limits its range in the U.S. Primrose-willow is extremely weedy and invasive. Any disturbance, such as clearing, ditching, or cattle grazing in a wetland provides ample opportunity for this weedy wildflower to invade and overwhelm the native vegetation.

Yellow

Ludwigia maritima R.M. Harper

Ludwigia peruviana (L.) H. Hara

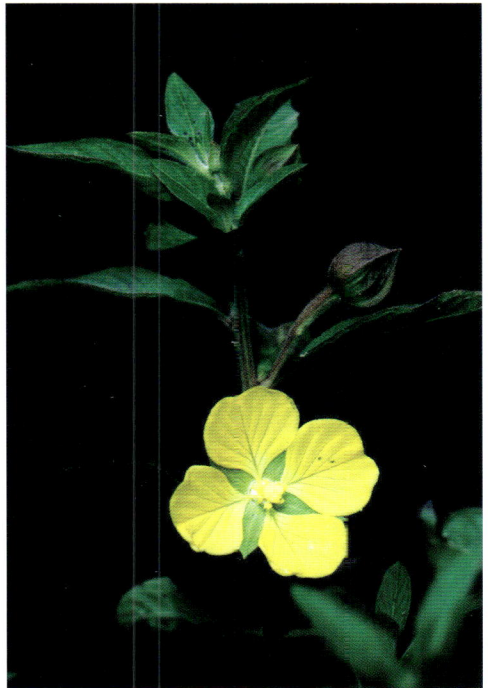

Yellow

Ludwigia suffruticosa Walter
Headed Seedbox
Onagraceae, Evening-primrose Family

Habit: Herbaceous perennial, 30-80 cm tall, from underground runners. Stems erect, branching from base, smooth.

Leaves: Alternate; stalkless or essentially so; blades narrowly lance-shaped to linear, to 4 cm long and 1 cm wide, smaller on upper stems.

Inflorescences: Dense, somewhat rounded, terminal clusters.

Flowers: 4 yellow-green sepals that become reddish, petals absent or minute, yellow.

Fruit: A 4-sided capsule, 3-4 mm long and wide.

Habitat and Distribution: Common; margins, shores, depressions, savannahs, and wet pinelands; throughout Florida, west to southern Mississippi, and north to southeastern North Carolina.

Comment: Flowers can be found in all warm months. Although easily identified by the dense terminal cluster of flowers and fruits together with the narrow leaves, Headed Seedbox is not an especially attractive wildflower.

Ludwigia virgata Michx.
Savannah Seedbox
Onagraceae, Evening-primrose Family

Habit: Herbaceous perennial, to 1 m tall. Stems erect, branched, smooth to hairy.

Leaves: Alternate; stalkless; blades oblong to narrowly linear-oblong, to 3 cm long, usually smooth, tips rounded.

Inflorescences: Solitary flowers, on stalks 6-10 mm long, in leaf axils, on upper wand-like branches.

Flowers: Yellow, with 4 petals, each 1-1.6 cm long.

Fruit: A cubical, 4-sided capsule, 5-7 mm long, often narrowly winged, smooth or nearly so.

Habitat and Distribution: Frequent; bogs and wet, sandy pinelands; central peninsula Florida, west to adjacent Alabama, and north to Virginia.

Comment: Yellow flowers of Savannah Seedbox occur in the warm months, but fall quickly. Nevertheless a few flowers are usually present on the plants. The plants are often found among sparse vegetation.

Yellow

Ludwigia suffruticosa Walter

Ludwigia virgata Michx.

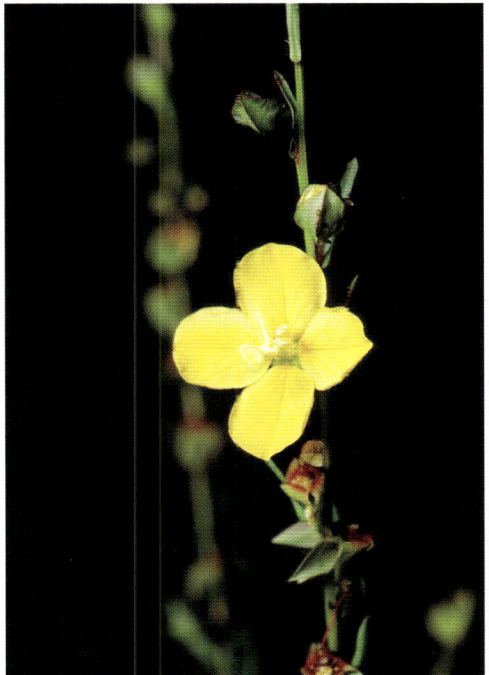

397

Yellow

Macfadyena unguis-cati (L.) A.H. Gentry
Cat's-claw Vine
Bignoniaceae, Trumpet Creeper Family

Habit: Woody perennial vine, to 15 m long. Stems high climbing.

Leaves: Opposite; stalked; blades with 2 ovate to lance-shaped leaflets, 2-7 cm long, and a 3-pronged tendril, tip of prongs hooked.

Inflorescences: Flowers solitary or in few-flowered clusters, in leaf axils.

Flowers: Yellow, funnelform, to 8 cm long, with 5 spreading lobes.

Fruit: A long, slender, flat capsule, to 50 cm long; seeds with thin papery wings.

Habitat and Distribution: Infrequent, but can be locally common; throughout Florida; widespread in warm temperate and tropical climates; native of continental tropical America.

Comment: The very large showy flowers occur throughout the year southward and in the summer in warm temperate areas. Widely cultivated, most escaped plants are found close to development.

Melilotus indicus (L.) All.
Sour-clover or Alfalfilla
Fabaceae (Leguminosae), Pea Family

Habit: Herbaceous annual, to 60 cm tall. Stems erect, branched, with few hairs.

Leaves: Alternate; stalked; blades with 3 broadly elliptic leaflets, often wider at tips, 1-2.5 cm long, smooth.

Inflorescences: Long cylindrical spikes from upper leaf axils.

Flowers: Yellow, slender, pea-shaped, 2.5-3 mm long, fragrant.

Fruit: A flat, round, 1-seeded pod, yellowish or reddish, 1.5-2.5 mm long, surface bumpy.

Habitat and Distribution: Infrequent to occasional; agricultural and other disturbed and waste areas; throughout Florida, north to North Carolina, west to Texas, and from Arizona into southern Oregon; native of Mediterranean region.

Comment: Flowering in spring and summer Sour-clover has a distinctive sweet fragrance. The small flowers, even clustered on the spike, are not really eye-catching. Sour-clover is cultivated as a green-manure or forage crop. Distribution is extremely scattered.

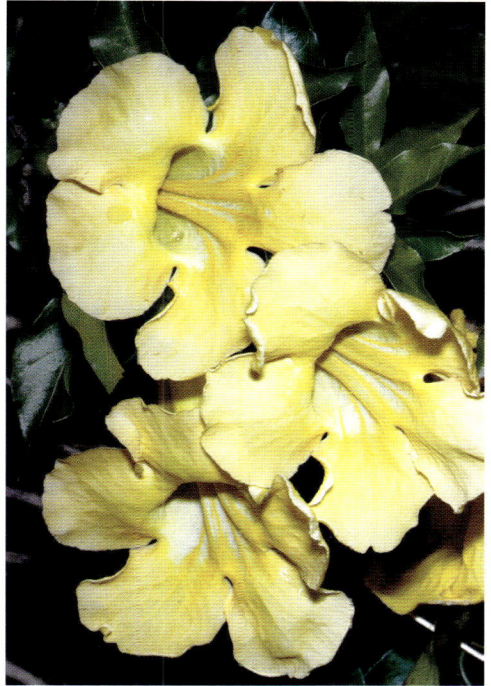

Macfadyena unguis-cati
(L.) A.H. Gentry

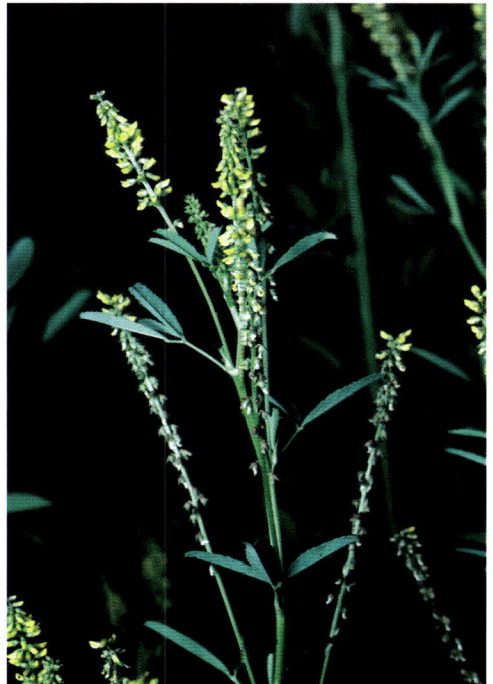

Melilotus indicus (L.) All.

Yellow

Melothria pendula L.
Creeping Cucumber or Meloncito
Cucurbitaceae, Gourd Family

Habit: Herbaceous perennial vine, to 4 m long, from a perennial root. Stems slender, mat-forming, creeping or climbing using tendrils at leaf junctions.

Leaves: Alternate; stalked; blades rounded, with 3 to 5 palmate lobes, 2-8 cm long and wide, margins with fine teeth, base heart-shaped.

Inflorescences: Flowers solitary or in small clusters in leaf axils, male and female flowers separate on the same vine.

Flowers: Yellow or somewhat greenish, small, with 5 petals, to 3 mm long.

Fruit: A long-stalked berry, resembling a cucumber or a tiny watermelon, 1-3 cm long, green or turning blackish.

Habitat and Distribution: Common; virtually all habitats lacking water at or on the surface, especially in disturbed habitats, not found in some cultivated crop lands; throughout Florida, west to Texas and Mexico, and north to Oklahoma, Missouri, Indiana, and Virginia.

Comment: As a wildflower, Creeping Cucumber flowers are not usually noticed unless you have a magnifying lens. The flowers, occurring in the warm months, are quite small. The spreading and matting weedy habit of the vine is easily noticed as it covers bare spots and climbs among low vegetation. It is a common weed of home landscapes. The very small fruits that look remarkably like a tiny watermelons are what most frequently provokes comments. The fruits can be eaten raw or pickled, but be cautious as the seeds are reported to be purgative.

Mentzelia floridana Nutt. ex Torr. & A. Gray
Poorman's Patch or Stick-leaf
Loasaceae, Stick-leaf Family

Habit: Herbaceous perennial, to 1 m tall. Stems sprawling, brittle, many-branched.

Leaves: Alternate; stalked; blades triangular or 3-lobed, 2-10 cm long, with barbed hairs, margins scalloped.

Inflorescences: Flowers solitary in upper leaf axils.

Flowers: Yellow, with 5 broad petals, each 1.5-1.8 cm long.

Fruit: An ellipsoid capsule, to 1.5 cm long, with stiff hairs.

Habitat and Distribution: Common; coastal hammocks, shell middens, dunes, and sandy disturbed areas; peninsula Florida south into the Keys; Bahamas.

Comment: Flowering all year, but not really known as a wildflower due to the unique barbed hairs on leaves that cause them to easily attach to clothing. The leaves can plaster themselves so tightly to clothing that they appear to be patches. Once attached the leaves are quite difficult to remove.

Melothria pendula L.

Mentzelia floridana
Nutt. ex Torr. & A. Gray

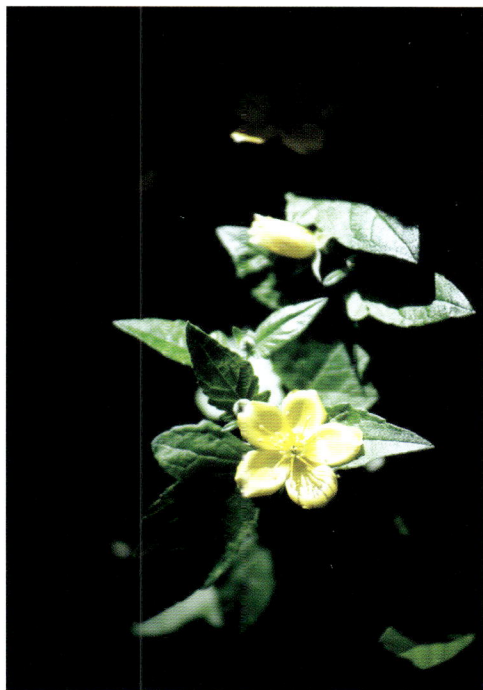

Yellow

Momordica charantia L.
Wild Balsam-apple
Cucurbitaceae, Gourd Family

Habit: Herbaceous annual vine, to 10 m long. Stems creeping or climbing using tendrils at leaf junctions.

Leaves: Alternate; stalked; blades rounded, with 5 to 7 palmate lobes, 6-7 cm long and wide, margins often with teeth, base heart-shaped.

Inflorescences: Male and female flowers separate on the same or separate plants, both from leaf axils, male flowers solitary or in small clusters, female flowers solitary.

Flowers: Bell-shaped, yellow, about 2 cm long, with 5 petals.

Fruit: A golden-yellow berry, ovoid, warty, 4-12 cm long, opening by 3 valves to expose shiny red seeds.

Habitat and Distribution: Common; roadsides, gardens, thickets, fence rows, and other disturbed habitats; Texas and central peninsula Florida southward; throughout the West Indies and the warm temperate and tropic zones of both the Old and New Worlds; native of the Old World tropics.

Comment: The large, yellow, bell-shaped flowers bloom spontaneously throughout the year. The large flowers and especially the fruits are very showy. The bright shiny seeds of open fruits also add to the wildflower display. The fruits and seeds are POISONOUS, not edible.

Nelumbo lutea Willd.
American Lotus, Yellow-lotus, Water-chinquapin, Pond-nut
Nelumbonaceae, Lotus Family

Habit: Herbaceous aquatic perennial, from underground runners. Stems (underground runners) spongy, to 1 cm in diameter.

Leaves: Alternate; long-stalked, stalk attached in center of blade; blades at or above water surface, round, 20-60 cm in diameter, flat if on surface, depressed in center if elevated.

Inflorescences: Solitary flower on a long stalk from the leaf axil, sometimes extending as much as 1 m above the water surface.

Flowers: Yellow, cup-shaped, 7-25 cm wide, with several to many petals.

Fruit: Of solitary, hard nuts, each contained in a hole in a hard, black, flat-topped receptacle, nuts about 1 cm in diameter, receptacle about 10 cm in diameter.

Habitat and Distribution: Infrequent to occasional, can be locally common; still waters of ponds, lakes, marshes, and slow moving streams; central peninsula Florida, west to Texas, and north to Oklahoma, Iowa, Minnesota, Ontario, and New York.

Comment: American Lotus is often found in large masses. The pale yellow flowers are a spectacular sight in summer, especially in very large populations. The very curious receptacles are highly prized for dried flower arrangements, with or without the nuts.

Momordica charantia L.

Nelumbo lutea Willd.

Yellow

Nuphar lutea (L.) Sm.
Spatter-dock, Cow-lily, Yellow Pond-lily
[*Nuphar advena* (Aiton) W. T. Aiton]
Nymphaeaceae, Water-lily Family

Habit: Herbaceous aquatic perennial, from underground runners. Stems (underground runners) branched.

Leaves: Alternate; long-stalked; blades at or above water surface, arrowhead-shaped to round with narrow notch, 25-40 cm long, basal lobes rounded.

Inflorescences: Solitary on a long stalk from the leaf axil, floating or held above the water surface.

Flowers: Roundish to cup-shaped, 3-5 cm across, yellow, petals many, fleshy, small.

Fruit: Ovoid, greenish, yellowish- or reddish-tinged, slightly ribbed, 2-5 cm long and broad.

Habitat and Distribution: Common; still waters of ponds, lakes, marshes, ditches, canals, and slow moving streams; throughout Florida, west to Texas and north to Wisconsin and Maine; Mexico; Cuba.

Comment: Spatter-dock flowers occur in the spring and summer. The unusual flowers are attractive, but not especially showy. The hard fruits are more of an oddity than attractive.

Oenothera biennis L.
Evening-primrose
Onagraceae, Evening-primrose Family

Habit: Herbaceous biennial, 1.3-2 m tall. Stems erect, leafy, with stiff hairs.

Leaves: Alternate; basal rosette with stalks, upper stalkless; blades lance-shaped, to 16 cm long, usually hairy, margins of basal rosette deeply cut, upper blades toothed.

Inflorescences: Solitary flowers from leaf axils at stem tips.

Flowers: Yellow, with 4 petals, each 1.5-3 cm long, above a long tube.

Fruit: A cylindrical capsule, 1.4-4 cm long, 4-5 mm wide.

Habitat and Distribution: Occasional to frequent; can be locally common; dry fields, woods, and disturbed areas; central peninsula Florida, west to Texas, and north to Wisconsin, Ontario, Quebec, and Newfoundland.

Comment: Usually in a mass, the tall erect stems topped with large yellow flowers that open late in the afternoon during summer and fall provide a display of wildflowers that is hard not to notice.

Nuphar lutea (L.) Sm.

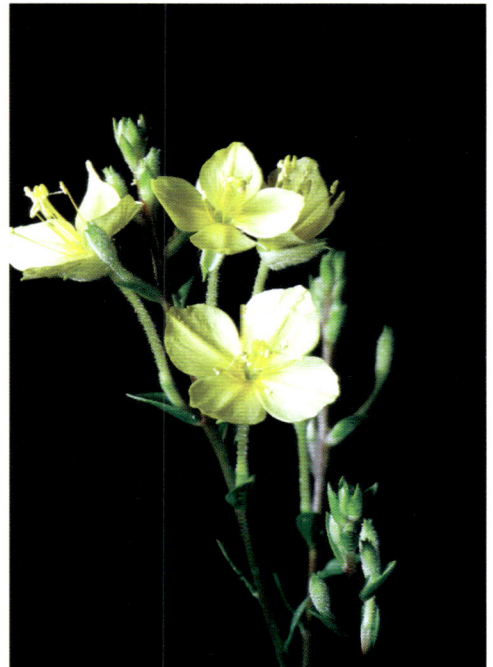

Oenothera biennis L.

Yellow

Oenothera fruticosa L.
Sundrops
Onagraceae, Evening-primrose Family

Habit: Herbaceous perennial, to 80 cm tall. Stems erect, usually branched, hairy.

Leaves: Alternate; stalkless or nearly so; blades narrowly elliptic, linear, or ovate, to 11 cm long, margins smooth or toothed.

Inflorescences: Solitary flowers from leaf axils at stem tips.

Flowers: Yellow, with 4 broad petals, each 1-2 cm long, above a long tube.

Fruit: A rounded to oblong, somewhat 4-sided capsule, 5-9 mm long, 3-4.5 mm wide, hairy.

Habitat and Distribution: Occasional; dry open woods, pinelands, meadows, fields, roadsides, and other disturbed sites; northern peninsula Florida, west to Louisiana, and north to Oklahoma, Missouri, and Virginia.

Comment: The large, showy Sundrops flowers are seen during the day in spring and summer. Frequently cultivated for the showy flowers.

Oenothera laciniata Hill
Cutleaf Evening-primrose
Onagraceae, Evening-primrose Family

Habit: Herbaceous biennial, to 80 cm tall. Stems spreading, branching from base, slightly hairy.

Leaves: Alternate; basal rosette with stalks, upper stalkless; blades elliptic, sometimes broader towards tip, 2.5-8 cm long, margins unevenly lobed or deeply cut, hairy.

Inflorescences: Solitary flowers from leaf axils at stem tips.

Flowers: Yellow to reddish-yellow, with 4 petals, 8-18 mm long, above a long tube.

Fruit: A cylindrical capsule, 2-4 cm long, 3-4 mm broad, hairy, usually slightly curved.

Habitat and Distribution: Common; fields, roadsides, sandy open clearings, and other disturbed sites; throughout Florida, west to Texas, and north to South Dakota and Maine; south into Ecuador.

Comment: Flowers occur near evening in all warm months. The flowers are large enough to be noticed, especially in the open sandy soils where they grow. Cutleaf Evening-primrose is an especially prolific weed in gardens and open crop lands.

Oenothera fruticosa L.

Oenothera laciniata Hill

Yellow

Opuntia humifusa (Raf.) Raf.
Prickly-pear Cactus
Cactaceae, Cactus Family

Habit: Shrubby succulent perennial, forming mats or clumps, usually to 30 cm tall. Stem segments round to broadly elliptical, occasionally elongate, fleshy, flattened, green, 4-16 cm long, difficult to separate.

Leaves: Alternate, quickly deciduous; stalkless; blades narrowly conical, 5-11 mm long, spaced along surface in spine bearing areas (areoles); areoles with 0-2 spines; spines gray, 2-3 cm long, round, not flattened.

Inflorescences: Solitary to several flowers from the areoles at the tips of last year's segments.

Flowers: Yellow often with a red center, cup-shaped, 4-9 cm long and wide, with many waxy petals.

Fruit: An ovoid berry, wider at the tip, 2-6 cm long, fleshy, spiny, becoming reddish or purple.

Habitat and Distribution: Common; dry, sandy, soils of dunes, sandhills, prairies, hammocks, and other open habitats; throughout Florida, west to Mississippi, and north to Minnesota, Ontario, and Massachusetts.

Comment: Prickly-pear Cactus blooms during any warm month southward and in the summer and early fall northward. The flowers are large and showy, especially so in the dry open areas in which they are found. Fruits are edible, but special techniques must be used to avoid the spines and small barbed hairs at the areoles.

Orontium aquaticum L.
Golden Club or Never-wet
Araceae, Arum Family

Habit: Herbaceous perennial, 15-60 cm tall, from a deep thick underground runner (rhizome).

Leaves: Alternate, in a basal cluster; long-stalked; blades rounded to elliptic, 10-40 cm long, 4-16 cm wide, often an iridescent green, sheathed at base.

Inflorescences: Dense flowers on upper 5-10 cm of a cylindrical flowering stalk.

Flowers: Small, yellow.

Fruit: Blue-green berries, each 6-10 mm diameter.

Habitat and Distribution: Infrequent; bogs, streams, and wet ditches; from central Florida west to Louisiana and north to Massachusetts.

Comment: Golden Club flowers in the winter southward and spring northward. The golden spike in a bed of iridescent leaves is absolutely striking. Soils washing into the water and along the waterway often bury the runner very deep.

Opuntia humifusa (Raf.) Raf.

Orontium aquaticum L.

Yellow

Oxalis corniculata L.
Creeping Wood-sorrel, Agrito, Jocoyote
Oxalidaceae, Wood-sorrel Family

Habit: Herbaceous perennial, 5-40 cm long, from a slender taproot. Stems creeping, rooting at joints, often reddish, lightly hairy.

Leaves: Alternate; long-stalked; blades with 3 heart-shaped leaflets, each to 1 cm long, green or reddish.

Inflorescences: 1- or 2- to sometimes several-flowered, in loose clusters on long stalks from leaf axils.

Flowers: Yellow, 5 petals, each 5-11 mm long.

Fruit: A cylindrical, beaked capsule; 0.8-1 cm long, usually hairy.

Habitat and Distribution: Common; lawns, roadsides, pastures, fields, greenhouses, and other disturbed habitats; throughout Florida, west to California, and north to Canada; pantropical; native of Europe.

Comment: Flowers can be found in any warm month. The creeping stems can become matted. Germinating easily from seeds, this extremely weedy species can pop up in virtually any disturbed soil. A pernicious greenhouse weed, Creeping Wood-sorrel travels easily and frequently in pots.

Oxalis dillenii Jacq.
Southern Yellow Wood-sorrel
[*Oxalis florida* Salisb.]
Oxalidaceae, Wood-sorrel Family

Habit: Herbaceous perennial, 10-25 cm tall. Stems erect or reclining, branching from the base, slender, hairy.

Leaves: Alternate; long-stalked; blades with 3 heart-shaped leaflets, each 10-16 mm wide, green.

Inflorescences: 2- to 9-flowered, in loose clusters on long stalks from leaf axils.

Flowers: Yellow, 5 petals, each 4-10 mm long.

Fruit: A cylindric, beaked capsule, 1-2.5 cm long, smooth to hairy.

Habitat and Distribution: Occasional to common; lawns, fields, pastures, woodlands, bottomlands, and disturbed habitats; throughout Florida, west to Texas, and north to New Jersey; becoming a cosmopolitan weed; introduced to western Europe.

Comment: Southern Yellow Wood-sorrel flowers in most warm months. The plant appears to be a cluster of green leaves with small yellow flowers poking out.

Oxalis corniculata L.

Oxalis dillenii Jacq.

Yellow

Oxalis stricta L.
Yellow Wood-sorrel, Yellow Oxalis, Chanchaquilla
Oxalidaceae, Wood-sorrel Family

Habit: Herbaceous perennial, 10-50 cm tall, from white underground runners. Stems erect to lying flat, almost smooth to hairy.

Leaves: Alternate; long-stalked; blades with 3 heart-shaped leaflets, each 0.7-2 cm broad, purplish.

Inflorescences: 2- to 7-flowered, in loose clusters on long stalks from leaf axils.

Flowers: Yellow, 5 petals, each 4-9 mm long.

Fruit: A cylindrical, beaked capsule, 8-1.5 cm long, hairy.

Habitat and Distribution: Common; woodlands, floodplains, and disturbed sites; panhandle Florida, west to Texas, and north to central and eastern U.S.; becoming a cosmopolitan weed; introduced into Europe.

Comment: Yellow Wood-sorrel blooms in the warm months. A typical weedy Wood-sorrel, but slightly taller. The flower is very attractive, but is not very large, and occurs close to the ground, so one must look carefully to find it.

Pectis prostrata Cav.
Chicken Weed
Asteraceae (Compositae), Aster or Sunflower Family

Habit: Herbaceous annual, to 30 cm long, from a taproot. Stems lying flat, matted, branching from base, hairy.

Leaves: Opposite; stalkless; blades linear to linear oblong, to 3 cm long, 2-5 mm wide, numerous, scattered, round, imbedded oil glands underneath, margins with bristles near base.

Inflorescences: Heads solitary or clustered in leaf axils.

Flowers: Rays 5, yellow, tubular, to 2 mm long.

Fruit: A linear nutlet, to 4 mm long, with scales and bristles at the tip.

Habitat and Distribution: Infrequent to occasional; dry soils of pinelands, lawns, roadsides, deserts, arroyos, washes, and disturbed sites; throughout Florida, and in Texas, New Mexico, and Arizona; Mexico south into northern South America.

Comment: The tiny flower heads can be found in summer and fall. Chicken Weed is most commonly encountered as a flat mat in dry turf. The flowers are matted into the foliage making them very difficult to see. Identification depends on looking at the oil glands under the blades with a magnifying lens.

Oxalis stricta L.

Pectis prostrata Cav.

Yellow

Phoebanthus tenuifolius (Torr. & A. Gray) S.F. Blake
Pineland False Sunflower
Asteraceae (Compositae), Aster or Sunflower Family

Habit: Herbaceous perennial, to 1 m tall, from tuberous underground runners.

Leaves: Opposite on lower stem, alternate on upper; stalkless or nearly so; blades linear, to 8 cm long, sandpapery, margins smooth.

Inflorescences: Heads solitary at stem and branch tips.

Flowers: Rays yellow, each 3-4.5 cm long.

Fruit: A thick, 4-sided nutlet, hairy, tipped with scales.

Habitat and Distribution: Rare; sandy pinelands and sandhills; Calhoun, Gulf, Franklin, and Liberty Counties along the Apalachicola River in central Panhandle Florida; endemic.

Comment: The very showy flowers of Pineland False Sunflower can be found in spring and summer. The large heads, as the common name implies, resemble Sunflowers. This False Sunflower is listed as Threatened by the State of Florida.

Physalis angulata L.
Cutleaf Ground-cherry
Solanaceae, Nightshade Family

Habit: Herbaceous annual, 0.3-1 m tall. Stems erect, smooth, angled, loosely branched.

Leaves: Alternate; long-stalked; blades ovate to somewhat lance-shaped, 4-13 cm long, margins usually deeply toothed.

Inflorescences: Solitary, long-stalked flowers from leaf axils.

Flowers: Yellowish, bell shaped, 0.6-1.2 cm long, with 5 lobes, nodding, anthers blue.

Fruit: A round, yellow berry, about 1 cm diameter, enclosed by joined green papery sepals, 1.5-3 cm long.

Habitat and Distribution: Frequent; sandy soils of woodlands, sandhills, fields, thickets, roadsides, and disturbed habitats; throughout Florida, west to Texas and north to Kansas, Missouri, Illinois, and Pennsylvania; Mexico; Central America; and West Indies.

Comment: Cutleaf Ground-cherry has light green yellowish flowers that bloom in any warm month. As a wildflower it is easily overlooked. What usually catches our attention is the large papery covering of the fruit. The papery covering is not showy, but is quite unusual.

Phoebanthus tenuifolius (Torr. & A. Gray) S.F. Blake

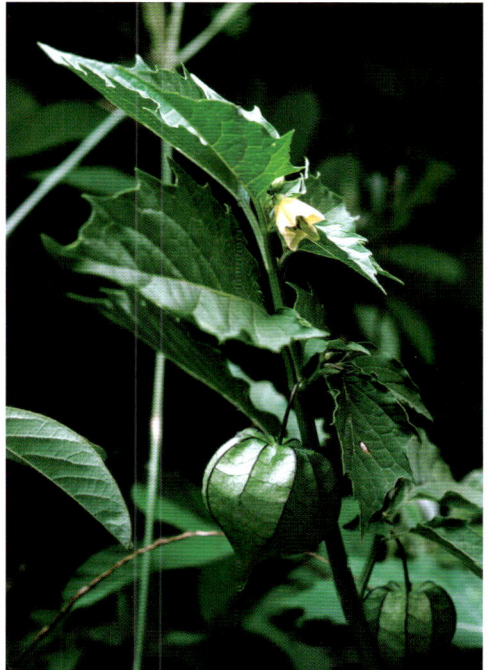

Physalis angulata L.

Yellow

Physalis arenicola Kearney
Pineland Ground-cherry or Sand Ground-cherry
[*Physalis arenicola* var. *ciliosa* (Rydb.) Waterf.; *Physalis ciliosa* Rydb.]
Solanaceae, Nightshade Family

Habit: Herbaceous perennial, to 30 cm tall. Stems erect to spreading, branched, hairy.

Leaves: Alternate; stalked; blades ovate, often somewhat heart-shaped, 1.5-8 cm long, margins scalloped to smooth.

Inflorescences: Solitary, long-stalked flowers from leaf axils.

Flowers: Yellowish, bell shaped, 1.5-2 cm long, with 5 lobes, nodding, anthers yellow.

Fruit: A round, yellow berry, about 9 mm in diameter, enclosed by connected green, papery sepals, to 3 cm long.

Habitat and Distribution: Frequent; sandy soils of pinelands, sandhills, scrub, open woodlands, thickets, and disturbed sites; peninsula to central panhandle Florida, and north into Georgia.

Comment: Pineland Ground-cherry has light green foliage, yellowish flowers and is usually diffusely spreading. Although the flowers can be present in any warm month, they are seldom noticed. As with most Ground-cherries, that large papery calyx covering the fruit, even though green, is what strikes the attention.

Physalis walteri Nutt.
Sticky Ground-cherry
[Kunze;L. var.(Kunze) Waterf.; L. var.(M.A. Curtis) Rydb.; L. subsp.(M.A. Curtis) Waterf.]
Solanaceae, Nightshade Family

Habit: Herbaceous perennial, usually 15-30 cm tall, from underground runners. Stems reclining to erect, covered with branched, star-shaped hairs.

Leaves: Alternate; stalked; blades rounded to elliptic, 3-7 cm long, lower surface covered with branched, star-shaped hairs, margins smooth.

Inflorescences: Solitary, long-stalked flowers from leaf axils.

Flowers: Yellow, bell-shaped, 1-2 cm long, nodding, anthers yellow.

Fruit: A round, yellow berry, 7-10 mm in diameter, enclosed by connected green papery sepals, to 2-4 cm long.

Habitat and Distribution: Common; sandy soils of sandhills, coastal sand dunes, and open woodlands; throughout Florida, west to Texas, and north to Virginia; south through Mexico and Central America into South America.

Comment: Flowering can occur in any warm month. Sticky Ground-cherry, as a wildflower, is not showy. In fact, the flowers are easily overlooked. The large papery calyx covering the fruit is often noticed and sometimes compared to a Chinese lantern.

Physalis arenicola Kearney

Physalis walteri Nutt.

Yellow

Pinguicula lutea Walter
Yellow Butterwort
Lentibulariaceae, Bladderwort Family

Habit: Herbaceous perennial, to 30 cm tall. Stems 1-8, leafless, with sticky hairs.

Leaves: In a basal rosette; stalkless; blades elliptic to oblong, 1-6 cm long, to 2 cm wide, yellowish-green, sticky on the upper surface, margins often rolled upward.

Inflorescences: Terminal, solitary flowers on a long stalk.

Flowers: Yellow, bell-shaped, spurred, to 3.5 cm long, 2-lipped, 5-lobed, lobe tips notched.

Fruit: A round capsule, 3-4.5 mm diameter.

Habitat and Distribution: Frequent to locally common; wet soils of pinelands, bogs, savannahs, and wet roadsides; throughout Florida, west to Louisiana, and north to North Carolina.

Comment: In spring the eye-catching bright yellow flowers can easily be seen among the dark soils of the boggy areas it inhabits. The sticky leaves catch and absorb insects. Yellow Butterwort is listed as Threatened by the State of Florida, although its distribution and frequency probably does not warrant such treatment.

Piriqueta caroliniana (Walter) Urb.
Piriqueta or Pitted Stripeseed
[*Piriqueta cistoides* (L.) Griseb. subsp. *caroliniana* (Walter) Arbo]
Turneraceae, Turnera Family

Habit: Herbaceous perennial, 15-50 cm tall. Stems slender, smooth to hairy, hairs variable with star-shaped and/or long spreading yellowish-brown hairs.

Leaves: Alternate; stalkless or nearly so; blades linear to oblong-elliptic to lance-shaped, 2-8 cm long, with fine star-shaped hairs, margins smooth to blunt toothed.

Inflorescences: Flowers solitary, on long stalks from the upper leaf axils.

Flowers: Yellow to orange-yellow, 5 petals, each 1.5-2.2 cm long, stigmas 3, branched and bushy.

Fruit: A round capsule, 5-7 cm in diameter.

Habitat and Distribution: Frequent; sandy soils of pinelands, sandhills, open margins, swales, marshes, and clearings; peninsula to central panhandle Florida, and north to North Carolina; Cuba; Hispaniola; Bahamas.

Comment: Deep yellow to bright orange-yellow flowers occur in all warm months. Plants tend to bloom in flushes within a general area. The flowers are so eye-catching that wildflower enthusiasts seldom pass them without comment.

Pinguicula lutea Walter

Piriqueta caroliniana (Walter) Urb.

419

Yellow

Pityopsis graminifolia (Michx.) Nutt.
Silk-grass
[*Chrysopsis graminifolia* (Michx.) Elliott; *Heterotheca graminifolia* (Michx.) Shinners]
Asteraceae (Compositae), Aster or Sunflower Family

Habit: Herbaceous perennial, 30-80 cm tall, from underground runners. Stems erect, often branching, with silvery hairs.

Leaves: Alternate; stalkless; blades slender, grass-like, 10-35 cm long, with 3 to 5 or more parallel veins, with silvery silky hairs.

Inflorescences: Heads in a many-branched arrangement at the top of stems.

Flowers: Rays yellow, 8-10 mm long, narrow.

Fruit: A narrow, ribbed, hairy nutlet, 2.3-3 mm long, with tuft of bristles 4.5-6 mm long at tip.

Habitat and Distribution: Common; dry sandy soils of pinelands, sandhills, dunes, old fields, open woods, and roadside banks; throughout Florida, west to Texas, and north to Oklahoma, Ohio, and Delaware; Guatemala; southern Mexico; Bahamas.

Comment: The flowers can be seen from summer into fall. Although not large and showy, Silk-grass as a wildflower is very appealing due to its environment, the many flowers, and silky foliage. Silk-grass is quite common in open, very disturbed habitats in which not much else flowers. The silky foliage with parallel veins is so grass-like that young plants frequently fool plant enthusiasts.

Pityopsis oligantha (Chapm. ex Torr. & A. Gray) Small
Few-flowered Silk-grass
[*Chrysopsis oligantha* Chapm. ex Torr. & A. Gray; *Heterotheca oligantha* (Chapm. ex Torr. & A. Gray) Harms]
Asteraceae (Compositae), Aster or Sunflower Family

Habit: Herbaceous perennial, 20-50 cm tall, from underground runners. Stems usually 1, erect, with silvery hairs and with dense stalked glandular hairs.

Leaves: Alternate, overwinter in basal rosette; stalkless; blades slender, grass-like, 10-30 cm long, with 3 to 5 or more parallel veins, with silver, silky, and glandular hairs.

Inflorescences: Fewer than 8 heads in a branched, terminal arrangement.

Flowers: Rays yellow, 9-13 mm long, narrow.

Fruit: A narrow, ribbed, hairy nutlet, 4-5 mm long, with a tuft of bristles 6-7 mm long at tip.

Habitat and Distribution: Occasional; open areas in wet flatwoods, bogs, and cypress pond margins; central and western panhandle Florida, and adjacent Georgia, Alabama, and Mississippi.

Comment: Few-flowered Silk-grass has slightly larger heads than Silk-grass, but fewer heads. This species flowers in spring. Unlike Silk-grass, Few-flowered Silk-grass grows exclusively in wet habitats. The young growth disturbingly resembles a grass.

Pityopsis graminifolia
(Michx.) Nutt.

Pityopsis oligantha
(Chapm. ex Torr. & A. Gray) Small

Yellow

Platanthera ciliaris (L.) Lindl.
Yellow-fringed Orchid or Orange Plume
[*Blephariglotis ciliaris* (L.) Rydb.; *Habenaria ciliaris* (L.) R. Br.]
Orchidaceae, Orchid Family

Habit: Herbaceous terrestrial perennial, to 1 m tall. Stem stout, leafy.

Leaves: Alternate; stalkless; blades 2-5, lance-shaped, 6-30 cm long.

Inflorescences: Flowers in a terminal, spike-like arrangement.

Flowers: Yellow to orange, about 4 cm long, with deeply fringed lip and slender spur 2-3.5 cm long.

Fruit: A cylindrical capsule.

Habitat and Distribution: Frequent; open wet soils of meadows, ditches, seeps, bogs, seepage slopes, floodplains, savannahs, prairies, bottomlands, and flatwoods; central peninsula Florida, west to Texas, and north to Arkansas, Missouri, Illinois, southern Michigan, and Massachusetts.

Comment: The spectacular flowers appear from late July into late September. These brilliant wildflowers are a showstopper. Yellow-fringed Orchid is listed as Threatened by the State of Florida, although it is quite frequent. Distribution is scattered giving the impression that it is less frequent than it actually is.

Polygala cymosa Walter
Tall Milkwort or Tall Pine Barren Milkwort
Polygalaceae, Milkwort Family

Habit: Herbaceous biennial, 0.4-1.2 m tall. Stems usually solitary, branched near the top below the flowers, smooth.

Leaves: Alternate; stalkless; blades in basal rosette, very slender, linear to narrowly lance-shaped, 3-14 cm long, stem leaves much smaller, linear.

Inflorescences: Flowers in dense, few- to many-branched, cylindrical clusters at stem tips.

Flowers: Yellow or greenish, 3-4 mm long.

Fruit: A rounded capsule, to 1 mm wide.

Habitat and Distribution: Frequent; wet pinelands, savannas, ponds, depressions, marshes, pine barrens, bogs, swamps, ditches, and other wet areas; central peninsula Florida, along the coastal plain west to Louisiana, and north to Delaware.

Comment: The showy flowers can be seen from spring into fall and appear on almost leafless stiffly erect stems. Basal leaves remain through flowering. The tall stems often exceed nearby vegetation.

Platanthera ciliaris (L.) Lindl.

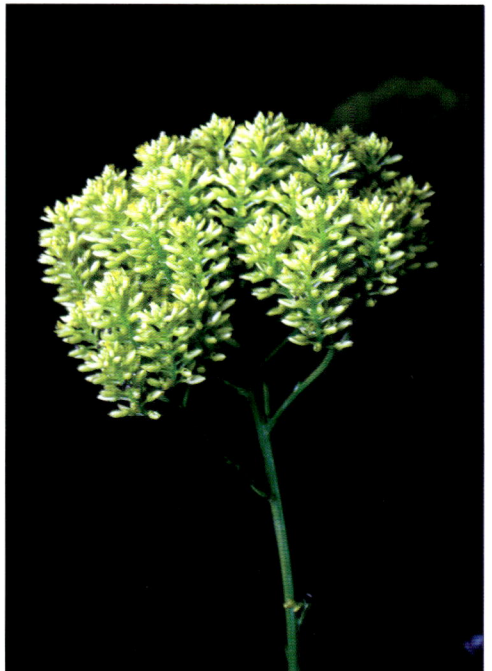

Polygala cymosa Walter

Yellow

Polygala nana (Michx.) DC.
Wild Bachelor's-button or Candyroot
Polygalaceae, Milkwort Family

Habit: Herbaceous annual or biennial, 2-15 cm tall. Plant compact, low growing.

Leaves: Alternate; mostly basal; stalkless; blades spatulate, 1-5.5 cm long, somewhat succulent.

Inflorescences: Flowers in terminal, almost round clusters that become cylindrical.

Flowers: Yellow or greenish, to 5.5-7.5 mm long.

Fruit: A rounded capsule, to 1 mm wide.

Habitat and Distribution: Frequent to common; wet soils of pinelands, swales, ditches, pond and bog margins, and moist sandy areas; throughout Florida, west to Texas, and north to North Carolina.

Comment: The dense clusters of lemon-yellow to somewhat greenish flowers occur from spring into summer. The colors are subdued and the heads are often somewhat nestled among the foliage. Upon drying the flowers become bluish green.

Polygala ramosa Elliott
Short Milkwort or Low Pinewoods Milkwort
Polygalaceae, Milkwort Family

Habit: Herbaceous annual, 10-30 cm tall. Stems solitary, branched below flowers, smooth.

Leaves: Alternate; stalkless; blades spatulate on lower stem, becoming narrower upwards, 1-2 cm or sometimes to 7 cm long, often withered when plant is flowering.

Inflorescences: Terminal, cylindrical, spike-like clusters on multiple branches, branches form a flat-topped arrangement.

Flowers: Yellow, 3 mm long.

Fruit: A rounded capsule, about 1 mm wide.

Habitat and Distribution: Frequent; moist to wet soils of pinelands, bogs, savannas, ponds, depressions, ditches, and swales; throughout Florida, west to Texas, and north to southern New Jersey.

Comment: The bright yellow flowers in large arrangements seen from summer into fall never fail to catch the eye. As a wildflower it is easy to spot. When dry the flowers turn a dark bluish green.

Polygala nana (Michx.) DC.

Polygala ramosa Elliott

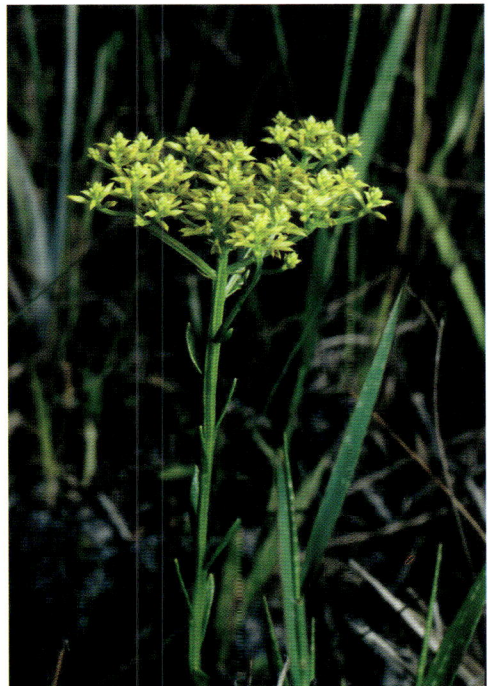

Yellow

Polygala rugelii Shuttlew. ex Chapm.
Yellow Milkwort
Polygalaceae, Milkwort Family

Habit: Herbaceous annual, biennial, or occasionally perennial, 25-75 cm tall. Stem usually several, leafy, often branched below flowers.

Leaves: Alternate, mostly in a basal rosette; stalkless; blades spatulate, 3-6 cm long, becoming narrower and shorter upward along stems.

Inflorescences: A somewhat rounded cluster at stem and branch tips.

Flowers: Yellow, 5-8 mm long, tips lobed.

Fruit: A rounded capsule, to 1 mm wide.

Habitat and Distribution: Common; moist to wet soils of flatwoods, savannas, ditches, depression margins, bogs, and other wet areas; peninsula and eastern panhandle Florida; endemic.

Comment: The bright yellow flowers in clusters are eye-catching and can be seen from spring into fall. When dry the flowers turn bluish green or yellowish green.

Polygonum punctatum Elliott
Dotted Smartweed
Polygonaceae, Buckwheat Family

Habit: Herbaceous perennial, 0.3-1 m tall. Stems leaning down and rooting at the joints, sometimes erect, often branched, joints covered by a papery tube, tube has bristles at top.

Leaves: Alternate; short-stalked; blades lance-shaped, 6-15 cm long, smooth or with short hairs.

Inflorescences: Terminal, spike-like arrangements at stem and branch tips.

Flowers: White to greenish-white with a yellowish cast, to 3 mm long, dotted with glands.

Fruit: A black, glossy, 3-angled nutlet, 2-3 mm long.

Habitat and Distribution: Common, often locally dense; virtually all habitats; throughout Florida, and temperate and subtropical North America; Bermuda; West Indies; tropical America.

Comment: Dotted Smartweed flowers during all the warm months. Often a few spikes of flowers can be seen in even the smallest depressions that occasionally hold water. Frequently growing in very dense populations, deep mats of old stems can push the plant to several feet above the surface.

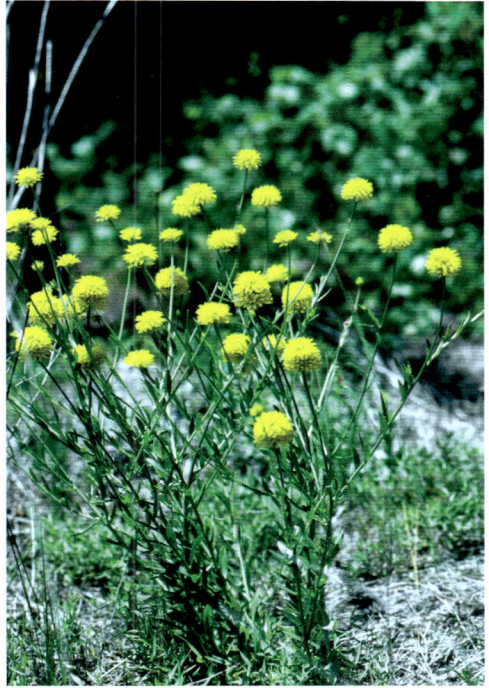

Polygala rugelii
Shuttlew. ex Chapm.

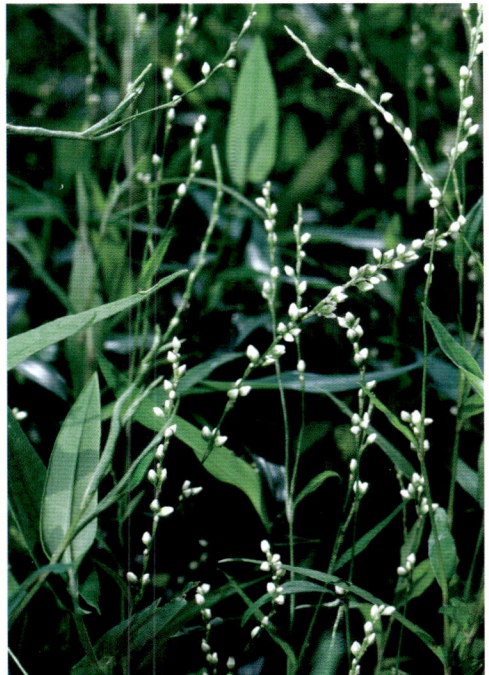

Polygonum punctatum Elliott

Yellow

Polymnia uvedalia (L.) L.
Bear's-foot
Asteraceae (Compositae), Aster or Sunflower Family

Habit: Herbaceous perennial, 1-3 m tall. Stems angled, purplish, hairy on upper portion.

Leaves: Opposite on lower stem, sometimes alternate on upper stem; blades taper to a leafy stalk; blades broad, triangular, 10-30 cm long and wide, with 3 to 5 lobes, hairy, rough.

Inflorescences: Heads terminal and axillary on long-stalked, leafy branches.

Flowers: Rays yellow, 1.5-2.5 cm long.

Fruit: An ovoid, flattened, blackish nutlet, 4-6 cm long, with shallow grooves.

Habitat and Distribution: Infrequent to frequent; moist woods, slits in limerock, meadows, and pastures; central peninsula Florida, west to Texas, and north to Oklahoma, Missouri, Illinois, Pennsylvania, and Delaware.

Comment: The summer and fall flowers are almost lost among the plentiful large leaves. As a wildflower the large leafy green plant with the yellow sprinkles of flowers among the leaves is quite attractive.

Portulaca oleracea L.
Common Purslane or Little Hogweed
Portulacaceae, Purslane Family

Habit: Herbaceous annual, to 45 cm long. Stems lying flat, mat forming, smooth, succulent, often reddish.

Leaves: Alternate or nearly opposite; stalkless; blades spatulate, to 5 cm long, thick, succulent.

Inflorescences: Flowers solitary or in clusters from leaf axils.

Flowers: Yellow, petals usually 5, 3-4.6 mm long.

Fruit: A rounded capsule, 3.5-4.2 mm in diameter, splitting open around the middle; seeds dark red or black.

Habitat and Distribution: Frequent or locally common; slopes, dunes, gardens, salt marshes, fields, crops, sod farms, turf grasses, and other disturbed sites; throughout Florida, the U.S. and southern Canada; temperate and tropical regions of the world; thought to be native to the Old World tropics.

Comment: Common Purslane flowers can occur in any warm month. The large flowers are showy as wildflowers. The fleshy prostrate plants spreading over open ground are curious and often noted, even without blooms. Cultivated forms are offered for sale. A more erect, thicker, very succulent form is grown in gardens as a pot herb.

Polymnia uvedalia (L.) L.

Portulaca oleracea L.

Yellow

Pyrrhopappus carolinianus (Walter) DC.
False Dandelion
Asteraceae (Compositae), Aster or Sunflower Family

Habit: Herbaceous annual or biennial, 0.3-0.7 m tall, from a taproot. Stems erect, sparingly branched, smooth or with few hairs above, sap milky.

Leaves: Alternate, basal rosette present; stalkless or the basal leaves with stalks; blades lance-shaped and widest above the middle or elliptic, 8-25 cm long, margins sometimes smooth or more often deeply lobed, stem leaves smaller.

Inflorescences: Heads at tips of stems and branches.

Flowers: Rays yellow, sometimes very pale, 2-2.5 cm long.

Fruit: A cylindric nutlet, tapering to both ends, 4-4.5 mm long, hairy, 5-grooved, and with a tuft of bristles 8-10 mm long at the tip of a long stalk.

Habitat and Distribution: Common; open dry soils of fields, meadows, pastures, lawns, roadsides, and other disturbed sites; central peninsula Florida, west to Texas, and north to Oklahoma, Kansas, Missouri, Indiana, and Delaware.

Comment: The heads of False Dandelion are quite photographic when the plants bloom in spring and summer. After the blooms have gone the many fruits produced by the heads each have a tuft of bristles resembling a powder puff. These fruiting heads are also attractive. A gentle puff of wind will often release the seeds, each with a parachute of bristles, so that the air will be thick with floating seeds. The flowers and fruits closely resemble Dandelion (*Taraxacum officinale*).

Raphanus raphanistrum L.
Wild Radish or Jointed Charlock
Brassicaceae (Cruciferae), Mustard Family

Habit: Herbaceous winter annual, 30-80 cm tall, from a taproot. Stems covered with stiff hairs.

Leaves: Alternate; stalked; blades larger on lower parts of plant, smaller and narrower on upper parts; lower leaves often dissected, upper leaves lobed, margins of all blades toothed.

Inflorescences: Flowers stalked, in an open, elongated assortment at stem and branch tips, flowering progressively upward.

Flowers: Yellow or white, less commonly pink or purple, 2-3 cm wide, with 4 petals.

Fruit: A narrow pod, 2-6 cm long and 3-5 mm diameter, with constrictions between seeds.

Habitat and Distribution: Frequent, locally common; disturbed sites, particularly cultivated fields; throughout Florida and throughout the temperate regions of the world; native to the Mediterranean region of Europe.

Comment: Flowering in spring, Wild Radish is familiar to many as a weed of roadsides and gardens. In mass it makes a showy wildflower.

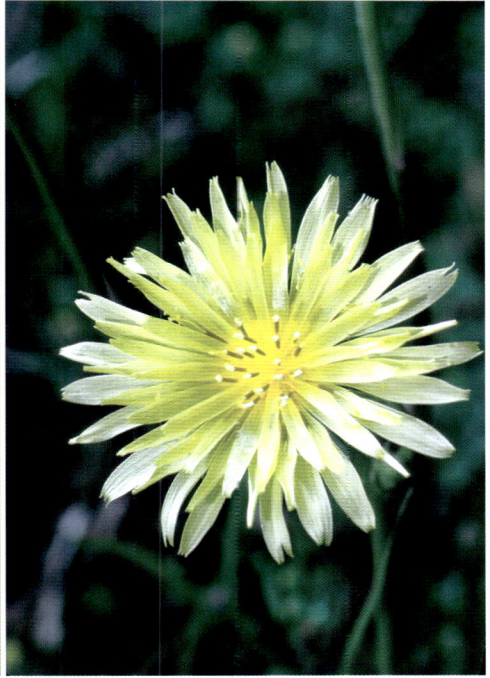

Pyrrhopappus carolinianus (Walter) DC.

Raphanus raphanistrum L.

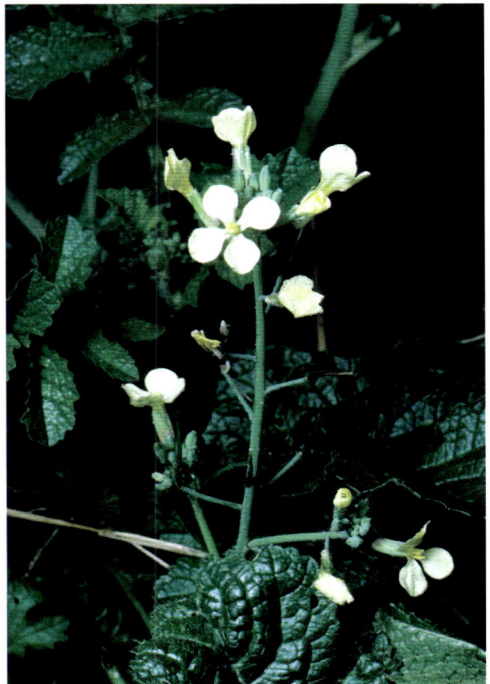

Yellow

Rhexia lutea Walter
Yellow Meadow-beauty
Melastomataceae, Melastome Family

Habit: Herbaceous perennial, 10-60 cm tall. Stems glandular hairy, 4-sided, usually branched.

Leaves: Opposite; stalkless; blades elliptic to lance-shaped and wider above the middle, to 2.8 cm long, 3-nerved, sparsely hairy.

Inflorescences: Open, branched clusters at stem tips and in upper leaf axils.

Flowers: Yellow, petals 4, to 13 mm long.

Fruit: An urn-shaped, ridged capsule, to 9 mm long, smooth.

Habitat and Distribution: Occasional; wet soils of pinelands, bogs, and savannahs; northern peninsula Florida, west to Texas, and north to eastern North Carolina.

Comment: The large showy flowers occur in the spring and summer. The color is distinctive and identifies the only Rhexia with yellow blooms.

Rhizophora mangle L.
Red Mangrove
Rhizophoraceae, Mangrove Family

Habit: Shrub or tree, to 20 m tall, with bowed stilt roots from the trunk and branches.

Leaves: Opposite; stalked; blades elliptic or oblong, sometimes broadly so, 4-15 cm long, evergreen, leathery.

Inflorescences: Clusters of 2 to 8 flowers in leaf axils.

Flowers: Yellow, petals 4, slender, triangular, to 1 cm long, hairy.

Fruit: A cone-shaped, leathery berry, with a single seed, seed germinates in the fruit to produce a 20-30 cm long club-shaped first root.

Habitat and Distribution: Frequent to locally common; shallow salty or brackish coastal waters, frequently tidal; central peninsula Florida (Levy County in the west coast and Volusia County on the east coast southward); coasts of the West Indies; tropical America, West Africa; the Pacific Islands.

Comment: Blooming all year, flowers of Red Mangrove are hardly spectacular. However, the unusual germinating fruits are very curious and provoke comments and photos. Red Mangrove is one of our most valuable plants. The curious fruits fall from the plant and float upright. If the primitive root makes contact with soil it will grow and produce a new plant. In this way the Red Mangrove can start new populations. The maze of stilt roots collects sand thereby extending land. These land builders also provide breeding areas and nurseries for a multitude of sea life.

Rhexia lutea Walter

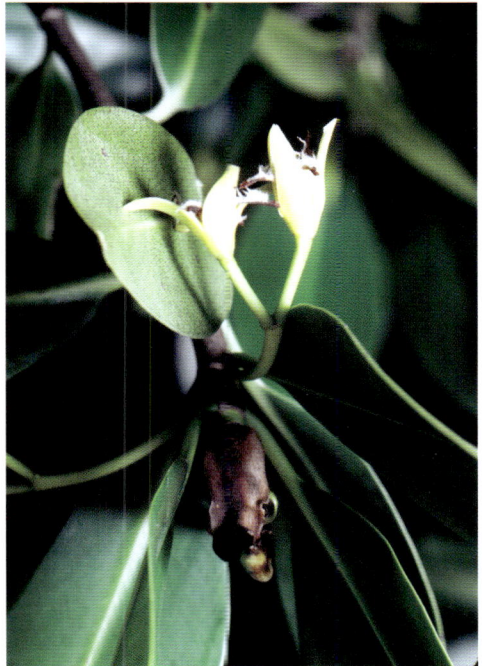

Rhizophora mangle L.

Yellow

Rhododendron austrinum (Small) Rehder
Florida Flame Azalea
[*Azalea austrina* Small]
Ericaceae, Heath Family

Habit: Shrub, to 3 m or more tall. Twigs hairy.

Leaves: Alternate, deciduous; stalked, stalks with stalked glandular hairs; blades elliptic to somewhat broader above the middle, to 11 cm long, appearing after flowers begin to open.

Inflorescences: Terminal clusters on shoots of the previous year.

Flowers: Yellow and orange, funnel-shaped, with 5 lobes, stamens long exserted.

Fruit: A cylindric, dry capsule, 1.5-2.5 cm long, 5-lobed, with stalked glandular hairs.

Habitat and Distribution: Occasional; hammocks, floodplain forests, bluffs, slopes, and stream banks; Baker County on the Georgia border in north peninsula, and Leon and Liberty Counties westward into panhandle Florida, north into southwest Georgia, and southern and central Alabama.

Comment: Spring flowering, Florida Flame Azalea deserves its common name as the blooms are spectacular. It can be cultivated in north peninsula Florida south of its natural range. Florida Flame Azalea is listed by the State of Florida as Endangered.

Rhynchosia difformis (Elliott) DC.
Twining Rhynchosia or Twining Snoutbean
[*Rhynchosia lewtonii* (Vail) Small; *Rhynchosia tomentosa* (L.) Hook & Arn., misapplied]
Fabaceae (Leguminosae), Pea Family

Habit: Herbaceous perennial vine, 0.5-3 m long. Stems climbing or trailing, with downward-pointing hairs.

Leaves: Alternate; stalked; blades with 3 leaflets, sometimes the lowest blades with 1 leaflet, each rounded, 1.5-5 cm long, hairy, resin dotted.

Inflorescences: Dense clusters in leaf axils.

Flowers: Yellow, typically pea-shaped, 5.5-11 cm long.

Fruit: A pod, 1.4-2.1 cm long, 6-8 mm wide, hairy.

Habitat and Distribution: Infrequent to occasional; dry woods, sandhills, sandy woods, old fields, roadsides, and clearings; central peninsula Florida, west to Texas, and north to Virginia.

Comment: The flowering clusters can be seen in spring and summer. The yellow flowers are showy enough to catch most wildflower watchers' eyes.

Rhododendron austrinum
(Small) Rehder

Rhynchosia difformis (Elliott) DC.

435

Yellow

Rhynchosia michauxii Vail
Oneleaf Rhynchosia or Michaux's Snoutbean
Fabaceae (Leguminosae), Pea Family

Habit: Herbaceous perennial, to 1 m tall or long. Stems trailing, zig-zagging, leafy, with gray hairs.

Leaves: Alternate; stalked; blades with 1 leaflet, kidney-shaped, 2-4.5 cm long.

Inflorescences: Dense clusters in leaf axils.

Flowers: Yellow, typically pea-shaped, 8-11 mm long.

Fruit: A pod, 1-1.6 cm long, 5-8 mm wide, hairy.

Habitat and Distribution: Frequent; dry, sandy soils of open woods, sandhills, coastal beaches, and roadsides; peninsula and eastern panhandle Florida, rarely westward in the panhandle and north into Georgia.

Comment: Although flowering all year, the yellow flowers can barely be seen as they are nestled in the green calyx that surrounds the petals.

Rorippa teres (Michx.) Stuckey
Terete Yellow-cress or Southern Marsh Yellow-cress
[*Rorippa walteri* (Elliott) C. Mohr]
Brassicaceae (Cruciferae), Mustard Family

Habit: Herbaceous annual or biennial, 10-30 cm tall. Stems erect to spreading, branched.

Leaves: Alternate; stalked; blades oblong, 2-8 cm long, finely dissected, lobes with blunt teeth.

Inflorescences: Dense terminal and axillary clusters.

Flowers: Yellow, petals 4, tiny, to 1 mm long.

Fruit: A 2-parted, cylindrical, slender capsule, 0.8-1.5 cm long.

Habitat and Distribution: Frequent; moist to wet soils of hammocks, fields, swamps, lake, pond, and stream margins, and disturbed sites; throughout Florida, west to Texas and Oklahoma, and north to North Carolina; south into Central America.

Comment: Although flowering in winter and spring, the multitude of tiny blooms is seldom noticed. More frequently the finely dissected leaves are likely to attract attention in open wet habitats.

Rhynchosia michauxii Vail

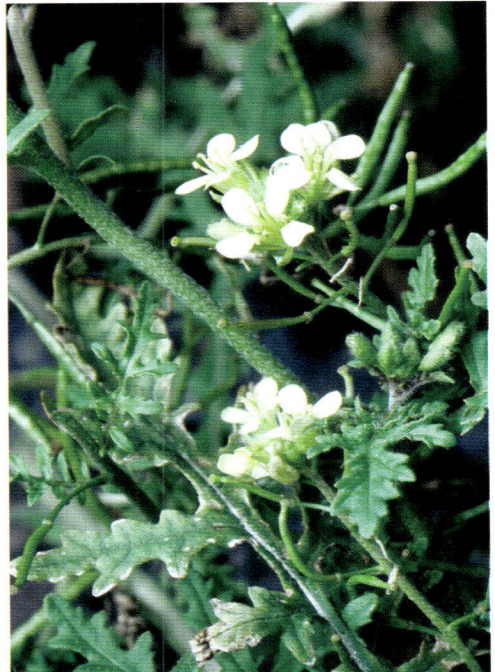

Rorippa teres (Michx.) Stuckey

Yellow

Rudbeckia hirta L.
Black-eyed Susan
[*Rudbeckia divergens* T.V. Moore; *Rudbeckia floridana* T.V. Moore]
Asteraceae (Compositae), Aster or Sunflower Family

Habit: Herbaceous annual, biennial, or perennial, 0.3-1 m tall. Stems several, emerging from basal crown, hairy, rough.

Leaves: Alternate, numerous at base; stalkless; blades elliptic, lance-shaped, or rounded, 5-18 cm long, 1-8 cm wide, hairy, rough, smaller on upper stems.

Inflorescences: Large terminal heads.

Flowers: Rays deep yellow to orange, 3.5-5 cm long, disc dark brown and domed.

Fruit: A 4-sided, blackish nutlet, 2.5-2.7 mm long, lacking a tuft of bristles at top.

Habitat and Distribution: Common; fields, flatwoods, sandhills, meadows, pastures, roadsides, and open disturbed areas; throughout Florida, the U.S., and southern Canada, more frequent in the midwest and eastern U.S.; northern Mexico; native of the midwestern U.S.

Comment: Black-eyed Susan flowers in the summer and fall. It is a very popular cultivated plant with several cultivated forms available. The very large flowers are striking and easily seen in the wild.

Rudbeckia mohrii A. Gray
Mohr's Coneflower
Asteraceae (Compositae), Aster or Sunflower Family

Habit: Herbaceous perennial, to 1.1 m tall. Stems thinly hairy.

Leaves: Alternate; stalkless; blades very narrow, grass-like, 3- or more-nerved, to 20 cm long, smooth.

Inflorescences: Heads terminal, large.

Flowers: Rays yellow, 1.5-3.5 cm long, spreading downward, disc domed to spherical, dark purple or reddish-brown.

Fruit: A 4-sided, smooth nutlet, 3.4-4.2 mm long, with a crown of scales on top.

Habitat and Distribution: Occasional; wet flatwoods, ditches, depressions, and cypress swamps; eastern and western panhandle Florida and southwest Georgia.

Comment: The large-flowered Mohr's Coneflower blooms in summer and fall. This distinctive showy wildflower can be easily identified by the large purple cone in the center surrounded by drooping rays.

Rudbeckia hirta L.

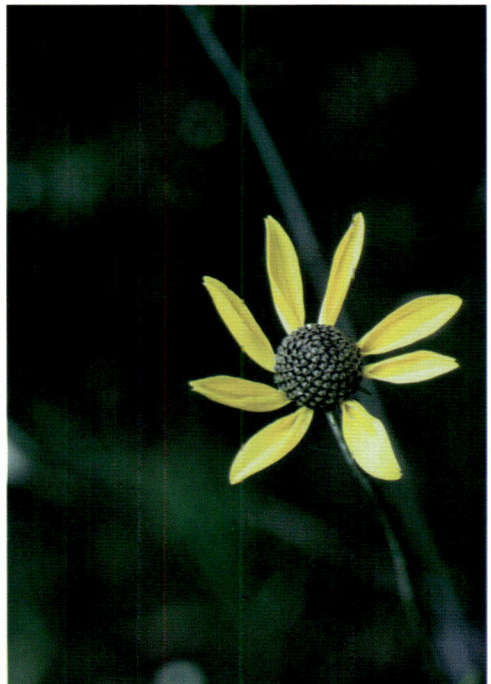

Rudbeckia mohrii A. Gray

Yellow

Salix caroliniana Michx.
Coastal Plain Willow or Carolina Willow
Salicaceae, Willow Family

Habit: Tree, to 12 m tall. Smaller branches hairy when young, becoming smooth.

Leaves: Alternate; stalked; blades lance-shaped, narrowly to broadly so, to 12 cm long, to 2.5 cm wide, glossy green above, waxy gray beneath, margins with small teeth, emerging after tree begins to flower, leafy stipule often present circling stem at axil.

Inflorescences: Drooping cylindrical spike-like clusters, 3-5 cm long, 0.8-1 cm in diameter, male and female flowers in separate spikes on the same plant.

Flowers: Yellowish-green, both sexes minute.

Fruit: A flask-shaped capsule, to 7 mm long.

Habitat and Distribution: Common; virtually all wet habitats; throughout Florida, west to Texas, and north to Kansas, Missouri, Illinois, Indiana, District of Columbia, and Maryland; Cuba.

Comment: The spring spikes of flowers are quite attractive, especially so, as they appear with the emerging light green leaves. Coastal Plain Willow is quite weedy, growing quickly from the wind dispersed seeds that seem to blanket all habitats.

Salix nigra Marshall
Black Willow
Salicaceae, Willow Family

Habit: Tree, to 20 m tall. Smaller branches hairy when young, becoming smooth.

Leaves: Alternate; stalked; blades lance-shaped, to 18 cm long, to 1.5 cm wide, glossy green above, light green beneath (occasionally a bit gray waxy), margins toothed, emerging after tree begins to flower, leafy stipule often present circling stem at axil.

Inflorescences: Drooping cylindrical spike-like clusters, 1.5-7.5 cm long, 0.8-1 cm in diameter, male and female flowers in separate spikes on the same plant.

Flowers: Yellowish-green, both sexes minute.

Fruit: A flask-shaped capsule, 3-4 mm long.

Habitat and Distribution: Frequent to common; virtually all wet habitats; north peninsula Florida, west to Texas, and north to Minnesota, Ontario, and New Brunswick; northern Mexico.

Comment: Similar to Coastal Plain Willow, Black Willow flowers in the spring. The spikes of light yellow flowers are very attractive set against the new light green of the leaves. A weedy species, it springs up quickly with disturbance of wet soils. Coastal Plain Willow and Black Willow can be quickly separated by examining the lower leaf surface. Coastal Plain Willow has a waxy gray undersurface and Black Willow has a light green undersurface.

Salix caroliniana Michx.

Salix nigra Marshall

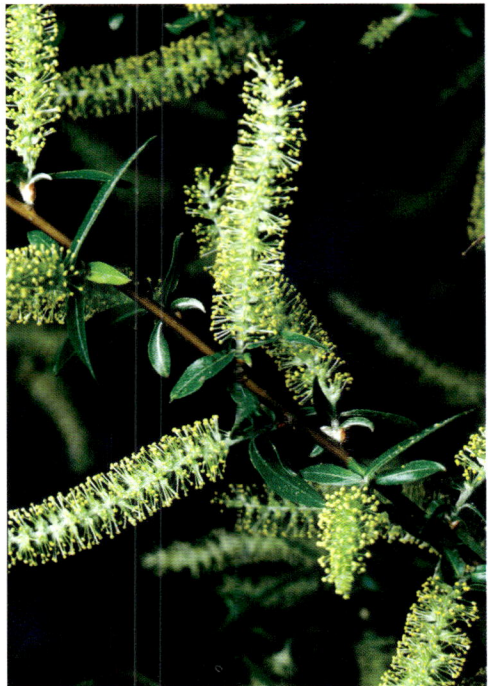

Yellow

Sarracenia flava L.
Trumpet-leaf or Yellow Pitcher-plant
Sarraceniaceae, Pitcher-plant Family

Habit: Herbaceous carnivorous perennial, 0.3-0.9 m tall, from an underground runner.

Leaves: Alternate; stalkless; of 2 types, scale-like cover the runner, trumpet-like are aerial; aerial blades, erect, to 0.9 m tall, hollow, green at base, widening toward the top, opening to a yellow, trumpet-shaped, hooded mouth, with a wing-like flange along the tube.

Inflorescences: Flowers nodding, terminal, on a long (to 0.6 m tall) stalk.

Flowers: Yellow, petals 5, fiddle-shaped, each 5-8 cm long.

Fruit: A round capsule, 1-2.5 cm in diameter.

Habitat and Distribution: Frequent, locally common; bogs, savannas, swamps, seepage slopes, Titi thickets, and wet flatwoods; north peninsula Florida, west to Mississippi, and north to Virginia.

Comment: The large, bright yellow to yellowish-green flowers can be spectacular when seen in mass during the spring. The large trumpet-shaped leaves are also showy. The inside of the aerial leaf is smooth and slick to enable the capture of insects. Trumpet-leaf is cultivated and a few specialty nurseries carry it.

Sarracenia minor Walter
Hooded Pitcher-plant
Sarraceniaceae, Pitcher-plant Family

Habit: Herbaceous carnivorous perennial, 12-40 cm tall, from an underground runner.

Leaves: Alternate; stalkless; of 2 types, scale-like cover the runner, trumpet-like are aerial; aerial blades erect, hollow, green at base, widest in the middle, opening to whitish or white spotted hooded mouth, wing-like flange along the tube.

Inflorescences: Flowers terminal on a long (to 55 cm tall) stalk.

Flowers: Light yellow, petals 5, strap-shaped, 2.5-4 cm long.

Fruit: A round capsule, to 1.5 cm in diameter.

Habitat and Distribution: Frequent, locally common; bogs, wet flatwoods, savannas, ditches, and wet rights-of way; southern central peninsula, north to central panhandle Florida, southwest Georgia, and North Carolina.

Comment: In appropriate habitats, the spring flowers of Hooded Pitcher-plant are sensational. The large, yellow, nodding flowers can be seen in masses covering several acres, particularly in open areas around cypress depressions that grade into wet flatwoods. The leaves are slick inside the tube. When insects venture into the tube, not being able to climb out, they frequently fly upward into the white-spotted hood which appears to offer escape only to be knocked back. This species is listed by the State of Florida as Threatened. It is quite common and persists very well after disturbance. Hooded Pitcher-plant is available for cultivation from specialty nurseries.

Sarracenia flava L.

Sarracenia minor Walter

Yellow

Senecio anonymus A.W. Wood
Southern Ragwort or Small's Ragwort
[*Senecio smallii* Britton]
Asteraceae (Compositae), Aster or Sunflower Family

Habit: Herbaceous perennial, 30-80 cm tall. Stems many, emerging from base, densely hairy at base, but otherwise smooth with age.

Leaves: Alternate; stalked; blades elliptic, deeply lobed or merely toothed, 4-30 cm long, 1-5 cm wide, stem leaves becoming smaller upwards.

Inflorescences: Heads in terminal branched clusters, numerous, 30-150 per plant.

Flowers: Rays yellow to orange, to 8 mm long, disc yellow, 2-3 mm wide.

Fruit: An angled nutlet, 1.5-1.8 mm long, slightly wider towards tip, tips with a crown of bristles.

Habitat and Distribution: Rare southward, more common northward, especially in the mountains; fields, meadows, pastures, roadsides, savannas, and dry woodlands; panhandle Florida, west to Mississippi, and north Kentucky, Tennessee, and Pennsylvania.

Comment: The large clusters of spring flowers are very showy. Ragworts are easily cultivated from nutlets.

Senecia glabellus Poir.
Butterweed or Yellowtop
Asteraceae (Compositae), Aster or Sunflower Family

Habit: Herbaceous annual, 0.2-1.5 m tall. Stems branching, ribbed, smooth.

Leaves: Alternate; stalked; blades lance-shaped and wider towards tip to elliptic, deeply lobed and toothed, 3-25 cm long, 2-7 cm wide, stem leaves becoming smaller upwards.

Inflorescences: Few to numerous heads in terminal branched clusters.

Flowers: Rays yellow-orange, 5-13 mm long, toothed, disc yellow-orange, 2.5-4 mm wide.

Fruit: A ribbed, ellipsoid nutlet, 1.4-3 mm long, with or without a tuft of bristles at tip.

Habitat and Distribution: Common; wet soils of swamps, depressions, woods, ponds, lakes, fields, pastures, roadsides, bottomlands, and disturbed sites; throughout Florida, west to Texas, and north to South Dakota, Missouri, Ohio, and North Carolina.

Comment: Blooms of Butterweed are common in the cool weather of spring. Often the spectacularly bright mass of flowers is amplified by masses of plants in some areas. It can be a photographer's delight.

Senecio anonymus A.W. Wood

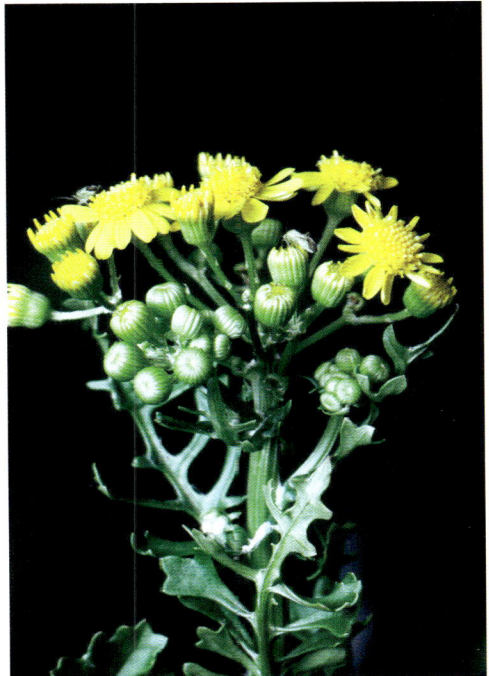

Senecio glabellus Poir.

Yellow

Senna marilandica (L.) Link
Wild Senna
[Cassia marilandica L.]
[Cassia bahamensis Mill., misapplied; Cassia chapmanii Isely]
Fabaceae (Leguminosae), Pea Family

Habit: Herbaceous, erect, perennial, to 2 m tall, from a woody rootstock.

Leaves: Alternate; stalked, with a rounded gland on the upper surface near the base; blades divided into 6 to 10 pairs of elliptic to lance-shaped leaflets, each 4 to 5 cm long, pointed.

Inflorescences: Clusters from leaf axils or terminal.

Flowers: Yellow, 1.1 cm wide, with 5 petals.

Fruit: A pod, straight or curved, 6 to 11 cm long, 8-11 mm wide.

Habitat and Distribution: Occasional; moist, open margins of sandy bottomlands, open sandy fields, and open woods; central peninsula Florida, north to southeast New York and Wisconsin, west to Texas and Nebraska.

Comment: Summer through fall flowering. Sometimes cultivated for the showy flowers, it is most often used as a nectar plant. Wild Senna is used medicinally, but large amounts are harmful.

Senna mexicana (Jacq.) H.S. Irwin & Barneby var. chapmanii (Isley) H.S. Irwin & Barneby
Bahama Senna
[Mill., misapplied;Isley]
Fabaceae (Leguminosae), Pea Family

Habit: Spreading shrub to 2 m tall, or erect tree to 8 m tall.

Leaves: Alternate; stalked; blades divided into 6, 8 or 10, elliptic to lance-shaped leaflets, each 2-4.5 cm long, a stalkless dome-shaped gland on the upper surface between the lowest pair of leaflets.

Inflorescences: Clusters from leaf axils or terminal.

Flowers: Yellow, 2-2.3 cm wide, with 5 petals.

Fruit: A pod, 6-10 cm long, 5 mm wide, straight or somewhat curved.

Habitat and Distribution: Rare; coastal pinelands, hammocks, and dunes; Dade and Monroe Counties of southernmost peninsula Florida; Cuba and the Bahamas.

Comment: Blooms occur all year. The golden-yellow flowers occurring in several to many clusters on a spreading shrub are quite eye-catching.

Senna marilandica (L.) Link

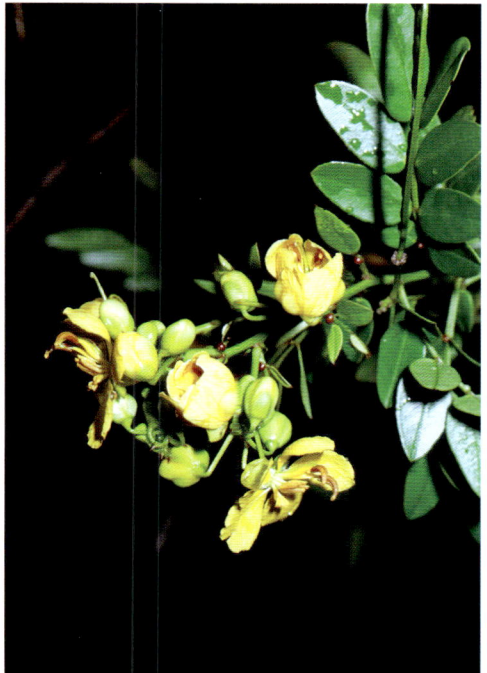

Senna mexicana (Jacq.) H.S.
Irwin & Barneby var. *chapmanii*
(Isley) H.S. Irwin & Barneby

Yellow

Senna occidentalis (L.) Link
Coffee Senna, Septic Weed, Bricho, Low Senna
Fabaceae (Leguminosae), Pea Family

Habit: Herbaceous annual, 0.1-2 m tall. Stems usually smooth.

Leaves: Alternate; stalked, with a rounded to conical gland on the upper surface; blades divided into 3 to 6 pairs of lance-shaped to oval leaflets, each pointed, 3.5-8 cm long, with hairy margins.

Inflorescences: Flowers solitary or in few-flowered clusters in leaf axils.

Flowers: Yellow, 2-3.6 cm wide, with 5 petals.

Fruit: A pod, narrow, 8-14 cm long and 5-9 mm broad, straight or slightly curved.

Habitat and Distribution: Frequent; disturbed areas, pastures, row crops, and old fields; throughout Florida, west to Texas, and north to North Carolina and Arkansas; a native of the subtropics and tropics of the New World, West Indies, Mexico south into South America; naturalized in the Old World Tropics.

Comment: Blooming occurs in the hot months of summer into fall. The flowers are attractive, but considering the weediness of the plant are not highly prized as a wildflower. Coffee Senna commonly occurs in open agricultural fields. The seeds and consequently the plants persist. The seeds have been roasted and used for coffee. The flowers and roots have been used medicinally.

Sesbania macrocarpa Muhl.
Hemp Sesbania
[*Sesbania exaltata* (Raf.) Cory]
Fabaceae (Leguminosae), Pea Family

Habit: Herbaceous annual, 1-4 m tall. Stems smooth.

Leaves: Alternate; stalked; blades divided, with 20-70 leaflets, each narrowly oblong, 1-3 cm long.

Inflorescences: Long-stalked, few-flowered clusters from leaf axils.

Flowers: Yellow to blood red, typically pea-shaped, 1.1-1.6 cm long, upper petal with purplish marks.

Fruit: A pod, very narrow, becomes 4-sided, 10-20 cm long and 3-4 mm wide, with about 30 seeds.

Habitat and Distribution: Frequent to common; disturbed sites, ditches, creek bottoms, fields, swales, and marshes; throughout Florida, north to New York, and west to Arkansas, Kansas, and California.

Comment: The few flowers found on a single plant at any time in the summer and fall are attractive but not especially showy, except when the blood red color form is present. The blood red color form always seems to attract attention. Hemp Sesbania is a tall gangly plant that is extremely weedy. Farmers and native habitat managers both struggle to control its spread which can crowd out other plants. Hemp Sesbania, also known as Colorado River Hemp, was an important fiber plant for the western Indians.

Yellow

Senna occidentalis (L.) Link

Sesbania macrocarpa Muhl.

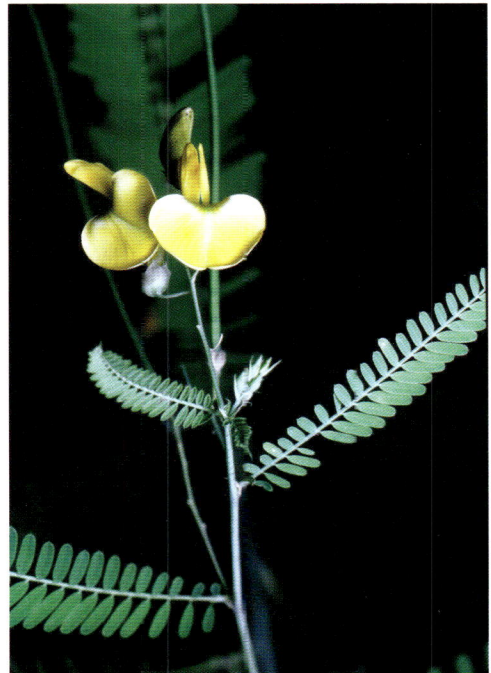

449

Yellow

Sesbania vesicaria (Jacq.) Elliott
Bag Pod or Bladder Pod
[*Glottidium vesicarium* (Jacq.) Harper]
Fabaceae (Leguminosae), Pea Family

Habit: Herbaceous annual, 1-3 m tall. Stems smooth to sparsely hairy.

Leaves: Alternate; stalked; blades divided with 20-52 leaflets, each elliptic-oblong, 1-4 cm long.

Inflorescences: Long-stalked, few-flowered clusters from leaf axils.

Flowers: Yellow, red, or almost black, typically pea-shaped, 0.8-1 cm long.

Fruit: A pod, elliptical, 2-8 cm long, 1.5-2 cm wide, flat, usually with 2 seeds.

Habitat and Distribution: Frequent; moist thickets, disturbed sites, marshes, pastures, swamps, fields, and ditches; throughout Florida, west to Oklahoma and Texas, and north to North Carolina.

Comment: The long gangly plant, flowering from the spring into the fall, is not very exciting, unless the weediness is affecting land management activities. Bag Pod, like Hemp Sesbania, can provide significant weed problems for agriculture and native habitats. The flowers when tinged with red often are noticed as an unusual wildflower.

Sida elliottii Torr. & A. Gray
Narrow-leaf Sida
[*Sida rubromarginata* Nash]
Malvaceae, Mallow Family

Habit: Herbaceous perennial, 30-50 cm tall. Stems simple or with a few branches, star-shaped hairs.

Leaves: Alternate; stalked; blades linear, 2-7 cm long, margins toothed.

Inflorescences: Solitary flowers in leaf axils.

Flowers: Yellow-orange, petals 5, each 1-1.5 mm long.

Fruit: A cluster of 7-12, 1-seeded capsules, each with 2 beaks.

Habitat and Distribution: Occasional; sandy openings in pinelands, woods, fields, and sandhills; throughout Florida, west to Mississippi, and north Missouri, Tennessee, and Virginia.

Comment: Flowering in the summer, Narrow-leaf Sida has attractive blooms of some size, but usually is overlooked. The plant appears to be weedy and has only a few flowers open at any one time.

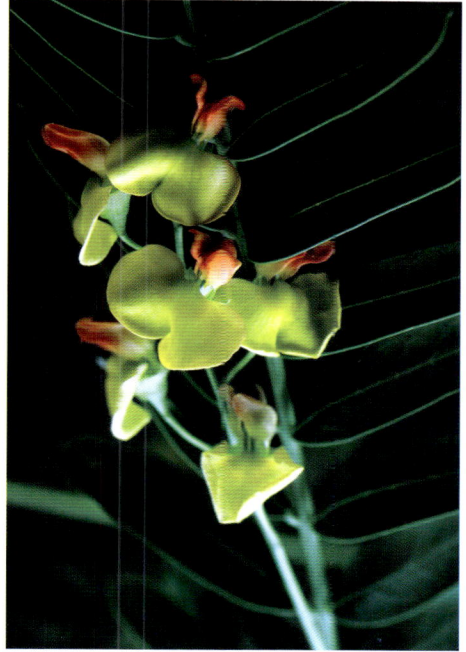

Sesbania vesicaria (Jacq.) Elliott

Sida elliottii Torr. & A. Gray

Yellow

Silphium asteriscus L.
Starry Rosin-weed or Southern Rosin-weed
Asteraceae (Compositae), Aster or Sunflower Family

Habit: Herbaceous perennial, 0.6-1.5 m tall. Stems rough, leafy.

Leaves: Opposite or alternate; stalked; blades broadly elliptic to lance-shaped to oval, 4 15 cm long, rough hairy, margins with coarse teeth or smooth.

Inflorescences: Few to many terminal heads.

Flowers: Rays yellow, wide, each to 3 cm long, with 3 teeth, disc yellow.

Fruit: An oval, flat, winged, smooth nutlet, 6-9 mm long.

Habitat and Distribution: Occasional to infrequent; dry woodlands, sandhills, old fields, pastures, and open areas; central peninsula Florida, west to Mississippi, and north to Arkansas, Missouri, Ohio, and Virginia.

Comment: Starry Rosin-weed occurs in open habitats so that the large showy flowers are easily seen and photographed while blooming in summer and fall.

Silphium compositum Michx.
Rosin-weed
Asteraceae (Compositae), Aster or Sunflower Family

Habit: Herbaceous perennial, 0.7-3 m tall. Stems smooth, branches many, loose.

Leaves: Alternate; stalked; blades broadly ovate, oblong or elliptic, 10-50 cm long, often hairy on the midrib underneath, margins deeply lobed or dissected, usually sharp teeth.

Inflorescences: Terminal heads.

Flowers: Rays yellow, wide, each to 2 cm long, disc yellow.

Fruit: An oval, flat, winged, smooth nutlet, 6-9 mm long, wings with 2 teeth at tip.

Habitat and Distribution: Occasional to frequent; dry, sandy soils of woodlands, thickets, and especially flatwoods; central peninsula Florida, north to Alabama, Tennessee, and Virginia.

Comment: The summer and fall flowers are easily seen in open flatwoods. An especially nice wildflower, as the many large blooms are accompanied by large green leaves situated around the base.

Silphium asteriscus L.

Silphium compositum Michx.

Yellow

Solidago odora Aiton var. *chapmanii* (Torr. & A. Gray) Cronquist
Chapman's Goldenrod
[*Solidago chapmanii* Torr. & A. Gray]
Asteraceae (Compositae), Aster or Sunflower Family

Habit: Herbaceous perennial, to 1.6 m tall. Stems hairy.

Leaves: Alternate; stalkless; blades rounded or somewhat elliptic, 3-9 cm long, smooth, margins rough hairy.

Inflorescences: Numerous heads in terminal clusters, branches bend downwards.

Flowers: Rays yellow, each 2-4 mm long.

Fruit: A narrow, hairy nutlet, 1.5-1.8 mm long, short hairy, with a tuft of bristles at tip, 2-3 mm long.

Habitat and Distribution: Common; sandy, dry soils of sandhills and open hammocks; peninsula west to central panhandle Florida.

Comment: Chapman's Goldenrod is easily recognized in the open sandy habitats of Florida by the rounded leaves and erect hairy stems. It flowers, as most Goldenrods do, in summer and fall. Sweet Goldenrod (*Solidago odora* var. *odora*) differs only in the lance-shaped blades and smooth or vertically lined stems. Sweet Goldenrod occurs from central peninsula Florida, west to Texas, and north to Missouri, Ohio, Vermont, and New Hampshire.

Solidago sempervirens L.
Seaside Goldenrod
Asteraceae (Compositae), Aster or Sunflower Family

Habit: Herbaceous perennial, 0.4-2 m tall, from a thick branched rootstock. Stems several, erect, stout, smooth, branched, leafy.

Leaves: Alternate, from a basal rosette; with winged stalks; blades elliptic to lance-shaped, 5-20 cm long, fleshy, basal leaves larger, gradually smaller upwards.

Inflorescences: Numerous heads in branched terminal or axillary clusters.

Flowers: Rays yellow, each to 5 mm long.

Fruit: A narrow, hairy nutlet, 1.8-2.5 mm long, with tuft of bristles at tip, 3.5-4.5 mm long.

Habitat and Distribution: Common; coastal pinelands, dunes, swales, and brackish marshes; throughout Florida, west to Texas, and north to Newfoundland; West Indies.

Comment: Seaside Goldenrod is one of many summer and fall flowering goldenrods. The bright yellow spires of blooms are eye-catching.

Solidago odora Aiton
var. *chapmanii*
(Torr. & A. Gray) Cronquist

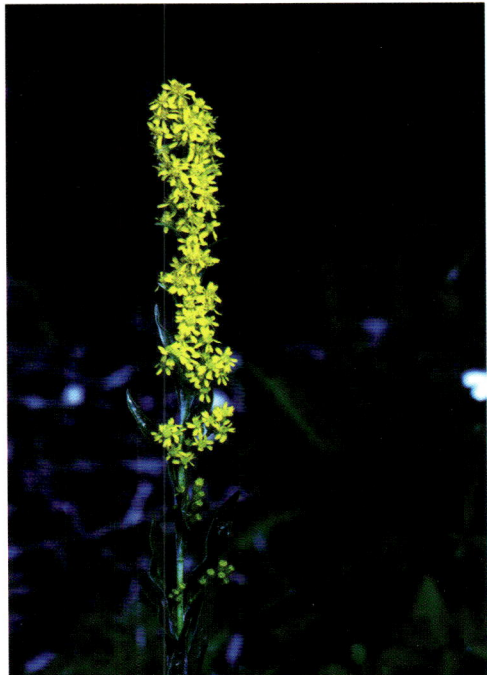

Solidago sempervirens L.

Yellow

Solidago stricta Aiton
Smooth Goldenrod
Asteraceae (Compositae), Aster or Sunflower Family

Habit: Herbaceous perennial, 0.7-2 m tall, from a short rootstock and long underground runners. Stems erect, slender, smooth, usually unbranched.

Leaves: Alternate, with a basal rosette; stalkless; blades narrowly elliptical or somewhat lance-shaped, 5-30 cm long, smooth, basal leaves largest, quickly become smaller on stem.

Inflorescences: Numerous heads in long, slender, spike-shaped, branched, terminal or axillary clusters.

Flowers: Rays yellow, each 2-5 mm long.

Fruit: A narrow, hairy nutlet, 1.8-2.5 mm long, with tuft of bristles at tip, 3.5-4.5 mm long.

Habitat and Distribution: Common; wet flatwoods, bogs, depression, freshwater marshes, meadows, tidal marshes, and wet other wet margins and banks; throughout Florida, west to Texas, and north to New Jersey; south into southern Mexico; West Indies.

Comment: Tall and wandlike, Smooth Goldenrod is usually found in large clusters of plants. Flowering is in summer and fall. As with most goldenrods the golden flowers can be spectacular.

Sonchus asper (L.) Hill
Spiny-leaved Sow-thistle, Achicoria Dulce, Prickly Sow-thistle
Asteraceae (Compositae), Aster or Sunflower Family

Habit: Herbaceous annual, 0.3-2 m tall. Stems erect, simple to many, smooth, with milky sap.

Leaves: Alternate; stalkless; blades lance-shaped, 6-30 cm long, spiny, with deep lobes, base clasping with rounded lobes.

Inflorescences: Terminal heads on many branches.

Flowers: Rays yellow, very narrow, about 1 cm long.

Fruit: A nutlet, to 2.5-3 mm long, with 3-5 longitudinal ribs.

Habitat and Distribution: Frequent; roadsides, fields, pastures, most open disturbed areas; throughout Florida; cosmopolitan; native to Europe.

Comment: Blooming all year, but mainly in spring, as a wildflower the flowers are not especially noticeable. The plant frequently looks ratty. Usually, the large thistle-like spiny leaves are noted, especially if someone is pricked by them.

Solidago stricta Aiton

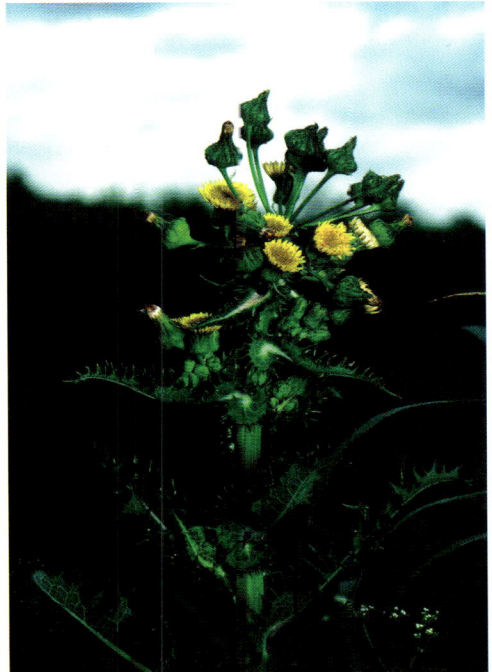

Sonchus asper (L.) Hill

Yellow

Sonchus oleraceus L.
Common Sow-thistle
Asteraceae (Compositae), Aster or Sunflower Family

Habit: Herbaceous annual, 0.3-2 m tall. Stems erect, simple to many, smooth, with milky sap.

Leaves: Alternate; stalkless; blades lance-shaped, to 30 cm long, spiny, with deep lobes, base clasping, with pointed lobes.

Inflorescences: Terminal heads on many branches.

Flowers: Rays yellow, very narrow, about 1 cm long.

Fruit: A nutlet, 2.5-3 mm long, with a ribbed and wrinkled surface.

Habitat and Distribution: Frequent; roadsides, fields, pastures, most open disturbed areas; throughout Florida; cosmopolitan; native to Europe.

Comment: Blooming all year, but mainly in spring, as a wildflower the flowers are not especially noticeable. The plant frequently looks ratty. Usually, the large thistle-like spiny leaves are noted, especially if someone is pricked by them.

Sophora tomentosa L.
Necklace-pod, Yellow Sophora, Silver Bush, Coast Sophora
Fabaceae (Leguminosae), Pea Family

Habit: Shrub, to 3 m tall. Stems densely grayish hairy.

Leaves: Alternate; stalked; blades once-divided, with 11-21 leaflets, each rounded to somewhat elliptic, 2.5-6 cm long, densely gray hairy, leathery.

Inflorescences: Terminal, spike-like.

Flowers: Yellow, typically pea-shaped, 1.8-2.3 cm long.

Fruit: A slender, leathery pod, 6-15 cm long, with deep constrictions between seeds, gray hairy.

Habitat and Distribution: Infrequent to frequent; coastal strands, flats, and hammocks; central peninsula Florida southward, coastal Texas; West Indies, throughout the warm coastal areas of the world.

Comment: The large yellow flowers of Necklace-pod are quite showy and ornamental. The grayish-hairy foliage is also attractive. It is offered for sale by some nurseries.

Sonchus oleraceus L.

Sophora tomentosa L.

Yellow

Sphagneticola trilobata (L.) Pruski
Creeping Oxeye or Trailing Wedelia
[*Wedelia trilobata* (L.) A. Hitchc.]
Asteraceae (Compositae), Aster or Sunflower Family

Habit: Herbaceous perennial, to 40 cm tall and 2 long. Stems creeping to erect, rooting at the joints, fleshy, hairy.

Leaves: Opposite; usually stalkless; blades broadly elliptic, diamond-shaped to lance-shaped, 3-12 cm long, frequently with 3-5 lobes, often fleshy, margins toothed.

Inflorescences: Solitary heads on long stalks from leaf axils.

Flowers: Ray flowers yellow to orange-yellow, 0.6-1.5 cm long, with few teeth, disc yellow.

Fruit: A rounded, bumpy nutlet, 3-5 mm long, with a papery crown at tip.

Habitat and Distribution: Occasional, locally common; virtually any moist disturbed area and in coastal habitats; peninsula Florida; Bermuda; West Indies; Mexico south into South America; Asia; Indian Ocean Islands; Pacific Islands; Australia; native to the New World tropics.

Comment: The bright yellow flowers can be found in any warm month. Extensively cultivated as a patio plant and for ground cover, it easily escapes and can become very weedy. Creeping Oxeye is invasive and can smother native species. It is easily propagated by cuttings.

Stillingia aquatica Chapm.
Corkwood
Euphorbiaceae, Spurge Family

Habit: Shrub, 0.6-1.8 m tall. Stems single, tapering in diameter upward, branches reddish or purplish.

Leaves: Alternate; short-stalked; blades narrowly lance shaped, 3-8 cm long, margins with blunt teeth.

Inflorescences: Terminal, spike-like, male and female flowers separate in the same spike, male flowers on upper spike, female flowers below the male.

Flowers: Yellow, green, or reddish, minute, dense, no petals.

Fruit: A rounded capsule, about 1 cm in diameter.

Habitat and Distribution: Occasional, locally common; flatwoods ponds, depressions in woodlands, marshes, ditches, canals, usually in standing water; throughout Florida, west to Mississippi, and north to South Carolina.

Comment: Corkwood flowers can be found in any warm month. The flower spikes are not eye-catching. The effect is of a shrub that looks somewhat unusual. Corkwood's fame is that its wood is lighter than cork and is used for fishing.

Sphagneticola trilobata (L.) Pruski

Stillingia aquatica Chapm.

Yellow

Stylosanthes biflora (L.) BSP.
Pencil-flower
Fabaceae (Leguminosae), Pea Family

Habit: Herbaceous perennial, 15-60 cm long. Stems few to many, lying flat, ascending or sprawling.

Leaves: Alternate; stalked; blades with 3 leaflets, each ovate to somewhat oblong, 1-3 cm long.

Inflorescences: Flowers solitary or crowded in leaf axils.

Flowers: Yellow to orange, typically pea-shaped, 5-7 mm long.

Fruit: A two-parted pod, 4-5 mm long, minutely hairy, with tiny curved beak at tip, 0.5-1 mm long.

Habitat and Distribution: Infrequent to occasional; dry sandy soils of sandhills, hammocks, roadsides, turf, old fields, and other dry sites and waste places; throughout Florida, west to Texas, and north to Missouri and New Jersey.

Comment: The ever so small flowers can be found in summer if one looks carefully down, especially in dry open turf. The flowers are a bright orange-yellow.

Stylosanthes hamata (L.) Taub.
Southern Pencil-flower or Cheesy-toes
Fabaceae (Leguminosae), Pea Family

Habit: Herbaceous annual or perennial, 15-50 cm long. Stems laying flat to ascending.

Leaves: Alternate; stalked; blades with 3 leaflets, each elliptic, 0.7-1.5 cm long.

Inflorescences: Flowers solitary or crowded in leaf axils.

Flowers: Yellow to orange-yellow, typically pea-shaped, to 6 mm long.

Fruit: A two-parted pod, to 6 mm long, minutely hairy, with small hooked beak at tip, 2-2.5 mm long.

Habitat and Distribution: Rare in Florida to common elsewhere; roadsides, pinelands, and virtually any habitat with dry soils; southern peninsula Florida; West Indies; Central America into northern South America.

Comment: Consistently flowering all year, the yellow flowers are very attractive, but small. Southern Pencil-flower is more frequent than reported in Florida, but is very difficult to separate from the common Pencil-flower.

Yellow

Stylosanthes biflora (L.) BSP.

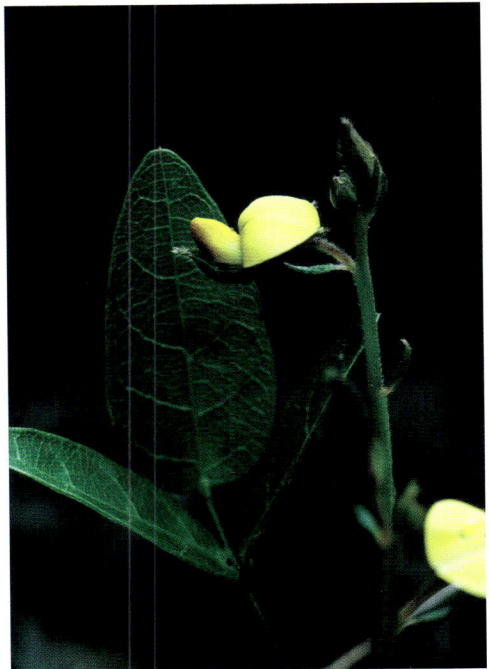

Stylosanthes hamata (L.) Taub.

463

Yellow

Symplocos tinctoria (L.) L'Her.
Horse-sugar or Sweetleaf
Symplocaceae, Sweetleaf Family

Habit: Shrub or small tree, to 6 m tall.

Leaves: Alternate; deciduous in spring, although some falling in fall; stalked; blades elliptic, 8-13 cm long, leathery, hairy beneath, sometimes with toothed edges toward tips.

Inflorescences: Spherical, puff-like clusters of flowers on twigs of the previous year.

Flowers: Yellow, petals 5, stamens long and numerous, appear before the leaves.

Fruit: A green, cylindrical drupe, 8-12 mm long.

Habitat and Distribution: Occasional, locally common; upland hammocks, bottomlands, wetland margins, and moist thickets; central peninsula Florida, west to Texas, and north to Oklahoma, Arkansas, Tennessee, and Delaware.

Comment: The yellow flowers with long stamens are very enchanting as they seem to pop out of the twigs in spring and rarely in the fall. The flowers are followed by cylindrical fruits that stick out at right angles from the twigs. The leaves are prized as browse by animals. Biting into the blade at or near the midrib produces a sweet taste.

Taraxacum officinale Weber ex F.H. Wigg.
Dandelion
[*Leontodon taraxacum* L.]
Asteraceae (Compositae), Aster or Sunflower Family

Habit: Herbaceous winter annual or perennial, to 30 cm tall, from a deep taproot. Stems solitary or few, hollow, not branched, hairy, sap milky.

Leaves: Alternate, in a basal rosette; stalked or stalkless; blades lance-shaped and widest above the middle or elliptic, 8-30 cm long, margins slightly to deeply lobed, lobes point back toward the base.

Inflorescences: Heads solitary, terminal.

Flowers: Rays yellow to orange-yellow, to 2.3 cm long.

Fruit: A cylindric nutlet, tapering to both ends, 2.7-3.1 mm long, 5-grooved, and with a tuft of long bristles at the tip of a long stalk.

Habitat and Distribution: Occasional to common, often locally common; lawns, roadsides, pastures, open pinewoods, fields, and open disturbed sites; throughout Florida, cosmopolitan in temperate climates; native to Eurasia.

Comment: The large heads flower from spring into fall. Dandelion is an eye-catching wildflower. Although not tall it stands out in the low grasses in which it commonly grows. The fruiting heads resembling a puff ball are also attractive. A gentle puff of wind will often release the seeds, each with a parachute of bristles, so that the air will be thick with floating seeds. The young leaves are eaten as greens. Extremely weedy, it is a common and despised problem in lawns.

Symplocos tinctoria (L.) L'Her.
photograph courtesy of John Tobe

Taraxacum officinale
Weber ex F.H. Wigg.

Yellow

Tetragonotheca helianthoides L.
Pineland-ginseng or Squarehead
Asteraceae (Compositae), Aster or Sunflower Family

Habit: Herbaceous perennial, 30-90 cm tall, from a taproot. Stems many, emerging from base, erect to spreading, branching above, glandular hairy.

Leaves: Opposite; stalkless; blades elliptic to rounded, 10-17 cm long, 2.5-13 cm wide, margins with teeth.

Inflorescences: Heads solitary, terminal.

Flowers: Rays yellow, 1.5-3 cm long, disc yellow; 4 bracts around head, ovate.

Fruit: A rounded nutlet, 4-5 mm long, slightly 4-sided, with no bristles at top.

Habitat and Distribution: Frequent; sandy open woods, sandhills, and thickets; central peninsula Florida, west to Mississippi, and north to Tennessee and Virginia.

Comment: Pineland-ginseng blooms in spring and summer. The large flower heads with the showy rays are certainly attractive, but the definitive feature is the four bracts from which it gets the common name, Squarehead.

Thaspium barbinode (Michx.) Nutt.
Bearded Meadow-parsnip
Apiaceae (Umbelliferae), Carrot Family

Habit: Herbaceous perennial, 0.3-1.2 m tall. Stems slender, erect, stiffly hairy around upper joints.

Leaves: Alternate; stalked; principal blades twice-divided, leaflets/segments lance-shaped to ovate, margins toothed or cut.

Inflorescences: Tiny flat-topped clusters arranged on long stalks in terminal, larger, flat-topped clusters.

Flowers: Yellow, about 3 mm wide, petals tiny.

Fruit: 2 1-seeded nutlets, each ribbed, smooth, roundly elliptic, 4-6 mm long, ribs winged.

Habitat and Distribution: Rare southward, infrequent elsewhere; rich woods, prairies, and along stream banks; Jackson and Okaloosa Counties in Panhandle Florida, west to Mississippi, and north to Oklahoma, Kansas, Minnesota, Ontario, and New York.

Comment: The very large arrangements of very tiny flowers can be seen in spring. The flowers are not very appealing, but the arrangement is interesting.

Tetragonotheca helianthoides L.

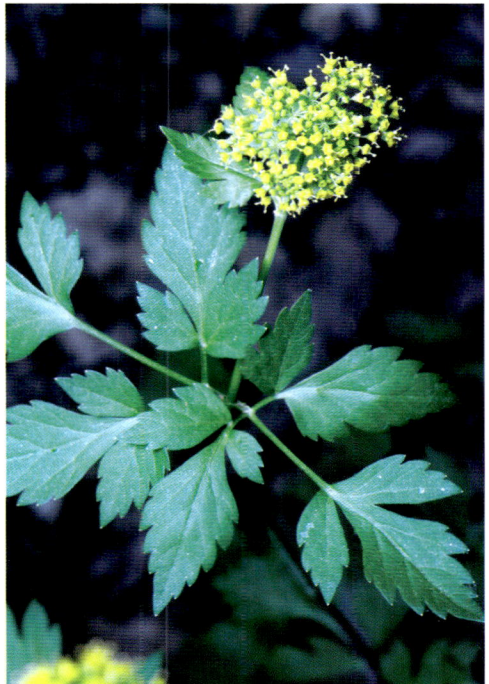

Thaspium barbinode
(Michx.) Nutt.

Yellow

Thunbergia alata Bojer ex Sims
Black-eyed Susan Vine
Acanthaceae, Acanthus Family

Habit: Herbaceous perennial vine, to 4 m long. Stems slender, twining, climbing, hairy.

Leaves: Opposite; winged stalk; blades triangular to triangular-ovate, 3-8 cm long, hairy, base heart- or arrowhead-shaped.

Inflorescences: Flowers solitary or few, on stalks from leaf axils.

Flowers: Yellow, orange-yellow, or cream, with dark centers, narrow tube expanding spreading to 5 lobes, 2-4 cm long, 2.5-3.5 cm wide.

Fruit: A rounded, beaked capsule, 8-10 mm in diameter.

Habitat and Distribution: Frequent; fields, trails, thickets, gardens, home sites, and other disturbed habitats; peninsula Florida, Texas; native to tropical east Africa, now widely spreading worldwide into warm climates.

Comment: Black-eyed Susan Vine flowers in all the warm months. The blooms are striking. Unfortunately, the vine reproduces readily from seeds and often becomes quite weedy. Several cultivated forms with different colored flowers are offered for sale. The vines can be killed by freezing temperatures.

Tribulus cistoides L.
Puncture Vine, Burr Nut, Large Yellow Caltrop
Zygophyllaceae, Caltrop Family

Habit: Herbaceous perennial, usually to 0.5 m long, sometimes to 3 m long, from a taproot. Stems lying flat, spreading from base, hairy.

Leaves: Opposite; stalked; blades with 5 to 7 pairs of leaflets, each oblong to elliptic, 1-1.9 cm long, hairy.

Inflorescences: Flowers solitary on stalks about 2-3 cm long.

Flowers: Yellow, petals 5, each 1-2.5 cm long.

Fruit: Burr-like, rounded, to 1.5 cm in diameter, hard, spiny.

Habitat and Distribution: Infrequent to frequent; sandy soils of open disturbed areas, often coastal, roadsides, turf, vacant lots, fields; throughout Florida, west to Louisiana, and north to Georgia; West Indies and Mexico, south into South America; native to tropical America.

Comment: The large, bright yellow, showy flowers can mostly be seen from winter into early fall. The common name, Puncture Vine or Burr Nut, refers to the spines on the fruits that can puncture tires and feet and can be fatal to livestock.

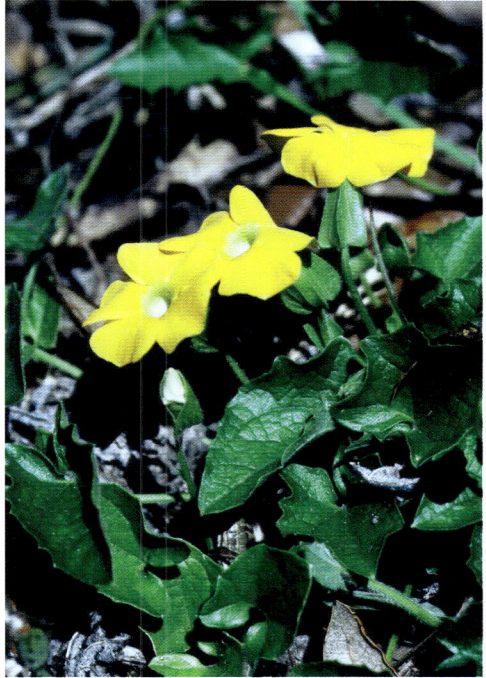

Thunbergia alata Bojer ex Sims

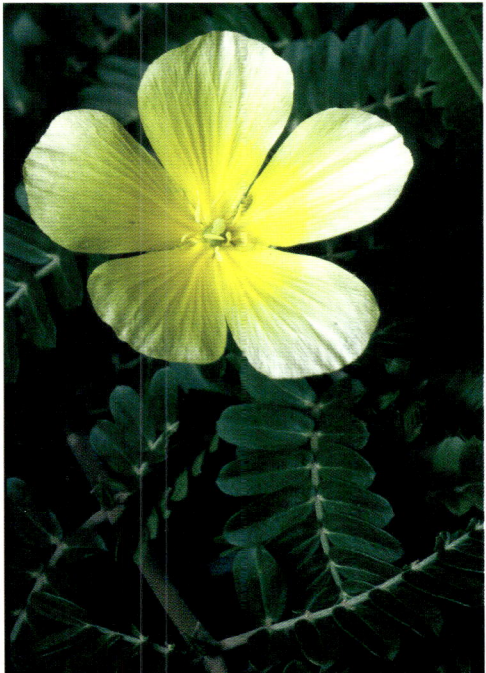

Tribulus cistoides L.

Yellow

Trifolium campestre Schreb.
Hop Clover, Field Clover, Large Hop Clover
[*Trifolium procumbens* L., misapplied]
Fabaceae (Leguminosae), Pea Family

Habit: Herbaceous annual, 5-40 cm long. Stems lying flat or ascending, hairy or sometimes smooth.

Leaves: Alternate; stalked; blades with 3 leaflets, each oval, oblong or somewhat lance-shaped, 0.6-1.5 cm long, margins with teeth toward the tips.

Inflorescences: Flowers in axillary and terminal, round, head-like clusters.

Flowers: Yellow turning brownish, lined, narrow, 2.5-5.5 mm long.

Fruit: An oblong pod, to 3 mm long, 1-seeded.

Habitat and Distribution: Frequent, locally common; sandy soils of open woods, roadsides, fields, lawns, and other disturbed areas; central peninsula Florida; west to Texas, north to Kansas, Missouri, and Virginia, and in the Pacific Northwest; native of Eurasia, now widespread in both the Old and New Worlds.

Comment: The small spring flowers of Hop Clover are attractive, but they quickly turn brown so that heads are not very photogenic.

Trifolium dubium Sibth.
Small Hop Clover
Fabaceae (Leguminosae), Pea Family

Habit: Herbaceous annual, 5-30 cm tall. Stems ascending or reclining, hairy or sometimes smooth.

Leaves: Alternate; stalked; blades with 3 leaflets, each narrowly rounded, 0.6-1.5 cm long, margins with teeth toward tips.

Inflorescences: Flowers in axillary and terminal, round, head-like clusters.

Flowers: Yellow turning brownish, not lined, narrow, 3.5-4 mm long.

Fruit: An oblong pod, 2.5-3 mm long, 1-seeded.

Habitat and Distribution: Occasional, locally common; lawns, roadsides, turf, fields, and other disturbed areas; northern peninsula Florida, west to Texas, and north to Kansas, southern Canada, and Massachusetts, and in the Pacific Northwest; native to Europe, widespread.

Comment: Like Hop Clover, Small Hop Clover, has heads of yellow flowers in spring that rapidly turn brown. It is often sold with the common name Shamrock.

Trifolium campestre Schreb.

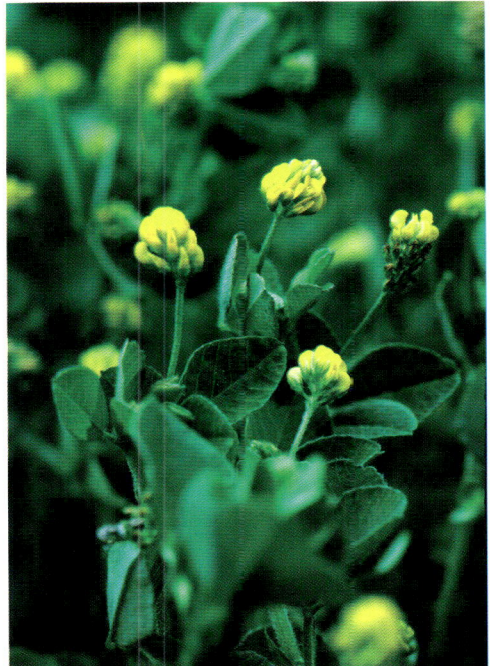

Trifolium dubium Sibth.

Yellow

Utricularia cornuta Michx.
Horned Bladderwort
Lentibulariaceae, Bladderwort Family

Habit: Herbaceous perennial, to 10-35 cm tall, from an underground stem. Plants terrestrial or on floating mats.

Leaves: Alternate; stalked; blades subterranean, thread-like, with bubble-like marginal bladders that trap small insects.

Inflorescences: Flowers solitary to several, terminal on leafless stems.

Flowers: Yellow to orange, 1.5-2.5 cm long, 2-lipped, lower lip larger and 3-lobed, with spur, 7-12 cm long.

Fruit: A round capsule, 2.5-3.5 mm in diameter.

Habitat and Distribution: Frequent; very wet soils of pond margins, bogs, or shallow ditches; throughout Florida, west to Texas, and north to Minnesota, Ontario, and Newfoundland.

Comment: The somewhat large flowers occur from spring into fall. They seem out of place on the thin stems. The tiny bladders are difficult to find except by delicately washing a handful of soil from the underground stem.

Utricularia inflata Walter
Floating Bladderwort
Lentibulariaceae, Bladderwort Family

Habit: Herbaceous aquatic perennial, emergent to 25 cm tall above water surface, rooted on the bottom or free floating.

Leaves: Alternate; stalked; blades of two types: 1) underwater, thread-like, finely branched, with rounded bladders, 2) 4-10 floating, spongy blades, at the water surface, radiating like spokes, keeping plant afloat, bladders on fine, thread-like divisions at tips.

Inflorescences: Several to many flowers, terminal on leafless stems.

Flowers: Yellow, 1.5-2.5 cm long, 2-lipped, lower lip slightly larger, 3-lobed, with spur, to 8 mm long.

Fruit: A round capsule, 3-6 mm in diameter.

Habitat and Distribution: Common; ditches, lakes, swamps, canals, sloughs, ponds, and other still waters; throughout Florida, west to Texas, and north to New Jersey.

Comment: The peculiar floating, spoke-like leaves are commonly noticed, even if the showy winter and spring flowers are not present.

Utricularia cornuta Michx.

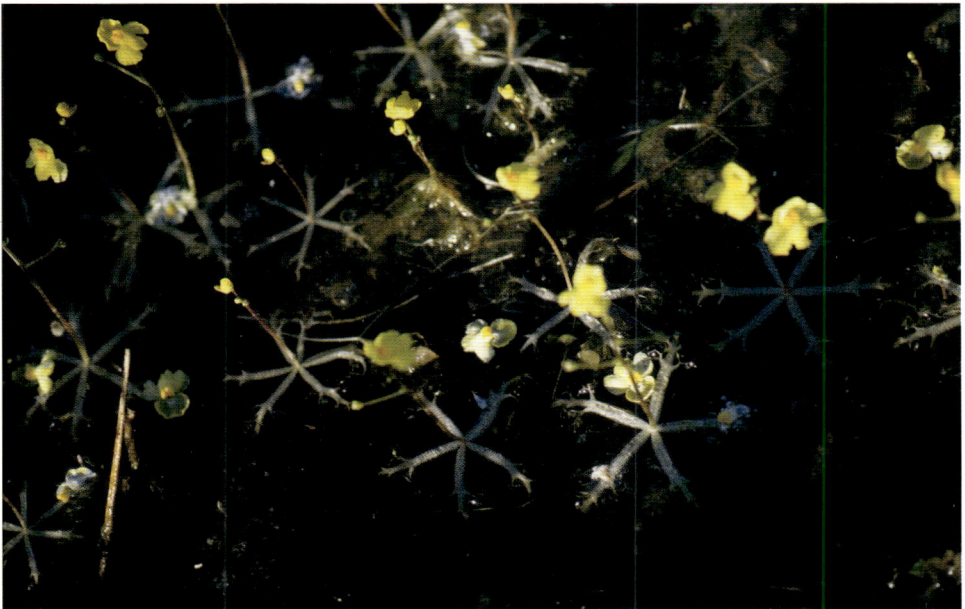

Utricularia inflata Walter
Photograph courtesy of UF/IFAS Center for Aquatic and Invasive Plants

473

Yellow

Utricularia subulata L.
Wet Sand Bladderwort or Zigzag Bladderwort
Lentibulariaceae, Bladderwort Family

Habit: Herbaceous perennial, to 18 cm tall. Plants terrestrial, branches filiform, under ground.

Leaves: Alternate; stalked; blades thread-like, underground with bladders or on soil surface and lacking bladders.

Inflorescences: Few to many flowers, terminal on leafless stem.

Flowers: Yellow, 5-12 mm long, 2-lipped, lower lip broader, 3-lobed, with spur, to 5 mm long.

Fruit: A round capsule, 1.5-2 mm in diameter.

Habitat and Distribution: Common; wet pinelands, seepage slopes, pond and lake margins, ditches, and sandy wet clearings; throughout Florida, west to Texas and Arkansas, and north to Nova Scotia and Massachusetts; West Indies; Central and South America; tropical Africa; Madagascar; Thailand; Borneo.

Comment: Usually only one or two flowers are seen at any one time during the spring to fall flowering season. The very tiny, thin stems seem to be too small to support the flower.

Uvularia perfoliata L.
Bellwort or Straw-bell
Liliaceae, Lily Family

Habit: Herbaceous perennial, 10-80 cm tall, from white underground runners, forming colonies. Stems unbranched or branched once.

Leaves: Alternate; stalkless; blades elliptic or occasionally more rounded, 3-9 cm long, base wrapping completely around stem.

Inflorescences: Flowers solitary or few, terminal.

Flowers: Dull yellow, slender, bell-shaped, 2-3 cm long, nodding.

Fruit: A rounded capsule, to 1 cm in diameter.

Habitat and Distribution: Rare southward, more frequent north and west, especially in the mountains and piedmont; rich woodlands, hammocks, coves, and bottomlands; Leon and Gadsden Counties in panhandle Florida, west to Arkansas and Oklahoma, and north to Ohio, Ontario, Quebec, and New Hampshire.

Comment: As a spring blooming wildflower Bellwort is not very showy. More noticeable are the colonies of plants with the stem seemingly growing through the blade.

Utricularia subulata L.

Uvularia perfoliata L.

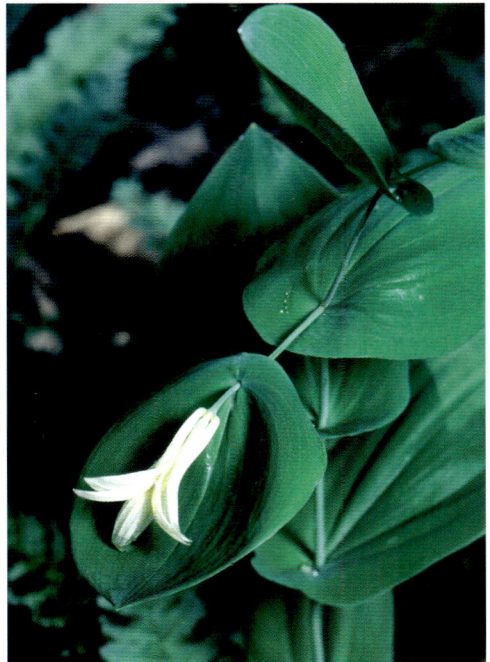

Yellow

Verbascum thapsus L.
Common Mullein, Flannel Plant, Velvet Plant
Scrophulariaceae, Figwort or Snapdragon Family

Habit: Herbaceous biennial or perennial, 0.3-1.8 m tall. Stem usually single, stout, leafy, woolly with star-shaped hairs.

Leaves: Alternate, basal rosette; lower stalked, upper stalkless; blades elliptic to lance-shaped, 5-30 cm long, light green, velvety, smaller on upper stem, margins smooth or blunt-toothed.

Inflorescences: Dense, cylindrical spikes at stem tips.

Flowers: Yellow, 1.5-2.5 cm wide, petals 5, anther filaments densely white hairy.

Fruit: A rounded capsule, 6-10 cm long.

Habitat and Distribution: Infrequent to occasional; roadsides, pastures, rocky banks, open woodlands, and in other disturbed sites; north peninsula Florida, throughout North America; native of Europe.

Comment: Common Mullein is a very attractive wildflower when blooming from summer into fall or even when simply vegetative. The tall gray-green woolly plants are always noticed, especially so along roadsides where they are often the only tall weed.

Verbascum virgatum Stokes
Purple-stamen Mullein or Wand Mullein
Scrophulariaceae, Figwort or Snapdragon Family

Habit: Herbaceous biennial or perennial, 0.6-1.2 m tall. Stems usually unbranched, leafy, hairy with star-shaped hairs.

Leaves: Alternate, basal rosette; lower stalked, upper stalkless; blades elliptic or wider towards tip, 7-25 cm long, hairy, margins toothed.

Inflorescences: Dense, cylindrical, terminal spikes.

Flowers: Yellow, 2-3 cm wide, petals 5, anther filaments densely purple hairy.

Fruit: A rounded capsule, 6-8 cm diameter.

Habitat and Distribution: Infrequent to occasional; old fields, roadsides, and in other disturbed open areas; central peninsula Florida, west to California, and north to North Carolina.

Comment: The yellow summer flowers are certainly worth a photograph, especially with the purple stamen hairs. Again, like Common Mullein the gray-green vegetation is also attractive. Purple-stamen Mullein usually grows among low grassy vegetation and sticks out like a sore thumb.

Verbascum thapsus L.

Verbascum virgatum Stokes

Yellow

Vicia grandiflora Scop.
Large-flowered Vetch
Fabaceae (Leguminosae), Pea or Bean Family

Habit: Herbaceous annual, 30-60 cm tall. Stems sprawling, ascending, or slightly climbing, finely hairy.

Leaves: Alternate; short-stalked; blades with 6-14 leaflets, leaflets 1-2.5 cm long, elliptic-oblong to oblong, terminal leaflet modified into a tendril.

Inflorescences: Axillary clusters of 2.

Flowers: Pale yellow, sometimes tinged with violet, typically pea-shaped, 2.5-3 cm long.

Fruit: A slender, somewhat flattened, 3.5-5 cm long pod, black when mature.

Habitat and Distribution: Occasional; in fields, roadsides, and disturbed areas, from eastern panhandle Florida, west to Louisiana, Arkansas and Missouri, and north to Georgia, Tennessee, and Maryland; native of Europe.

Comment: The large yellow flowers of Large-flowered Vetch are easily noticed. This low climber is easily recognized by the terminal tendril and the large yellow flowers.

Vigna luteola (Jacq.) Benth.
Wild Cowpea or Yellow Vigna
Fabaceae (Leguminosae), Pea Family

Habit: Herbaceous perennial vine, to 3 m long. Stems trailing or twining, hairy or smooth.

Leaves: Alternate; stalked; blades with 3 leaflets, each ovate to narrowly lance-shaped, 2-8 cm long.

Inflorescences: Flowers clustered at tip of long stalks from leaf axils.

Flowers: Yellow, typically pea-shaped, 1.4-1.8 mm long.

Fruit: A narrow, oblong pod, to 7 cm long, hairy.

Habitat and Distribution: Frequent; wet margins, beaches, tidal flats, mangrove hammocks, thickets, and open disturbed sites; throughout Florida, west to Texas, and north to North Carolina; Bermuda; West Indies; Mexico south into South America; Old World tropics.

Comment: The yellow flowers occur in the warm months. The flowers are not particularly showy, but stand out against the green foliage. Wild Cowpea is frequently found in ditches and along roadside swales.

Vicia grandiflora Scop.

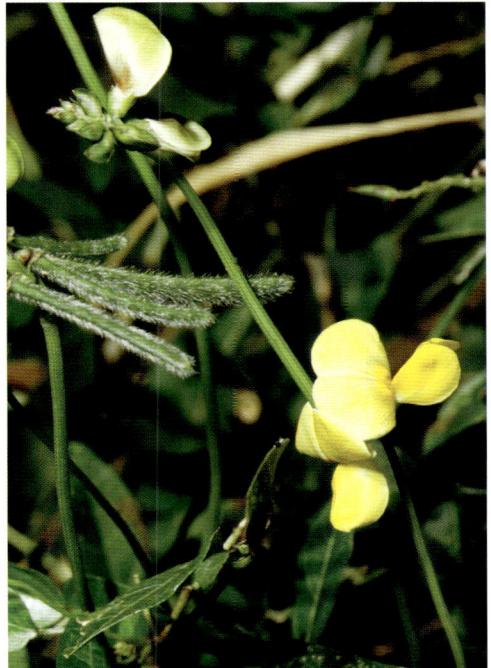

Vigna luteola (Jacq.) Benth.

Yellow

Xyris caroliniana Walter
Carolina Yellow-eyed-grass
Xyridaceae, Yellow-eyed-grass Family

Habit: Herbaceous perennial, to 1.1 m tall. Stems twisted, smooth, 1-ridged at top, sheath shorter than leaves.

Leaves: Alternate, all basal; stalkless; blades very narrow, grasslike, to 50 cm long, 5 mm wide.

Inflorescences: Terminal elliptic cones, 1.5-3 cm long.

Flowers: Yellow or white, petals 3, 8-9 mm long, opening in the afternoon, lateral sepals exserted from cone.

Fruit: An oblong capsule; seeds translucent, 0.8-1 mm long.

Habitat and Distribution: Common; sands of pinelands, sandhills, savannahs, and scrub; throughout Florida, west to Texas, and north to New Jersey.

Comment: In summer and fall, flowers poke out of the cone-like heads. The flowers are not especially showy, but the cones with an open flower or two are eye-catching. When the head has finished blooming the scales open, resembling a pine cone.

Xyris elliottii Chapm.
Elliott's Yellow-eyed-grass
Xyridaceae, Yellow-eyed-grass Family

Habit: Herbaceous perennial, to 60 cm tall. Stems thin, twisted or straight, 1- to several-ridged at top, sheath shorter than leaves.

Leaves: Alternate, all basal; stalkless; blades linear to narrowly linear, to 30 cm long and 1-2 mm wide, flat.

Inflorescences: Terminal ovoid cones, 0.6-1.5 cm long.

Flowers: Yellow, petals 3, about 5 mm long, opening in the morning, lateral sepals not exserted from cone.

Fruit: An oblong capsule; seeds translucent, to 0.5-0.6 mm long.

Habitat and Distribution: Common; wet flatwoods, depressions, pond and lake margins, savannahs, ditches, and other wet areas; throughout Florida, west to Mississippi, and north to South Carolina.

Comment: Flowering from spring into fall, Elliott's Yellow-eyed-grass can still be identified even when not in flower by the large clump of narrowly linear blades. The pale yellow flowers are not particularly attractive. The cone-like heads usually garner most of the attention.

Xyris caroliniana Walter

Xyris elliottii Chapm.

Yellow

Xyris jupicai Rich.
Common Yellow-eyed-grass
Xyridaceae, Yellow-eyed-grass Family

Habit: Herbaceous perennial, to 80 cm tall. Stems straight, becoming flattened and 1- or 2-ridged at top, sheath shorter than leaves.

Leaves: Alternate, all basal; stalkless; blades broadly linear or linear, grasslike, to 60 cm long, to 1 cm wide.

Inflorescences: Terminal ovoid to oblong cones, 0.5-1.5 cm long.

Flowers: Yellow, petals 3, to 3 mm long, opening in the morning.

Fruit: An oblong capsule; seeds translucent, 4-5 mm long.

Habitat and Distribution: Common; pond, lake, and swamp margins, bogs, wet pine flatwoods, ditches, savannahs, and other low, wet areas; throughout Florida, west to Texas and Arkansas, and north to Tennessee and New Jersey; native of tropical America.

Comment: Flowering all year, Common Yellow-eyed-grass is perhaps the most frequently encountered species in this group. The petals are quite small, so like most Yellow-eyed-grasses the cone-like heads and large grass-like leaves are most often noticed.

Youngia japonica (L.) DC.
Asiatic False Hawksbeard
Asteraceae (Compositae), Aster or Sunflower Family

Habit: Herbaceous annual or biennial, 0.1-0.5 m tall, from a taproot. Stems 1-5, sap milky.

Leaves: Alternate, basal rosette; stalked or stalkless; blades oval, oblong, lance-shaped, 6-12 cm long, deeply lobed, margins toothed.

Inflorescences: Terminal many-flowered clusters.

Flowers: Rays yellow, to 6 mm long.

Fruit: A spindle-shaped nutlet, 1.5-2.5 mm long, ribbed, ribs hairy, with white tuft of bristles at tip, to 3 mm long.

Habitat and Distribution: Common; fields, lawns, gardens, roadsides, and most open disturbed habitats; throughout Florida, west to Texas and Arkansas, and north to Pennsylvania and New York; West Indies; Mexico into South America; Europe; Africa; Pacific Islands; Australia; native to southeastern Asia.

Comment: The very small light yellow flowers can be found all year. Asiatic False Hawksbeard often attracts attention as it germinates and quickly forms a basal rosette in open disturbed soils, such as gardens. When blooms occur, it quickly becomes apparent that this is a weed.

Xyris jupicai Rich.

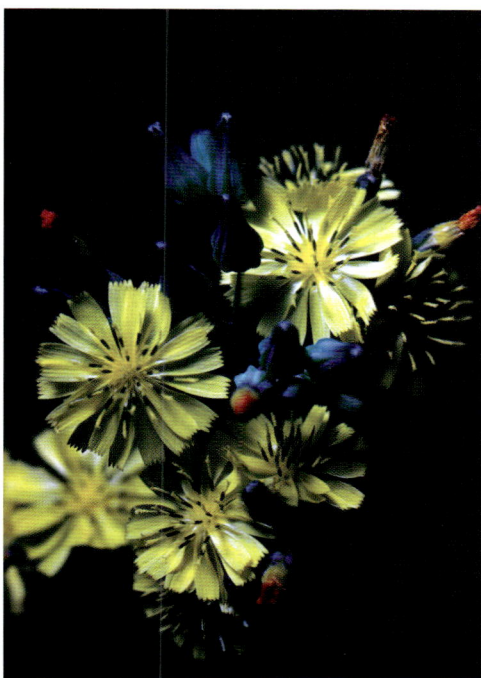

Youngia japonica (L.) DC.

Yellow

Zizia aurea (L.) Koch
Golden Alexanders
Apiaceae (Umbelliferae), Carrot Family

Habit: Herbaceous perennial, 20-80 cm tall, from thickened roots.

Leaves: Alternate; stalked; blades divided once, twice, or three times into 3 triangular segments, each ovate to lance-shaped, 2-6 cm long, margins sharply and finely toothed, upper leaves smaller.

Inflorescences: Tiny clusters arranged on long stalks in terminal larger, round, flat-topped clusters.

Flowers: Yellow, minute, 3-3.5 mm wide.

Fruit: 2 1-seeded nutlets, each ribbed, ellipsoid, smooth, laterally flattened, 3-4 mm long.

Habitat and Distribution: Rare southward, infrequent northward; fields, floodplains, meadows, wet woods, swamp forests, and creek bottoms; Holmes County in north central panhandle Florida, west to Texas, and north to Montana, Saskatchewan, and New Brunswick.

Comment: The tiny golden yellow flowers occur in spring and early summer. As a wildflower the delicate arrangement of the many small flowers certainly makes Golden Alexanders photogenic.

Zizia aurea (L.) Koch

White

Achillea millefolium L.
Common Yarrow
Asteraceae (Compositae), Aster or Sunflower Family

Habit: Herbaceous perennial, to 1.2 m tall, from underground runners. Stems slightly to densely hairy, 1 to several from base.

Leaves: Alternate, basal rosettes; leaves on stems stalkless, leaves at base stalked; blades 2-15 cm long, very finely divided, feather-like, giving frilly appearance.

Inflorescences: Dense, domed, terminal clusters of flower heads.

Flowers: White, whitish yellow or rarely pink, very small, rays 1-3 mm long.

Fruit: An achene, narrow, smooth, 2-3 mm long.

Habitat and Distribution: Frequent; open, dry, primarily disturbed sites, meadows, old fields; throughout the United States from central peninsula Florida northward; a circumboreal species.

Comment: Flowering in spring the plant is cultivated for the gray-hairy, fern-like foliage. Many different forms are available from most nurseries. Dried leaves and flowers are used medicinally.

Agarista populifolia (Lam.) Judd
Pipewood or Pipestem
[*Leucothoe populifolia* (Lam.) Dippel]
Ericaceae, Heath Family

Habit: Shrub, 1-4 m tall. Stems ascending. Branches shelving downwards, becoming shorter towards tip of stem giving plant a pyramidal shape.

Leaves: Alternate; short-stalked, stalks minutely hairy; blades lance-shaped to somewhat oval, 4-10 cm long, evergreen, with fine netted veins, bases usually rounded, tip pointed, and margins smooth to wavy to slightly toothed near tip.

Inflorescences: Short, spike-like clusters from leaf axils.

Flowers: White, pendulent, on stalks 7-10 mm long, urn-shaped, 7-10 mm long, with 5 small lobes.

Fruit: A roundish capsule, 4-5 mm long and 5-6 mm wide.

Habitat and Distribution: Infrequent; wet woods and low, swampy sites, often along creek floodplains; from central peninsula Florida, north along the Coastal Plain to southeastern South Carolina.

Comment: This shrub has been featured in regional landscaping magazines. The pyramidal shape, evergreen foliage, and profuse white flowers in spring are very attractive. The shelving branching pattern retains the shape of the shrub and does not allow pruning.

Achillea millefolium L.

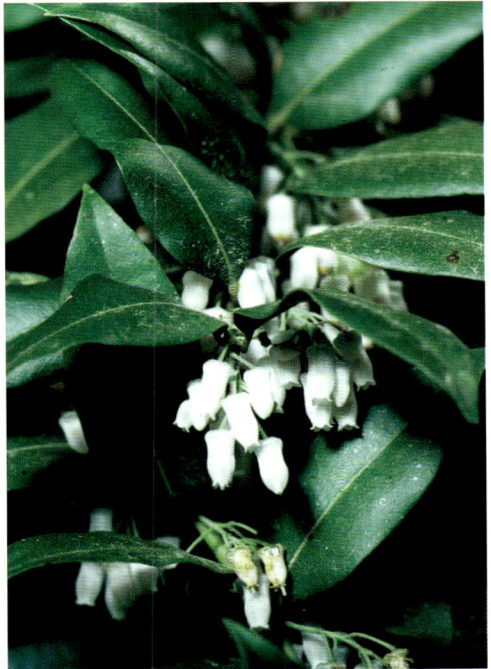

Agarista populifolia (Lam.) Judd

White

Aletris obovata Nash
White Colicroot or Southern Colicroot
Liliaceae, Lily Family

Habit: Herbaceous perennial, to 80 cm tall. Stems slender, erect, mostly leafless.

Leaves: Alternate, basal rosette; stalkless; blades 2-8 cm long, narrowly elliptic and broadest toward the tips.

Inflorescences: Spike-like, terminal arrangements.

Flowers: White, 5-6 mm long, cylindrical to spindle-shaped.

Fruit: A capsule, 3-chambered, 6-8 mm long, somewhat rounded.

Habitat and Distribution: Frequent; moist open pinelands; from central peninsula Florida west to central panhandle Florida, and north to southeast South Carolina.

Comment: Flowers are seen in spring and early summer. Plants are not particularly showy. The tall nearly leafless stems usually poke above the surrounding vegetation providing spires of small white flowers.

Aleurites fordii Hemsl.
Tung Tree
[*Vernicia fordii* (Hemsl.) Airy-Shaw]
Euphorbiaceae, Spurge Family

Habit: Tree, to 10 m tall. Twigs relatively thick, sap milky.

Leaves: Alternate; long-stalked, two glands on upper side of stalk just below blade; blades rounded, to 30 cm long and wide, 5-veined, heart-shaped base, sharp-pointed tip, with two lobes on young plants, margins smooth.

Inflorescences: Somewhat large, open, branched clusters, on stalks from developing leaf axils at stem tips.

Flowers: White, red-veined at base, about 4 cm wide, 5-8 petals.

Fruit: A capsule, rounded, 3-sided, 4-8 cm wide, with 3-5 large, 3-sided seeds.

Habitat and Distribution: Rare, can be locally common; disturbed habitats, usually near former cultivation; central peninsula Florida, to and throughout Louisiana, and north into south Georgia.

Comment: The very showy flowers appear before the developing leaves in spring. Tung Tree was cultivated in orchards as a crop for the valuable oil obtained from the seeds. It is also cultivated as an ornamental. The seeds are POISONOUS.

Aletris obovata Nash

Aleurites fordii Hemsl.

White

Alternanthera philoxeroides (Mart.) Griseb.
Alligator-weed
Amaranthaceae, Amaranth Family

Habit: Herbaceous perennial, 0.3-1.5 m long, sometimes forming a large, hard, woody rootstock. Stems to 2 cm thick, hollow, creeping, or trailing, forming mats, nodes frequently swollen, pinkish, rooting at the joints.

Leaves: Opposite; stalkless; blades elliptic, 3-11 cm long, tapered to base.

Inflorescences: Terminal or from leaf axils, round heads to 1.5 cm in diameter, on long stalks.

Flowers: White, papery.

Fruit: A small, papery 1-seeded bladder.

Habitat and Distribution: Frequent to locally common; usually aquatic, in waterways, ponds, lakes, swamps, and sometimes in crops and gardens; throughout Florida, north to Virginia, and west to Texas; Central America; native to South America.

Comment: Alligator-weed blooms in any warm month. The hollow stems with opposite leaves, pinkish joints, and balls of papery white flowers on long stalks make this plant readily identifiable. Occasionally it can be found in irrigated crops and gardens, in which case the base often becomes hard and woody. This weedy exotic plant is one of the first successful subjects of biological control efforts. It is being repressed by three insects imported from its native range. When populations grow thick these insects inevitably find them and reduce the plants to fragments.

Amelanchier arborea (Michx. f.) Fernald
Serviceberry
Rosaceae, Rose Family

Habit: Small tree, 5-16 m tall.

Leaves: Alternate; stalked; blades ovate to elliptic-ovate, 2-13 cm long, densely hairy underneath when young, less hairy with age, with fine teeth on margins.

Inflorescences: Loose, terminal clusters of 5-15, usually hanging flowers.

Flowers: White or sometimes pinkish, 5 strap-shaped petals, 1-1.7 cm long.

Fruit: An apple-like pome, round, 5-8 mm in diameter, red to purplish, usually dry, tasteless, dried flower parts at tip.

Habitat and Distribution: Occasional; upland woods, usually well-drained; south Georgia and Florida panhandle, west to northeast Texas, and north to Minnesota, Michigan, Ontario, Quebec, and Maine.

Comment: Serviceberry flowers in spring just as the leaves are appearing. This subcanopy species is not especially noticeable, but can be an attractive specimen tree.

Alternanthera philoxeroides
(Mart.) Griseb.

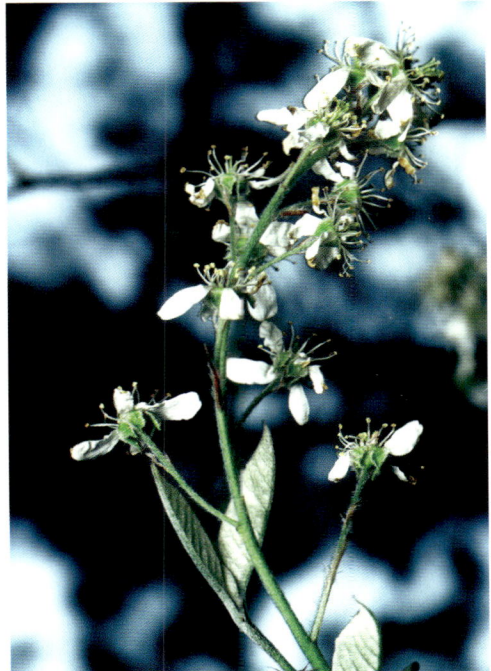

Amelanchier arborea
(Michx. f.) Fernald

White

Angelica dentata (Chapm. ex Torr & A. Gray) J.M. Coult. & Rose
Smooth Angelica
Apiaceae (Umbelliferae), Carrot Family

Habit: Herbaceous perennial, to 1 m tall. Stems smooth, slender.

Leaves: Alternate; stalked; mostly basal; blades somewhat leathery, finely divided into narrow, angular segments with toothed edges, veins run to the tip of the tooth.

Inflorescences: Long-stalked, dome-shaped, 5-8 cm wide clusters from upper leaf axils.

Flowers: White, small.

Fruit: Rounded, smooth, 5-6 mm long, with wide thin wings on each edge, and 3 ribs on each surface.

Habitat and Distribution: Occasional; dry pinelands and bogs; from central panhandle Florida to southwest Georgia.

Comment: The attractive clusters of flowers seen in summer and fall help with identification, but the small winged fruits are most distinctive.

Ardisia crenata Sims
Coral-berry
Myrsinaceae, Myrsine Family

Habit: Small shrub, usually to 1 m, occasionally to 3 m tall.

Leaves: Alternate, evergreen; stalked; blades elliptic, leathery, to 12 cm long, margins wavy and blunt-toothed.

Inflorescences: Terminal, branched clusters, on distinctive branches.

Flowers: White, about 7 mm wide.

Fruit: Round, with a central stone, about 7 mm in diameter, bright red.

Habitat and Distribution: Occasional, but locally common; moist hammocks; central and north peninsula and panhandle Florida; native from Japan to northern India.

Comment: Coral-berry is a very colorful ornamental. The spring flowers are attractive, but the multitude of red fruits held just below the shiny green leaves are spectacular. A white-fruited form, although rare, is occasionally seen. The fruits are edible but somewhat dry and definitely not good. Unfortunately, this handsome ornamental is listed by the Florida Exotic Pest Plant Council as a Category I invasive nuisance. It has escaped into native habitats and, although slow growing, is crowding and shading out native species.

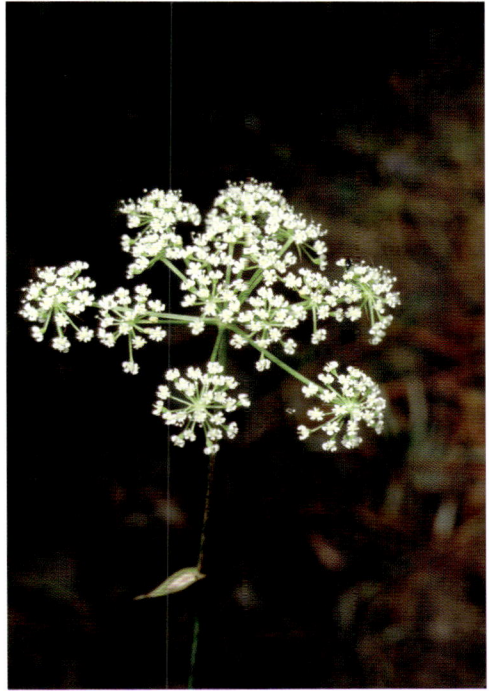

Angelica dentata
(Chapm. ex Torr. & A. Gray) J.M.
Coult. & Rose

Ardisia crenata Sims

495

White

Ardisia escallonioides Schiede & Deppe ex Schldl. & Cham.
Marlberry
Myrsinaceae, Myrsine Family

Habit: Shrub or small tree, to 7 m tall. Bark whitish, scaly, twigs thick.

Leaves: Alternate, evergreen; stalked; blades elliptic or broadest toward tip, 4-18 cm long, leathery, margins smooth.

Inflorescences: Dense, terminal, branched clusters.

Flowers: White with purplish lines and spots, 6-8 mm wide, petals 5, arching outward.

Fruit: Round, with a central stone, black, to 9 mm diameter.

Habitat and Distribution: Frequent; hammocks, thickets, and pinelands; central and southern peninsula Florida; Mexico, Guatemala, Cuba, Hispaniola, British Honduras, Bahamas.

Comment: The large masses of flowers occur mainly in spring, but can be found at any time. The evergreen shiny leaves and black shiny fruits make Marlberry appealing all year.

Argemone albiflora Hornem.
Carolina-poppy
Papaveraceae, Poppy Family

Habit: Herbaceous annual or biennial, 0.3-1.5 m tall, from a taproot. Stems erect, solitary, spiny, branched, sap yellow.

Leaves: Alternate; stalkless, clasping stem; blades leathery, deeply lobed at base of stem, less so on upper stem, 3-20 cm long, with toothed and spiny margins.

Inflorescences: Terminal flowers on branches and stems.

Flowers: White, 5-12 cm wide, with 4-6 broad crinkled petals and yellow anthers in the center.

Fruit: A spiny, elliptic capsule, 2.5-4 cm long.

Habitat and Distribution: Occasional, can be locally common; dry soils, primarily along rights-of-way and other open disturbed areas; throughout Florida, west to Texas, and north to Arkansas, Missouri, and North Carolina.

Comment: Carolina-poppy is a spectacular wildflower. The large white flowers with the yellow center can easily be seen at some distance. The forbidding spines of this plant make it easily identifiable.

White

Ardisia escallonioides Schiede &
Deppe ex Schdl. & Cham.

Argemone albiflora
Hornem.

White

Argusia gnaphalodes (L.) Heine
Sea lavender
[*Mallotonia gnaphalodes* (L.) Britton; *Tournefortia gnaphalodes* (L.) R. Br. ex Roem. & Schult.]
Boraginaceae, Borage Family

Habit: Shrub, to 3 m tall. Branches densely hairy.

Leaves: Alternate; stalkless; blades linear, 3-10 cm long, thick, succulent, densely hairy.

Inflorescences: Dense flowers in a short, arching spike from upper leaf axils.

Flowers: White, tubular, 3-7 mm long.

Fruit: Dry, brown, smooth, ovoid, somewhat corky, 5-6 mm long.

Habitat and Distribution: Rare; open sandy beach areas, dunes, rocky shelves; coastal east peninsula and coastal south peninsula Florida; the West Indies, Bermuda, Mexico, Central America.

Comment: The fragrant flowers of Sea-lavender bloom in winter and spring. The densely hairy foliage makes this an appealing ornamental. Sea-lavender is listed by the State of Florida as Endangered.

Asclepias cinerea Walter
Carolina Milkweed
Asclepiadaceae, Milkweed Family

Habit: Herbaceous perennial, 30-70 cm tall. Stems smooth or nearly so, slender, solitary, with white milky sap.

Leaves: Opposite; stalkless; blades very narrow, 3-9 cm long, 1-2 mm wide.

Inflorescences: Flat-topped clusters, terminal or from upper leaf axils.

Flowers: Whitish with a light lavender to purple tint, petals 5-7 mm long, lobes reflexed, upright central crown 4-5 mm in diameter.

Fruit: An elongate capsule, broader in the middle and tapering toward both ends, smooth, 8-12 cm long and 5-10 mm wide, erect.

Habitat and Distribution: Frequent; dry pinelands, sandhills; north peninsula Florida, west to Mississippi, and north to South Carolina.

Comment: This small spring and summer flowering plant has attractive flowers, but unless blooming, is difficult to find among the other low herbaceous plants.

Argusia gnaphalodes (L.) Heine

Asclepias cinerea Walter
photograph courtesy of John Tobe

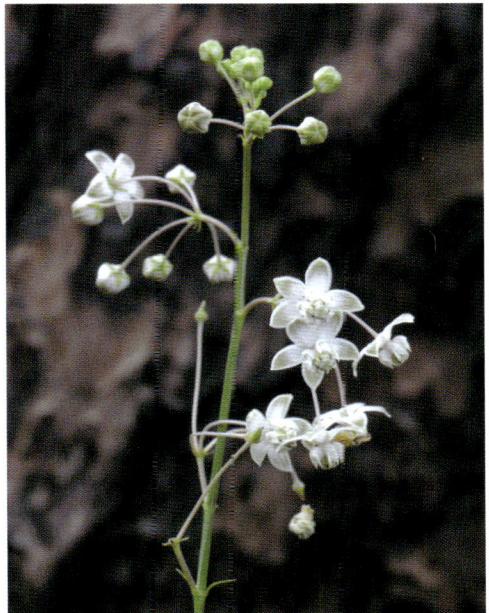

White

Asclepias perennis Walter
Swamp Milkweed
Asclepiadaceae, Milkweed Family

Habit: Herbaceous perennial, 30-50 cm tall. Stems several or solitary, almost smooth.

Leaves: Opposite; stalked; blades thin, elliptic to lance-shaped, 6-12 cm long, 1-3 cm wide.

Inflorescences: 1-4, stalked, disk-like or hemispheric clusters, in axils of the upper leaves.

Flowers: White and sometimes tinged pink, 5 outer petals, 2.5-4 mm long, lobes reflexed, upright central crown 2-3 mm wide.

Fruit: A capsule, wide in center, tapering toward ends, 4-8 cm long, 1-2.5 cm wide, smooth, drooping.

Habitat and Distribution: Frequent; wet woods, swamps, along rivers, around ponds; from central peninsula Florida, north to South Carolina, Missouri, Illinois, and Indiana, west to Texas.

Comment: Swamp Milkweed flowers in spring and summer. Some habitats have Swamp Milkweed as a ground cover which can provide a showy display when in bloom.

Asimina incarna (Bartr.) Exell
Woolly Pawpaw
Annonaceae, Custard-apple Family

Habit: Shrub, to 1.5 m tall.

Leaves: Alternate, deciduous; short-stalked; blades elliptic to oblong, 4-10 cm long, leathery, woolly when young, sparsely hairy when mature.

Inflorescences: Solitary or to 4 flowers, from leaf axils, just prior to new season's growth, nodding.

Flowers: White, outer 3 petals 4-7 cm long, inner 3 petals to 2 cm long and yellow-roughened on inside base.

Fruit: A berry, oblong, green-yellow, 4-8 cm long.

Habitat and Distribution: Occasional; pinelands and sandhills; central peninsula and eastern panhandle Florida to southeastern Georgia.

Comment: The large spring flowers of Flag Pawpaw can be spectacular. The flowers, sometimes in abundance, hang from the bare stems. Since the plants occur in sparsely vegetated habitats they are easily spotted. The pulp of the fruit is edible and is prized by various mammals.

Asclepias perennis Walter

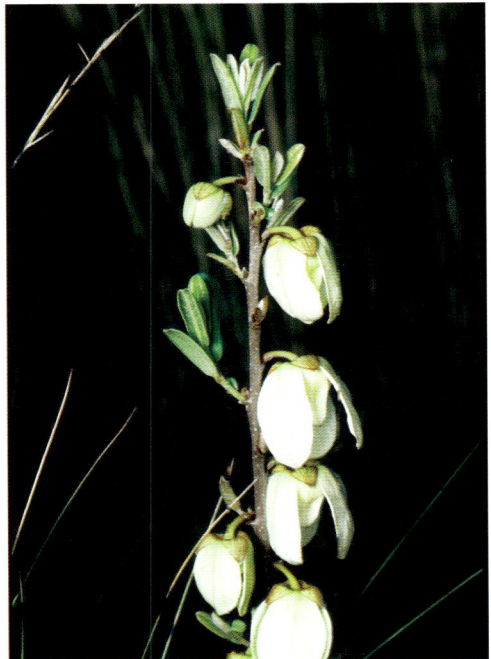

Asimina incarna (Bartr.) Exell

White

Asimina reticulata Shuttlew. ex Chapm.
Flatwoods Pawpaw
Annonaceae, Custard-apple Family

Habit: Shrub, to 1 m tall.

Leaves: Alternate, deciduous; short-stalked; blades oblong, 3-10 cm long, leathery, sparsely covered with orange hairs above, densely reddish hairy below when young.

Inflorescences: From leaf axils, 1 or occasionally 2 per axil, just before new growth, nodding.

Flowers: White, outer petals 3-4 cm long, inner petals 2-2.5 cm long with purplish-roughened bases.

Fruit: A berry, elliptic, yellow-green, 2.5-9 cm long.

Habitat and Distribution: Common; scrub, sandhills, pastures, and dry pine flatwoods; southern to northern peninsula Florida.

Comment: Spring flowers are large and showy. Flatwoods Pawpaws are common in pastures on dry sandy soils. Although a prolific weed, the plants are wonderful wildflowers. The pulp of the fruit is edible. It is difficult to find ripe fruit due to its' being relished by most mammals.

Aster reticulatus Pursh
Pine Barren Aster or White-top Aster
[*Oclemena reticulata* (Pursh) G.L. Nesom]
Asteraceae (Compositae), Aster or Sunflower Family

Habit: Herbaceous perennial, to 1.2 m tall. Stems few to many, forming clumps, erect, hairy.

Leaves: Alternate; stalkless; blades elliptic to oval, 3-8 cm long, hairy on both surfaces, veins prominent underneath.

Inflorescences: Heads in open branched terminal arrangements.

Flowers: White rays, 8-20, narrow, strap-like, 1-2 cm long, drooping, with yellow discs.

Fruit: An achene, elliptic, to 3 mm long, hairy, with whitish to tan tuft of bristles 7 mm long on top.

Habitat and Distribution: Frequent; moist and low pinelands; central peninsula Florida north on the coastal plain into southern South Carolina.

Comment: Spring blooming, Pine Barren Aster is frequently seen throughout pine woods.

White

Asimina reticulata
Shuttlew. ex Chapm.

Aster reticulatus Pursh

503

White

Aster tortifolius Michx.
Dixie White-topped Aster
[Sericocarpus tortifolius (Michx.) Nees]
Asteraceae (Compositae), Aster or Sunflower Family

Habit: Herbaceous perennial, to 1 m tall. Stems densely hairy, slender.

Leaves: Alternate, no basal rosette; stalkless; blades rounded to somewhat lance-shaped, broader near tip, 1-3 cm long, densely hairy.

Inflorescences: Heads in flat-topped terminal arrangement.

Flowers: White with white discs, with 3-7 thin rays, to 8 mm long.

Fruit: An achene, to 3 mm long, hairy, with white tuft of bristles 6-8 mm long on top.

Habitat and Distribution: Frequent; sandy soils of dry pinelands, hammocks, and old fields; throughout Florida, west to Louisiana, and north to North Carolina.

Comment: Dixie White-topped Aster flowers in the summer and fall and occasionally in all warm months southward. The mass of flowers is noticeable, but not really showy.

Baccharis glomeruliflora Pers.
Silverling Groundsel Bush
Asteraceae (Compositae), Aster or Sunflower Family

Habit: Shrub, to 3 m tall. Stems many-branched.

Leaves: Alternate; stalked; blades grayish or dull green, elliptic to oval and broader at tip, 3-6 cm long, with few large teeth at tip.

Inflorescences: Heads in axillary stalkless clusters, male and female flowers separate and on separate plants, only disc flowers present.

Flowers: White or cream, 4-6 mm long.

Fruit: An achene, to 1.5 mm long, ribbed, with tuft of white to tan hair-like bristles on top, male bristles short, female bristles 7-10 mm long.

Habitat and Distribution: Common; wet woods, hydric hammocks, swamps, fresh and brackish marshes; throughout Florida, and north to North Carolina; West Indies.

Comment: Blooming in fall and winter, the female flowers are showy in mass. Silverling Groundsel Bush is very similar to Groundsel Bush, differing in that flowers are not in spreading, stalked, terminal clusters.

Aster tortifolius Michx.

Baccharis glomeruliflora Pers.

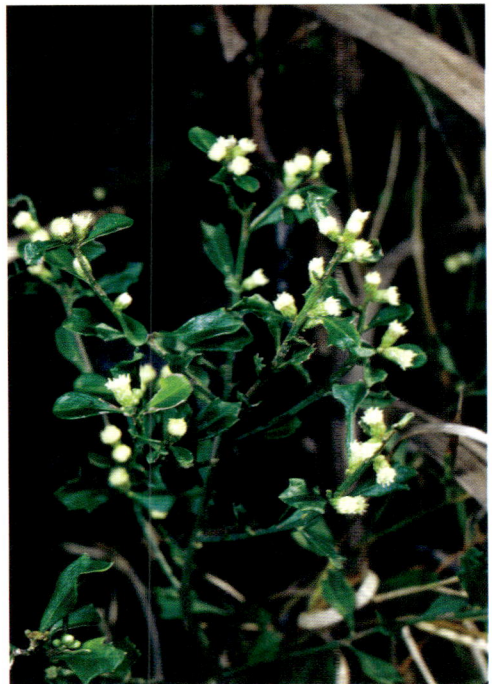

White

Baccharis halimifolia L.
Groundsel Bush
Asteraceae (Compositae), Aster or Sunflower Family

Habit: Shrub, to 4 m tall. Stems many-branched.

Leaves: Alternate; stalked; grayish or dull green, elliptic to oval and broader at tip, 3-6 cm long, with few large teeth at tip.

Inflorescences: Heads stalked, terminal, male and female flowers separate and on separate plants, only disc flowers present.

Flowers: White or cream, 4-6 mm long.

Fruit: An achene, to 1.5 mm long, ribbed, with tuft of white to tan hair-like bristles on top, male bristles short, female bristles 7-10 mm long.

Habitat and Distribution: Common; brackish and fresh marshes, beaches, hammocks, old fields, roadsides, woodland margins, and other disturbed areas; throughout Florida, west to Texas, and north to Arkansas and Massachusetts.

Comment: The female flowers in fall and winter are very showy. Female plants cultivated along edges can be spectacular. Groundsel Bush can scarcely be separated from Silverling Groundsel Bush except by the flower clusters being in axillary stalkless clusters in the latter and stalked terminal clusters in the former.

Baptisia alba (L.) Vent.
White Wild Indigo
[*Baptisia lactea* (Raf.) Thieret]
Fabaceae (Leguminosae), Pea or Bean Family

Habit: Herbaceous perennial, to 2 m tall. Stems erect, spreading, smooth, usually waxy-gray.

Leaves: Alternate; stalked; blades with 3-leaflets, leaflets short-stalked, broadly elliptic to somewhat rounded, 2-6 cm long.

Inflorescences: Terminal, stalked, spike-like arrangements.

Flowers: White petals, 2-2.5 cm long.

Fruit: An inflated pod, 2-4 cm long, black at maturity.

Habitat and Distribution: Occasional; pinelands, hammocks, sand ridges, water course banks, and roadsides; central peninsula Florida, north to North Carolina, Michigan, and Minnesota, west to Alabama and Texas.

Comment: White Wild Indigo is a striking spring blooming wildflower. The large assemblies of bright white flowers stand out among all other vegetation. It is being offered by some nurseries.

Baccharis halimifolia L.

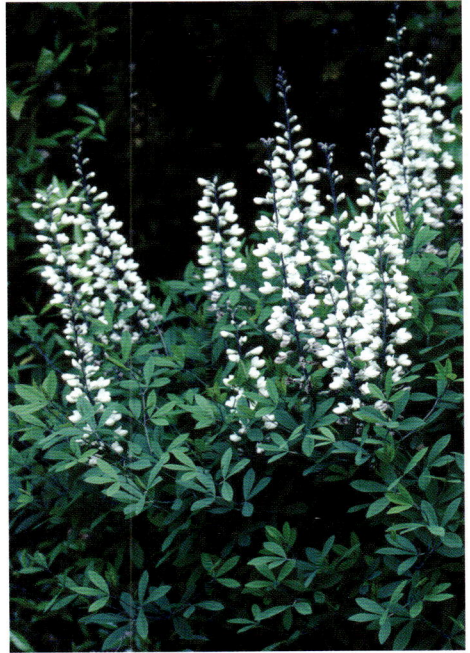

Baptisia alba (L.) Vent.

White

Befaria racemosa Vent.
Tarflower
[*Bejaria racemosa* Vent.]
Ericaceae, Heath Family

Habit: Shrub, 0.4-2.5 m tall. Stems woody, branched, hairy with distinctive stiff, long, white hairs.

Leaves: Alternate; lacking stalks or nearly so; blades narrowly elliptic to nearly oval, 2-6 cm long, evergreen, leathery, somewhat powdery.

Inflorescences: Loose clusters at stem and branch tips.

Flowers: White to pinkish, with usually 7 strap-or club-shaped petals 2-3 cm long, fragrant, sticky.

Fruit: A roundish capsule, 6-8 mm diameter, 7-parted.

Habitat and Distribution: Frequent to common; sandhills, scrub, and wet to dry pine flatwoods; peninsula Florida north along the Coastal Plain into southeastern Georgia.

Comment: Tarflower is all but invisible in its native habitats, except during the summer flowering season. The large bright white flowers are eye-catching. The sticky petals catch more than the eye as small insects stick to the surfaces.

Bidens alba (L.) DC.
Common Beggar-ticks
[*Bidens pilosa* L., misapplied]
Asteraceae (Compositae), Aster or Sunflower Family

Habit: Herbaceous annual or short-lived perennial, to 2 m tall. Stems smooth and mostly erect.

Leaves: Opposite; stalked; blades simple or with 3, 5 or 7 leaflets, leaflets elliptic to oval, each 2-10 cm long, with toothed edges.

Inflorescences: Heads terminal on stalks.

Flowers: White with yellow-orange discs, usually with 5 rays to 1.5 cm long and 0.6-1 cm wide.

Fruit: Achenes linear, flattened to 4 sided, 7-16 mm long and 1 mm wide, with 2 stiff bristles 2-3 mm long on top, bristles downwardly barbed.

Habitat and Distribution: Common; disturbed habitats, roadsides; throughout Florida, west to Texas, and north to North Carolina; pantropical.

Comment: Common Beggar-ticks can flower in any warm month. This plant is extremely weedy. It is propagated by the seeds that move everywhere humans and fur bearing animals travel by means of the bristles catching on fur and clothing. Fruits of this weed were found in the Spanish treasure ship Atocha that sank over 300 years ago and were successfully germinated. Young leaves are touted to contain helpful amounts of vitamin C and are edible though not particularly tasty. Killed by cold it is constantly reintroduced along the margins of its range.

Befaria racemosa Vent.

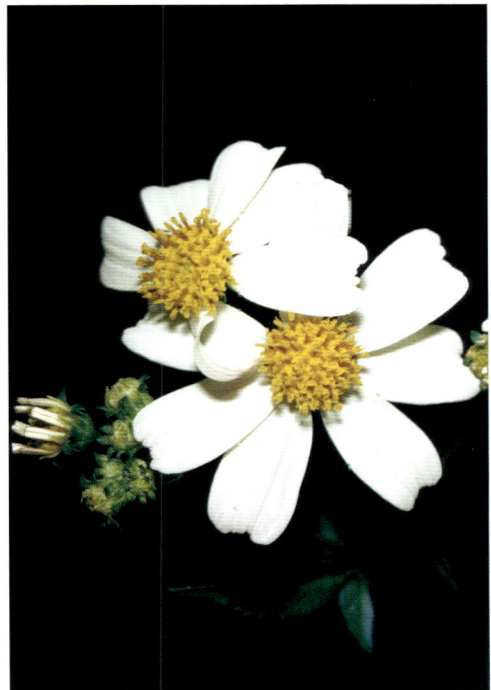

Bidens alba (L.) DC.

White

Blutaparon vermiculare (L.) Mears
Beach Carpet
[*Philoxerus vermicularis* (L.) Sm.]
Amaranthaceae, Amaranth Family

Habit: Herbaceous perennial, to 2 m long. Stems lying flat to ascending, succulent.

Leaves: Opposite; stalkless; blades narrow, succulent, 1-5 cm long, smooth, hairy in axils.

Inflorescences: Terminal, congested spikes, 1-3 cm long.

Flowers: Silvery white, 3-5 mm long.

Fruit: Inflated, tiny, to 1 mm long.

Habitat and Distribution: Frequent; coastal sites, flats, swales, and dunes; central and southern peninsula Florida, west to Texas; south through Mexico into Panama, Columbia into Brazil, the West Indies, west coast of tropical Africa.

Comment: Flowering spikes can be seen any time during the year. The silvery stems and spikes are attractive and easily seen due to the open habitat in which they occur.

Bourreria succulenta Jacq.
Strongback
[*Bourreria ovata* Miers; *Bourreria revoluta* Kunth]
Boraginaceae, Borage Family

Habit: Shrub or small tree, to 10 m tall. Bark reddish-brown.

Leaves: Alternate; stalked; blades somewhat rounded, often broadest towards tip, to 12 cm long, leathery, margins smooth.

Inflorescences: Terminal branched clusters.

Flowers: White, about 1 cm long, lobes rounded.

Fruit: Nearly round, with a central stone, orange-red, 1-1.5 cm in diameter, flesh thin.

Habitat and Distribution: Occasional; hammocks and thickets; Dade and Monroe Counties in south peninsula Florida; Cuba, Bahamas.

Comment: The clusters of large flowers that can be seen all year are showy as are the orange-red fruits that follow. Strongback is a good ornamental.

White

Blutaparon vermiculare (L.)Mears

Bourreria succulenta Jacq.

White

Byrsonima lucida (Mill.) DC.
Locustberry
Malpighiaceae, Malpighia Family

Habit: Shrub, to 4 m tall, occasionally a small tree to 8 m tall.

Leaves: Opposite, evergreen; stalked; blades spatulate, to 5 cm long, smooth, margins smooth.

Inflorescences: Terminal branched clusters.

Flowers: White to red, petals 5-7 mm long, rounded at tip, narrowed at base.

Fruit: Rounded, 3-lobed, drupe 9-12 mm in diameter.

Habitat and Distribution: Rare; rocky pinelands, hammocks, rocky flats and depressions; Dade and Monroe Counties in south Peninsula Florida; the West Indies.

Comment: Blooming throughout the year, Locustberry is an excellent wildflower and ornamental. The clusters of flowers are eye-catching, especially with many clusters on a many-branched shrub. The fruits are edible and said to taste like cranberries.

Burmannia capitata (Walt.) Mart.
Southern Bluethread
Burmanniaceae, Burmannia Family

Habit: Herbaceous annual, 5-20 cm tall. Stems very slender, threadlike, rarely branched.

Leaves: Alternate; stalkless; few, very small, scale-like, 3-5 mm long.

Inflorescences: Dense, terminal, cap-like clusters with as many as 20 flowers.

Flowers: Cream, sometimes tinged with blue, 3-5 mm long and about 1 mm broad.

Fruit: A capsule, to 3 mm long, 3-lobed.

Habitat and Distribution: Occasional, often overlooked; low wet woodlands and bogs; throughout Florida, west to Texas, and north to North Carolina.

Comment: This tiny warm season bloomer is seldom noticed. Almost never seen except when in flower, it is treated as a novelty.

Byrsonima lucida (Mill.) DC.

Burmannia capitata (Walt.) Mart.

White

Cakile lanceolata (Willd.) O.E. Schulz
Southern Sea rocket
Brassicaceae (Cruciferae), Mustard Family

Habit: Herbaceous annual, to 60 cm long. Stems prostrate to ascending, smooth, succulent.

Leaves: Alternate; stalked; blades elliptic to oblong, succulent, 3-15 cm long, margins undulating to having rounded, blunt teeth.

Inflorescences: Flowers scattered along upper stem.

Flowers: White to slightly purple, about 1 cm across, petals 4.

Fruit: A capsule, slender, 1-3 cm long, 2-segmented.

Habitat and Distribution: Common; coastal - dunes, strand, marshes, and flats; Mississippi south into the Florida Keys; the West Indies, Mexico south into northern South America.

Comment: Flowers appear in spring and summer. Southern Sea-rocket is difficult to spot when not blooming, as many similar succulent plants occur in the coastal habitats.

Calystegia sepium (L.) R. Br.
Hedge Bindweed
Convolvulaceae, Morning-glory Family

Habit: Herbaceous perennial vine, 0.1-2 m long, from underground runners. Stems trailing or twining, branching.

Leaves: Alternate; long-stalked; blades arrow-shaped or triangular, 4-10 cm long, smooth.

Inflorescences: Flowers solitary or paired in leaf axils.

Flowers: White, funnel-shaped, 5-7 cm wide, petals 5, fused, with 2 leaf-like bracts around base of tube.

Fruit: A rounded capsule, 2-4 seeded.

Habitat and Distribution: Occasional; salt and fresh water marshes, rights-of-way, and other open disturbed areas; throughout Florida, and most of North America; cosmopolitan in temperate regions; a native of Eurasia.

Comment: The large showy flowers in spring and summer are easily noticed. Hedge Bindweed is normally weedy and is to be expected in such areas.

Cakile lanceolata (Willd.) O.E. Schulz

Calystegia sepium (L.) R. Br.

White

Capsella bursa-pastoris (L.) Medik.
Shepherd's Purse
Brassicaceae (Cruciferae), Mustard Family

Habit: Herbaceous annual or winter annual, 30-90 cm tall. Stem single from base then branching.

Leaves: Alternate, mostly basal; stalkless, but narrowed to base; blades 3-14 cm long, basal leaves deeply lobed, becoming less lobed to nearly unlobed upwards on stem.

Inflorescences: Long, spike-like arrangements from upper leaf axils.

Flowers: White, petals 4, 2-4 mm long, broadest at tip.

Fruit: A 2-parted wedge-shaped capsule, erect, to 8 mm long.

Habitat and Distribution: Occasional; fields and other disturbed sites; central peninsula Florida, throughout the United States; cosmopolitan; a native of Europe.

Comment: This weedy little spring bloomer can be attractive. It is recognized by the heart-shaped, two-sided fruits.

Catharanthus roseus (L.) G. Don
Madagascar Periwinkle
[Vinca rosea L.]
Apocynaceae, Dogbane Family

Habit: Herbaceous perennial, 20-80 cm tall. Stems erect, somewhat bushy.

Leaves: Opposite; short-stalked; blades dark green, glossy, oblong-elliptic, 2-8 cm long.

Inflorescences: Short stalks with 1-3 flowers from leaf axils.

Flowers: White, rose purple, or pink, with 5 broad, pointed petals, each lobe 1.5-2.5 cm long, and a slender tubular throat, 2-3.5 cm long.

Fruit: 2-parted, dry, each part slender and 1.5-3.5 cm long.

Habitat and Distribution: Frequent; dry, open disturbed habitats, waste places, pinelands, scrub land, and other open sites; central and southern peninsular Florida, and rarely panhandle Florida, north to the Carolinas; pantropical; a native of Madagascar.

Comment: Madagascar Periwinkle is very popular as a cultivated plant. Blooming in all warm months, it is reliable for color in hot weather. Further spread northward into the southern margins of temperate areas is to be expected.

Capsella bursa-pastoris
(L.) Medik.

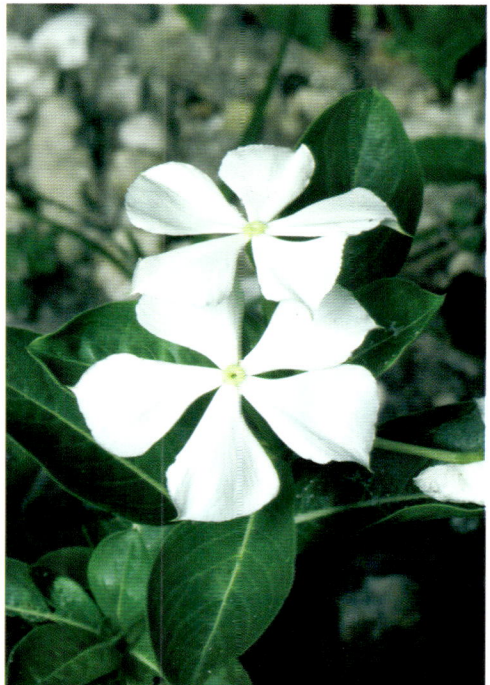

Catharanthus roseus (L.) G. Don

White

Ceanothus americanus L.
New Jersey Tea or Redroot
Rhamnaceae, Buckthorn Family

Habit: Shrub, 0.2-1 m tall. Branches low, bushy.

Leaves: Alternate, deciduous; short-stalked; blades lance-shaped to rounded, 2-8 cm long, hairy, clearly 3-veined beneath, margins with teeth.

Inflorescences: Stalked, dense clusters, terminal or from leaf axils.

Flowers: Creamy white, about 5 mm wide, petals 5, rounded at tips, narrowed at base.

Fruit: 3-lobed, capsule-like, slightly fleshy, 4-6 mm in diameter, 3-seeded.

Habitat and Distribution: Frequent; dry sites - hammocks, sandhills, mixed forests, and clearings; central peninsula Florida, west to Texas, and north to Manitoba, Quebec, and Maine.

Comment: New Jersey Tea is a very attractive shrub, much more so when in flower in spring. The dried leaves were used as a tea substitute by early settlers. Roots have been used medicinally.

Ceanothus microphyllus Michx.
Small-leaf Redroot
Rhamnaceae, Buckthorn Family

Habit: Shrub, 30-60 cm tall. Branches low, bushy, but not dense.

Leaves: Alternate, deciduous; short-stalked; blades elliptic to rounded, very small, 3-10 mm long.

Inflorescences: Small, round clusters, terminal on branches and twigs.

Flowers: White, about 3 mm wide, petals 5, rounded at tips, narrowed at base.

Fruit: 3-lobed, capsule-like, 4-5 mm in diameter, 3-seeded.

Habitat and Distribution: Frequent; dry pinelands and sandhills; central peninsula Florida, north to Alabama and Georgia.

Comment: This small-leaved shrub is not particularly attractive, even though the small spring flowers are notable.

Ceanothus americanus L.

Ceanothus microphyllus Michx.
photographs courtesy of John Tobe

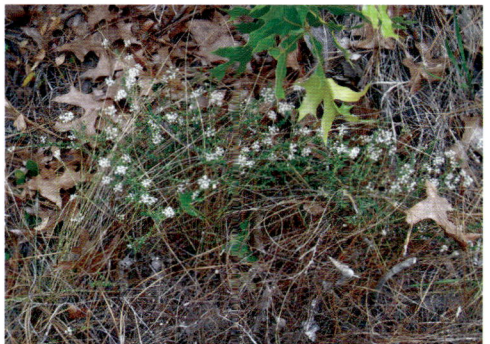

White

Celosia trigyna L.
African Cock's Comb
Amaranthaceae, Amaranth Family

Habit: Herbaceous annual, 0.3-1.2 m tall. Branches spreading, vine-like.

Leaves: Alternate; stalked; blades broadly elliptic, up to 8 cm long.

Inflorescences: Long-stalked, dense, tight spikes, from leaf axils or terminal.

Flowers: Greenish white, star-shaped, to 3 mm long.

Fruit: A capsule, breaks open around middle, like a bowl with a lid to release seeds.

Habitat and Distribution: Infrequent; disturbed areas; central peninsula Florida, widely scattered elsewhere in Florida; a native of Africa.

Comment: African Cock's Comb is an extremely popular summer flowering ornamental. Due to its popularity it will probably be more frequently found as an escape in warmer climates.

Cephalanthus occidentalis L.
Button Bush
Rubiaceae, Madder Family

Habit: Shrub to 3 m tall or, exceptionally, a small tree to 16 m tall.

Leaves: Opposite or whorled; stalked, a papery structure or line is on each side of the stem between the leaf stalks; blades elliptic to somewhat rounded, 4-20 cm long, margins smooth.

Inflorescences: Long-stalked, spherical heads, 2-3 cm in diameter, terminal or from leaf axils.

Flowers: Creamy white, with many yellow tipped anthers exserted 5-8 cm outward.

Fruit: In dark brown clusters resembling the flower heads, elongate, angled; nutlets 4-8 mm long.

Habitat and Distribution: Common; wet sites - banks of streams, lakes, ponds, marshes, and swamps; throughout Florida, north to Nova Scotia, New Brunswick, Quebec, and Minnesota, west to Texas and California; Mexico, the West Indies.

Comment: Flowering is generally summer through fall, but is all year in warmer climates. The exceptionally showy balls of flowers make this shrub an attractive landscape plant. It can be used in wet to moist soils. Some pruning of older plants is desirable. The fruits are eaten by several birds. Parts of this shrub are used medicinally. Horses are poisoned by the foliage and twigs, but some other animals are apparently not harmed.

Celosia trigyna L.

Cephalanthus occidentalis L.

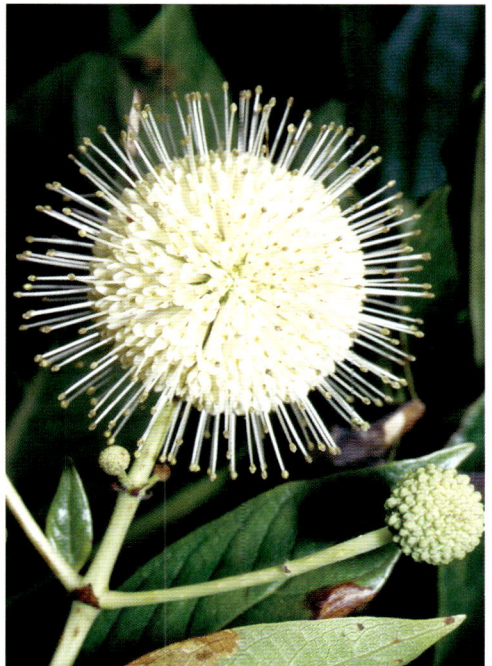

White

Cerastium glomeratum Thuill.
Mouse-ear Chickweed
Caryophyllaceae, Pink Family

Habit: Herbaceous winter annual, 5-30 cm tall. Stems mat forming, with sticky hairs.

Leaves: Opposite; stalkless; blades spatulate, 0.4-3 cm long, with bases clasping stems.

Inflorescences: Terminal, congested clusters.

Flowers: White, about 1 cm wide, petals 5, two-lobed.

Fruit: A cylindric capsule, 5-9 mm long.

Habitat and Distribution: Frequent; fields, roadsides, and other disturbed habitats; central peninsula Florida, west to Texas and California, and north to British Columbia, South Dakota, Illinois, and Nova Scotia.

Comment: The spring flowers of Mouse-ear Chickweed are delicate, but attractive when viewed closely. This is a common turf weed and moves easily by means of the sticky stems and small seeds.

Chamaelirium luteum (L.) A. Gray
Devil's-bit or Fairy-wand
Liliaceae, Lily Family

Habit: Herbaceous perennials, separate male and female plants, male plants to 0.7 m tall, female plants to 1.2 m tall, from short, thick, underground runners. Stems simple.

Leaves: Alternate, evergreen, larger leaves basal, decrease in size upwards on stem; stalkless; blades mostly elliptic, 5-20 cm long, tapered to base.

Inflorescences: Slender spikes, 2-15 cm long.

Flowers: White, male 3-4 mm long, female 2-3 mm long.

Fruit: An ellipsoid capsule, 7-11 mm long and 5-6 mm in diameter.

Habitat and Distribution: Occasional; rich woodlands, flatwoods, bluffs, moist thickets, and bogs; central peninsula Florida, west to Louisiana and Arkansas, north to Illinois, Ohio, Ontario, and Massachusetts.

Comment: The flowers in spring and summer are very difficult to see as are the plants. The flowers are small and the plants blend into the other low vegetation. However, the very long spikes can catch the eye.

Cerastium glomeratum Thuill.

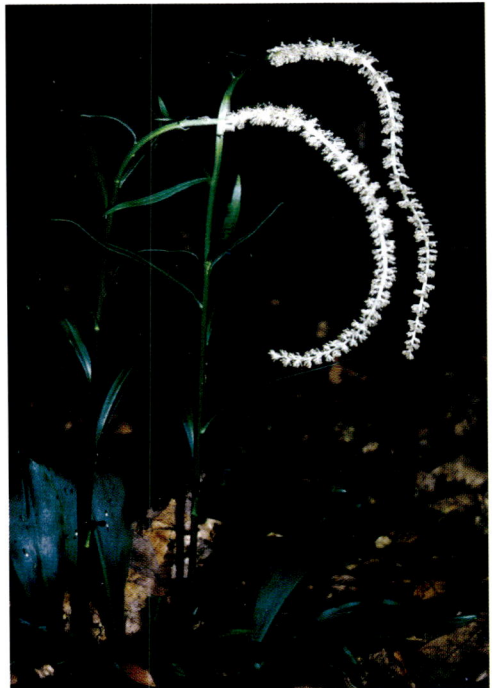

Chamaelirium luteum (L.) A. Gray

White

Chaerophyllum tainturieri Hook.
Wild Chervil
Apiaceae (Umbelliferae), Carrot Family

Habit: Herbaceous annual, 20-90 cm tall. Stems erect, solitary, stiffly hairy.

Leaves: Alternate; stalked; blades finely divided, 2-12 cm long, smooth to hairy.

Inflorescences: Small, terminal clusters of 3-10 flowers.

Flowers: White, small, with 5 petals.

Fruit: Narrow, broadest in middle, 4-8 mm long, ribbed.

Habitat and Distribution: Occasional; wet hammocks, open disturbed sites, and prairies; from central peninsula Florida, west to Texas, Arizona, Kansas, and Missouri, and north to Virginia.

Comment: Even when flowering in spring, Wild Chervil is difficult to identify. The fruits are distinctive and provide the best characteristic for identification.

Chamaesyce hypericifolia (L.) Millsp.
Upright Spurge
[*Euphorbia hypericifolia* L.]
Euphorbiaceae, Spurge Family

Habit: Herbaceous annual, to 60 cm tall. Stems smooth, erect or ascending, often reddish, with milky sap.

Leaves: Opposite; short-stalked; blades oblong, 1-4 cm long, 0.3-1.7 cm wide, margins with teeth towards tip.

Inflorescences: Small, round, stalked clusters, from leaf axils.

Flowers: White, tiny.

Fruit: A smooth capsule, about 1.3 mm long, 3-parted.

Habitat and Distribution: Common; disturbed sites; throughout Florida, north to Georgia, west to Texas; south through Mexico into Venezuela and Columbia; West Indies.

Comment: Upright Spurge blooms all year. The clusters of tiny white flowers are attractive when viewed closely. This plant is a common wildflower of disturbed soils, frequently seen in gardens and cultivated fields.

White

Chaerophyllum tainturieri Hook.

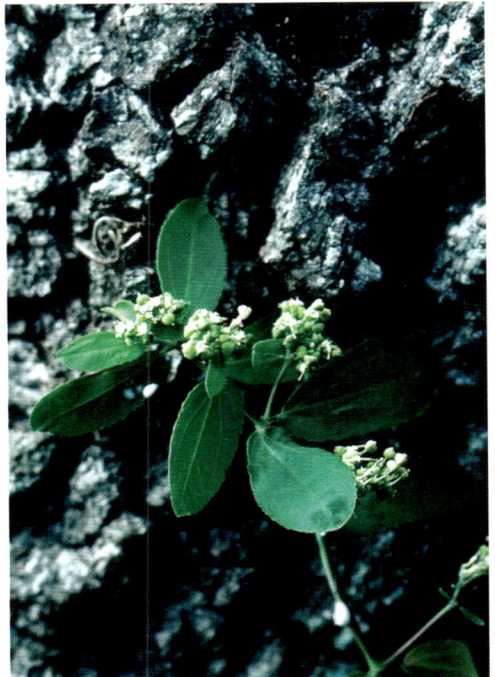

Chamaesyce hypericifolia (L.) Millsp.

525

White

Chaptalia tomentosa Vent.
Pineland Daisy or Sun-bonnets
Asteraceae (Compositae), Aster or Sunflower Family

Habit: Herbaceous perennial, to 40 cm tall. Stems one to few, hairy.

Leaves: Alternate, mostly in basal rosettes; stalkless, narrowed to base; blades lance-shaped and broader at tip to oval, 4-18 cm long and 1-4 cm wide, dense soft white hairs beneath, with minutely toothed margins.

Inflorescences: Heads solitary, terminal, erect or nodding.

Flowers: White with a white or cream disk, with 13-21 rays, 3-6 mm long, 3-toothed at tips, usually pink beneath.

Fruit: A grayish brown, elliptical achene, about 4-6 mm long tapering to a distinct beak, with pale hair-like bristles 4-6 mm long at top.

Habitat and Distribution: Frequent; moist and wet flatwoods, bogs, and sandy forested habitats; throughout Florida, west to east Texas, and north to North Carolina.

Comment: Pineland Daisy blooms in spring and is characteristic in wet pinelands.

Chionanthus virginicus L.
Fringe Tree
Oleaceae, Olive Family

Habit: Tall shrub or small tree, to 10 m tall.

Leaves: Opposite; stalked; blades elliptic to oval, to 20 cm long, smooth above and smooth or hairy below.

Inflorescences: Showy, fringe-like clusters below newest growth.

Flowers: White or cream, nodding, with 4 strap-like petals 1.5-3 cm long.

Fruit: Elliptical or oval with a central stone, 1-1.8 cm, 6-10 mm diameter, blue.

Habitat and Distribution: Frequent; moist woods, thickets, seepage areas, along stream banks, sandhill and scrub margins; central peninsula Florida, west to Texas and Oklahoma, and north to Missouri, Ohio and New Jersey.

Comment: The large showy clusters of Fringe Tree blooms appear in early spring just as the leaves are beginning to come out. Flowers on a plant all function as either male or female. Fringe Tree is sometimes called Old-man's-beard. It is one of the plants that early settlers used as an ornamental. It is still frequently collected from the wild and used in rural settings. When used as a specimen tree, Fringe Tree is spectacular in flower.

Chaptalia tomentosa Vent.

Chionanthus virginicus L.

White

Cicuta maculata L.
Water Hemlock
[Cicuta mexicana J.M. Coult. & Rose]
Apiaceae (Umbelliferae), Carrot Family

Habit: Herbaceous perennial, 1-2.5 m tall. Stems smooth, hollow, purplish or with purple stripes, from clustered often tuberously-thickened roots.

Leaves: Alternate; long-stalked; blades divided 2 or 3 times, leaflets odd-numbered, lance-shaped, each to 10 cm long and 3.5 cm wide, thin, margins toothed.

Inflorescences: Groups of small, domed clusters that form a large clump, from leaf axils or terminal.

Flowers: Whitish green, tiny, with 5 petals.

Fruit: Elliptical, flattened, to 3 mm long, with corky ribs.

Habitat and Distribution: Frequent; low, wet sites, swamps, marshes, banks; throughout Florida, north to Virginia, Prince Edward Island, and Quebec, west to the Dakotas, Texas, and eastern Mexico.

Comment: The hollow stems with purplish-stripes and large clusters of tiny white flowers are quite attractive. Flowers can be found in summer and fall. The plant is deadly POISONOUS when ingested. The flower clusters and general habit of growth can easily be confused with Elderberry which has opposite leaves.

Citrus aurantium L.
Sour Orange
Rutaceae, Citrus Family

Habit: Small tree, to 10 m tall. Branches many, with spines.

Leaves: Alternate; stalked, stalks often winged; blades oval, 6-12 cm long.

Inflorescences: Flowers solitary or in small clusters from leaf axils.

Flowers: White, about 2.5 cm wide.

Fruit: A rounded berry, 7-9 cm in diameter, segmented, many-seeded, outer rind normally roughened.

Habitat and Distribution: Frequent; disturbed woods; peninsula Florida into the central panhandle; a native of southeast Asia, now pantropical.

Comment: The spring flowers are attractive, especially against the backdrop of the dense somewhat shiny green leaves. The flowers are extremely fragrant and are noticed at long distances. Sour Orange has been used as a rootstock for other types of Citrus due to its resistance to root rot disease. Other types of Citrus grown on this rootstock are sensitive to freezing and, when frozen, the Sour Orange rootstock often survives. Seeds from mature plants can spread the species, especially as the fruits are eaten by livestock. The fruit is edible but very sour. The juice is used as an orangeade when sugar is added. The peel is a source for marmalade and is candied. The peel is also used for an essential oil and is a source for a perfume. Some medicines have been made from the peel and its oil. The peel has been used as a source for vitamins. The plant is also a source for honey.

Cicuta maculata L.

Citrus aurantium L.

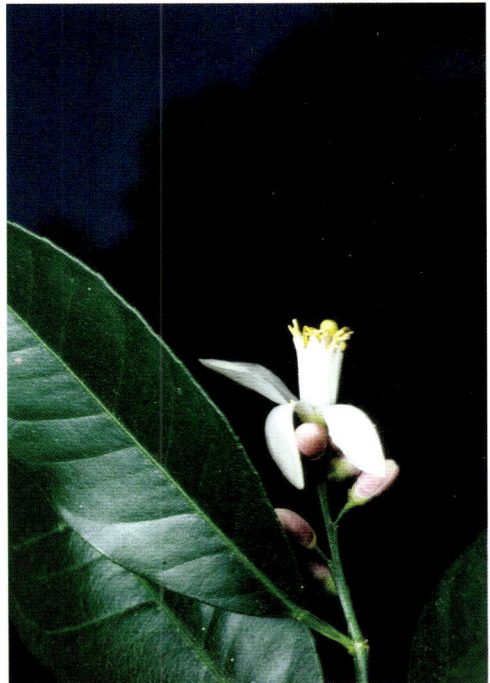

White

Clerodendrum indicum (L.) Kuntze
Sky Rocket or Turk's Turban
Verbenaceae, Verbena Family

Habit: Shrub, 1-4 m tall. Stems smooth, stout, hollow, branching in whorls.

Leaves: Opposite or whorled in 3s to 6s; stalked; blades oblong to elliptic, to 23 cm long, smooth.

Inflorescences: Few-flowered, branched clusters from upper leaf axils and terminal.

Flowers: White or yellowish, 2.5-3.4 cm long, arching downward, with elongated slender tube, with 5 recurved triangular lobes, the outer sepals green turning red, 1.1-1.7 cm long, 5-lobed.

Fruit: With a central 4-parted stone, 4-lobed, 8-12 mm long, red, purple, or black.

Habitat and Distribution: Frequent; disturbed wooded and open habitats; throughout Florida, west to Texas and north to South Carolina; widely distributed in the warm temperate and tropical parts of the world; a native of the East Indies.

Comment: The always showy flowers and fruits of this erect slender shrub can be seen throughout the warm months of the year and never fail to provoke comment. The plants frequently escape into untended thickets and the tall stems poke out through the surrounding vegetation to the amazement of passerby.

Clethra alnifolia L.
Sweet Pepper-bush
Clethraceae, White-alder Family

Habit: Shrub, to 3 m tall. Young twigs hairy, older twigs smooth.

Leaves: Alternate, deciduous; short-stalked; blades elliptic or lance-shaped, wider near tip, pointed at both ends, 4-11 cm long, margins toothed..

Inflorescences: Dense, terminal, frilly spikes, 10-20 cm long.

Flowers: White (rarely pink), with 5 narrow petals, each 5-8 mm long.

Fruit: Capsules 3-parted, erect, roundish, 2-3 mm long, hairy.

Habitat and Distribution: Frequent to locally common; prairies, bays, swamps, banks, bogs, and wet pine woods; from north peninsula Florida, west to Texas, and north to southern Maine.

Comment: Summer blooming, Sweet Pepper-bush is used as a flowering shrub. The flowers are quite fragrant and can be mashed to form a lather. Leaves turn a shade of orange in the fall.

Clerodendrum indicum (L.)
Kuntze

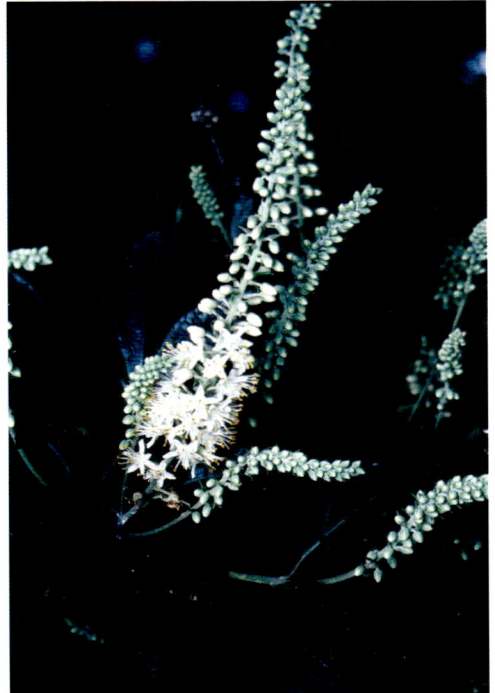

Clethra alnifolia L.

White

Cliftonia monophylla (Lam.) Britton ex Sarg.
Black Titi or Buckwheat Tree
Cyrillaceae, Titi Family

Habit: Shrub or small tree, to 10 m tall.

Leaves: Alternate, evergreen; short-stalked or stalkless; blades elliptic, 3-10 cm long, leathery, smooth, lateral veins difficult to see.

Inflorescences: One to several, spike-like arrangements near the tip of last year's twigs.

Flowers: White, 5-8 mm long, with 5 petals.

Fruit: 3- to 5-lobed, 3- to 5-winged, 3- to 5-seeded.

Habitat and Distribution: Frequent and locally common; acid swamps, bogs, stream banks, and wet ditches; northern peninsula Florida, north into Georgia, and west into Louisiana.

Comment: The Black Titi is prized as a honey tree by bee keepers when it flowers in the spring. The masses of fragrant, white flowers can be sensational when viewing large populations. Flowers appear before new growth for the season. The winged fruits are distinctive.

Cnidoscolus stimulosus (Michx.) Engelm. & A. Gray
Spurge-nettle, Stinging-nettle, Bull-nettle, Tread-softly
Euphorbiaceae, Spurge or Poinsettia Family

Habit: Herbaceous perennial, to 1 m tall, from a long, deep root. Stems usually branched and covered with stinging hairs, sap milky.

Leaves: Alternate; long-stalked; blades deeply 3- to 5-lobed, 10-20 cm wide, with jaggedly toothed edges and stinging hairs.

Inflorescences: Few-flowered clusters in the axils of the upper leaves or terminal, male and female flowers separate, but in the same cluster, the central flower usually female.

Flowers: Bright white sepals, 5, no petals, 1-2 cm wide.

Fruit: A 3-seeded capsule; seeds dark brown.

Habitat and Distribution: Common; dry disturbed sites, scrub, sandhills, sandy woods, hammocks and old fields; from Virginia south throughout Florida, and west to Louisiana.

Comment: Blooming from late spring into early fall northward and all year southward, Spurge-nettle is very weedy in most dry sites including turf and is commonly found in lawns. The flowers are so bright white that they almost seem disconnected from the plant in some levels of daylight. The contents of the stinging hairs produce a small pustule when injected into the skin by even a slight touch. Some individuals are very allergic and much of their bodies can become swollen by contact with even one hair. The root is edible and is used medicinally.

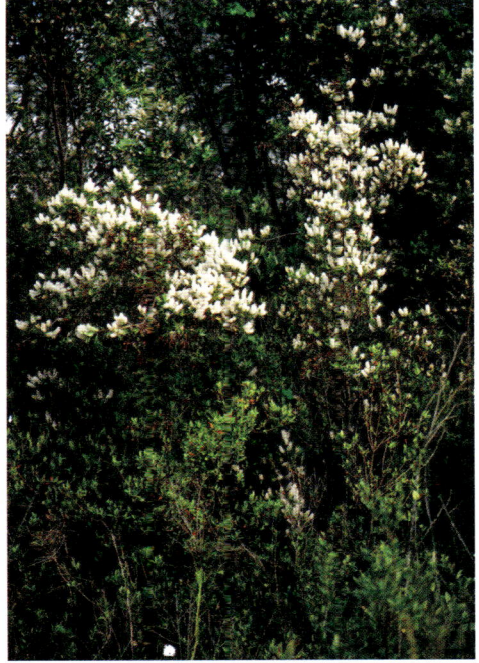

Cliftonia monophylla (Lam.) Britton ex Sarg.

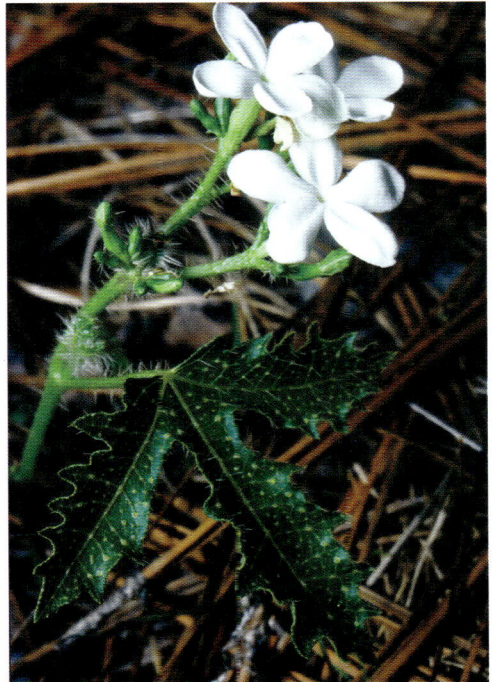

Cnidoscolus stimulosus
(Michx.) Engelm. & A. Gray

White

Coccoloba uvifera (L.) L.
Seagrape
Polygonaceae, Buckwheat Family

Habit: Shrub or small tree, to 15 m tall. Branches spreading, bark smooth.

Leaves: Alternate, evergreen, sheathing stem at base; stalked; blades round or somewhat heart-shaped, 4-27 cm long and wide, thick, sometimes with reddish veins, margins smooth.

Inflorescences: Long, slender spikes from leaf axils.

Flowers: White or cream to greenish, 63 mm wide.

Fruit: Round, 1-2 cm in diameter, purplish when ripe, hanging in grape-like clusters.

Habitat and Distribution: Frequent; coastal hammocks and strands, dunes, beaches, and rocky outcrops; the Florida Keys to northern peninsula Florida along the coast; Bermuda, the West Indies, Mexico, Central America, South America.

Comment: Very popular as an ornamental, Seagrape blooms from spring into fall. The long strings of flowers followed by the grape-like fruits are ornamental as is the spreading habit with the round leaves. Fruits are juicy, somewhat harsh to the taste, and frequently used for jelly. The wood is used in cabinetry.

Conopholis americana (L.) Wallr.
Squawroot or Cancer-root
Orobanchaceae, Broomrape Family

Habit: Herbaceous parasitic perennial, 5-25 cm tall and 2-4 cm diameter, often on oak roots. Stems unbranched, usually clumped, stout, covered with yellowish-brown, overlapping leaves.

Leaves: Alternate; stalkless; blades yellowish brown, scale-like.

Inflorescences: Numerous flowers in a terminal spike, spike 4-20 cm long.

Flowers: Pale yellow to yellowish, 1-1.5 cm long, curved downward, upper lip notched, lower lip 3-parted.

Fruit: A rounded capsule, 0.8-1.4 cm long.

Habitat and Distribution: Occasional; dry oak hammocks and woodlands; central peninsula and eastern panhandle Florida, north through Alabama to Michigan and Maine.

Comment: As a curiosity, the spring blooming Squawroot is noticed, but certainly not in any demand as a wildflower.

Coccoloba uvifera (L.) L.

Conopholis americana (L.) Wallr.

White

Cornus florida L.
Flowering Dogwood
Cornaceae, Dogwood Family

Habit: Small to medium sized tree, 5-15 m tall.

Leaves: Opposite, deciduous; stalked; blades broadly elliptic, 5-12 cm long, margins smooth.

Inflorescences: Disc-like clusters at branch tips.

Flowers: Green, small, surrounded by 4 large, white or pink, rounded, notched, petal-like bracts, 4-6 cm long.

Fruit: Drupe ellipsoid, 8-18 mm long, 4-8 mm in diameter, and red when ripe.

Habitat and Distribution: Frequent to common; well drained, moist soils of hammocks, deciduous woodlands, thickets, and fence rows; central peninsula Florida, through the northeastern states into Maine, southern Ontario, Michigan, Illinois, Missouri, and south into Oklahoma, Kansas, Texas, and Mexico.

Comment: This understory tree is one of the most popular landscape plants in the south. Flowering is in spring and the Dogwood season can be easily traced from Florida northward by following the distinct gradually moving line of flowering plants. The tree has attractive bark and foliage most of the growing season. The very showy, petal-like structures appearing with the flowers in the spring can be spectacular. In the fall the foliage turns shades of orange and red and the accompanying clusters of red fruit add to the display. The wood has a variety of uses.

Cornus foemina Mill.
Swamp Dogwood
[*Cornus stricta* Lam.; *Svida stricta* (Lam.) Small]
Cornaceae, Dogwood Family

Habit: Shrub, to 6 m tall. Twigs smooth.

Leaves: Opposite, deciduous; stalked; blades broadly elliptical, lance-shaped, oval, 4-10 cm long, smooth or with few, short, flattened hairs beneath, margins smooth.

Inflorescences: Terminal and axillary domed clusters.

Flowers: White, small.

Fruit: Drupe ellipsoid, to 4-6 mm long, blue.

Habitat and Distribution: Frequent; rich, swampy wooded sites; throughout Florida, west to Texas, north to southeastern Virginia, and inland into Indiana, Illinois, and Missouri.

Comment: This spring blooming shrub is occasionally used as a hedge. The flower clusters are not especially large or showy, but are attractive. The blue clusters of fruits add to the appeal in summer, but are not large. Unlike Flowering Dogwood, the leaves of Swamp Dogwood do not show fall color.

536

Cornus florida L.

Cornus foemina Mill.

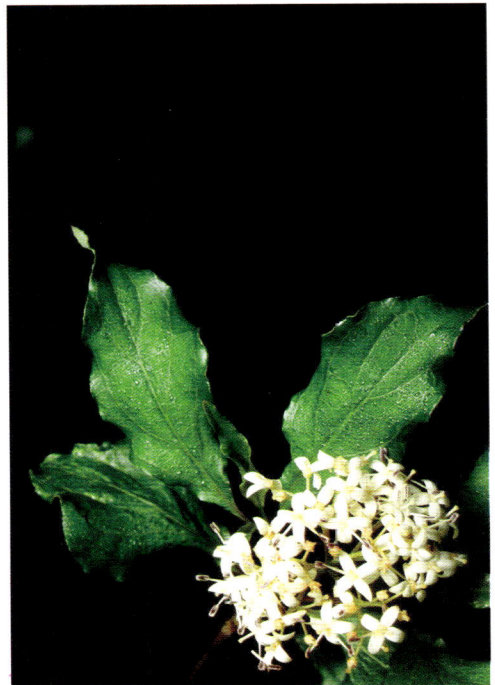

White

Crataegus michauxii Pers.
Summer Haw
[*Crataegus flava* Aiton, misapplied; *Crataegus lacrimata* Small]
Rosaceae, Rose Family

Habit: Shrub or small tree, to 5 m tall. Stems thorny with spines 2-6 cm long, young branches woolly.

Leaves: Alternate, deciduous; glandular-stalked; blades paddle-shaped, wider toward tip, sometimes shallowly lobed, with black-tipped teeth along edges.

Inflorescences: Solitary flowers from leaf axils or in small clusters at tips of shoots.

Flowers: White, with 5 separate, rounded petals, 5-8 mm long.

Fruit: An apple-like pome, round, 8-17 mm diameter, red, orange, or yellow, with dried flower parts at tip.

Habitat and Distribution: Frequent; dry open sandy woods, sandhills, scrub, pinelands, old fields, fence rows, and rights-of-way; from northern peninsula Florida, west to Mississippi, and north to Tennessee and North Carolina.

Comment: Summer Haw is an especially ornamental shrub. The spring flowers are profuse and showy. The shrub/small tree usually has drooping branches and a rounded crown. The paddle-shaped leaves are noticeable and turn shades of yellow in fall. The fruits reach maturity in late summer and fall. Trees with a multitude of fruits are striking. The fruits are edible and often used for jelly.

Crinum americanum L.
String-lily
Amaryllidaceae, Amaryllis Family

Habit: Herbaceous perennial, to 90 cm tall, from a thick bulb, with above ground runners.

Leaves: Alternate, in a basal rosette; stalkless; blades strap-shaped, 60-150 cm long and to 5 cm wide, margins with teeth.

Inflorescences: 2-6, terminal flowers on a leafless stem.

Flowers: White, fragrant, with 6 long slender, outward arching lobes 5-14 cm long.

Fruit: A globose capsule, about 3 cm thick, 3-lobed.

Habitat and Distribution: Frequent; swamps, marshes, ditches, lakes, shores, and wet woods; throughout Florida, north to Georgia, and west to Texas.

Comment: String-lily flowers in spring and summer. The large fragrant flowers are very showy. The shiny, spirally-arranged, strap-shaped leaves are also attractive and make this a very useful landscape specimen.

Crataegus michauxii Pers.

Crinum americanum L.

White

Croton argyranthemus Michx.
Silver Croton
Euphorbiaceae, Spurge Family

Habit: Herbaceous or almost shrubby perennial, 10-60 cm tall, from underground runners. Stems herbaceous, covered with brownish star-shaped hairs.

Leaves: Alternate; stalked; blades narrow, elliptic to lance-shaped, 1-5 cm long, with silvery scales beneath, often brown-flecked, margins smooth.

Inflorescences: Small terminal clusters, male and female flower separate in the same cluster, male above the female.

Flowers: White to yellow, minute.

Fruit: A capsule, to 5 mm long, 3-seeded.

Habitat and Distribution: Frequent; dry pinelands, scrub, and sandhills; from central peninsula Florida, west to Texas, and north to Georgia.

Comment: Silver Croton blooms in spring and summer. The flowers are scarcely noticeable, but the silvery leaves are appealing.

Croton glandulosus L.
Tropic Croton
Euphorbiaceae, Spurge Family

Habit: Herbaceous annual, 10-60 cm tall, from a tap root. Stems branched, reddish, covered with tiny, star-shaped hairs.

Leaves: Alternate; stalked; blades narrow, oblong, 1-9 cm long, margins with teeth, a saucer-shaped gland on each side at base.

Inflorescences: Terminal clusters, male and female flowers separate in the same cluster, male above female.

Flowers: White, small.

Fruit: An erect capsule, 4-6 mm long, 3-seeded.

Habitat and Distribution: Frequent; dry sandy soils of pinelands, scrub, beaches, and disturbed habitats; throughout Florida, west to Texas, and north to Virginia.

Comment: Tropic Croton blooms all year. The tiny flowers of this little ungainly plant are difficult to see.

Croton argyranthemus Michx.

Croton glandulosus L.

White

Croton linearis Jacq.
Pineland Croton or Granny Bush
Euphorbiaceae, Spurge Family

Habit: Shrub, to 2 m tall.

Leaves: Alternate; short-stalked; blades very slender, to 7 cm long, yellow to silvery underneath, two glands at base.

Inflorescences: Terminal clusters, male and female flowers separate in the same cluster, male above female.

Flowers: White to yellow, minute.

Fruit: A capsule, to 5 mm long, yellow hairy, 3-seeded.

Habitat and Distribution: Occasional; open coastal pinelands and sandhills; east coast of southern peninsula Florida, and the Florida Keys; the West Indies.

Comment: The whitish flowers of Pineland Croton can be seen any time during the year. They are scarcely noticeable, and one must look carefully.

Cuscuta campestris Yuncker
Field Dodder
Convolvulaceae, Morning-glory Family

Habit: Herbaceous parasitic annual. Stems thin, often yellow or orange, twining around and attaching to the host plant.

Leaves: Leafless.

Inflorescences: Globular clusters around stems.

Flowers: White, tubular, 2.5 mm long, calyx lobes as long as flower tube, tips rounded, petals spreading, with tips bent inward.

Fruit: A round capsule, 2.5-3.5 mm diameter.

Habitat and Distribution: Occasional; mostly parasitic on herbaceous plants, in old fields and rights-of-way; throughout temperate North America, including northern Mexico; the West Indies.

Comment: Field Dodder blooms in any warm month. The characteristic parasitic stems can be found clinging to almost any herbaceous plant in open weedy habitats.

Croton linearis Jacq.

Cuscuta campestris Yuncker

White

Cynoglossum virginianum L.
Wild Comfrey
Boraginaceae, Borage Family

Habit: Herbaceous perennial, 30-80 cm tall, from a taproot. Stems bristly.

Leaves: Alternate; lower stalked, upper stalkless; blades elliptic becoming more rounded upward, 10-20 cm long, hairy, mostly basal, becoming fewer and smaller upward.

Inflorescences: Small terminal clusters.

Flowers: White or usually light blue, 7-12 mm wide, funnel-shaped, with short, broad tube, and 5 rounded, spreading lobes.

Fruit: A capsule, 4-lobed, 7 mm long, covered with bristles, wrinkled between the bristles, 4-seeded.

Habitat and Distribution: Rare to infrequent southward, frequent northward; dry deciduous woods, thickets, meadows, bluff forests, and along roadsides; Gadsden and Liberty Counties in eastern panhandle Florida, west to Texas, and north to British Columbia, Quebec, and New Brunswick.

Comment: Flowering in spring the flowers are noticeable due to the lack of leaves along the upper stem. The plant itself is not much taller than the lower vegetation in which it grows. Wild Comfrey is listed as Endangered by the state of Florida.

Cyrilla racemiflora L.
Titi or Leatherwood
[*Cyrilla arida* Small; *Cyrilla parvifolia* Raf.; *Cyrilla racemiflora* L. var. *parvifolia* Sarg.]
Cyrillaceae, Titi Family

Habit: Shrub, to 10 m tall, frequently forming root shoots.

Leaves: Alternate, tardily deciduous; stalkless or nearly so; blades elliptic, frequently wider toward tip, 1-10 cm, somewhat leathery, smooth, lateral veins easily seen.

Inflorescences: Spike-like arrangements at the tip of last season's twigs.

Flowers: White, 2-3.5 mm long, with 5 petals.

Fruit: A dry, 2- or 3-lobed capsule, 2-2.8 mm long, 2- or 3-seeded.

Habitat and Distribution: Occasional, usually locally common; swamps, low woods, wet flatwoods, bogs, bottomlands, stream banks, and rarely in scrub; central peninsula Florida, west to Texas, and north to Virginia; southern Mexico, Belize, northern South America, West Indies.

Comment: The flowers occur in spring and early summer just after the growth for the current season. Titi usually grows in masses and the numerous spikes of flowers make a dramatic show. Bee keepers prize the flowers for honey.

544

Cynoglossum virginianum L.

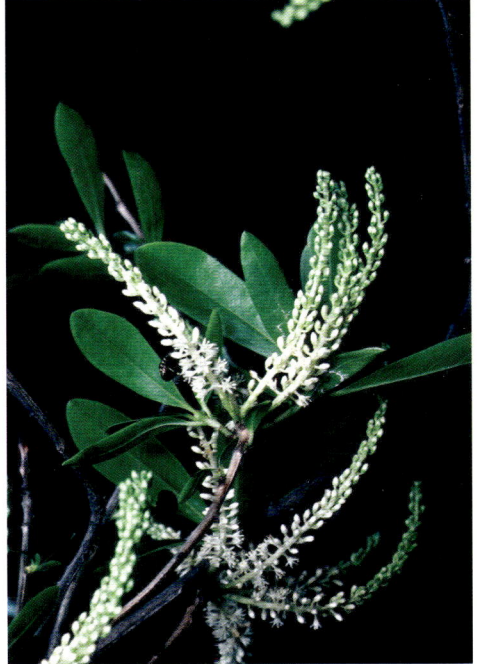

Cyrilla racemiflora L.

White

Dalbergia ecastophyllum (L.) Taub.
Coin Vine
Fabaceae (Leguminosae), Pea or Bean Family

Habit: Shrubby perennial, to 5 m tall. Stems viney, climbing or sprawling.

Leaves: Alternate; stalked; blades with only one leaflet, leaflet broadly ovate to rounded, 6-13 cm long, leathery.

Inflorescences: Branched clusters from leaf axils.

Flowers: White, sometimes with a yellow tinge, 7-9 mm long.

Fruit: A rounded, flat, winged pod, 2-3 cm long.

Habitat and Distribution: Frequent; coastal hammocks, strand, dunes, shell mounds, and mangroves; central and south peninsula Florida; pantropical.

Comment: Coin Vine mostly flowers in spring and summer, but blooms all year southward. The flowers are attractive, but the unusual fruit gets this vine most frequently noticed.

Dalea pinnata (J.F. Gmel.) Barneby
Summer Farewell
Fabaceae (Leguminosae), Pea or Bean Family

Habit: Herbaceous perennial, 30-80 cm tall, in clumps. Stems branching, smooth, slightly woody.

Leaves: Alternate; stalked; blades once-divided, with 3-9 needle-like leaflets, leaflets 5-8 mm long.

Inflorescences: Somewhat flattened or domed terminal heads with numerous leaf-like bracts.

Flowers: White, 8-9 mm long, with 5 petals (4 stamen-like) and 5 stamens.

Fruit: Pods 2 mm long, with 1 or 2 seeds.

Habitat and Distribution: Frequent; dry pinelands, sandhills, and scrub; throughout Florida, in the Coastal Plain from Mississippi into North Carolina.

Comment: Summer Farewell flowers in the summer and fall. The heads of flowers, resembling the Aster Family, persist. This and other species of this genus are frequent wildflowers in dry sandy habitats.

Dalbergia ecastophyllum
(L.) Taub.

Dalea pinnata
(J.F. Gmel.) Barneby

White

Datura stramonium L.
Jimson Weed
Solanaceae, Nightshade or Potato Family

Habit: Herbaceous annual, 0.5-1.5 m tall, from a tap root. Stem mostly smooth, green, often tinged with purple, branching above first flower.

Leaves: Alternate; stalked; blades elliptic to somewhat rounded, 7-20 cm long, with wavy margins or with large rough jagged teeth on the margins.

Inflorescences: Solitary flowers in the forks of branching stems.

Flowers: White, funnel-shaped, 6-10 cm long, with 5 lobes ending as long narrow tips.

Fruit: An erect capsule, roundish or oval, 3-5 cm diameter, spiny.

Habitat and Distribution: Occasional to infrequent; fields, roadsides, lots, agricultural areas, and other disturbed sites; throughout Florida, north to Massachusetts, west to California; now widely distributed throughout the temperate, warm temperate, and tropical regions of the world.

Comment: The large showy flowers can be seen in the warm months of the year. As an ornamental Jimson Weed is popular. One form of the plant has no spines on the fruit and another form has lavender flowers. The plant has a rank odor and is DEADLY POISONOUS if ingested. Occasionally, it is grown as a source of an alkaloidal drug.

Daucus carota L.
Wild Carrot or Queen Anne's-lace
Apiaceae (Umbelliferae), Carrot Family

Habit: Herbaceous biennial, 0.4-1.2 m tall, from a thick taproot. Stems solitary, usually hairy, often branched.

Leaves: Alternate; stalked; blades to 20 cm long, finely divided into many narrow segments, delicate.

Inflorescences: Terminal and axillary, flat, disc-like clusters, clusters lacy in appearance and become concave when older.

Flowers: White, tiny, central flower of cluster is maroon.

Fruit: Capsules oblong, 2-4 mm long, with winged ribs and prickles.

Habitat and Distribution: Infrequent; fields, roadsides, and disturbed areas; widely distributed in North America, from central peninsula Florida north to southern Canada, and west to the West Coast; West Indies; a native of Eurasia.

Comment: Wild Carrot is spring flowering. The clusters of white flowers and the frilly leaves provide an attractive wildflower in old fields and roadsides in cooler parts of the U.S.

Datura stromonium L.

Daucus carota L.

White

Dichondra carolinensis Michx.
Ponyfoot
Convolvulaceae, Morning-glory Family

Habit: Herbaceous perennial, to about 1 m long, from above ground runners. Stems thin, creeping, mat forming.

Leaves: Alternate; long-stalked; blades rounded to kidney shaped, 0.5-3 cm wide, margins smooth.

Inflorescences: Solitary or paired flowers on long stalks from leaf axils.

Flowers: White or greenish, with 5 spreading petals, 2-3 mm long.

Fruit: A capsule, 2-lobed, to 2.5 mm long.

Habitat and Distribution: Common; moist soils of pinelands, hammocks, roadsides, and lawns; throughout Florida, west to Texas, and north to Arkansas and Virginia; Bermuda, Bahamas.

Comment: Although Ponyfoot flowers in most warm months, one has to look very closely to find the flowers. The tiny flowers are usually just at or just below the height of the leaves. The creamy white to greenish color almost renders them invisible unless crawling on hands and knees.

Diodia virginiana L.
Button Weed
Rubiaceae, Madder Family

Habit: Herbaceous perennial, to 1.5 m long. Stems lying flat, smooth or sparsely hairy, reddish green, weakly branching.

Leaves: Opposite; stalkless; blades narrowly lance-shaped to narrowly elliptic, 2-9 cm long, band of tissue between leaf bases with few bristles on upper edge.

Inflorescences: Usually solitary in leaf axils.

Flowers: White, 1 cm long, with 4 narrow, pointed petals.

Fruit: Ellipsoid, two-parted, 6-9 mm long, with 6 vertical ridges, smooth or hairy, crowned with two or sometimes 3 pointed teeth.

Habitat and Distribution: Common; wet woods and pinelands, marshes, swamps, wet margins, and banks; throughout Florida, west to Texas and north to Missouri, Illinois, and New Jersey.

Comment: This low-growing plant often forms a ground cover along wet banks and blankets shallow water. Button Weed flowers in all warm months and, in mass, becomes showy.

Dichondra carolinensis Michx.

Diodia virginiana L.

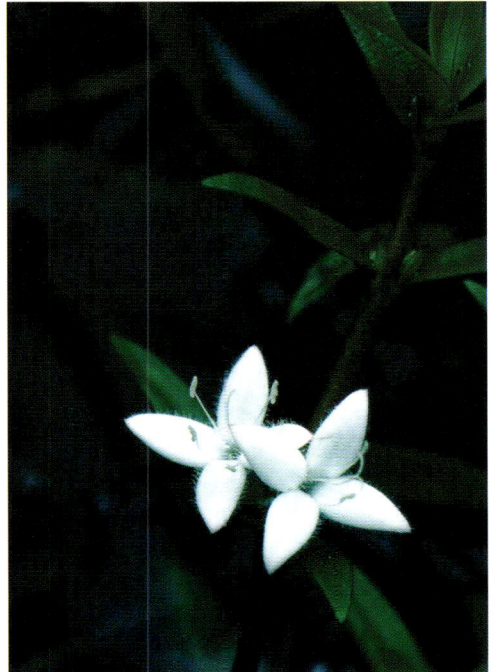

White

Dionaea muscipula J. Ellis
Venus' Fly Trap
Droseraceae, Sundew Family

Habit: Herbaceous perennial, to 40 cm tall, from short runners encased in the fleshy leaf bases.

Leaves: Alternate, in basal rosette; with fleshy, winged stalks; blades to 2.5 cm long, nearly round, with 2 hinged kidney shaped lobes, folded lengthwise, upper surface with a few sensitive hairs that trigger folding of the lobes, edges with bristles to 8 mm long.

Inflorescences: Cluster of a few flowers on a leafless stalk.

Flowers: White, with 5 white petals, about 1 cm long.

Fruit: An oval capsule, 3-4 mm long.

Habitat and Distribution: Infrequent to rare; endemic to wet sandy ditches, savannahs, and bog margins in the coastal plain of the Carolinas; planted and persisting in roadside seepage areas in Liberty and Franklin Counties in the Florida panhandle.

Comment: Flowering in spring, Venus' Fly Trap is a well known insectivorous plant with the amazing characteristic of trapping insects within the folds of the leaves once the sensitive hairs on the leaf surface are touched. In 1975 plants were planted along roadsides in the Florida panhandle and have persisted.

Drosera brevifolia Pursh
Dwarf Sundew
[*Drosera annua* E.L. Reed; *Drosera leucantha* Shinners]
Droseraceae, Sundew Family

Habit: Herbaceous perennial, to 15 cm tall.

Leaves: Alternate, basal; long-stalked; blades spoon-shaped to round, 7-16 mm long, thick, with sticky hairs.

Inflorescences: Few flowers, along one side of the terminal 2-6 cm of a leafless stalk.

Flowers: White, sometimes pinkish, 5 petals, each 5-8 mm long.

Fruit: A rounded capsule, 3.5-4 mm long, partially enclosed by sepals. Seeds black, egg shaped, pitted, less than 0.5 mm long.

Habitat and Distribution: Occasional, can be locally common; moist pinelands, bogs, pond and swamp margins, and sandy roadside ditches; peninsula Florida west to Texas, and north to Kentucky, Tennessee, and Virginia.

Comment: Dwarf Sundew blooms throughout the spring and early summer. The plants can be red in color. The basal cluster of spoon-shaped leaves with sticky hairs and white to pink flowers are often noticed, even though quite small. Insects adhere to the sticky hairs and are gradually absorbed into the blades.

Dionaea muscipula J. Ellis

Drosera brevifolia Pursh

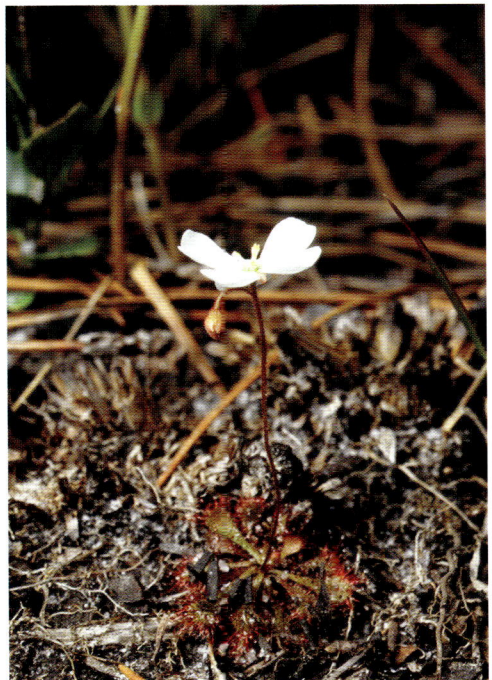

White

Drosera intermedia Hayne
Spoonleaf Sundew, Water Sundew, Narrow-leaved Sundew
Droseraceae, Sundew Family

Habit: Herbaceous perennial, to 20 cm tall.

Leaves: Alternate, in basal rosette and along stem; long-stalked; blades oblong-spatulate to oval, to 2 cm long, thick, with sticky hairs.

Inflorescences: Flowers along one side at the tip of a leafless stalk arcing upward above the leaves.

Flowers: White to pinkish, about 1 cm across, with 5 petals.

Fruit: A rounded capsule, about 5 mm long, partially enclosed by sepals. Seeds reddish-brown to black, 0.4-1 mm long, densely covered with tiny bumps.

Habitat and Distribution: Occasional; sandy acid soils of bogs, marshes, pond margins, and wet ditches; central Florida, west to Texas and north to Ohio, Illinois, Minnesota, Ontario, and Newfoundland.

Comment: Spring flowers are easily seen, but not showy. The small green plants with long, spoon-shaped, leaves with sticky hairs are difficult to see among all the other wetland species. As with all the Sundews, insects land on the sticky hairs, become stuck, and can be slowly dissolved. Spoonleaf Sundew is listed as Threatened by the State of Florida.

Echites umbellata Jacq.
Rubber Vine or Devil's Potato-root
Apocynaceae, Dogbane Family

Habit: Twining, semi-woody vine or sub-shrub, to 2 m long.

Leaves: Opposite; stalked; blades elliptic to oblong, 4-10 cm long, leathery, margins smooth and folded under.

Inflorescences: Loose, stalked clusters from leaf axils.

Flowers: White, 4-6 cm long, tubular, swollen at the middle, with 5 pinwheel-like petals at mouth.

Fruit: Paired, long, thin, cylindric, rigid pods, 10-25 cm long; seeds about 5 mm long, with a tuft of hairs.

Habitat and Distribution: Occasional; pinelands and coastal thickets; central peninsula Florida southward; the West Indies, Mexico south into northern South America.

Comment: The large showy flowers occur in all warm months. The vines can be trained over a trellis and on fences.

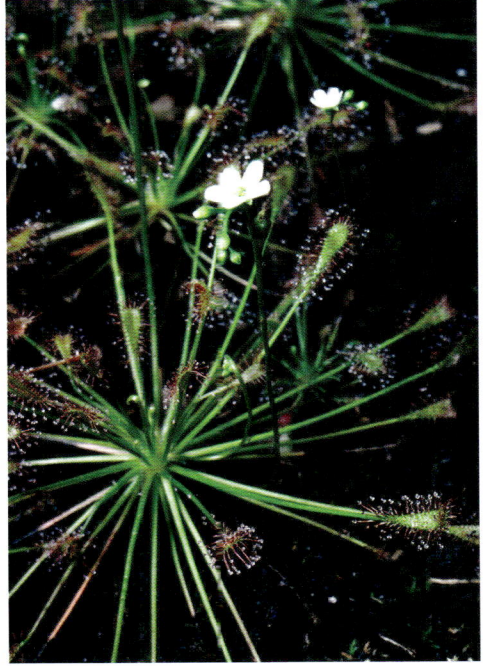

Drosera intermedia Hayne

Echites umbellata Jacq.

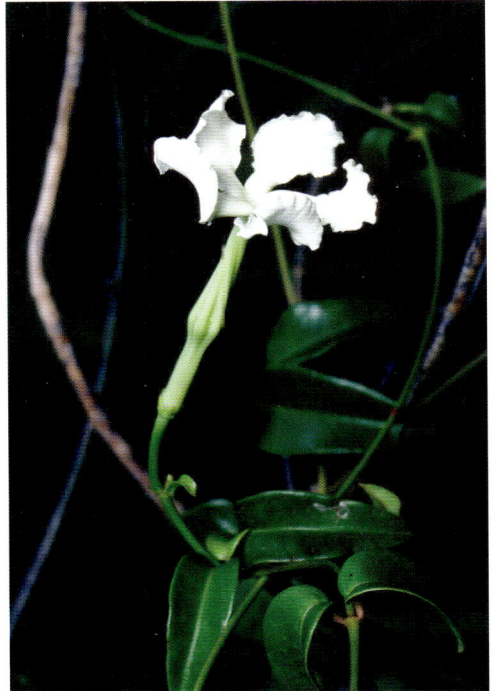

White

Eclipta prostrata (L.) L.
Eclipta, False Daisy, Yerba-de-tago
[*Eclipta alba* (L.) Hassk.]
Asteraceae (Compositae), Aster or Sunflower Family

Habit: Herbaceous annual, to 1 m tall or long, from a taproot. Stems most often sprawling and low growing, sometimes rooting at the joints, hairy.

Leaves: Opposite; stalkless or nearly so; blades lance-shaped, 3-13 cm long, with toothed edges.

Inflorescences: Terminal heads on axillary stalks, heads about 5 mm wide.

Flowers: Rays white, numerous, 1-2 mm long, thin.

Fruit: A brown achene, about 2 mm long, fairly thick, rugose, 3- or 4-sided.

Habitat and Distribution: Common; fresh water meadows, marshes, floodplains, low woods, and bogs; throughout Florida, throughout the United States, north into Canada; pantropical, and warm temperate regions of the world.

Comment: Flowers in all warm seasons. Eclipta can be weedy in almost any wet soil, commonly found in irrigated gardens. The rough 3- or 4-sided seeds are distinctive, but quite small.

Erigeron quercifolius Lam.
Southern Fleabane
Asteraceae (Compositae), Aster or Sunflower Family

Habit: Herbaceous perennial, 10-80 cm tall. Stems hairy, especially at base.

Leaves: Alternate; stalkless; blades mostly basal, widest toward ends, mostly 3-8 cm long, toothed and/or lobed, hairy, stem leaves clasping.

Inflorescences: Heads terminal on loose branches on upper stems.

Flowers: Rays white, blue or pinkish, 100-200, discs yellow, 1-2 cm wide.

Fruit: An achene, to 0.7 mm long, stiff hairs on top.

Habitat and Distribution: Common; sandy open disturbed sites, open pine woods, and open hardwoods; throughout Florida, west to Louisiana, and north to southern Virginia.

Comment: Southern Fleabane mostly flowers in spring and early summer and sporadically in other warm months. It is extremely common along moist roadsides and in open moist to wet fields. Providing a show in mass this plant should be used more commonly as a planted wildflower.

Eclipta prostrata (L.) L.

Erigeron quercifolius Lam.

White

Erigeron vernus (L.) Torr. & A. Gray
Marsh Fleabane
Asteraceae (Compositae), Aster or Sunflower Family

Habit: Herbaceous biennial or short-lived perennial, 10-60 cm tall. Stems smooth, almost leafless, hollow.

Leaves: Alternate; stalkless or short-stalked; basal leaves rounded, blades often broader at the tip, 2-8 cm long, fleshy.

Inflorescences: Heads in terminal more or less flat-topped clusters.

Flowers: White or pinkish with yellow discs, about 2 cm wide, with less than 40 rays.

Fruit: An achene, about 1.5 mm long, yellowish to reddish, with tuft of bristles at top.

Habitat and Distribution: Common; savannahs, bogs, pine flatwoods, pond margins, and ditches; throughout Florida, west to Louisiana, and north to Virginia.

Comment: White heads of flowers on nearly leafless stems can commonly be seen along roadsides in spring. Isolated blooms can occur in other warm months.

Eriocaulon compressum Lam.
Softhead Pipewort
Eriocaulaceae, Pipewort Family

Habit: Herbaceous perennial, to 50 cm tall, from white septate roots.

Leaves: Alternate, in a basal rosette; blades 3-35 cm long and 3-6 mm broad, tapering from base to tip, spongy with visible air spaces at base.

Inflorescences: Terminal, round heads.

Flowers: Heads white, compact, to 2 cm across, soft, with many flowers.

Fruit: A 2- or 3-seeded capsule; seeds 0.7 mm long, red, ellipsoid, smooth.

Habitat and Distribution: Common; wet soil or shallow water of bogs, marshes, ponds, and flatwoods; throughout Florida, west to Texas, and north to New Jersey.

Comment: Plants often occur in large populations. In summer the round white heads provide a distinctive and somewhat unusual display for a wildflower.

Erigeron vernus (L.) Torr. & A. Gray

Eriocaulon compressum Lam.

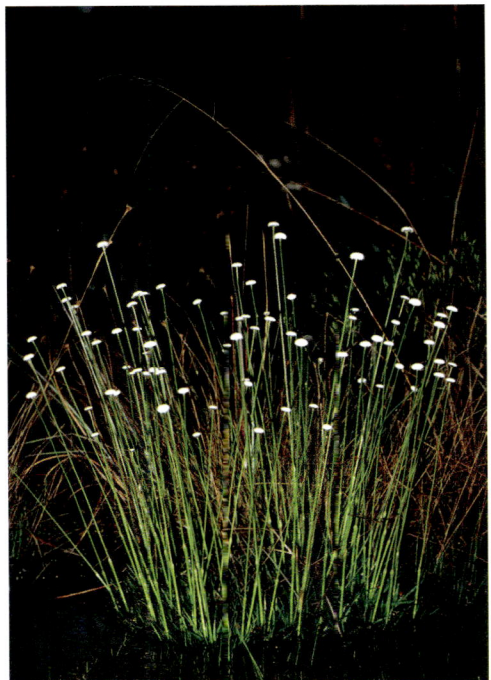

White

Eriogonum longifolium Nutt. var. *gnaphalifolium* Gand.
Scrub Buckwheat
[*Eriogonum floridanum* Small]
Polygonaceae, Buckwheat Family

Habit: Herbaceous perennial, 0.6-1 m tall. Stem single, branched or unbranched, covered with soft, white, silvery hairs.

Leaves: Alternate, mostly basal; stalkless; blades narrowly elliptical, basal leaves 8-20 cm long, stem leaves much smaller, all blades green and hairy above, whitish and hairy below.

Inflorescences: Branched, terminal, drooping clusters.

Flowers: Greenish white, small, 5-6 mm long, lacking petals, hairy.

Fruit: A nutlet, to 6 mm long, with white silky hairs.

Habitat and Distribution: Occasional; sandhills, scrub, and dry pinelands; central and northern peninsula Florida; endemic.

Comment: Scrub Buckwheat flowers from late winter into fall. The white woolly hairs are more noticeable than the flowers. Not particularly attractive this rare plant is listed by the Federal Government and the State of Florida as Endangered.

Eriogonum tomentosum Michx.
Wild Buckwheat or Dog's Tongue
Polygonaceae, Buckwheat Family

Habit: Herbaceous perennial, 0.4-1.2 m tall. Stems somewhat woody, branched, hairy, tan-colored.

Leaves: In whorls of 3-5 on stem, basal rosette present; basal leaves stalked, stem leaves stalkless; blades oblong to spatulate, 7-12 cm long, white or tan hairy beneath.

Inflorescences: Spreading, branched clusters from leaf axils.

Flowers: White to pinkish, small, 3-4 mm long, lacking petals, hairy.

Fruit: A nutlet, slightly flattened, 5 to 6 mm long, 3-ribbed.

Habitat and Distribution: Frequent; dry pinelands and sandhills; central peninsula Florida, west to Alabama, and north to South Carolina.

Comment: Wild Buckwheat can be seen blooming from spring into early winter. The tan stems and lower surfaces of leaves are attractive and noticeable. It can be locally common.

Eriogonum longifolium Nutt. var.
gnaphalifolium Gand.

Eriogonum tomentosum Michx.

White

Eryngium yuccifolium Michx.
Rattlesnake Master or Button Snakeroot
[*Eryngium synchaetum* (A. Gray ex J.M. Coult. & Rose) J.M. Coult. & Rose]
Apiaceae (Umbelliferae), Carrot Family

Habit: Herbaceous perennial, 0.3-1.6 m tall, from a cluster of thickened roots. Stem solitary, upper stem with branches.

Leaves: Alternate; stalkless; blades parallel-veined, grass-like; basal leaves narrow, 15-80 cm long and 1-3 cm wide, stiff, margins spiny, tips sharply pointed; stem leaves are smaller and sometimes wider.

Inflorescences: Terminal, loose clusters, of compact, spherical heads, heads about 1-2.5 cm long and wide.

Flowers: White or greenish petals.

Fruit: Oblong, rounded, 2-parted, scaly, 4-8 mm long and to 2.5 mm wide.

Habitat and Distribution: Frequent; moist to dry sandy soils of pinelands, prairies, and open woods; throughout Florida, west to Texas, Oklahoma, and Kansas, and north to Minnesota and Connecticut.

Comment: Members of this genus are food for black swallowtail caterpillars. The basal leaves with their tiny spines are quite distinctive and provide an attractive accent to native gardens. The flowers are profuse, although not large, in heads with a toothed bract. Flowering can be anytime during the warm months. Purple flowered relatives are Corn Snakeroot (*Eryngium aquaticum*) and Fragrant Eryngium (*Eryngium aromaticum*).

Eupatorium album L.
White Thoroughwort
Asteraceae (Compositae), Aster or Sunflower Family

Habit: Herbaceous perennial, to 1 m tall, from crown or thick rhizome. Stems erect, usually solitary, solid, hairy.

Leaves: Opposite; stalkless; blades elliptic to somewhat rounded, 5-12 cm long, hairy, 3 prominent parallel veins and resinous-dotted beneath, margins toothed.

Inflorescences: Heads terminal on spreading branches, only disc flowers present, heads cylindrical, 6-8 mm long and 2-3 mm in diameter.

Flowers: Tips of bracts and flowers white.

Fruit: An achene, 3-4 mm long, resinous-dotted, with tuft of bristles on top.

Habitat and Distribution: Occasional; dry open woodlands, pinelands, sandhills, and old fields; central peninsula Florida, west to Mississippi, and north to Ohio and southern New York.

Comment: Flowering from late June into frost, the plant is showy due to the large mass of the small white heads.

White

Eryngium yuccifolium Michx.

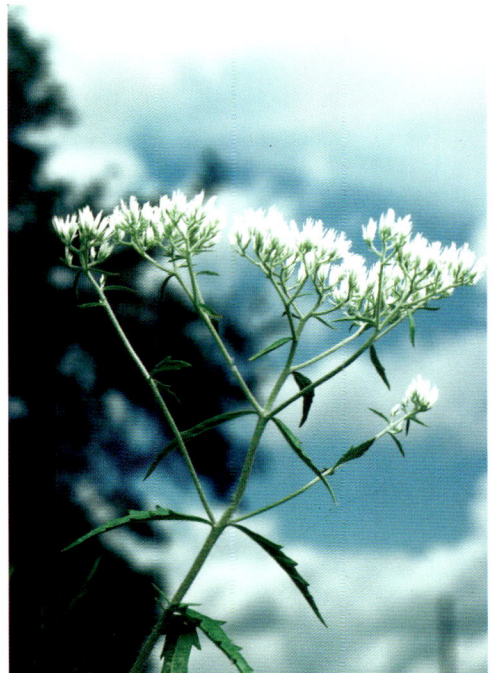

Eupatorium album L.

White

Eupatorium capillifolium (Lam.) Small
Dog-fennel
Asteraceae (Compositae), Aster or Sunflower Family

Habit: Herbaceous perennial, 1-3.5 m tall, from a thick woody crown. Stems hairy, with many small branches, branches often drooping.

Leaves: Opposite on lower stem, alternate on upper stem; stalked; very delicate and finely divided, 2-7 cm long and less than 1 mm broad, feathery in appearance, with glandular dots; uppermost leaves on stems not divided.

Inflorescences: Heads in lax, elongate, drooping, spray-like clusters, feathery in appearance, only disc flowers present, heads cylindrical, 2-3 mm long and 1 mm in diameter.

Flowers: Creamy white, tiny.

Fruit: An achene, 1-1.6 mm long with a tuft of bristles on top.

Habitat and Distribution: Common; sandy soils of dry and wet habitats, pinelands, sandhills, marshes, old fields, pastures, and disturbed sites; throughout Florida, west to Texas and Arkansas, and north to New Jersey.

Comment: Dog-fennel blooms from summer into frost or late fall. This species is facultative and will grow in many environmental circumstances. Its facultative abilities and small wind-blown seeds contribute to this species' ability to invade almost any disturbed area and become a terrible weed.

Eupatorium jucundum Greene
Fairy Snakeroot
[*Ageratina juncunda* (Greene) Clewell & Wooten]
Asteraceae (Compositae), Aster or Sunflower Family

Habit: Herbaceous perennial, 0.4-1.2 m tall. Stem hairy, woody at base, branches paired and flexuous.

Leaves: Opposite; short-stalked; blades triangular to oval, 1.5-6 cm long, three-nerved, margins toothed.

Inflorescences: Heads in terminal flat-topped clusters, only disc flowers present.

Flowers: White, less than 20 per head, each 4 mm long.

Fruit: An achene, 2-3 mm long, smooth or short hairy near top.

Habitat and Distribution: Frequent; dry sandy soils of open habitats, hammocks, and pine woods; throughout the peninsula and into the central panhandle of Florida and north into southeast Georgia.

Comment: Fairy Snakeroot flowers in summer and fall. It provokes comments as a wildflower due to the flat-topped masses of white.

Eupatorium capillifolium
(Lam.) Small

Eupatorium jucundum Greene

White

Eupatorium mohrii Greene
Mohr's Thoroughwort
[*Eupatorium recurvans* Small]
Asteraceae (Compositae), Aster or Sunflower Family

Habit: Herbaceous perennial, 3 to 1.2 m tall, from a thickened rhizome. Stems erect, hairy.

Leaves: Opposite, or the upper alternate; stalkless; blades lance-shaped to somewhat elliptic, 1.5-10 cm long, three-nerved, margins few-toothed, dotted underneath, often curved downward.

Inflorescences: Heads in terminal flat-topped clusters, only disc flowers present.

Flowers: White, about 5 per head, each 3 mm long.

Fruit: An achene, 1.5-2.8 mm long, resinous dotted, with tuft of hairs at tip.

Habitat and Distribution: Common; flatwoods, pond margins, ditches, and sandhills; throughout Florida, north to Virginia, and west to Louisiana.

Comment: Flowering in summer and fall, Mohr's Thoroughwort is common and provides a wildflower display that provokes much comment. It often is seen in the open areas near and around water.

Eupatorium semiserratum DC.
Smallflower Thoroughwort
Asteraceae (Compositae), Aster or Sunflower Family

Habit: Herbaceous perennial, 0.7-1.2 m tall, from a short thick rhizome. Stems erect, usually solitary, hairy.

Leaves: Opposite, simple or sometimes with clusters of small leaves; short-stalked; blades elliptic, firm, 4.5-8 cm long, three-nerved, margins toothed, blades twist into a vertical position.

Inflorescences: Terminal flat-topped clusters of cylindrical heads, only disc flowers present.

Flowers: Cream, heads 2.5-3.5 mm long. cylindrical, 2.5-3.5 mm long.

Fruit: An achene, 1.5-2.5 mm long.

Habitat and Distribution: Occasional; sandy wet soils, pine barrens and flatwoods, bogs, and swamp forests, Florida panhandle, west to Texas, and north to Arkansas, Missouri, Tennessee, and Virginia.

Comment: This summer and fall blooming plant is relatively easy to identify with its vertical blades.

Eupatorium mohrii Greene

Eupatorium semiserratum DC.

White

Eupatorium serotinum Michx.
Late Boneset
Asteraceae (Compositae), Aster or Sunflower Family

Habit: Herbaceous perennial, 0.8-2 m tall, from a rhizome. Stems erect, hairy, forming colonies, upper stems branched.

Leaves: Opposite; long-stalked; blades lance-shaped to rounded, to 20 cm long, 3- or 5-nerved, hairy and resinous-dotted beneath, margins toothed.

Inflorescences: Dense, terminal, flat-topped clusters, heads cylindrical to flared, only disc flowers present.

Flowers: White, heads 3-5 mm long and about as wide.

Fruit: An achene, 1.7-2.8 mm long, resinous-dotted, with a tuft of bristles on top.

Habitat and Distribution: Frequent; moist open hardwoods, wet hammocks, tidal marshes, and pond margins; throughout Florida, north to New York, Illinois, and Kansas, west to Oklahoma, Texas, and northern Mexico.

Comment: Late Boneset flowers in late summer and fall. The tall plants with 3- or 5-nerved, hairy leaves are distinctive, but the flowers are not particularly showy.

Froelichia floridana (Nutt.) Moq.
Cotton-weed
Amaranthaceae, Amaranth Family

Habit: Herbaceous annual, 0.4-1.8 m tall. Stems stiffly erect, hairy.

Leaves: Opposite, mostly basal; short-stalked; blades oblong to spatulate, 3-12 cm long, woolly beneath.

Inflorescences: Few to several, compact, erect spikes, often stalked.

Flowers: White, small, about 5 mm long, cottony.

Fruit: About 5 mm long, thin-walled, with one or two spiny ridges; seeds shiny, about 1.5 mm long.

Habitat and Distribution: Common; sandy soils of sandhills, pinelands, dry hammocks, fields, roadsides, and other dry sites; throughout Florida, west to Texas, and north to Delaware, Maryland, and New Jersey.

Comment: The flowers in summer and fall appear as cottony spikes on tall willowy stems. Commonly seen in dry sandy sites the plants often provoke comment.

Eupatorium serotinum Michx.

Froelichia floridana (Nutt.) Moq.

White

Galactia elliottii Nutt.
White Milk-pea
Fabaceae (Leguminosae), Pea or Bean Family

Habit: Perennial vine, 0.5-2 m long, from underground runners. Stems prostrate or twining and climbing, rooting at nodes.

Leaves: Alternate; stalked; blades with 7-9 elliptic leaflets, leaflets 2-5 cm long, sparsely to densely hairy on both surfaces.

Inflorescences: Long-stalked clusters of 2-6 flowers in leaf axils.

Flowers: White or tinged pink, typically pea-shaped, upper petal 11-15 mm long, folded.

Fruit: A flat pod, 3-6 cm long, hairy.

Habitat and Distribution: Common; dry pinelands, sandhill, scrub, and sandy water margins; peninsula to central panhandle Florida, north to southeastern South Carolina.

Comment: Flowering from spring into early summer, White Milk-pea is a wonderful wildflower. The vines can become a dense ground cover or climb into and over low vegetation. The evergreen leaves and bright white flowers are eye-catching. It should be utilized more frequently.

Galium aparine L.
Catchweed
Rubiaceae, Madder Family

Habit: Herbaceous annual, 0.1-1.5 m long, from a taproot. Stems reclining, often forming mats, angles lined with backward pointing bristles.

Leaves: In whorls of 6 or 8; stalkless; blades narrow, 1-8 cm long, with bristly edges.

Inflorescences: Small, stalked, axillary clusters.

Flowers: White, with 4 narrow petals.

Fruit: Dry, 2-4.5 mm long and 3-5 mm wide, brown or black, bristly with hooked hairs.

Habitat and Distribution: Occasionally, can be locally common; hammocks, meadows, woodlands, prairies, seashores, and other disturbed shady places; central peninsula Florida, north throughout North America into Alaska and Newfoundland; circumpolar.

Comment: Flowering in early spring in shady moist sites Catchweed can become a light green ground cover. The bristly plant gets its common name from its ability to catch on clothing or fur. The brittle stem breaks and the plant fragment can be transported to another site.

Galactia elliottii Nutt.

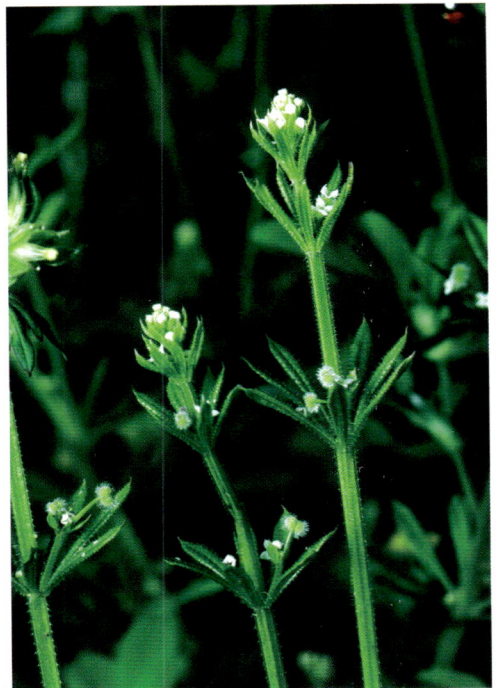

Galium aparine L.

White

Gaura angustifolia Michx.
Southern Gaura
Onagraceae, Evening-primrose Family

Habit: Herbaceous perennial, 0.7-2 m tall. Stems slender, hairy, branched.

Leaves: Alternate, single or clustered at joints; stalkless; blades narrowly elliptic or lance-shaped, 1-12 cm long, margins minutely toothed.

Inflorescences: Terminal, spike-like branches, flowers stalked.

Flowers: White to pink, with 3 or 4 slender, spatula-shaped petals 3-6 mm long.

Fruit: Capsules 5-10 mm long, 2-3 mm broad, sides 3- to 4-angled, stalked.

Habitat and Distribution: Common; sandy soils of pinelands, sandhills, roadsides, open woods, fields, and disturbed sites; throughout Florida, west to Mississippi, and north to North Carolina.

Comment: Flowers in spring and summer are easily overlooked. Often seen along roadsides the plant looks quite weedy.

Gaylussacia mosieri Small
Woolly-berry
Ericaceae, Heath Family

Habit: Shrub, to 1.5 m tall. Stem with long gland-tipped hairs.

Leaves: Alternate; stalked; blades elliptic, 3-6 cm long, thick, somewhat hairy on both surfaces when young, margins smooth.

Inflorescences: Spike-like, spreading, hairy arrays on last year's growth.

Flowers: White, urn-shaped, 8-9 mm long.

Fruit: A berry, 8-10 mm in diameter, glandular hairy.

Habitat and Distribution: Frequent; hydric hammocks, wet flatwoods, bogs, titi and cypress swamps; central peninsula Florida, north along the coastal plain into east Georgia, west into Louisiana.

Comment: The flowers in spring are sprinkled among the leaves. The fruits, although somewhat seedy, are edible as are most Huckle-berries.

Gaura angustifolia Michx.

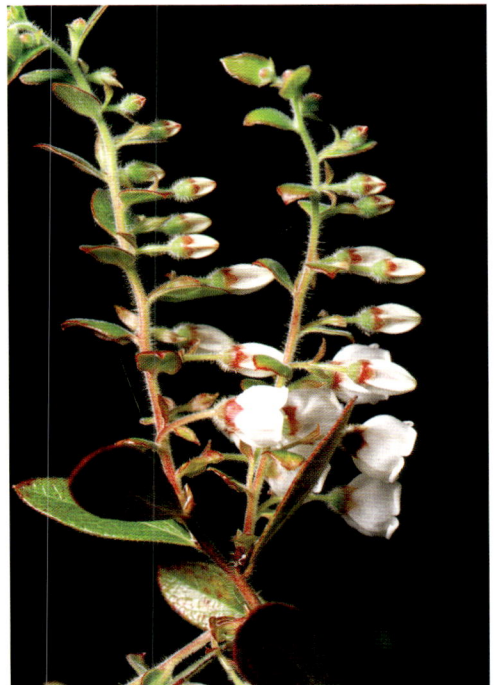

Gaylussacia mosieri Small

White

Gentiana pennelliana Fernald
Wiregrass Gentian
Gentianaceae, Gentian Family

Habit: Herbaceous perennial, to 2 m tall.

Leaves: Opposite; stalkless; blades linear spatulate to linear.

Inflorescences: Flowers solitary in leaf axils.

Flowers: White, spotted with blue-green on the inside, lobes rounded, toothed.

Fruit: Capsule, about 1.5 cm long.

Habitat and Distribution: Rare; wet flatwoods, slash pine plantations, and roadside ditches; Bay, Calhoun, Franklin, Gadsden, Gulf, Liberty, Wakulla, and Walton Counties in central panhandle Florida.

Comment: Flowers can be seen from October into February. The flowers are showy and the plants are very attractive. Wiregrass Gentian is endemic and listed by the State of Florida as Endangered.

Gnaphalium obtusifolium L.
Rabbit's-tobacco, Fragrant Cudweed, Cat-foot
[*Pseudognaphalium obtusifolium* (L.) Hilliard & B.L. Burtt]
Asteraceae (Compositae), Aster or Sunflower Family

Habit: Herbaceous annual or winter-annual, 20-80 cm tall. Stems leafy, covered with woolly white hairs.

Leaves: Alternate; stalkless; blades narrowly elliptic, to 8 cm long and 8 mm wide, densely white-woolly hairy beneath.

Inflorescences: Heads in much-branched, flat or round-topped, terminal clusters, only disc flowers present.

Flowers: White, 5-7 mm long, papery.

Fruit: An achene, less than 1 mm long, smooth.

Habitat and Distribution: Frequent; flatwoods, sandhills, old fields, pastures, and disturbed areas; throughout Florida, west to Texas, and north to Nebraska, Wisconsin, Ontario, and Nova Scotia.

Comment: Rabbit's-Tobacco flowers in summer and fall. For decades kids have used this plant as a staple in play cigarettes made with newspaper and dried leaves.

White

Gentiana pennelliana Fernald

Gnaphalium obtusifolium L.

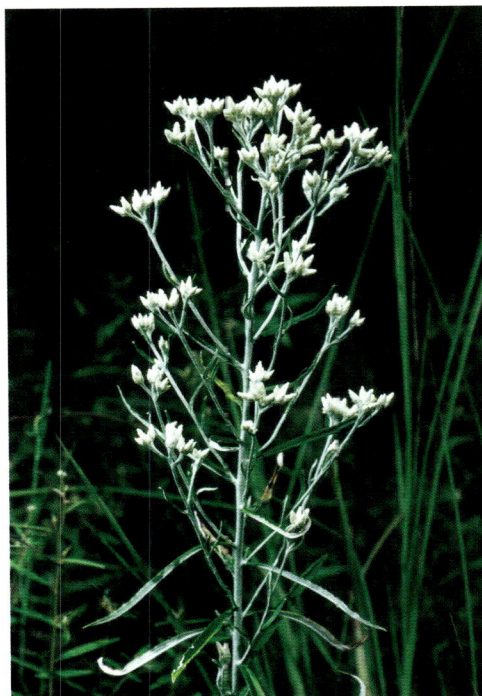

575

White

Gnaphalium purpureum L.
Purple Cudweed
[*Gamochaeta purpurea* (L.) Cabrera]
Asteraceae (Compositae), Aster or Sunflower Family

Habit: Herbaceous annual or biennial, 10-40 cm tall, often from a basal rosette. Stems ascending, long hairy.

Leaves: Alternate; stalkless; blades broader at tip, to 10 cm long, smaller upwards, densely hairy underneath, slightly hairy on top.

Inflorescences: Heads in stalked clusters from leaf axils on upper plant, only disc flowers present.

Flowers: White, 3-5 mm long, papery.

Fruit: An achene, 0.5-0.7 mm long, dotted.

Habitat and Distribution: Common; sandy soils of crops, fields, pastures, roadsides, and disturbed areas; throughout Florida, north throughout North America into southern Canada, south through Mexico and Central American into South America.

Comment: This common crop and field weed, flowering in spring and summer, often occurs in significant numbers, but is not much of a scenic wildflower.

Gomphrena serrata L.
Globe Amaranth
[*Gomphrena decumbens* Jacq.; *Gomphrena dispersa* Standl.]
Amaranthaceae, Amaranth Family

Habit: Herbaceous annual or perennial, to 10 cm tall. Stems hairy, diffuse, low growing, branched and spreading or ascending.

Leaves: Opposite; stalked; blades elliptic, 1.5-5 cm long, hairy.

Inflorescences: Rounded heads, terminal or in upper axils.

Flowers: White, yellow, purple, or red, composed of small, papery, cottony-haired flowers.

Fruit: Nutlike, roundish, 1-2 mm long.

Habitat and Distribution: Common; dry sandy soils of open disturbed sites; throughout Florida and west to Texas; Mexico into Central and South America, West Indies.

Comment: Flowering all warm months this low-growing plant can be seen in disturbed areas and most frequently in cultivation. The various color forms are a favorite of gardeners.

Gnaphalium purpureum L.

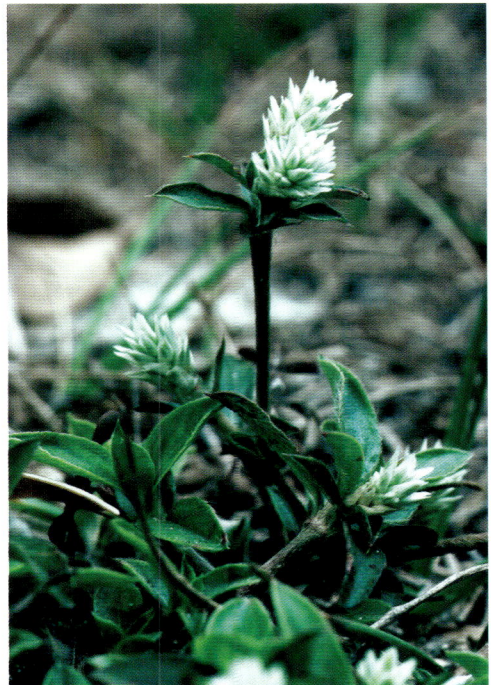

Gomphrena serrata L.

White

Gordonia lasianthus (L.) J. Ellis
Loblolly-bay
Theaceae, Tea Family

Habit: Shrub, small or large tree, to 25 m tall. Bark deeply furrowed on older trees.

Leaves: Alternate, evergreen; stalked; blades thick, glossy, elliptic, 8-16 cm long, margins bluntly or sharply toothed.

Inflorescences: Flowers solitary in leaf axils at stem tips.

Flowers: White with yellow centers, 4-8 cm across, petals 5, rounded, with fringed edges.

Fruit: A woody, rounded capsule, about 1 cm diameter.

Habitat and Distribution: Frequent; bay forests, bogs, wet flatwoods, cypress depressions, and swamps; central peninsula Florida, west to Louisiana, and north to North Carolina.

Comment: Flowers for several weeks during the summer. Flowers usually occur one at a time on each branch and only a few open on the tree during the same period. The flowers appear like jewels among the dark green leaves during the hot days of summer. The trees are difficult to cultivate and have to be perfectly situated. Often short-lived in cultivation.

Gossypium hirsutum L.
Wild Cotton
Malvaceae, Mallow Family

Habit: Herbaceous or shrubby annual or perennial, to 4 m tall.

Leaves: Alternate; stalked; blades rounded, to 15 cm long, three- or five-lobed, sometimes not lobed.

Inflorescences: Flowers solitary, stalked, from leaf axils.

Flowers: White, fading to pinkish-purple, sometimes with a purple spot at base, funnel-shaped, to 6 cm long.

Fruit: A capsule, rounded, 2-3 cm long; seeds covered with hairs.

Habitat and Distribution: Occasional; coastal hammocks, beaches, and disturbed areas along the coast; peninsula Florida, Texas; the West Indies, coastal Mexico, Cayman Islands, Hispaniola.

Comment: Wild Cotton can flower at any time during the year. The hairs on the seeds are the cotton of commerce. The seeds yield an oil. As a wild flower it is seldom planted, but certainly attractive.

Gordonia lasianthus (L.) J. Ellis

Gossypium hirsutum L.

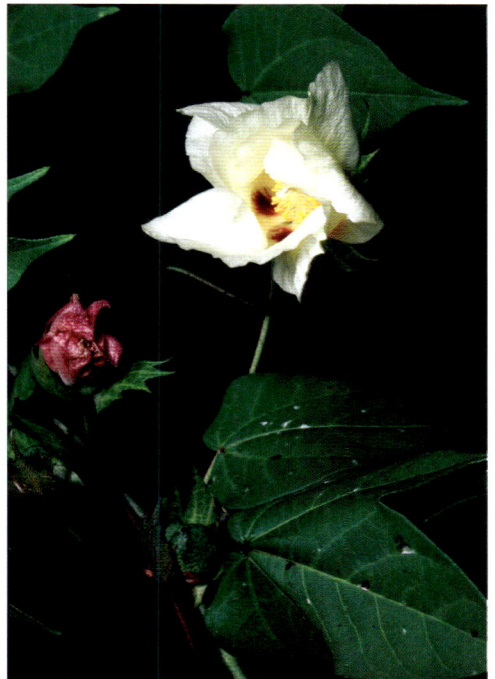

White

Habenaria quinqueseta (Michx.) Eaton
Michaux's Orchid
Orchidaceae, Orchid Family

Habit: Herbaceous perennial, 20-50 cm tall, from a rounded tuber. Stem erect, leafy, green.

Leaves: 3-7, alternate; stalkless; blades elliptic, 5-25 cm long and 2.5-6 cm broad, reduced on upper stem.

Inflorescences: 15-25 flowers in a loose, spike-like arrangement at stem tips.

Flowers: White or greenish, 2-3 cm wide, with 5 very slender, curving lobes, and a thin green spur usually 4-8 cm long.

Fruit: A broadly cylindric capsule, to 2 cm long and 1 cm wide.

Habitat and Distribution: Occasional, but locally common; rich, moist to occasionally wet hammocks and pine flatwoods; peninsula and eastern panhandle Florida, north to South Carolina, and west to Texas; Mexico, Central America, West Indies.

Comment: The attractive, distinctive, flowers that occur in fall and winter make Michaux's Orchid easy to identify. It is difficult to cultivate, but a very attractive wildflower.

Halesia carolina L.
Carolina Silver-bell
Styracaceae, Storax Family

Habit: Small tree, to 12 m tall.

Leaves: Alternate, deciduous; stalked; blades elliptic, 7-12 cm long, some star-shaped hairs on the lower surface, margins smooth or with teeth.

Inflorescences: Short, stalked clusters of 2-7 flowers, from axils of last year's leaves.

Flowers: White or tinged pink, pendent, bell-shaped, 1.2-1.5 cm long, with 4 lobes, hairless.

Fruit: Elliptic in length and round in cross section, 4-winged, 2-4.5 cm long, 1- to 3-seeded.

Habitat and Distribution: Occasional; forested slopes, banks, floodplains, hammocks; central peninsula Florida, west to Mississippi and north to Georgia.

Comment: Carolina Silver-bell blooms in spring. The many small flowers are attractive and appear before the leaves.

Habenaria quinqueseta
(Michx.) Eaton

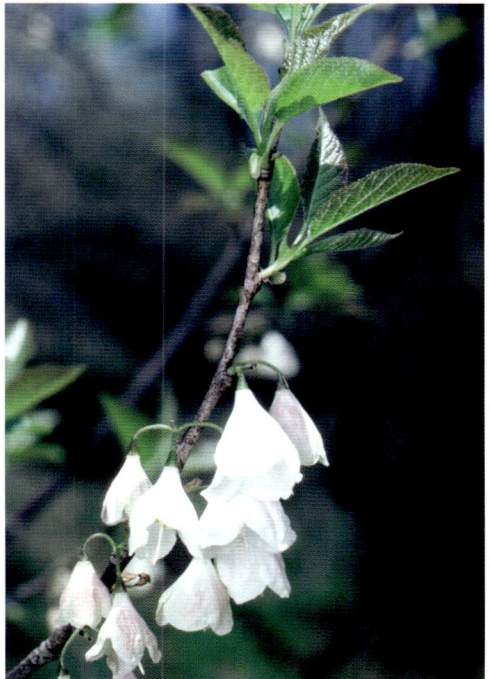

Halesia carolina L.

White

Halesia diptera J. Ellis
Two-winged Silver-bell
Styracaceae, Storax Family

Habit: Small tree, to 10 m tall.

Leaves: Alternate, deciduous; stalked; blades broadly elliptic, 9-17 cm long, both surfaces with star-shaped hairs, margins with teeth.

Inflorescences: Short, stalked clusters of 2-7 flowers, from axils of last year's leaves.

Flowers: White or tinged pink, pendent, bell-shaped, 1.5-3 cm long, with 4 lobes, hairy.

Fruit: Narrowly elliptic in length and somewhat flattened in cross section, 2-winged, 4-5 cm long, 1- to 3-seeded.

Habitat and Distribution: Occasional; floodplains, slopes, ravines, upland woodlands; panhandle Florida, north to southeast South Carolina, west to east Texas.

Comment: Two-winged Silver-bell flowers for about four weeks in the spring. Blooming before the leaves appear, the larger flowered plant is a great addition to the landscape.

Hedyotis procumbens (J.F. Gmel.) Fosberg
Fairy Footprints or Innocence
[*Hedyotis procumbens* (J.F. Gmel.) Standl.]
Rubiaceae, Madder Family

Habit: Herbaceous perennial, 5-40 cm long. Stems prostrate, creeping.

Leaves: Opposite; stalked; blades oval to nearly round, 0.5-1.8 cm long, with hairy edges.

Inflorescences: Flowers solitary or in small clusters in leaf axils.

Flowers: Bright white, narrow tube 5-7 mm long, with 4 pointed lobes 4-6 mm long.

Fruit: A 2-lobed capsule, 3-4 mm long and 4-5 mm broad.

Habitat and Distribution: Common; sandy soils of pinelands, sandhills, dunes, and disturbed areas; throughout Florida, west to Louisiana and north to South Carolina.

Comment: Flowering in all warm months, this small prostrate wildflower can provide a marvelous display just before dusk. The blooms are so bright white that they seem to resemble ones perception of Fairy Footprints in a lawn or when growing in other low vegetation.

White

Halesia diptera J. Ellis

Hedyotis procumbens
(J.F. Gmel.) Fosberg

White

Heliotropium polyphyllum Lehm.
Pineland Heliotrope
Boraginaceae, Borage Family

Habit: Herbaceous perennial, prostrate to erect, 10-50 cm long, 5 cm to 1 m tall. Stems hairy.

Leaves: Alternate; stalkless or nearly so; blades narrowly elliptic, 0.5-2.5 cm long, margins smooth.

Inflorescences: Flowers arranged along one side of terminal spike, becoming progressively smaller upward.

Flowers: White or yellow, tubular, 3-4 mm long.

Fruit: 4 seed-like nutlets, to 1.5 mm wide; seeds 4.

Habitat and Distribution: Frequent; flatwoods, open sites, often in sandy soils; throughout Florida; Bahamas, South America.

Comment: The bright flowers bloom all year, especially southward. Pineland Heliotrope tolerates poorer soils. It is a small but showy wildflower that should be used more often as an ornamental.

Hydrangea quercifolia W. Bartram
Oakleaf Hydrangea
Hydrangeaceae, Hydrangea Family

Habit: Shrub, to 3 m tall. Older stems with peeling, papery bark.

Leaves: Opposite; stalked; blades 25-30 cm long, broadly elliptic to oval, sometimes several large lobes, both surfaces with soft cottony hairs, margins toothed.

Inflorescences: Large, terminal, short-branched arrangements.

Flowers: Both sterile and fertile flowers present, sterile flowers with 4 large, white, showy, petal-like sepals, fertile flowers with 5 tiny, white petals.

Fruit: A capsule, about 3 mm long and wide, 8- to 10-ribbed; seeds brown, curved, ribbed, 0.5-1 mm long.

Habitat and Distribution: Occasional; rich, mixed woods, banks, and ravines; Mississippi River east to Marion County in north peninsula Florida, north to Tennessee.

Comment: Flowering in spring, the large showy arrangements of white sepals are spectacular. The sepals, even when dry, are very showy, and hang on the plant for many weeks. The stems with dry sepals are used frequently in dried arrangements. The large soft green leaves add to the attractiveness. The papery bark is attractive during the cold months of the year after the leaves fall. This shrub is easily grown, but does need a larger area to reach its potential.

Heliotropium polyphyllum Lehm.

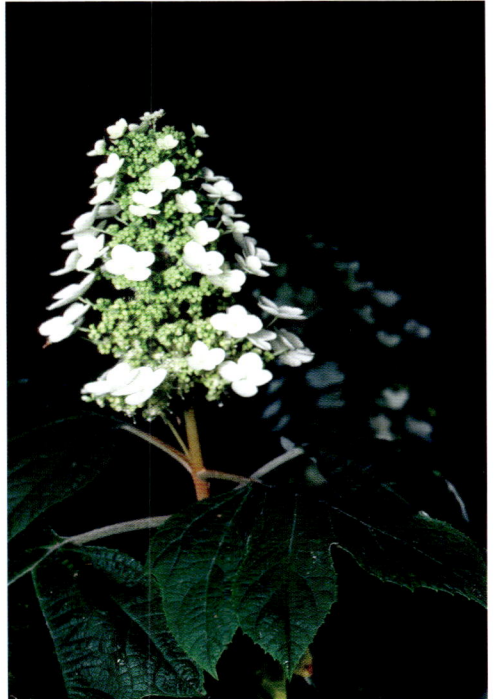

Hydrangea quercifolia W. Bartram

White

Hylocereus undatus (Haw.) Britton & Rose
Night-blooming Cereus
Cactaceae, Cactus Family

Habit: Herbaceous perennial, to 7 m long. Stems fleshy, thick, smooth, climbing, rooting along stem, branching; branches 3-angled, spiny, with 2-5 spines, each 3-4 mm long, at each joint/indentation.

Leaves: Lacking.

Inflorescences: Flowers solitary, from joints.

Flowers: White, to 30 cm long, covered with scales, petals many, stamens numerous.

Fruit: A round berry, red, about 8 cm long.

Habitat and Distribution: Occasional to frequent; disturbed habitats, frequently remaining after cultivation; central and south peninsula Florida; the West Indies, Mexico south into South America; a native of South America.

Comment: The very large showy night-blooming flowers of Night-blooming Cereus, also known as Queen-of-the-night, are spectacular. Large buds swell during the day and open at night during any warm month.

Hydrocotyle umbellata L.
Water Pennywort or Ombligo de Venus
Apiaceae (Umbelliferae), Carrot Family

Habit: Herbaceous perennial. Stems smooth, succulent, creeping or floating, rooting at nodes, often forming mats.

Leaves: Alternate; stalks 4-25 cm long and attached near center of blade; blades round, 2-7 cm wide, with scalloped edges.

Inflorescences: Long-stalked, disc-like clusters of 15-50 flowers from leaf axils, clusters 1-2 cm wide.

Flowers: White, small, with 5 petals.

Fruit: Flattened, roundish, ribbed, 2-3 mm wide.

Habitat and Distribution: Common; shallow water and moist to wet areas, including lawns; throughout Florida, north to Nova Scotia and Minnesota, west to Texas; California, Oregon; West Indies; Mexico south into South America.

Comment: The shiny umbrella-like (peltate) blades of this common creeper are frequently noticed. The long-stalked clusters of flowers are as long or longer than the leaves and when appearing with the shiny blades in spring and summer are attractive. The leaves often are massed into large concentrations and can be mistaken for a planned ground cover.

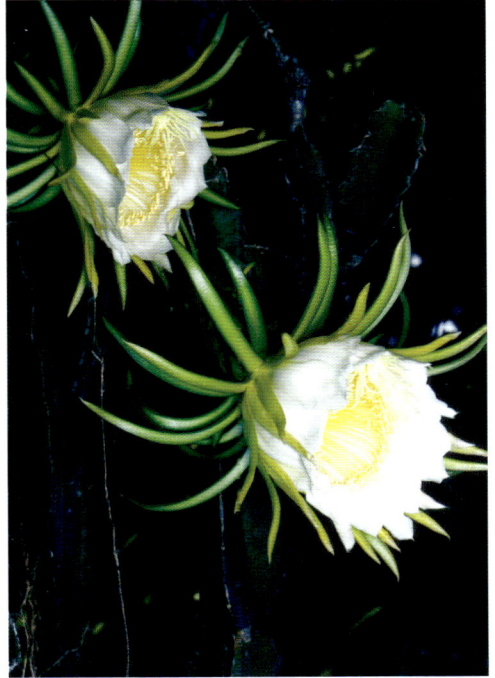

Hylocereus undatus
(Haw.) Britton & Rose

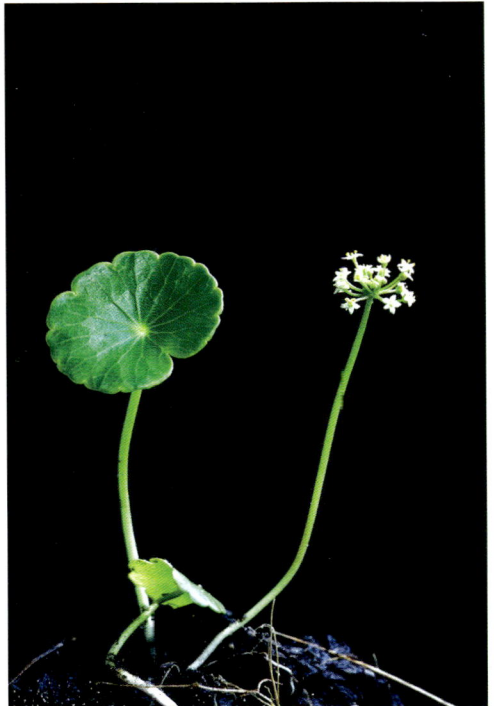

Hydrocotyle umbellata L.

White

Hymenocallis crassifolia Herb.
Coastal Plain Spider-lily
Amaryllidaceae, Amaryllis Family

Habit: Herbaceous perennial, to 60 cm tall, from a bulb.

Leaves: Alternate, in a basal rosette; stalkless; blades linear, to 50 cm long and 1-3 cm broad, leathery.

Inflorescences: 1-3 flowers in a terminal cluster.

Flowers: White, 5-8 cm across, with a united, funnel-shaped center and 6 long, slender petals and sepals.

Fruit: A 1- to 3-seeded capsule; seeds green and fleshy, roundish, 1.5-2 cm long.

Habitat and Distribution: Occasional; lake, river, and stream shores, swamps, bogs, marshes, brackish marshes, and wet hammocks; northern peninsula Florida, north to North Carolina.

Comment: The very showy spring flowers are delicate. Coastal Plain Spider-lily is very effective in landscapes. The dark green leaves are attractive during the rest of the year.

Hymenopappus scabiosaeus L'Her.
Old Plainsman or Blushing Sandweed
Asteraceae (Compositae), Aster or Sunflower Family

Habit: Herbaceous perennial, to 1 m tall, from a taproot. Stems solitary to several, erect, angular, upper branches hairy.

Leaves: Alternate; lower stalked, upper stalkless; blades once-dissected or -lobed, basal rosette present, stem leaves finely divided, 8-25 cm wide, woolly.

Inflorescences: Heads on long stalks, in more or less flat-topped terminal clusters, only disc flowers present.

Flowers: White, heads with petal like bracts arranged in 1 or 2 rows.

Fruit: An achene, narrowly oblong, 4-angled, 3-4 mm long, blackish brown.

Habitat and Distribution: Occasional to locally common; oak woodlands, sandy fields, prairies, and dry open areas; from central peninsula Florida, north to South Carolina, Indiana, Illinois, and Missouri, west to Arkansas, Texas, and northern Mexico.

Comment: Old Plainsman flowers in summer. The papery flower heads and frilly leaves are offered for cultivation in varieties.

Hymenocallis crassifolia Herb.

Hymenopappus scabiosaeus L'Her.

White

Hyptis alata (Raf.) Shinners
Musky Mint
Lamiaceae (Labiatae), Mint Family

Habit: Herbaceous perennial, 0.4-2.5 m tall. Stems square, hairy, rarely branched.

Leaves: Opposite; stalked; blades lance-shaped, 4-12 cm long, with toothed edges.

Inflorescences: Many flowers in rounded, stalked, paired heads, each 1.5-2.5 cm wide.

Flowers: White, with purple spots.

Fruit: A smooth, black nutlet, 1-1.7 mm long.

Habitat and Distribution: Common; wet ditches, flatwoods, prairies, margins, and other low areas; throughout Florida, west to Texas, and north to North Carolina; West Indies.

Comment: Blooming in warm months, Musky Mint could be utilized as a cultivated wildflower in masses. Obviously, the plant needs moist to wet soils.

Ilex ambigua (Michx.) Torr.
Sand Holly or Carolina Holly
Aquifoliaceae, Holly Family

Habit: Shrub or tree, to 6 m tall. Twigs smooth or hairy.

Leaves: Alternate, deciduous; stalked; blades elliptic, often broadly so, 2-5 cm long, margins smooth or with fine teeth at tips.

Inflorescences: Small clusters of flowers in leaf axils, male and female flowers separate and on separate plants.

Flowers: White, about 5 mm across, with 4-5 petals.

Fruit: A round berry, to 1 cm in diameter, red; seeds smooth.

Habitat and Distribution: Common; dry soils of hammocks, scrub, sandhills, and dunes; central peninsula Florida, west to Texas, Arkansas, and Oklahoma, and north to North Carolina.

Comment: Flowers occur from winter into spring. Flowers are small and not easily noticed, even in mass. The bright red fruits on the female plants are very ornamental. Shrubs are occasionally used as a patio plant or more commonly left where they were growing.

White

Hyptis alata (Raf.) Shinners

Ilex ambigua (Michx.) Torr.

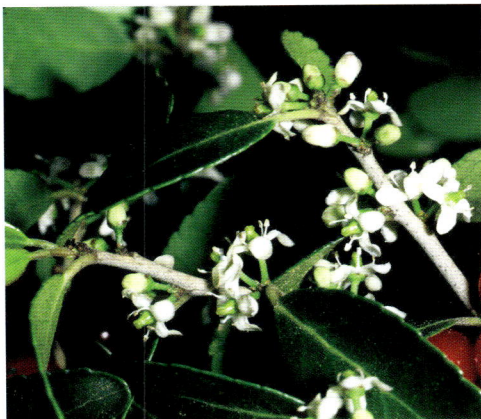

591

White

Ilex cassine L.
Dahoon Holly
Aquifoliaceae, Holly Family

Habit: Shrub or tree to 12 m tall. Bark grayish, branches smooth or with fine hairs.

Leaves: Alternate, evergreen; stalked; blades elliptic, often broadly so, 2-10 cm long, margins smooth or with few small teeth, lower surface pale and hairy on midrib.

Inflorescences: Small clusters of flowers in leaf axils, male and female flowers separate and on separate plants.

Flowers: White, small, 4-lobed.

Fruit: A berry, red, orange or yellow, round, 5-8 mm in diameter. Seeds ribbed.

Habitat and Distribution: Common; hammocks, low woods, wet flatwoods, wet margins, bogs, marshes, and swamps; throughout Florida, west to Texas, and north to North Carolina.

Comment: Dahoon Holly flowers in all the warm months. Flowers are small and not readily seen. The bright fruits of female plants are showy. The female plant is common and prized as an ornamental and can be easily grown on drier sites.

Ilex coriacea (Pursh) Chapm.
Large Gallberry, Sweet Gallberry, Bay-gall Bush
Aquifoliaceae, Holly Family

Habit: Shrub, to 5 m tall.

Leaves: Alternate, evergreen; stalked; blades elliptic to oval, 3-9 cm long, margins smooth or with a few small sharp teeth, lower surface with dark glands.

Inflorescences: Small clusters of flowers in leaf axils, male and female flowers separate and on separate plant.

Flowers: White, small, 5- to 9-lobed.

Fruit: A berry, black, shiny, round, 6-8 mm in diameter. Seeds smooth.

Habitat and Distribution: Occasional, locally common; wet pinelands, bogs, seepage areas, wet hammocks, and along streams; central peninsula Florida, north to Virginia, and west to Texas.

Comment: Large Gallberry blooms in the early spring. The dark green leaves are ornamental, but the small flowers and black fruits are difficult to see. This plant is an indicator of wetlands.

Photo courtesy of Ms. Jessie M. Harrris

***Ilex cassine* L.**

***Ilex coriacea* (Pursh.) Chapm.**

White

Ilex glabra (L.) A. Gray
Gallberry
Aquifoliaceae, Holly Family

Habit: Shrub, 0.5-3 m tall. Branches and twigs hairy when young, smooth at maturity.

Leaves: Alternate, evergreen; stalked; blades elliptic to somewhat rounded, often broader at the tip, 1-8 cm long, with few blunt teeth at tip.

Inflorescences: Small clusters of flowers in leaf axils, male and female flowers separate and on separate plants.

Flowers: Whitish green, small, with 5-8 petals.

Fruit: A berry, black when ripe, to 7 mm in diameter; seeds smooth.

Habitat and Distribution: Common; flatwoods, prairies, and pinelands; throughout Florida, west to Texas and north to Nova Scotia.

Comment: Gallberry flowers in winter and spring. The flowers are noticeable among the shiny green leaves. The black fruits on female plants are not very showy. Gallberry occurs in clones connected by extensive systems of underground runners. This plant can be used to provide some stability to sandy soils.

Ilex opaca Aiton
American Holly
Aquifoliaceae, Holly Family

Habit: Medium or large tree, to 15 m tall, cylindrical or conical in overall shape. Twigs slightly hairy, becoming smooth.

Leaves: Alternate, evergreen; stalked; blades leathery, oblong, elliptic to somewhat oval, 4-10 cm long, margins with spine-like teeth.

Inflorescences: Small clusters of flowers in leaf axils, male and female flower separate and on separate plants.

Flowers: White, small, with 4 petals.

Fruit: A berry, red, round, 0.7-1.2 cm in diameter; seeds ribbed and grooved.

Habitat and Distribution: Occasional to frequent; sandy soils of mixed deciduous woodlands and drier floodplains; central peninsula Florida, west to Texas, Ohio, and Missouri, and north to West Virginia, Pennsylvania, Maryland, New York, and Massachusetts.

Comment: American Holly is a spring blooming staple of the ornamental industry. It has been cultivated since early times along the east coast of the U.S. The shiny leaves of both male and female plants are attractive, but the red berries on the females are most popular. The leaves and berries are commonly used as Christmas decorations. It can be easily grown in most soils.

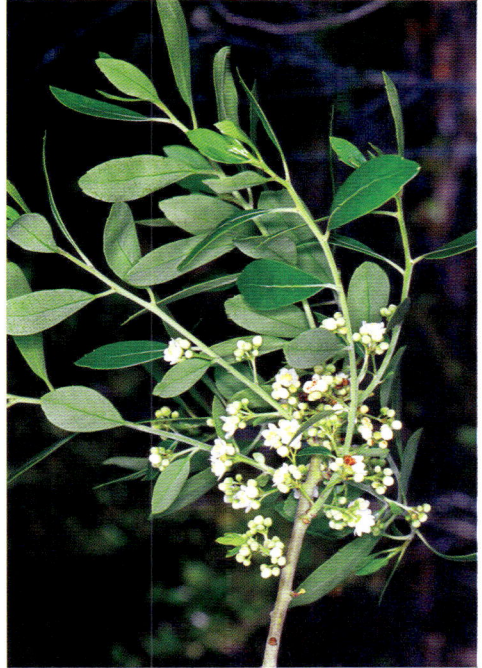

Ilex glabra (L.) A. Gray

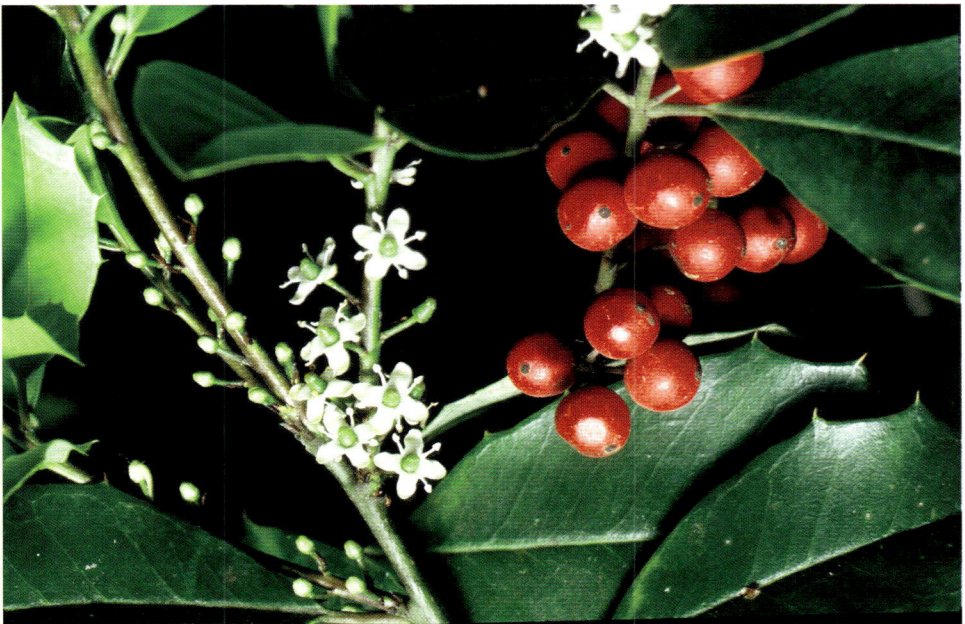

Ilex opaca Aiton

White

Ilex vomitoria Aiton
Yaupon Holly
Aquifoliaceae, Holly Family

Habit: Shrub or small tree, to 8 m tall. Young branches hairy, smooth with age.

Leaves: Alternate, evergreen; stalked; blades leathery, oval to elliptic, 1-3 cm long, margins with blunt, shallow teeth.

Inflorescences: Small clusters of flowers in leaf axils, male and female flowers separate and on separate plants.

Flowers: White, small, with 4 petals.

Fruit: A berry, red, clear, 4-7 mm in diameter; seeds ribbed.

Habitat and Distribution: Frequent to common; sandy woods, dunes, thickets, maritime forests, well-drained floodplains, banks, and forest margins; central peninsula Florida, west to Texas, Oklahoma, and Arkansas, and north to Virginia.

Comment: As a spring flowering species, Yaupon Holly, with its shiny green leaves, is a popular cultivated plant. Females with the shiny, clear red fruits are very attractive. It can be used as a specimen plant or most commonly as a hedge. Several cultivated forms, especially short round plants, are very popular.

Ipomoea alba L.
Moonflower
Convolvulaceae, Morning-glory Family

Habit: Herbaceous annual or perennial vine, to 5 m long, sometimes woody at base. Stems twining, smooth, high climbing, sap milky.

Leaves: Alternate; long-stalked; blades heart-shaped, frequently 3- to 5-lobed, to 20 cm long.

Inflorescences: Flowers solitary to several, in clusters from leaf axils.

Flowers: White with yellow radial stripes, 10-15 cm long, 8-10 cm wide, with a long slender tube, night blooming.

Fruit: A roundish capsule, to 3 cm long; seeds dark brown to black, 8-10 mm long, smooth, rounded.

Habitat and Distribution: Frequent; hammocks, mangroves, lake sides, rights-of-way, and disturbed sites; northern to southern peninsula Florida, Texas; a native of tropical America, now pantropical.

Comment: Moonflower blooms during all warm months. It has been extensively cultivated for the huge, fragrant, showy white flowers that open at dusk. This vine can be cultivated as an annual in temperate climates.

Ilex vomitoria Aiton

Ipomoea alba L.

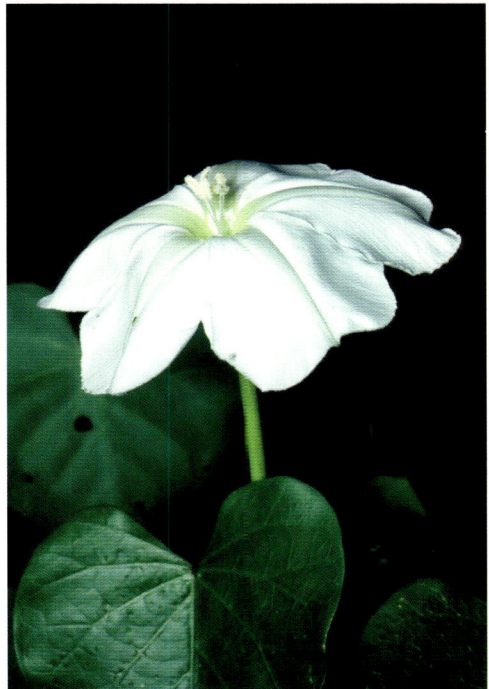

White

Ipomoea pandurata (L.) G. Meyer
Bigroot Morning-glory
Convolvulaceae, Morning-glory Family

Habit: Herbaceous perennial vine, to 5 m long, from a woody tuber-like root. Stems twining or trailing, smooth or with a few hairs.

Leaves: Alternate; long-stalked; blades heart-shaped and sometimes lobed, 4-12 cm long and almost as wide, often hairy underneath.

Inflorescences: 1-5 flowers in clusters from leaf axils.

Flowers: White, funnel-shaped, 5-8 cm long, with pinkish or purple throat.

Fruit: A rounded capsule, about 1 cm long; seeds with woolly hairs on the angles.

Habitat and Distribution: Frequent; dry soils of rights-of-way, fences, waste places, sandhills, and disturbed dry woods; central peninsula Florida, west to Texas, and north to Ontario and Connecticut.

Comment: Bigroot Morning-glory blooms in summer and fall. The large white flowers with the dark pink or maroon throat are very showy in the morning, but wilt during day. It can be cultivated with great effect on a trellis. Sometimes this plant is called: Man-of-the-earth, Wild-potato, or Man-root.

Ipomoea sagittata Poir.
Glades Morning-glory
Convolvulaceae, Morning-glory Family

Habit: Herbaceous perennial vine, from a creeping rootstock. Stems twining or trailing, smooth.

Leaves: Alternate; long-stalked; blades narrowly arrowhead-shaped, 4-10 cm long and 3-15 mm wide.

Inflorescences: Flowers solitary or in clusters of 2 or 3 from leaf axils.

Flowers: White or more commonly purplish-pink, funnel-shaped, 5-9 cm long and almost as wide.

Fruit: A round capsule, about 1 cm diameter; seeds with stiff hairs on the angles.

Habitat and Distribution: Frequent; low, wet habitats, fresh and brackish marshes, coastal beaches and dunes, and rights-of-way; throughout Florida, west to Texas, and north to North Carolina; the West Indies; introduced into the western Mediterranean.

Comment: A summer and fall blooming wildflower often seen in wetter habitats. The large flowers, opening in the morning, are spectacular in mass. Cultivating this vine on a trellis can yield an early morning treat.

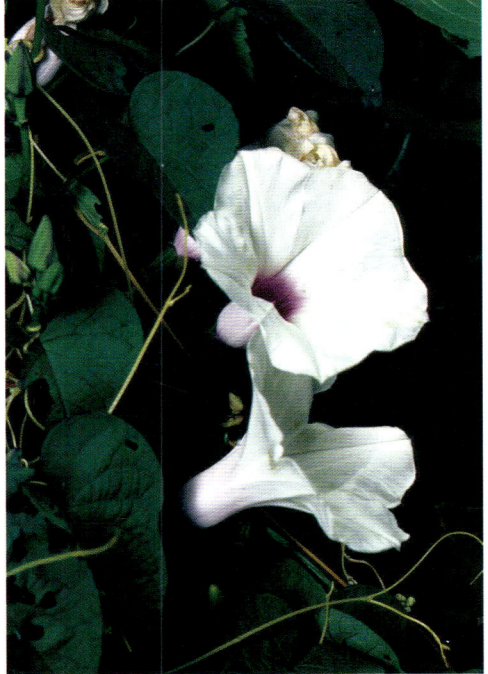

Ipomoea pandurata (L.) G. Meyer

Ipomoea sagittata Poir.

White

Iresine diffusa Humb. & Bonpl. ex Willd.
Bloodleaf or Juba's Bush
[Iresine celosia L., misapplied]
Amaranthaceae, Amaranth Family

Habit: Herbaceous annual, 1-3 m tall, from a thick taproot. Stems sprawling or erect, smooth, often pinkish/reddish, nodes swollen.

Leaves: Opposite; stalked; blades often pinkish/reddish, oval to somewhat rhombic, 3-15 cm long, smooth or hairy.

Inflorescences: Feather-like, diffuse, elongated clusters at ends of stems.

Flowers: White, pinkish, or greenish, minute, male and female flowers separate on the same plant.

Fruit: A tiny, 1-seeded, papery bladder; seeds round, about 1.5mm in diameter.

Habitat and Distribution: Common; marshes, hammocks, old fields, and disturbed areas; throughout Florida, north to North Carolina, and west to Texas; West Indies, central and western Mexico, south through Central American into South America.

Comment: The dense fine plumes of white to pinkish flowers can be seen in the warm months. Frequently the leaves and stems are reddish which is why this plant is called Bloodleaf. Both the flower plumes and leaves are colorful and ornamental.

Itea virginica L.
Virginia-willow, Virginia Sweetspire, Tassel-white
Escolloniaceae (sometimes a part of Saxifragaceae), Escallonia Family

Habit: Shrub, 1-3 m tall. Stems woody, often sprouting from roots, twigs hairy.

Leaves: Alternate, deciduous; short-stalked; blades elliptic to oval, often broader toward tips, pointed at both ends, 5-10 cm long and 1-4 cm broad, with distinctly dense minute teeth along edges.

Inflorescences: Terminal, bottlebrush-shaped clusters, 4-15 cm long and about 1 cm in diameter.

Flowers: White, 4-7 mm long, with 5 petals.

Fruit: A cylindrical capsule, tapering at tip, 3-7 mm long, hairy, grooved, splits along grooves at maturity; seeds about 1 mm long, oblong, surface reticulate.

Habitat and Distribution: Frequent; low woods, stream banks, lake banks, and swamps; throughout Florida, west to eastern Texas, and north to Oklahoma, Arkansas, Missouri, Illinois, Kentucky, Pennsylvania, and New Jersey.

Comment: Flowering in spring just as or before the leaves expand, the dense spires of white flowers are very attractive. Virginia-willow shrubs are frequently locally common, the root sprouts leading to large clusters of plants and subsequently showy masses of blooms.

Male

Female

Iresine diffusa Humb. & Bonpl. ex Willd.

Itea virginica L.

White

Jasminum multiflorum (Burm. f.) Andrews
Star Jasmine or Downy Jasmine
Oleaceae, Olive Family

Habit: Spreading or climbing shrub, to 2 m tall. Stems visibly hairy.

Leaves: Opposite, evergreen; stalked; blades ovate, rounded at base, pointed at tip, 3-7 cm long, hairy on both surfaces, margins smooth.

Inflorescences: Flowers in axillary or terminal clusters.

Flowers: White, 1.2-2.2 cm long, petals fused at base, 6- to 9-lobed, fragrant.

Fruit: A 2-lobed berry, infrequently develops.

Habitat and Distribution: Occasional; disturbed habitats; central and north peninsula Florida; a native of India; now throughout the world's tropics and subtropics.

Comment: Flowering all year in sun and partial shade, the white flowers with the dark green leaves provide an outstanding backdrop or specimen planting. Propagated by cuttings, this shrub can also be trained as a vine on a trellis. When seen as an escape the bright white flowers are eyecatching.

Kalmia hirsuta Walter
Wicky
Ericaceae, Heath Family

Habit: Shrub, 20-60 cm tall. Stems numerous, emerging from underground base; upper twigs hairy.

Leaves: Alternate, evergreen; stalkless; blades hairy, variable, broadly or narrowly elliptic with blunt or pointed ends, 5-15 mm long and 2-8 mm wide, margins smooth.

Inflorescences: 1-4 flowers clustered at base of new leaves.

Flowers: White or pale to deep pink, broadly bell-shaped, 1-1.5 cm diameter, with 5 short pointed lobes, and ring of dark pink spots inside.

Fruit: A round capsule with pointed tip, about 3 mm long, surrounded by sepals; seeds tiny, about 0.5 mm long.

Habitat and Distribution: Frequent; flatwoods, pine savannahs, and bog borders; from northern peninsula Florida, west to southeast Louisiana, and north to southeast South Carolina.

Comment: When in flower in spring and summer this small somewhat straggling shrub is quite showy. It is almost unnoticed in other seasons. The native habitat is subject to numerous fires suggesting that frequent pruning might be in order when cultivating. In nature soils are acidic and frequently flooded.

White

Jasminum multiflorum
(Burm. f.) Andrews

Kalmia hirsuta Walter

603

White

Kalmia latifolia L.
Mountain Laurel
Ericaceae, Heath Family

Habit: Large shrub, 3-10 m tall. Twigs hairy, mature stems smooth.

Leaves: Alternate, evergreen; stalks short and hairy; blades leathery, thick, narrow to oval, 5-12 cm long and 1.5-5 cm wide.

Inflorescences: Clusters of several flowers on ends of stems.

Flowers: White or pink, broadly bell-shaped, 2-3 cm diameter, with 5 short pointed lobes and a ring of purple spots inside.

Fruit: A roundish capsule, broader than long, 4-7 mm wide. Seeds tiny.

Habitat and Distribution: Common northward, infrequent southward; creek banks and swamps, rocky woods, ravine slopes, wooded bluffs; extreme north peninsula and panhandle Florida, north into New Brunswick, southern Ontario, and southern Indiana, west into Louisiana.

Comment: Mountain Laurel provides a very good show in spring. This shrub can be spectacular with a profusion of bright flowers against the shiny evergreen leaves. Normally the many stems are quite crooked and bent and the plant forms a dense green mound that can be covered with clusters of blooms.

Lachnocaulon anceps (Walter) Morong
Little White Bogbutton
Eriocaulaceae, Pipewort Family

Habit: Herbaceous perennial, 5-40 cm tall, from thin, dark, fibrous, branched roots. Stem thin, hairy.

Leaves: Alternate, in a basal rosette; stalkless; blades very slender, pointed, 2-6 cm long, lacking air spaces at base.

Inflorescences: Round terminal heads, 4-8 mm in diameter.

Flowers: White, bracts gray.

Fruit: A capsule, 3-parted; seeds 0.5-0.8 mm long, ridged.

Habitat and Distribution: Common; wet soil of flatwoods, bogs, marshes, and shores; throughout Florida, north to New Jersey, and west to Texas; Cuba.

Comment: These little Bogbuttons can be seen flowering through most of the summer. Masses of blooming plants can provide quite a show. The small clusters of basal rosettes are distinctive for the genus.

Kalmia latifolia L.

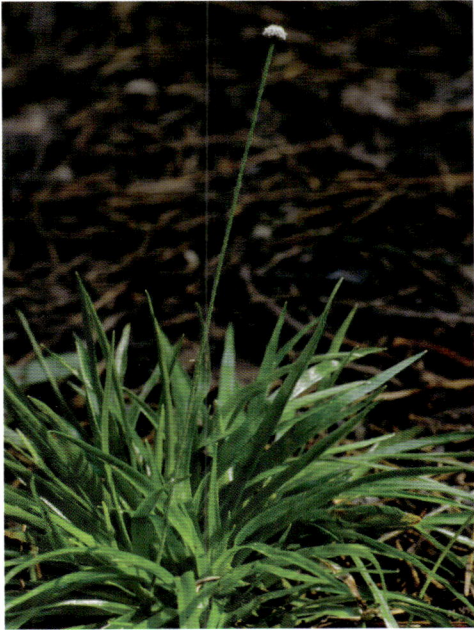

Photo courtesy of Ms. Jessie M. Harris

Lachnocaulon anceps (Walter) Morong

White

Lantana involucrata L.
White Lantana, White Sage, Wild Lantana, Button Sage
Verbenaceae, Vervain Family

Habit: Erect shrub, to 2 m tall. Branches thin, hairy.

Leaves: Opposite; stalked; blades aromatic, oval, often wider toward base, 1-5 cm long, covered with fine hairs, margins toothed.

Inflorescences: Stalked heads from leaf axils.

Flowers: White to pinkish to purple with a yellow eye, 4 to 8 mm long, petals fused at base.

Fruit: A fleshy, rounded drupe, 2-4 mm in diameter.

Habitat and Distribution: Infrequent; coastal sites - hammocks, sand dunes, beaches, disturbed areas; central and southern peninsula Florida; Bermuda, West Indies, Mexico south into northern South America, Galapagos Islands.

Comment: Like many tropical species White Lantana flowers all year. The white heads on bending branches are attractive, but not showy. This is a nice specimen plant for a garden.

Lepidium virginicum L.
Pepperweed
Brassicaceae (Cruciferae), Mustard Family

Habit: Herbaceous annual or biennial, 20-90 cm tall. Stem erect, branched, sparsely hairy.

Leaves: Alternate, deciduous; basal leaves (when biennial) deeply toothed and lobed, cut to midrib, to 15 cm long, tapered to base; stem leaves short-stalked, lower leaves similar to basal leaves, upper leaves becoming less toothed and with smooth margins near tip, 2-10 cm long.

Inflorescences: Terminal, cylindrical, spike-like arrangements.

Flowers: White, 4 petals, each to 2 mm long, broadest at tip.

Fruit: Round and flattened, 3-4 mm long, long-stalked, borne horizontally along flowering stalks.

Habitat and Distribution: Common; dry open areas, roadsides, old fields, and disturbed sites; throughout the United States; Newfoundland; introduced into the West Indies and Europe.

Comment: Spring flowering, Pepperweed can be an attractive wildflower when growing in masses. The fruits can be used to flavor salads.

Lantana involucrata L.

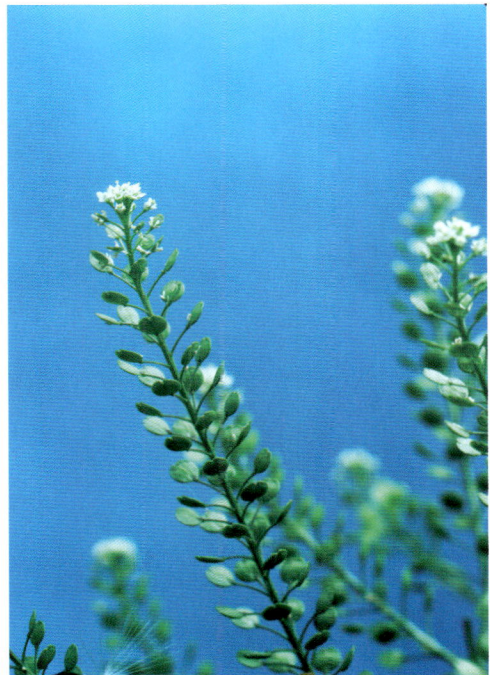

Lepidium virginicum L.

White

Licania michauxii Prance
Gopher-apple
Chrysobalanaceae, Cocco Plum Family

Habit: Low shrub, 10-40 cm tall, in large colonies from underground runners. Stems slender.

Leaves: Alternate, evergreen; short-stalked; blades oblong, 2-12 cm long, leathery, shiny above, veins evident, margins smooth.

Inflorescences: Terminal clusters.

Flowers: White, about 2 mm long, hairy, with 5 petals and a roundish tube.

Fruit: Ellipsoid, with a stone, 2-3 cm long, white, reddish, or purplish.

Habitat and Distribution: Frequent; dry woods and sandy soils, sandhills, scrub, dunes, and pinelands; throughout Florida, west to Mississippi, and north to South Carolina.

Comment: Gopher-apple can flower in any warm month. It can be dramatic as a wildflower when in large colonies. It is very underutilized as a landscape plant. The fruit is edible and is prized by wildlife, especially the Gopher Tortoise (*Gopherus polyphemis*).

Ligustrum lucidum W.T. Aiton
Chinese Privet, Glossy Privet, Wax-leaf Ligustrum
Oleaceae, Olive Family

Habit: Tall shrub or subcanopy tree, to 12 m tall. Branches smooth.

Leaves: Opposite, evergreen; stalked; blades oval, glossy, wider at base, distinct point at tip, 5-12 cm long and up to 5 cm broad, smooth, margins smooth.

Inflorescences: Branched, spreading, terminal clusters.

Flowers: White or cream, with slender tube 2.5-3.5 mm long and 4 spreading lobes 2-2.5 mm long.

Fruit: An oval, berrylike drupe, 4-9 mm long and 5 mm diameter, blue-black.

Habitat and Distribution: Infrequent, but can be locally common; shady disturbed habitats; panhandle and north and central peninsula Florida, coastal areas west to Louisiana and north along the coast to North Carolina; a native of China and Korea.

Comment: Flowering is in the spring in most of its range, intermittent in more tropical areas. The large clusters of flowers are quite showy against the glossy green leaves. The drooping clusters of bluish fruits are also attractive. Propagation is by seeds or cuttings. To keep this plant as a shrub requires pruning. Several cultivated forms have been named.

Licania michauxii Prance

Ligustrum lucidum W.T. Aiton

White

Lobelia paludosa Nutt.
White Swamp Lobelia
Campanulaceae, Bellflower Family

Habit: Herbaceous perennial, 20-90 cm tall. Stems smooth, leaves mostly basal, much smaller upwards.

Leaves: Alternate; tapering into stalks; basal leaves slender, often widest at tip, 3-25 cm long and up to 1.5 cm broad, margins smooth or wavy; upper leaves much smaller.

Inflorescences: Terminal, spike-like clusters.

Flowers: White or bluish, to 1.5 cm long, with smaller 2-lobed upper lip and larger 3-lobed lower lip.

Fruit: A capsule, to 3.5 mm wide.

Habitat and Distribution: Frequent; wet pinelands, bogs, and swamps; throughout peninsula Florida, west into the central panhandle, and north to adjacent Georgia.

Comment: The large flowers can be seen in all the warm months. Among the several herbaceous wildflowers in wetter habitats seen in the warmer months of the year, White Swamp Lobelia stands out with its spike of large flowers.

Lonicera japonica Thunb.
Japanese Honeysuckle
Caprifoliaceae, Honeysuckle Family

Habit: Perennial vine. Stems trailing or high climbing, semi-woody, hairy.

Leaves: Opposite, evergreen; short-stalked; blades oval, 2-8 cm long and 1-4 cm broad, hairy underneath, margins smooth, but variously toothed on young shoots.

Inflorescences: Flowers solitary or in pairs in leaf axils.

Flowers: White or pinkish fading to yellow, 2.5-5 cm long, with slender throat and 5 slender, unequal lobes, very fragrant.

Fruit: A round berry, 5 mm diameter, black; seeds black, shiny, reticulate, oblong, about 3 mm long.

Habitat and Distribution: Frequent, locally common; disturbed habitats, hammocks, pinelands, fields, roadsides; central peninsula Florida, west to Texas, and north to Missouri, Kansas, Massachusetts, and New York; native to Eastern Asia.

Comment: Japanese Honeysuckle flowers from spring into midsummer. This vine is a pretty wildflower, although many consider it an awful weed. It spreads by runners and seeds and can overgrow large areas destroying native vegetation. Propagation is by seeds, cuttings, and runners. Several named forms are available. Even when planted in a confined space, gardener beware.

Lobelia paludosa Nutt.

Lonicera japonica Thunb.

White

Lycopus americanus Muhl. ex W.P.C. Barton
American Water Hoarhound
Lamiaceae (Labiatae), Mint Family

Habit: Herbaceous perennial, 30-70 cm tall, from a long underground runner, not tuberously thickened. Stems mostly smooth, 4-angled.

Leaves: Opposite; very short-stalked; blades smooth to somewhat hairy underneath, glandular dotted, to 12 cm long, with toothed margins sometimes cut almost to midrib.

Inflorescences: Dense axillary clusters.

Flowers: White, tiny, with 4 lobes, each 2-4 mm long.

Fruit: A nutlet, 1-1.4 mm long.

Habitat and Distribution: Infrequent; wet woods, marshes, riverbanks, and pond margins; from Newfoundland, Quebec, and British Columbia, south to California and western panhandle Florida.

Comment: Flowering from late spring into frost. The clusters of flowers are so small that they are often overlooked.

Lyonia fruticosa (Michx.) G.S. Torr.
Stagger-bush or Poor-grub
Ericaceae, Heath Family

Habit: Shrub, 1-3 m tall. Young stems rusty hairy.

Leaves: Alternate, evergreen; short-stalked; blades oval to elliptic, 1-5 cm long, leathery, flat or slightly cup-shaped upwards, with rust-colored scales on lower surfaces.

Inflorescences: Axillary clusters on newer twigs.

Flowers: White, to 5 mm long, urn-shaped, with 5 tiny lobes.

Fruit: A capsule, to 5 mm long, with thick, lighter colored sutures between divisions.

Habitat and Distribution: Frequent in wetter pine flatwoods and bogs; occasional in scrub; throughout peninsula Florida, west into the eastern panhandle, and north to south central Georgia.

Comment: Spring flowers are not especially showy, but the total effect of the rusty new leaves and twigs, and curved stems makes a handsome display. Very infrequent in cultivation Stagger-bush is used along borders and as a hedge. The stems are used for fake ornamental shrubs.

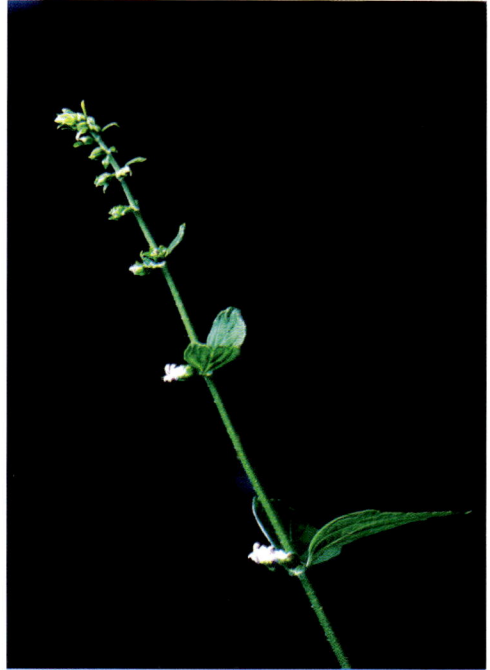

Lycopus americanus
Muhl. ex W.P.C. Barton

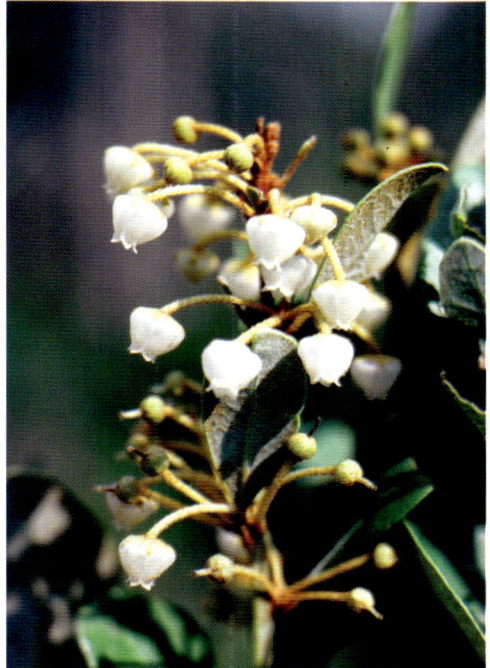

Lyonia fruticosa
(Michx.) G.S. Torr.
photograph courtesy of Walter Judd

White

Macbridea alba Chapm.
White Birds-in-a-nest
Lamiaceae (Labiatae), Mint Family

Habit: Herbaceous perennial, 30-50 cm tall. Stems erect with few or no branches.

Leaves: Opposite; short-stalked near base of plant, stalkless upwards; blades elliptic or often wider at tip, fleshy, margins smooth.

Inflorescences: Axillary clusters at stem tips.

Flowers: White, 2.5-3 cm long, two-lipped, upper lip forming a hood, lower lip 3-lobed; stamens 4, exserted.

Fruit: 4 distinct nutlets.

Habitat and Distribution: Rare; wet pine flatwoods, bogs, and savannahs; panhandle Florida in the Apalachicola River Basin in Bay, Franklin, Gulf, and Liberty Counties.

Comment: The flowers of White Birds-in-a-nest seen in summer are striking. The large white flowers viewed against the background greens are always eye-catching. White Birds-in-a-nest is listed by the State of Florida as Endangered, and by the U.S. Fish and Wildlife Service as Threatened.

Magnolia grandiflora L.
Southern Magnolia or Bullbay
Magnoliaceae, Magnolia Family

Habit: Large tree, to 30 m tall. Bark gray, smooth.

Leaves: Alternate; stalked; blades broadly elliptic to oval, 10-30 cm long and 4-15 cm broad, evergreen, leathery, glossy, dull reddish hairy underneath, margins smooth.

Inflorescences: Terminal on new branches after the leaves develop.

Flowers: White or cream, cup-shaped, 10-20 cm diameter, with 6-15 broad oval parts, very fragrant.

Fruit: Cone-like, oval, 6-10 cm long and 5-6 cm diameter; seeds bright red, 1-2 cm long, hanging from a thin thread when ripe.

Habitat and Distribution: Frequent to common; dry to wet hammocks, swamp forests, poorly and well-drained woods, and planted as an ornamental tree; central peninsula Florida, north to Virginia, and west to east Texas.

Comment: The large, fragrant flowers occur from late spring into summer. Southern Magnolia has been cultivated since the southern states were settled. It has, along with the Live Oak, become a symbol of the Old South. The wood was extensively harvested for making crates.

Macbridea alba Chapm.

Magnolia grandiflora L.

White

Magnolia macrophylla Michx. var. *ashei* (Weath.) D.L. Johnson
Ashe Magnolia
[*Magnolia ashei* Weath.]
Magnoliaceae, Magnolia Family

Habit: Shrub or small tree, usually to 3-5 m tall, rarely to 15 m tall. Stems mostly leaning. Branches crooked, young twigs hairy, shoots often short-lived, but frequently sprouting.

Leaves: Alternate, deciduous; stalked; blades somewhat spathulate, broadest just above middle or at tip, base rounded or lobed, very large, to 60 cm long and 30 cm wide, silvery hairy on margins and underneath.

Inflorescences: Flowers terminal, solitary.

Flowers: White, 25-30 cm wide, 9 spreading, petal-like segments.

Fruit: Cone-like, rounded, 4-5 cm long, 3 cm wide; seeds red, exerted on thin threads.

Habitat and Distribution: Rare; deciduous woodlands, ravines, and bluffs; six counties in panhandle Florida, from the Ochlockonee River to Santa Rosa County.

Comment: The fragrant spring flowers are spectacularly showy. Often two or more flowers occur at a time. The shrub is frequently multi-trunked. The distinctive habit with the uncommonly large leaves and flowers is a marvelous landscape plant. It provides a conspicuous contrast and much conversation. Listed by Florida as Endangered, but available from nurseries.

Magnolia virginiana L.
Sweetbay
Magnoliaceae, Magnolia Family

Habit: Medium to large tree, to 30 m tall. Bark whitish, smooth.

Leaves: Alternate, mostly evergreen southwards; stalked; blades elliptic, 6-15 cm long and 2-6 cm wide, leathery, silvery white underneath, margins smooth.

Inflorescences: Flowers terminal, solitary, on current year's growth after the leaves mature.

Flowers: White, very fragrant, cup-shaped, 3-8 cm diameter, with 6-15 oval parts.

Fruit: Cone-like, oval or round, 2-5 cm long; seeds red, 6-9 m long, hanging from thin thread when ripe.

Habitat and Distribution: Common; swamps, bayheads, wet woods, boggy habitats, and wet flatwoods; throughout Florida, north to Tennessee and Massachusetts, west to Arkansas and Texas.

Comment: Flowers appear in late spring and summer, often in profusion. Seldom cultivated and sometimes difficult to establish Sweetbay is a very attractive tree with layers of silvery leaves rippling in the wind and many sparkling white flowers.

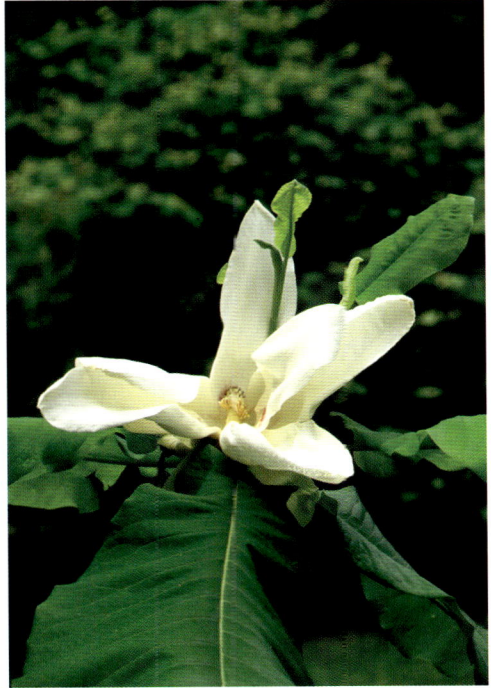

Magnolia macrophylla
Michx. var. *ashei* (Weath.) D.L.
Johnson

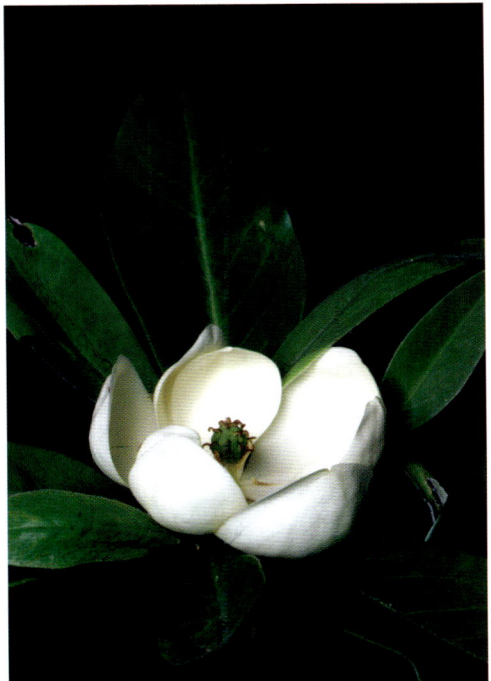

Magnolia virginiana L.

White

Marshallia tenuifolia Raf.
Barbara's Buttons
Asteraceae (Compositae), Aster or Sunflower Family

Habit: Herbaceous perennial, to 1 m tall. Stem usually branched at or below the middle, lower stem smooth and upper hairy.

Leaves: Alternate; stem leaves stalkless; blades slender, grass-like; basal rosette leaves stalked, lance-shaped and broader at tip.

Inflorescences: Heads, solitary, terminal, only disc flowers present.

Flowers: Whitish, pink or purple, heads 2-3 cm wide, glandular-dotted.

Fruit: An achene, 5-angled, 10-ribbed, 5 scales on top.

Habitat and Distribution: Frequent to locally common; grasslands, flatwoods, sandhills, open woods, and bogs; throughout Florida, west to Texas, and north to southern Georgia.

Comment: Barbara's Buttons flowers in summer and fall. The single heads resemble pink buttons and can be seen poking out of vegetation.

Mecardonia acuminata (Walter) Small
Axil-flower or White flowered Mecardonia
Scrophulariaceae, Figwort or Snapdragon Family

Habit: Herbaceous perennial, 10-70 cm tall. Stems 4-sided, erect or ascending, branched, hairless.

Leaves: Opposite; short- or non-stalked; blades elliptic to broader at tip, 1-5 cm long, firm, with glandular dots throughout, margins toothed towards tip.

Inflorescences: Long-stalked, solitary flowers from leaf axils.

Flowers: White with purple veins, 6-11 mm long, tubular, with 5 lobes and hairy throat.

Fruit: An oval capsule, 4-8 mm long; seeds cylindric.

Habitat and Distribution: Frequent; bogs, marshes, pond margins, ditches, streams, alluvial woods, wet prairies, and pine flatwoods; throughout Florida, north to Delaware and Maryland, west to east Texas and Missouri.

Comment: Flowering from summer into fall the diffuse branches with long-stalked white flowers are appealing. In moist to wet areas, a mass of these plants can be handsome.

Marshallia tenuifolia Raf.

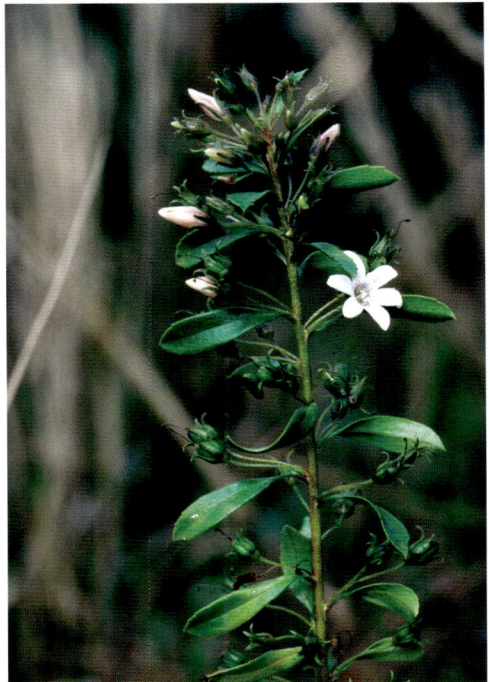

Mecardonia acuminata
(Walter) Small

White

Melaleuca quinquenervia (Cav.) S.T. Blake
Melaleuca Tree, Broadleaf Paperbark Tree, Punk Tree
Myrtaceae, Myrtle Family

Habit: Tree, to 12 m or more tall. Bark soft, white, peeling in thin layers. Branches drooping.

Leaves: Alternate; short-stalked; blades elliptic, 5-10 cm long, tapered at both ends, 3- to 7-veined, hairy when young, smooth and dull green when older, reddish dots on both surfaces.

Inflorescences: Dense terminal spikes, branches continue to grow past the flowers, producing normal leaves and leaving behind spikes of fruit surrounding the growing branch.

Flowers: White, with many stamens, giving bottlebrush-like appearance, petals 5.

Fruit: A short squarish or rounded capsule; seeds many, reddish brown, wedge-shaped, about 1 mm long.

Habitat and Distribution: Common; planted and commonly naturalized; swamps, low flatwoods, wetland margins, and other low habitats; central and southern peninsula Florida; Bahamas; a native of Australia, New Guinea, and New Caledonia.

Comment: Flowering all year, the silvery young leaves and many bottle-brush spikes provoke comments on the attractiveness. The white peeling bark is also unusual and ornamental. A cause of respiratory irritation to some persons, especially when in bloom, occasionally, causing a rash or headache. The bark can also cause dermatitis. Especially weedy, this tree has been placed in the highest category of invasive species, almost completely displacing native vegetation in some habitats. Sale of this tree is prohibited.

Melanthera angustifolia A. Rich.
Narrow-leaf Cat-tongue
Asteraceae (Compositae), Aster or Sunflower Family

Habit: Herbaceous perennial, to 1 m tall, often leaning on ground. Stems square, thin, branched, mottled with purple, covered with rough hairs.

Leaves: Opposite; nearly stalkless; blades very narrow, oblong to linear, to 12 cm long, sandpapery.

Inflorescences: Terminal, solitary heads, only disc flowers present.

Flowers: White, heads 1-2 cm wide, black-tipped anthers prominent.

Fruit: An achene, dark brown, broader at tip, 2.5-3 mm long, 4-angled, with few bristles on top.

Habitat and Distribution: Frequent to common; pinelands, hammocks, and disturbed sites; south peninsula Florida, West Indies, Mexico, and Guatemala.

Comment: Flowering in warm months, the plants are distinct with the prominent black-tipped anthers and very narrow, rough sandpapery blades. The blades are thought to feel like a rough cat's-tongue.

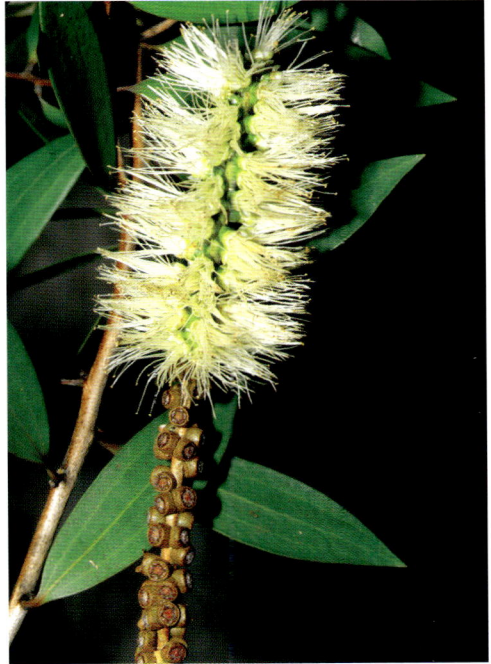

Melaleuca quinquenervia
(Cav.) S.T. Blake

Melanthera angustifolia A. Rich.

White

Melanthera nivea (L.) Small
Common Cat-tongue, Black Anthers, Small's Cat-tongue
[*Melanthera hastata* Michx.]
Asteraceae (Compositae), Aster or Sunflower Family

Habit: Herbaceous perennial, to 2 m tall. Stems square, many branched, mottled with purple, covered with rough hairs.

Leaves: Opposite; stalked; blades narrow to broad, oval to triangular, 5-16 cm long, often 3-lobed, 3-nerved, sandpapery.

Inflorescences: Terminal, solitary heads, only disc flowers present.

Flowers: White, heads 1-2 cm wide, black-tipped anthers prominent.

Fruit: An achene, dark brown, broader at tip, 2.5-3 mm long, 4-angled, with few bristles on top.

Habitat and Distribution: Frequent to common; dry or moist woods, thickets and on beaches; throughout Florida, west to Louisiana, and north to South Carolina; throughout tropical America.

Comment: Flowering in warm months, the plants are easy to identify with the prominent black-tipped anthers and triangular, 3-nerved, rough sandpapery blades. The blades are thought to feel like a rough cat's-tongue.

Melanthera parvifolia Small
South Florida Cat-tongue
Asteraceae (Compositae), Aster or Sunflower Family

Habit: Herbaceous perennial, to 0.6 m tall, often leaning on ground. Stems square, thin, many branched, mottled with purple, covered with rough hairs.

Leaves: Opposite; nearly stalkless; blades narrow to broad, oval to triangular, to 4 cm long, often 3-lobed, 3-nerved, sandpapery.

Inflorescences: Terminal, solitary heads, only disc flowers present.

Flowers: White, heads 1-2 cm wide, black-tipped anthers prominent.

Fruit: An achene, dark brown, broader at tip, 2.5-3 mm long, 4-angled, with few bristles on top.

Habitat and Distribution: Frequent to common; pinelands, hammocks, and disturbed sites; south peninsula Florida.

Comment: Flowering in warm months, the plants are distinct with the prominent black-tipped anthers and small, triangular, 3-nerved, rough sandpapery blades. The blades are thought to feel like a rough cat's-tongue.

Melanthera nivea (L.) Small

Melanthera parvifolia Small

White

Melilotus albus Medik.
Sweet-clover or Hubam
Fabaceae (Leguminosae), Pea or Bean Family

Habit: Herbaceous annual or biennial, 0.6-2.5 m tall. Stems branched.

Leaves: Alternate; stalked; blades with 3 oval leaflets, each 1-3 cm long, margins toothed.

Inflorescences: Stalked spikes of flowers from upper leaf axils.

Flowers: White, typically pea-shaped, with upper petal 4-5 mm long.

Fruit: An oval, flattened pod, 2.5-4 mm long, blackish, containing 1-2 seeds.

Habitat and Distribution: Frequent to locally common; fields, disturbed sites, pastures, river bottoms, and along roadsides; throughout Florida and North America, north to Nova Scotia, west to British Columbia, south to Mexico, and the West Indies; a native of western Asia and adjacent Europe.

Comment: Flowering in spring and summer this sweet-smelling wildflower can be locally very common. It has been extensively cultivated for soil improvement and forage. Sweet-clover is usually found in disturbed soils and therefore quite frequently in agricultural areas and on spoil piles.

Merremia dissecta (Jacq.) Hallier f.
Cutleaf Morning-glory or Alamo Vine
[*Ipomoea sinuata* Ort.; *Operculina dissecta* (Jacq.) House]
Convolvulaceae, Morning-glory Family

Habit: Herbaceous perennial vine. Stems trailing or climbing, often hairy.

Leaves: Alternate; long-stalked; blades 3-14 cm long and wide, with 5, 7, or 9 lobes and toothed edges.

Inflorescences: 1 or 2 flowers on long stalks from leaf axils.

Flowers: White with darker pink centers, funnel-shaped, 3-5 cm long and wide.

Fruit: A roundish capsule, 1-2 cm in diameter.

Habitat and Distribution: Frequent; primarily in open disturbed sites; throughout Florida, west to Texas, and north to Georgia; the West Indies, Mexico south into South America.

Comment: Often used as an ornamental, Cutleaf Morning-glory is cultivated on fences and trellises. The late spring into fall flowering period produces many handsome flowers each day. The deeply cut leaves are also provide interest.

Melilotus albus Medik.

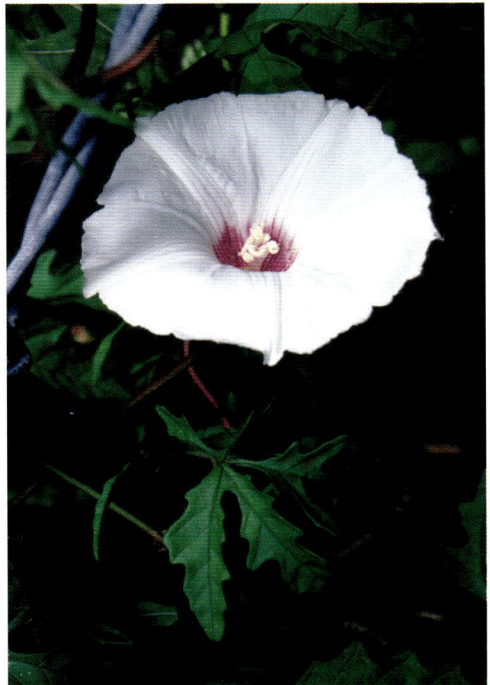

Merremia dissecta
(Jacq.) Hallier f.

White

Mikania scandens (L.) Willd.
Climbing Hempweed
Asteraceae (Compositae), Aster or Sunflower Family

Habit: Perennial vine, to 3 m long. Stems twining and climbing, sometimes sprawling, forming mats over low vegetation, hairy.

Leaves: Opposite; stalked; blades heart-shaped, 2-14 cm long, margins smooth to wavy and/or toothed, 3-nerved.

Inflorescences: Domed clusters of heads on stalks from leaf axils, only disc flowers present.

Flowers: Dull white or pinkish, 4 per head, each 5-6 mm long.

Fruit: An achene, black, 1.8-2.2 mm long, glandular, with bristles on top.

Habitat and Distribution: Common; usually in very wet habitats, wet thickets, woods, marshes, bogs, irrigated fields, and disturbed sites; throughout Florida, north to Maine, Illinois, Indiana, and Michigan, west to Texas and adjacent Mexico; Bahama Islands, Cuba.

Comment: Flowering in late summer into fall and all year southward, Climbing Hempweed is an extremely troublesome weed, especially in irrigated crops.

Mitchella repens L.
Partridge-berry or Twinberry
Rubiaceae, Madder Family

Habit: Herbaceous perennial, 15-30 cm long. Stems creeping, branched, forming mats.

Leaves: Opposite; short-stalked; blades oval, 0.6-2 cm long, margins smooth.

Inflorescences: Flowers stalked, terminal, paired.

Flowers: White, sometimes with a little pink, pairs joined at base, tubular, 9-14 mm long, with 4 spreading lobes, hairy inside.

Fruit: Red or rarely white, paired, roundish, 4-8 mm diameter.

Habitat and Distribution: Frequent to common; moist to dry woods; central peninsula Florida, west to east Mexico and east Texas, and north Minnesota, Ontario, and Nova Scotia.

Comment: Partridge-berry can make a great ground cover in shady situations. Attractive flowers occur from May into July making the ground cover even more desirable. The red fruits following the flowers, from late spring into fall, continue the ornamental aspect. This creeper with its shiny leaves and very attractive flowers and fruits can also be cultivated in hanging baskets or small ornamental areas.

Mikania scandens (L.) Willd.

Mitchella repens L.

White

Mitreola sessilifolia (J.F. Gmel.) G. Don
Miter-wort
[*Cynoctonum sessilifolium* (Walt.) J.F. Gmel.]
Loganiaceae, Logania Family

Habit: Herbaceous annual, 10-60 cm tall. Stems 4-angled, smooth, simple or branched above.

Leaves: Opposite; stalkless; blades ovate or elliptic, 0.5-2.5 cm long, margins smooth.

Inflorescences: Terminal, flowers along 1 side of spike-like, downward curving branches.

Flowers: White, tiny, with 5 petals.

Fruit: A capsule, to 3 mm long.

Habitat and Distribution: Frequent; wet habitats on the Coastal Plain; throughout Florida, west to Texas, and north to southeast Virginia.

Comment: Hardly an ornamental, the tiny-flowered Miter-wort blooms from summer into fall. The flowers are so small that even in reasonably dense populations the plant appears weedy.

Monarda punctata L.
Horsemint, Spotted Beebalm, Dotted-mint
Lamiaceae (Labiatae), Mint Family

Habit: Herbaceous perennial, biennial or annual, 0.3-1 m tall. Stems hairy, square, leafy.

Leaves: Opposite; stalked; blades lance-shaped to somewhat oblong, 2-9 cm long and 0.5-1.7 cm wide, margins with teeth.

Inflorescences: Clustered at stem tips or around upper leaf axils.

Flowers: White or yellow with pink spots, 1.5-2 cm long, 2-lipped, upper lip strongly arching, with whorl of pink, pointed bracts below flowers.

Fruit: 4 1-seeded nutlets, smooth, to 1.5 mm long.

Habitat and Distribution: Frequent; dry, sandy soils of flatwoods, fields, woods, and disturbed sites; central peninsula Florida, west to Texas, north to Missouri, Illinois, and Minnesota, and east to Vermont and New York.

Comment: Horsemint is a showy wildflower. Flowering from summer into fall, it always provokes comments. This small leafy shrub-like plant is striking in masses and easily cultivated.

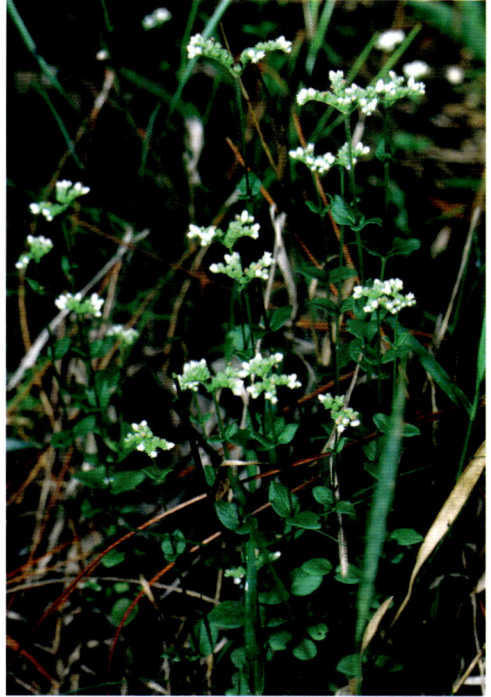

Mitreola sessilifolia
(J.F. Gmel.) G. Don

Monarda punctata L.

White

Monotropa uniflora L.
Indian Pipe
Ericaceae, Heath Family

Habit: Herbaceous perennial, to 20 cm tall, lacking chlorophyll. Stems erect, thick, fleshy, pale, often emerging in small clump, usually emerge white (occasionally pink), turn pink in age, and dry black.

Leaves: Alternate; stalkless; blades small, scale-like along stems.

Inflorescences: Solitary, terminal, nodding at first, become erect.

Flowers: White or pale pink, broadly tubular, 1.5-3 cm long.

Fruit: A capsule, 0.8-1.3 cm long.

Habitat and Distribution: Occasional to locally frequent; sandy to rich woods southward, rich woods northward; from Newfoundland to British Columbia, south through the United States to central peninsula Florida and Texas; Mexico south into Columbia.

Comment: Emerging and flowering in summer and fall, this strange-looking plant is parasitic on tree roots and never fails to provoke comment. The blackened drying plants can often be seen for several months after flowering. The entire plant resembles a pipe used by the original residents of this continent, therefore the common name.

Morus rubra L.
Red Mulberry
Moraceae, Mulberry Family

Habit: Small to large tree, 20-30 m tall.

Leaves: Alternate; stalked; blades rounded in general shape, lobed or unlobed, 4-15 cm long, margin with teeth, rough hairy on top, soft hairy underneath especially along veins.

Inflorescences: Cylindrical, pendulant spikes, either axillary or from just below the axil of the lowest leaf, male and female flowers in separate spikes, either on the same or separate trees.

Flowers: Whitish green, minute, male spikes 2-5 cm long, female spikes 1-2.5 cm long.

Fruit: An aggregate of fleshy nutlets, dull red to purple.

Habitat and Distribution: Frequent; rich woodlands, hammocks, floodplain forests; throughout Florida, west to central Texas, north to Kansas, Nebraska, and South Dakota, and east to Vermont.

Comment: Flowers in spring are followed by edible fruit in the late spring or early summer. The tree is used as an ornamental, more so than for fruit. Birds are very attracted to the fruit, widely spreading the seeds.

Monotropa uniflora L.

Morus rubra L.

White

Nemophila aphylla (L.) Brummitt
Baby Blue-eyes
[*Nemophila microcalyx* (Nutt.) Fisch. & C.A. Mey.; *Nemophila triloba* (Raf.) Thieret]
Hydrophyllaceae, Waterleaf Family

Habit: Herbaceous annual, 10-40 cm long or tall. Stems weak, branched.

Leaves: Alternate; long-stalked; blades once divided with 3 or 5 leaflets; leaflets rounded, often with teeth or lobes.

Inflorescences: Long-stalked, solitary flowers, from a stem node.

Flowers: White or bluish, small, to 5 mm wide.

Fruit: A capsule, 3-5 mm wide, hairy, rounded.

Habitat and Distribution: Infrequent; wet rich woods, floodplain forests, swamps; central panhandle Florida, west to east Texas, north to Arkansas and Missouri, and east to Kentucky and Virginia.

Comment: The spring flowers of this small diffuse plant are not easily seen. The weak spreading stem positions the flowers near the ground.

Nothoscordum bivalve (L.) Britton
False-garlic
[*Allium bivalve* (L.) Kuntze]
Amaryllidaceae, Amaryllis Family

Habit: Herbaceous perennial, 10-55 cm tall, from a bulb covered by a membrane.

Leaves: Basal; stalkless; blades slender, grass-like, 10-50 cm long and 2-8 mm wide.

Inflorescences: 6-12, in a terminal cluster on a leafless stalk.

Flowers: Whitish, with a purplish band on the outer midrib, 0.8-1.3 cm long, 6-parted.

Fruit: A capsule, oval, 4-7 mm long, 3-sided.

Habitat and Distribution: Occasional; grassy sites, roadsides, pastures, rocky outcrops, prairies, and disturbed areas; central peninsula Florida, north to Virginia, Ohio, Indiana, Illinois, and Nebraska, west to Texas; Mexico.

Comment: Flowering intermittently from spring into fall, the flowers are attractive. The plants are often noticed in grassy areas. Frequently confused with Wild Onion, False-garlic is easily separated by the lack of an odor.

Nemophila aphylla (L.) Brummitt

Nothoscordum bivalve (L.) Britton

White

Nymphaea odorata Ait.
White Water-lily or Fragrant Water-lily
Nymphaeaceae, Water-lily Family

Habit: Herbaceous aquatic perennial. Stem running underground.

Leaves: Alternate; very long-stalked; blades rounded, cleft with pointed lobes at base, 15-30 cm diameter, green on top and usually purple beneath.

Inflorescences: Solitary flowers, terminal on very long stalks.

Flowers: White or pinkish, 8-20 cm diameter, with many tapering petals and stamens.

Fruit: Round, 2-3 cm diameter, berry-like, with many seeds, stalks bend down after pollination and mature near the bottom.

Habitat and Distribution: Common to frequent; slow moving or standing fresh water ponds, lakes, streams, canals, and ditches; throughout Florida, west to southeast Texas, and north to Minnesota and Michigan, east to Newfoundland and Manitoba.

Comment: The large showy flowers can be spectacular when in mass. Individual flowers usually remain open for no longer than three days. Flowering spring through summer White Water-lily can be easily propagated by seeds or fragments of the rhizome (stem). A number of named cultivated forms are listed.

Nymphoides aquatica (Walt. ex J.F. Gmel.) O. Kuntze
Floating-heart, Banana-lily, Banana Plant
Menyanthaceae, Bogbean Family

Habit: Herbaceous aquatic perennial. Stem from underground runner (rhizome), thick, stocky.

Leaves: Alternate; long leaf stalks from rhizome, shorter leaf stalks from thin stem; blades rounded with cleft at base, to 20 cm long, to 15 cm wide, green above, purple and rough beneath with dots.

Inflorescences: Flowers on long stalks from leaf axils, in a cluster on secondary stalks, each secondary stalk from a common point, stalks sometimes bear a cluster of short tuberous roots near the top.

Flowers: White, 1-1.4 cm across, with 5 petals.

Fruit: A capsule, 5-14 mm long.

Habitat and Distribution: Frequent and locally common; quiet or standing water, ponds, ditches, canals, slow streams, swamps; throughout Florida, west to east Texas, and north to New Jersey.

Comment: The clusters of flowers are found from spring through summer. Floating-heart is a showy wildflower and is cultivated in aquariums. The tuberous roots just below the leaf blades are an oddity. Like White Water-lily the water surface can be densely covered with the flowers. The dense flowers sometimes look like white flakes strewn across the water.

Nymphaea odorata Ait.

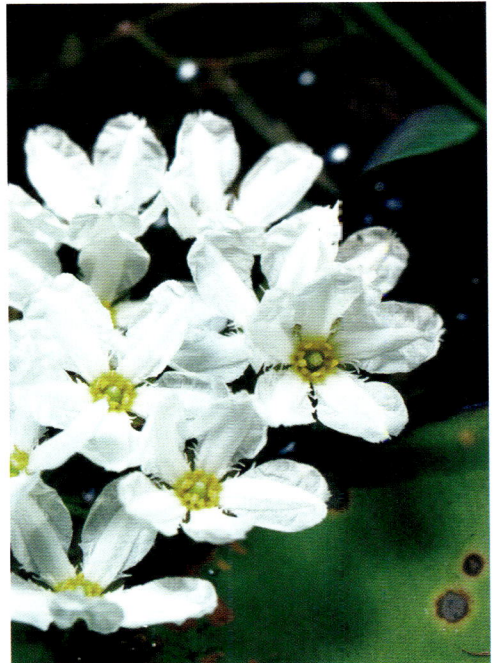

Nymphoides aquatica
(Walt. ex J.F. Gmel.) O. Kuntze

White

Oeceoclades maculata (Lindl.) Lindl.
Mottled Orchid or African Spotted Orchid
Orchidaceae, Orchid Family

Habit: Herbaceous perennial, to 40 cm tall. Stem a pseudobulb.

Leaves: 1 from pseudobulb; stalkless; leathery, strap-shaped, to 25 cm long and 5 cm wide, with silver spots mottled over surface.

Inflorescences: 5-15 flowers in a terminal, spike-like, loose arrangement.

Flowers: White with purple markings, not always opening.

Fruit: A capsule.

Habitat and Distribution: Rare, but locally frequent; dry hammocks; central and south peninsula Florida; West Indies, Central America, and native to Africa.

Comment: The flowers in summer and winter are attractive but not showy. The leaves are strikingly showy. Mottled Orchid is spreading northward.

Orobanche uniflora L.
Broomrape or Cancer Root
Orbanchaceae, Broomrape Family

Habit: Herbaceous perennial, 1-14 cm tall, parasitic on roots of several species of host plants. Stem often below surface, scaly, branched, bluish white.

Leaves: Alternate; stalkless; blades brown, scale-like, overlapping, at base of flowering stalk.

Inflorescences: Flowers solitary, terminal on long, hairy, flowering scapes.

Flowers: Whitish, 1.2-2.5 cm long, tubular, with 5 lobes.

Fruit: A capsule, 6-12 mm long.

Habitat and Distribution: Occasional to rare; rich, moist, deciduous woods; central peninsula Florida, throughout the United States, and north to New Brunswick and the Yukon.

Comment: The small almost invisible plant has a visible flower in spring. The flower almost seems to come from the ground. The flower is difficult to see.

White

Oeceoclades maculata (Lindl.)
Lindl.

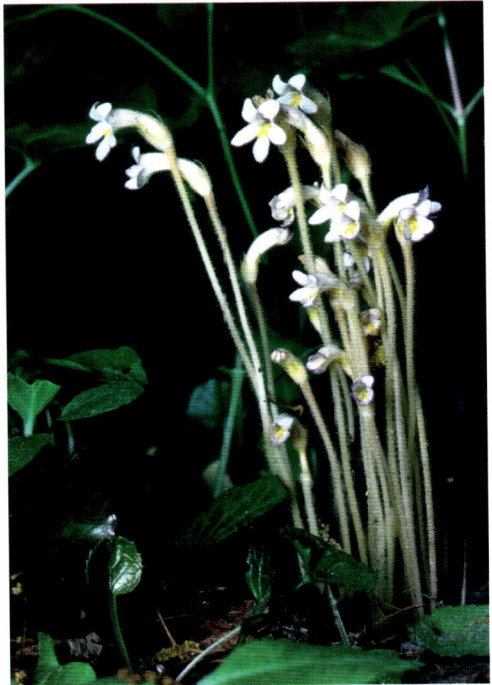

Orobanche uniflora L.

637

White

Osmanthus americanus (L.) Benth. & Hook. f. ex A. Gray
Wild-olive or Devil-wood
Oleaceae, Olive Family

Habit: Shrub or small tree, to 15 m tall. Bark pale.

Leaves: Opposite, evergreen; stalked; blades elliptic to lance-shaped to broader towards tip, 5-18 cm long, shiny, leathery.

Inflorescences: Axillary clusters.

Flowers: White, to 4 mm long, bell-shaped, with 4 petals.

Fruit: A drupe, oval to round, to 1.5 cm diameter, blue or purple.

Habitat and Distribution: Frequent; usually in dry wooded habitats, occasionally in wet forests; central peninsula Florida, west into Louisiana, north into southeast Virginia; Mexico.

Comment: The fragrant flowers can be found in spring, but are not showy. The shiny, dark green, evergreen leaves make this small tree valued as a cultivated specimen plant. Seeds do not germinate until the second year.

Oxydendrum arboreum (L.) DC.
Sourwood
Ericaceae, Heath Family

Habit: Medium to large tree, to 20 m tall. Branches smooth.

Leaves: Alternate; stalked; blades elliptic, 8-20 cm long and 2.5-7 cm broad, deciduous, with finely toothed edges.

Inflorescences: Upward turned branches of distinctive long-branched clusters from the axils/leaf scars of last year's leaves.

Flowers: White, urn-shaped, 5-8 mm long, 5-lobed.

Fruit: A capsule, oval, 5-7 mm long, 5-chambered, pendulent.

Habitat and Distribution: Occasional and scattered; wooded slopes, bluffs, ravines, hills, and stream banks; panhandle Florida north into New Jersey, Pennsylvania, Ohio, Indiana, and Illinois, west into Louisiana.

Comment: Flowering in spring, the downward sweeping clusters are quite distinctive. These clusters are no less attractive when dry, remaining on the tree for some time and are popular in dried flower arrangements. A prominent feature, in addition to the flowers, is the orange and red fall leaf color.

Osmanthus americanus (L.) Benth. & Hook. f. ex A. Gray

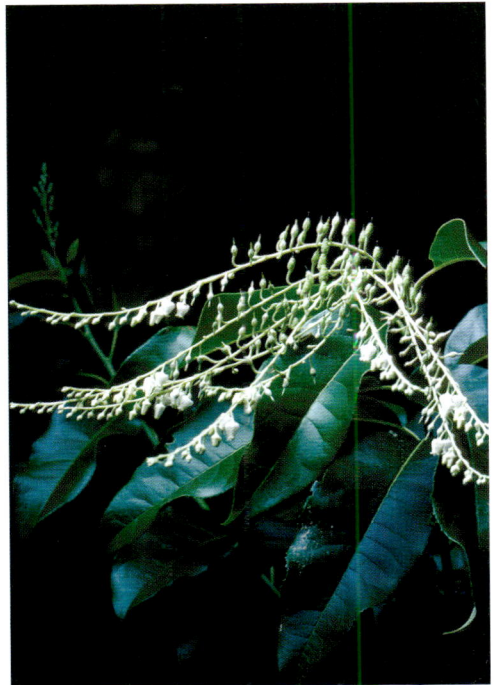

Oxydendrum arboreum (L.) DC.

639

White

Oxypolis filiformis (Walter) Britton
Water Dropwort
Apiaceae (Umbelliferae), Carrot Family

Habit: Herbaceous perennial, 0.5-1.8 m tall, from clustered, thickened roots. Stems erect, smooth, hollow, segmented, branched towards top.

Leaves: Alternate; stalk not evident; blades slender and round, 3-60 cm long, hollow.

Inflorescences: Terminal, long-stalked, domed-shaped clusters of 6-17 flowers.

Flowers: White, each 1.5 mm wide.

Fruit: Flattened, 4-7 mm long, with broad wings and ribs.

Habitat and Distribution: Common; open wet sites, ditches, ponds, and swamp margins; throughout Florida, west to Texas, and north to North Carolina; Bahama Islands, Cuba.

Comment: The flowers in summer and fall are attractive but not showy in domed masses. Always generating comment is the distinctive segmented stem and round leaves.

Oxypolis ternata (Nutt.) Heller
Piedmont Cowbane
Apiaceae (Umbelliferae), Carrot Family

Habit: Herbaceous perennial, 50-90 cm tall, from fibrous roots with tubers at the tips.

Leaves: Alternate; long-stalked; blades slender, 20 cm long and to 3 mm wide, simple or often dissected into 3 leaflets.

Inflorescences: Terminal, long-stalked, dome-shaped clusters of 8-12 flowers.

Flowers: White, each 3 mm wide.

Fruit: Flattened, 3-5 mm long, with broad wings and ribs.

Habitat and Distribution: Infrequent to occasional; low, wet sites, wet flatwoods, bogs; panhandle Florida north to southeast Virginia.

Comment: This thin, fall flowering, water loving plant is scarcely noticed, even in flower, as it blends into the other wetland foliage. The flower clusters and thin leaves are distinct features for identification.

Oxypolis filiformis
(Walter) Britton

Oxypolis ternata (Nutt.) Heller

White

Palafoxia feayi A. Gray
Feay's Palafoxia
Asteraceae (Compositae), Aster or Sunflower Family

Habit: Herbaceous or shrubby perennial, 1-3 m tall. Stems slender, branching, hairy, and herbaceous above, woody below.

Leaves: Upper leaves alternate, lower leaves opposite; short-stalked; blades elliptic, 2-7 cm long, evergreen, with rough hairs.

Inflorescences: Heads scattered towards or at the tops of stems, sometimes in flat-topped clusters, only disc flowers present.

Flowers: White, pink, or lavender, 17-30 flowers per head, each 3-6 mm long.

Fruit: An achene, 5-6 mm long, with scales on top.

Habitat and Distribution: Frequent; dry woods, sandhills, and scrub; central and south peninsula Florida.

Comment: Feay's Palafoxia is endemic in Florida. This slender plant blooms in fall and can have many flowers.

Palafoxia integrifolia (Nutt.) Torr. & A. Gray
Many Wings or Polypteris
[*Polypteris integrifolia* Nutt.]
Asteraceae (Compositae), Aster or Sunflower Family

Habit: Herbaceous or shrubby perennial, to 1.5 m tall. Stems slender, branching, hairy and herbaceous above, woody below.

Leaves: Upper leaves alternate, lower leaves opposite; short-stalked; blades lance-shaped to linear, 3-7 cm long, hairy.

Inflorescences: Heads scattered towards or at the tops of stems, sometimes in flat-topped clusters, only disc flowers present.

Flowers: White or pink-tinged, 13-26 flowers per head, each 9-14 mm long.

Fruit: An achene, 4-5 mm long, with scales on top.

Habitat and Distribution: Frequent; dry pine and oak woods, sandhills; peninsula to central panhandle Florida and adjacent Georgia.

Comment: Many Wings is a fall flowering, tall, slender wildflower restricted to sandhill habitats.

Palafoxia feayi A. Gray

Palafoxia integrifolia (Nutt.) Torr. & A. Gray

White

Parnassia caroliniana Michx.
Carolina Grass of Parnassus or Coastal Grass of Parnassus
Saxifragaceae, Saxifrage Family [Parnassiaceae, Grass-of-Parnassus Family]

Habit: Herbaceous perennial, 20-50 cm tall.

Leaves: Alternate, mostly basal; long-stalked; blades rounded, 2-6 cm long and wide.

Inflorescences: Terminal, very long-stalked, solitary flowers.

Flowers: White tinged with green, 5 petals, 1.5-2 cm long, with 9-19 visible veins.

Fruit: A capsule, 1-1.5 cm long.

Habitat and Distribution: Rare; swamps, wet flatwoods, bogs; eastern panhandle Florida to North Carolina.

Comment: This rare plant blooms in the fall. The flowers on thin erect stalks are a delicate white and somewhat difficult to see mixed against the other herbaceous species. The rounded basal rosette of leaves with flowers above are very attractive. Propagation is by seeds or division. It is listed by the State of Florida as Endangered.

Parnassia grandifolia DC.
Grass-of-Parnassus or Undine
Saxifragaceae, Saxifrage Family [Parnassiaceae, Grass-of-Parnassus Family]

Habit: Herbaceous perennial, 10-60 cm tall.

Leaves: Alternate, mostly basal; long-stalked; blades oval to nearly rounded, often kidney-shaped, 3-8 cm long and nearly as wide.

Inflorescences: Terminal, very long-stalked, solitary flowers.

Flowers: White tinged with green, 5 petals 1.2-2.5 cm long, with 7-9 visible veins.

Fruit: A capsule, 1.5 cm long.

Habitat and Distribution: Rare; moist woods, seepage areas, wet marls, boggy woods; Marion and Putnam Counties in northern peninsula and Franklin and Liberty Counties in eastern Panhandle Florida, west to eastern Texas, north to Missouri, West Virginia, and Virginia.

Comment: Grass-of-Parnassus flowers in fall. As a wildflower it is distinct when found, but quite rare. It is cultivated and can be planted in damp locations. Propagation is by seeds or divisions. Grass-of-Parnassus is listed by the State of Florida as Endangered.

Parnassia caroliniana Michx.

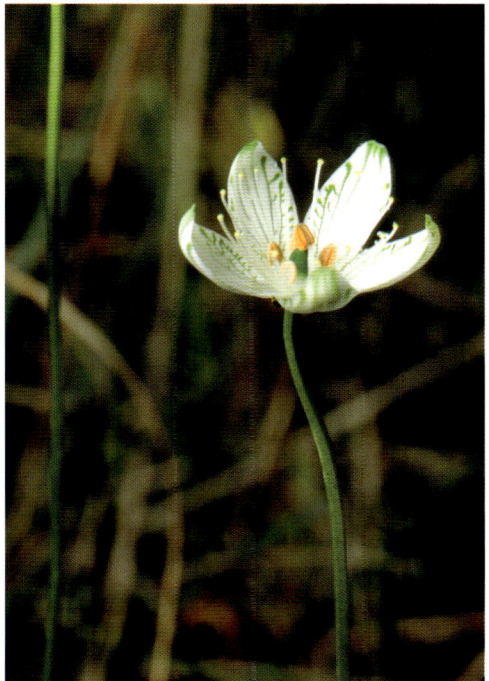

Parnassia grandifolia DC.

White

Paronychia baldwinii (Torr. & A. Gray) Fenzl ex Walp.
Baldwin's Whitlow-wort, Baldwin's Nailwort, Southern Whitlow-wort
[*Anychiastrum baldwinii* (Torr. & A. Gray) Small; *Anychiastrum riparium* (Chapm.) Small]
Caryophyllaceae, Pink Family

Habit: Herbaceous perennial, to 80 cm long, from a taproot. Stems freely branching, trailing.

Leaves: Opposite; stalkless; blades elliptic, 5-20 mm long, margins smooth.

Inflorescences: Loose clusters from leaf axils.

Flowers: No petals, 5 white sepals, to 1 mm long.

Fruit: A bladder-like utricle, thin-walled, round, to 0.4 mm long.

Habitat and Distribution: Frequent; dry sandy soils of scrubs, sandhills, dunes, river banks; central peninsula Florida, west to Louisiana, and north into southeast Virginia.

Comment: As a wildflower Baldwin's Whitlow-wort's thin trailing stems are more of a curiosity. The flowers in most of the warm months are so small that even in clusters they are difficult recognize.

Paronychia patula Shinners
Spreading Whitlow-wort or Pineland Nailwort
[*Siphonychia diffusa* Chapm.]
Caryophyllaceae, Pink Family

Habit: Herbaceous annual or biennial, 10-60 cm long. Stems lying flat, branched, minutely hairy.

Leaves: Opposite; stalkless; blades hairy, linear to somewhat broader near the tip, 5-25 mm long, margins smooth.

Inflorescences: Dense clusters from leaf axils.

Flowers: No petals, 5 white sepals, hairy, to 1.5 mm long with firm pointed tips.

Fruit: A bladder-like utricle, thin-walled, to 1 mm long.

Habitat and Distribution: Infrequent to occasional; dry sites, scrubs, sandhills; central peninsula Florida, west to Louisiana, and north to Georgia.

Comment: The thin spreading stems of Spreading Whitlow-wort blend with the sandy surroundings. Even when blooming in summer the dense clusters of small flowers are scarcely noticeable.

White

Paronychia baldwinii
(Torr. & A. Gray) Fenzl ex Walp.

Paronychia patula Shinners

White

Paronychia rugelii *(*Chapm.) Shuttlew. ex Chapm.
Sand Squares or Rugel's Nailwort
[*Gibbesia rugelii* (Chapm.) Small]
Caryophyllaceae, Pink Family

Habit: Herbaceous annual or biennial, 10-50 cm tall. Stems erect, branching, minutely hairy.

Leaves: Opposite; stalkless; blades hairy, linear to somewhat broader near the tip, 1-3 cm long, margins smooth.

Inflorescences: Small dense clusters from leaf axils.

Flowers: No petals, narrow pinkish sepals, to 2 mm long with a thin point at tip.

Fruit: A bladder-like utricle, thin-walled, less than 1 mm long.

Habitat and Distribution: Occasional; dry sites, sandhills, xeric flatwoods, old fields; central peninsula Florida, north to Alabama and Georgia.

Comment: Summer flowering, this delicate plant sticks upright from the sand. The entire plant appears to be squarish from above, hence the name, Sand Square. The square appearance is frequently noted as out of the ordinary by passersby.

Parthenocissus quinquefolia (L.) Planch.
Virginia Creeper or Woodbine
Vitaceae, Grape Family

Habit: Perennial woody vine, high climbing to well over 80 feet. Stems climbing or creeping by means of branched tendrils.

Leaves: Alternate; long-stalked; blades divided into 5 or 7 (sometimes 3) elliptic to oval leaflets, each with a short stalk, 6-15 cm long, sometimes hairy underneath, margins with teeth towards the tip.

Inflorescences: Large stalked clusters, terminal or opposite the leaves.

Flowers: Yellowish green, 5 petals, 2-3 mm long.

Fruit: A berry-like drupe, round, 5-9 mm diameter, blue or black, 1- to 3-seeded.

Habitat and Distribution: Common; virtually all but the wettest and driest habitats; throughout Florida, west to eastern Texas, and north to Nebraska, Ontario, and Maine.

Comment: Flowering is from spring into summer. Fruiting is from late spring into fall. Virginia Creeper is frequently used as an ornamental and has several named varieties. It is used to cover walls, fences, and arbors. As a ground cover the growth habit in the open is quite flat. Sometimes it is used to cover the lower woody portions of large trees. In the fall the foliage turns various shades of orange and red. Propagation can be by seeds or vegetatively by cuttings. It can become a troublesome weed.

Paronychia rugelii (Chapm.) Shuttlew. ex Chapm.

Parthenocissus quinquefolia (L.) Planch.

White

Penstemon multiflorus (Benth.) Chapm. ex Small
White Beard's-tongue or Many-flower Beard's-tongue
Scrophulariaceae, Snapdragon or Figwort Family [Veronicaceae, Speedwell Family]

Habit: Herbaceous perennial, 0.8-1.5 m tall, from basal rosette. Stems smooth.

Leaves: Opposite; stalked on the lower stem, stalkless towards the top; blades lance-shaped and wider towards the tip, to 20 cm long, margins sometimes toothed and wavy.

Inflorescences: Nodding, in successive whorls at stem tips.

Flowers: White, tubular, to 2.2 cm long, 5-lobed.

Fruit: A capsule, 7-9 mm long.

Habitat and Distribution: Frequent; dry habitats, scrubs, sandhills, xeric pinelands; throughout Florida and north to southern Georgia.

Comment: This member of the Snapdragon Family likes lots of sun or it will get leggy, but plant supports placed over the basal rosettes early in the year will help. It can bloom more than once during its blooming season from spring into fall. Cut flowers will last. As a wildflower it is very noticeable. The many flowers stay open for a long period and contribute to the summer vistas.

Photinia pyrifolia (Lam.) K.R. Robertson & J.B. Phipps
Red Chokeberry
[*Aronia arbutifolia* (L.) Ell.; *Pyrus arbutifolia* (L.) L. f.; *Sorbus arbutifolia* (L.) Heynh.]
Rosaceae, Rose Family

Habit: Perennial shrub, to 3.5 m tall, often from underground runners and forming large thickets. Stems thin, thornless, newer branches hairy.

Leaves: Alternate, deciduous; short-stalked; blades elliptic to ovate, 3-10 cm long, densely hairy underneath, with fine reddish gland-tipped teeth on margins.

Inflorescences: Dome-like clusters on twig tips.

Flowers: White, to 1.0 cm wide, with 5 roundish petals and numerous stamens.

Fruit: An apple-like pome, red, smooth, to 6-9 mm in diameter, dried flower parts at tip.

Habitat and Distribution: Frequent; moist to wet soils, along stream and pond margins, pinelands, bogs, and seepage sites; central peninsula Florida, west to Texas, and north to Arkansas, Kentucky, Pennsylvania, New York, Nova Scotia, and Newfoundland.

Comment: This willowy shrub blooms in spring. The white flower clusters are quite attractive. Perhaps more noticeable are the red fruits appearing in summer and persisting into the winter. The leaves with reddish margins add to the attractiveness.

Penstemon multiflorus
(Benth.) Chapm. ex Small

Photinia pyrifolia (Lam.) K.R. Robertson & J.B. Phipps

White

Physostegia godfreyi Cantino
Godfrey's False Dragon-head or Apalachicola Dragon-head
Lamiaceae (Labiatae), Mint Family

Habit: Herbaceous perennial, 0.6-1.0 m tall, from a rhizome. Stem erect, unbranched or with a few branches.

Leaves: Opposite; stalked on lower part of stem, stalkless upwards on stem; blades elliptic, sometimes broader towards tip, 3-6.5 cm long, margins smooth, wavy, or occasionally with a few teeth.

Inflorescences: Opposite or whorled, in long spike-like arrangements at stem tips.

Flowers: White to pale rose, 1-2 cm long, tubular, with purple veins, and purple splotches.

Fruit: Composed of four nutlets, 1.7-2 mm long, three-sided, warty.

Habitat and Distribution: Occasional; wet pine flatwoods, bogs, swamps, savannas, ditches; Bay, Calhoun, Franklin, Gulf, Liberty, and Walton Counties in the central panhandle of Florida; endemic.

Comment: The flowers found in summer can be extremely attractive due to the long spire of blooms. Godfrey's False Dragon-head can be used in a wild garden in a moist to wet area. It can be propagated by the rhizome or by seeds.

Pieris phillyreifolia (Hook.) DC.
Vine-wicky
Ericaceae, Heath Family

Habit: Perennial woody shrub, to 1 m, or vine, to 7 m, occasionally from underground runners. Stems erect or climbing upwards under bark of Pond-cypress or Atlantic White Cedar; branches erupting from beneath bark to produce leafy collections.

Leaves: Alternate; very short-stalked; blades elliptic or somewhat oblong or lance-shaped, leathery, evergreen, 1-7 cm long, margins rolled under, smooth or with a few blunt teeth at tip.

Inflorescences: 3- to 9-flowered arrays from leaf axils near stem tips.

Flowers: White, urn-shaped, 7-9 mm long, with tiny lobes.

Fruit: A capsule, somewhat rounded, 5-parted.

Habitat and Distribution: Occasional, but locally common; cypress ponds, white cedar swamps; from central peninsula Florida, north to South Carolina, and west to southern Alabama.

Comment: This strange vine can sometimes be found growing as a shrub. It can be seen blooming from winter into spring and the sprays of flowers are pretty. The strange viney habit of growing under soft bark with the bunches of leafy emergent branches is an unusual sight.

Physostegia godfreyi Cantino

Pieris phillyreifolia (Hook.) DC.

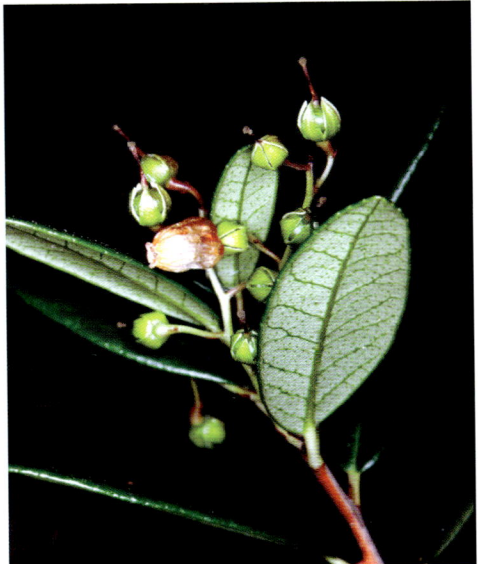

White

Pinguicula ionantha Godfrey
Godfrey's Butterwort or Panhandle Butterwort
Lentibulariaceae, Bladderwort Family

Habit: Herbaceous perennial, to 15 cm tall, often submersed.

Leaves: Alternate, in basal rosette; stalkless; blades oblong, to 8 cm long, upper surface covered with short glandular hairs, margins smooth and rolled upward.

Inflorescences: Solitary, terminal flowers on slender, leafless stalks covered with short, glandular hairs.

Flowers: White with purple throat, to 2 cm wide, 5 notched lobes, with spur to 5 mm long at base.

Fruit: A capsule, rounded, about 5 mm in diameter.

Habitat and Distribution: Rare; wet soils, bogs, flatwoods depressions, ditches, canals, shallow water; Bay, Franklin, Gulf, Liberty, and Wakulla Counties in eastern panhandle Florida; endemic.

Comment: The large flowers can be seen in spring. When part of a large population the effect can be dramatic. It is listed by Florida as Endangered and also by the US Fish and Wildlife Service as Threatened.

Pinguicula planifolia Chapm.
Chapman's Butterwort
Lentibulariaceae, Bladderwort Family

Habit: Herbaceous perennial, to 25 cm tall.

Leaves: Alternate, in basal rosette; stalkless; blades oblong to elliptic, to 8 cm long, upper surface covered with short glandular hairs, flat and often reddish.

Inflorescences: Solitary, terminal flowers on slender, leafless stalks covered with short, glandular hairs.

Flowers: Whitish to violet, to 3 cm wide, with 5 deeply notched lobes, with spur to 4 mm long at base.

Fruit: A capsule, rounded, to 5 mm wide.

Habitat and Distribution: Occasional; wet soils, shallow standing water, bogs, flatwoods, ditches, canals; panhandle Florida to southeastern Mississippi.

Comment: Chapman's Butterwort flowers in winter and spring. These large flowers are easily found in wet areas.

White

Pinguicula ionantha Godfrey

Pinguicula planifolia Chapm.
photographs courtesy of John Tobe

White

Pinguicula pumila Michx.
Small Butterwort
Lentibulariaceae, Bladderwort Family

Habit: Herbaceous perennial, to 10 cm tall.

Leaves: Alternate, in basal rosette; stalkless; blades' upper surface covered with glandular hairs, somewhat rounded and nearly oblong, 1-2.5 cm long, margins rolled upwards.

Inflorescences: Solitary, terminal flowers on slender, leafless stalks.

Flowers: Whitish to pink and purple, or rarely yellow, 1-1.5 cm wide, with 5 rounded and gently notched petals, with spur to 5 mm long at base.

Fruit: A capsule, rounded, 4-5 mm wide.

Habitat and Distribution: Frequent; moist, acidic soils - flatwoods, pastures, ditches; throughout Florida, west to southeast Texas, and north to southeast North Carolina; Bahamas.

Comment: Found blooming in winter and spring the Small Butterwort can be very showy in spite of the small flowers. Frequently growing in large masses in wet pastures the ground appears to be covered with flowers.

Platanthera blephariglottis (Willd.) Lindl.
White-fringed Orchid
[*Habenaria blephariglottis* (Willd.) Lindl., misapplied]
Orchidaceae, Orchid Family

Habit: Herbaceous perennial, 0.3-1 m tall, from fleshy roots. Stem erect, leafy.

Leaves: Alternate; stalkless; blades 2-4, slender, lance-shaped, 5-40 cm long and 1-5 cm wide.

Inflorescences: Many flowers in a spike-like arrangement at stem tip.

Flowers: White, 3 cm long, with main petal lobed and fringed, and slender spur 3-4 cm long.

Fruit: A capsule, ellipsoid, 2 cm long.

Habitat and Distribution: Occasional; open, moist to wet pine flatwoods, roadsides, and roadside ditches; central peninsula Florida, west to Texas, and north to North Carolina.

Comment: A spectacular wildflower that blooms in summer and early fall. Typical pine flatwoods habitats that are burned regularly, a necessity once common, are difficult to find.

Pinguicula pumila Michx.

Platanthera blephariglottis
(Willd.) Lindl.

White

Platanthera flava (L.) Lindl.
Southern Tubercled Orchid
[*Habenaria flava* (L.) R. Br. ex Spreng.]
Orchidaceae, Orchid Family

Habit: Herbaceous perennial, 10-60 cm tall, from fleshy roots. Stem erect, leafy.

Leaves: Alternate; stalkless; blades 1-4, slender, lance-shaped, 5-25 cm long and 1-4 cm wide.

Inflorescences: 10-40 flowers in a spike-like arrangement at stem tip.

Flowers: White to pale yellow or greenish, to 7 mm long, lip with prominent tubercle, and slender spur 8 mm long.

Fruit: A capsule, ellipsoid, to 8 mm long.

Habitat and Distribution: Occasional; wet hammocks, swamps, floodplains, roadside ditches, and seeps; central peninsula Florida, west to Texas and north to Missouri, Nova Scotia, and Maryland.

Comment: Most species in this genus are spectacular in flower. Southern Tubercled Orchid is noticeable when blooming in spring into early summer, but not much of a wildflower.

Platanthera nivea (Nutt.) Lindl.
Snowy Orchid
[*Habenaria nivea* (Nutt.) Spreng.]
Orchidaceae, Orchid Family

Habit: Herbaceous perennial, 20-90 cm tall, from fleshy tuberous roots. Stems erect, slender, rigid.

Leaves: Alternate; stalkless; blades 2-3, lance-shaped, 5-26 cm long and to 3 cm wide.

Inflorescences: 20-50 flowers in a spike-like arrangement at stem tip.

Flowers: White, to 10 mm long, lip pointing upward, and slender spur horizontal or curved upward, to 1.6 cm long.

Fruit: A capsule, cylindrical, to 1.2 cm long.

Habitat and Distribution: Frequent; wet flatwoods, moist meadows, prairies, seeps, and bogs; throughout Florida, west to Arkansas and Texas, and north to Delaware and New Jersey.

Comment: The delightfully fragrant, late spring to early fall blooms are magnificent. The white color and many flowers are striking. It is commonly found in the acid soils where one would find sundews and sedges. Listed as Threatened by the State of Florida.

Platanthera flava (L.) Lindl.

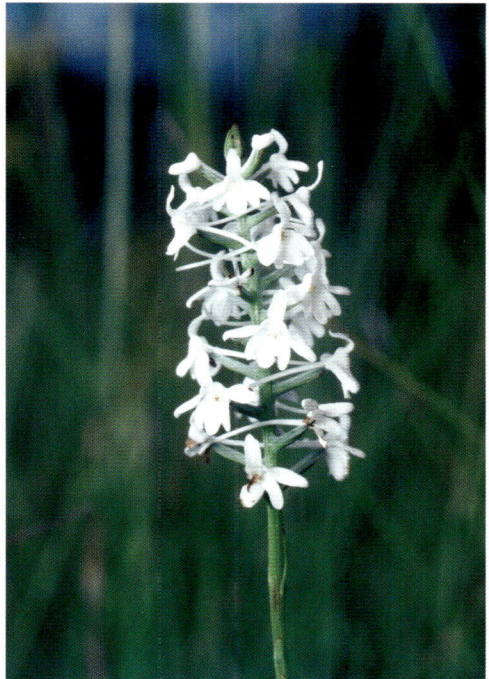

Platanthera nivea (Nutt.) Lindl.

White

Pluchea carolinensis (Jacq.) D. Don
Bushy Fleabane
[*Pluchea symphytifolia* (Mill.) Gillis, misapplied]
Asteraceae (Compositae), Aster or Sunflower Family

Habit: Perennial shrub, 1-4 m tall. Stems many-branched, glandular hairy.

Leaves: Alternate; stalked; blades elliptic to somewhat rounded, to 20 cm long, glandular and nearly smooth on top, densely hairy underneath, margins smooth or with small blunt teeth.

Inflorescences: Heads in large, long-stalked, flat-topped clusters at stem and branch tips, only disc flowers present.

Flowers: Pink to lavender, sometimes pale, heads 4.5-6 mm long and 5-10 mm wide.

Fruit: An achene, about 1 mm long, black, with bristles on top.

Habitat and Distribution: Occasional to frequent; hammocks, roadsides, and margins; west central and south peninsula Florida; West Indies, Bermuda, Mexico south into northern South America; Pacific Islands; native from the West Indies and Mexico into northern South America.

Comment: Bushy Fleabane blooms during the winter and early spring. Flowers are not very noticeable. The leaves produce a musty odor when bruised. The leaves are used medicinally.

Pluchea foetida (L.) DC.
White Stinking Fleabane
Asteraceae (Compositae), Aster or Sunflower Family

Habit: Herbaceous perennial, 0.3-1 m tall. Stems solitary or several, glandular, and cobwebby-hairy.

Leaves: Alternate; stalkless; blades oblong to elliptic to somewhat rounded, to 13 cm long, usually smooth and glandular on top and hairy underneath, margins toothed.

Inflorescences: Heads in usually long-stalked, flat-topped clusters from leaf axils, secondary clusters just below the terminal cluster of the main stem often extend higher, only disc flowers present.

Flowers: Creamy white, heads 5-8 mm high, 6-12 mm wide.

Fruit: An achene, to 1 mm long, pinkish, with bristles on top.

Habitat and Distribution: Frequent; low, wet sites, swamps, marshes, meadows, ponds, and ditches; throughout Florida, north to New Jersey, and west to Arkansas and Texas; West Indies.

Comment: The flowers of White Stinking Fleabane are most often seen in late summer into fall, and less frequently in any warm month southward. If the plants are brushed or touched a very distinctive odor emanates from the glands on the stems and leaves.

White

Pluchea carolinensis (Jacq.) D. Don

Pluchea foetida (L.) DC.

66666

White

Podophyllum peltatum L.
Mayapple or Mandrake
Berberidaceae, Barberry Family

Habit: Herbaceous perennial, 20-50 cm tall, from creeping rhizomes.

Leaves: 1 or 2, alternate; long-stalked; blades peltate (stalk attached like an umbrella, towards the center of the blade), rounded, 12-34 cm in diameter, divided into 5-9 segments, with toothed or lobed tips.

Inflorescences: A solitary flower, emerging at top of stem between the two leaves.

Flowers: White, nodding, 2-5 cm broad, with 6-9 firm broad petals.

Fruit: A berry, rounded, 3-5 cm long, red or yellow.

Habitat and Distribution: Rare in Florida, occasional to frequent elsewhere; deciduous woods, meadows, pastures, hammocks, and roadsides; Jackson County in central panhandle Florida, west to eastern Texas, north to Minnesota, Ontario, and Quebec.

Comment: Mayapple, often known as Mandrake, blooms in the early spring with fruit appearing in late spring. The pulp of the fruit is edible, but beware, the seeds and foliage are quite POISONOUS.

Pogonia ophioglossoides (L.) Ker-Gawl.
Rose Pogonia
Orchidaceae, Orchid Family

Habit: Herbaceous perennial, to 35, rarely 70 cm tall, from fibrous roots. Stem erect, smooth, single.

Leaves: Alternate; stalkless on mature plants, purple-stalked on seedlings; blades 1 or rarely 2, elliptic to somewhat rounded, 2-12 cm long, and 1-3 cm wide, sheathing the stem.

Inflorescences: 1-3 terminal flowers.

Flowers: White or usually rose, 2.5-4 cm across, lower lip with a white to purple bristled margin and yellow center.

Fruit: A capsule, ellipsoid, 2-3 cm long.

Habitat and Distribution: Frequent; bogs, marshes, prairies, seeps, roadside ditches, and wet pinelands; from central peninsula Florida, west to Texas, and north to Newfoundland and Manitoba.

Comment: Blooming in spring, the large bearded flowers have a leaf-like bract at their point of attachment with the stem that aids in identification. In wet pinelands they can be hard to spot unless the area has been recently burned. Sometimes the plants occur in dense clusters. Listed as Threatened by the State of Florida.

Podophyllum peltatum L.

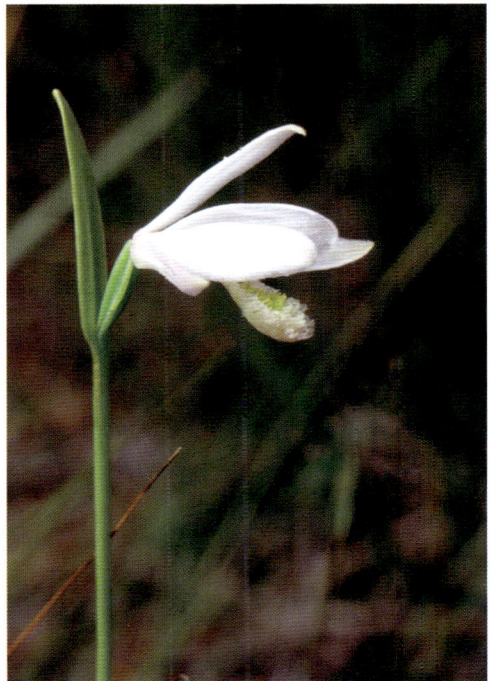

Pogonia ophioglossoides
(L.) Ker-Gawl.

White

Polanisia tenuifolia Torr. & A. Gray
Pineland Catchfly
[*Aldenella tenuifolia* (Torr. & A. Gray) Greene]
Capparaceae, Caper Family [Brassicaceae, Mustard Family]

Habit: Herbaceous annual, 20-80 cm tall.

Leaves: Alternate; stalked; blades 1-4 cm long, with 1, 2, or 3 leaflets, leaflets very slender or needle-like.

Inflorescences: Clusters at stem and branch tips.

Flowers: White, with 4 unequal petals, largest to 8 mm long.

Fruit: A capsule, very slender, 4-6 cm long.

Habitat and Distribution: Frequent; dry soils of sand hills, pinelands, scrubs; throughout Florida, west to Mississippi, and north to Georgia.

Comment: This peculiar thin plant is easily overlooked. The plants are clammy and sticky to the touch. The lopsided flowers are found in spring and summer.

Polygala balduinii Nutt.
White Bachelor's-button
[*Pilostaxis balduinii* (Nutt.) Small; *Pilostaxis carteri* (Small) Small]
Polygalaceae, Milkwort Family

Habit: Herbaceous annual or biennial, 10-70 cm tall. Stems erect, branched.

Leaves: Alternate; stalkless; blades at base broader at tip, those upwards elliptic, 0.5-2.5 cm long, basal leaves dying with age.

Inflorescences: Branched, roundish or cylindrical, 2-3 cm long clusters at top of stem.

Flowers: White to greenish, 3-4 mm long.

Fruit: A capsule, less than 1 mm wide.

Habitat and Distribution: Frequent; wet sites, bogs, marshes, swales, wet flatwoods, ditches, shores, marl prairies, swamps; throughout Florida, west to Mississippi, and north to Georgia; western Cuba.

Comment: Flowers are seen from spring into fall. This skinny plant is difficult to see when lacking flowers. When many plants are blooming in close proximity, especially along a shore, the effect of the white heads is picturesque.

White

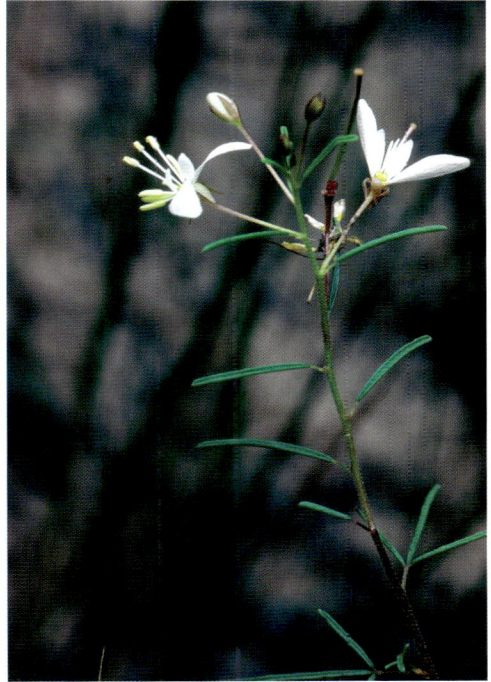

Polanisia tenuifolia Torr. &
A. Gray

Polygala balduinii Nutt.

White

Polygonella fimbriata (Elliott) Horton
Sandhill Wireweed
Polygonaceae, Buckwheat Family

Habit: Herbaceous to semi-woody perennial, 15-60 cm tall. Stems branching.

Leaves: Alternate, persistent; stalkless; blades very narrow, 1-3 cm long, sheathing at base, sheath fringed with hairs on margin.

Inflorescences: Spike-like clusters near or at stem tips.

Flowers: White to pink, tuft-like, sepals resemble petals, inner sepals fringed, to 2.5 mm long.

Fruit: Nutlets 3-sided, to 2 mm long and 1 mm wide.

Habitat and Distribution: Infrequent; sandhills and dry pinelands; Holmes County in panhandle Florida, north to Alabama and Georgia.

Comment: Sandhill Wireweed is a very attractive wildflower. Flowers occur in profusion in summer and fall.

Polygonella gracilis Meisn.
Wireweed
Polygonaceae, Buckwheat Family

Habit: Herbaceous annual, 0.3-1.7 m tall. Stems slender, branching.

Leaves: Alternate, shed early; stalkless; blades narrow, wider toward tips, 2-6 cm long, sheathing at base, sheath smooth on margin.

Inflorescences: Spike-like clusters.

Flowers: White or pink, sepals resemble petals, inner sepals smooth, to 2 mm long.

Fruit: Nutlets 3-sided, to 2.2 mm long and 1-1.3 mm wide.

Habitat and Distribution: Frequent; dry pinelands, dunes, scrubs, and sandhills; throughout Florida, west to Louisiana, and north to South Carolina.

Comment: Wireweed is so-called due to the slender stems and tendency to drop its leaves early in the summer. Flowering occurs from summer into fall. The tall skinny plants and white flowers are frequently noticed in spite of their slender nature.

Polygonella fimbriata
(Elliott) Horton

Polygonella gracilis Meisn.

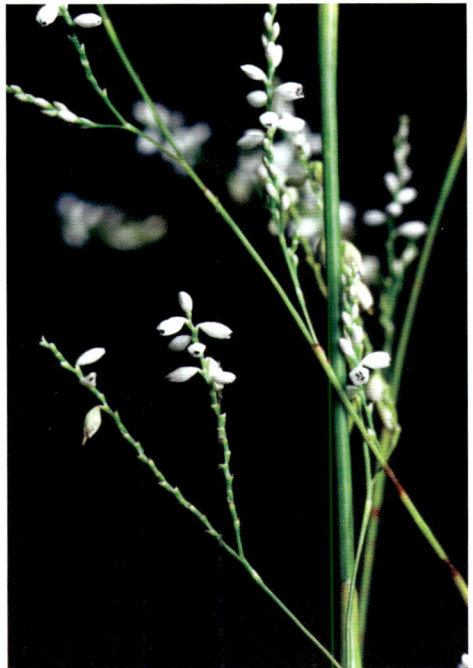

White

Polygonum hydropiperoides Michx.
Mild Water-pepper or False Water-pepper
Polygonaceae, Buckwheat Family

Habit: Herbaceous perennial, 0.6-1 m tall, from underground runners, often mat forming. Stems erect at tips, stems and leaf sheaths with appressed hairs, joints sheathed, sheaths with stiff hairs, upper margin of sheath with stiff hairs.

Leaves: Alternate, from a sheathing base; stalked; blades lance-shaped, 4-13 cm long, hairy.

Inflorescences: Axillary or terminal spikes.

Flowers: White, pink or green, to 3.5 mm long.

Fruit: Nutlets 3-sided, to 3 mm long, shiny dark brown.

Habitat and Distribution: Common; virtually all wet habitats; Nova Scotia across southern Canada to British Columbia, south throughout the United States; Mexico and the West Indies, through Central America into South America.

Comment: Flowering during all the warm months, Mild Water-pepper can provide a show when in masses. It freezes to the ground and comes back from the underground runners.

Polygonum virginianum L.
Jumpseed
[*Antenoron virginianum* (L.) Roberty & Vautier; *Tovara virginiana* (L.) Raf.]
Polygonaceae, Buckwheat Family

Habit: Herbaceous perennial, 0.3-1.2 m tall, from underground runners. Stems with swollen joints, joints sheathed, sheaths with stiff hairs, upper margin with long stiff hairs.

Leaves: Alternate, from a sheathing base; stalked; blades elliptic or rounded, 5-16 cm long, margins smooth.

Inflorescences: Terminal, long, widely-spaced, spike-like arrangements.

Flowers: White, greenish, or pinkish.

Fruit: Nutlets 2-sided, about 4 mm long, glossy brown.

Habitat and Distribution: Occasional, but locally common; forested floodplains, hydric hammocks, rich woods and thickets; north peninsula Florida, north to New Hampshire, Quebec, and Ontario, south and west to Texas and northern Mexico.

Comment: Jumpseed flowers in summer and fall. It is best identified by the long, whip-like flower spike. The small inconspicuous flowers are widely spaced. The fruit has two small hooked appendages at the tip and seems to "jump" and attach onto suitable clothing or fur.

Polygonum hydropiperoides Michx.

Polygonum virginianum L.

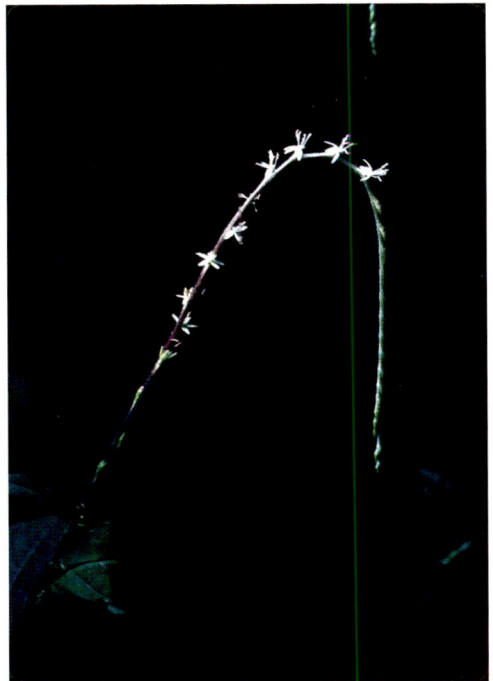

White

Polypremum procumbens L.
Rustweed
Buddlejaceae, Butterfly Bush Family (Loganiaceae, Logania Family; Tetrachondra-ceae, Tetrachondra Family)

Habit: Herbaceous perennial, 10-30 cm long. Stems spreading or ascending, mat forming, older plants rusty.

Leaves: Opposite; stalkless; blades linear, sharply pointed, 1-2.5 cm long.

Inflorescences: Solitary flowers in forks and at branch tips.

Flowers: White, 4 petals, to 3 mm long.

Fruit: A capsule, 1.5-2.5 mm long, ovoid, notched at tip. Seeds yellow, squarish.

Habitat and Distribution: Common; dry areas, disturbed sites, roadsides, sandy soils of floodplains, and pond margins; throughout Florida, west to Texas, and north to Missouri, Pennsylvania, and New Jersey; south into the West Indies and Columbia.

Comment: The small, commonly rust-colored plants are easy to recognize in dry soils. The tiny bright white flowers occur in most warm months.

Prunus americana Marshall
Wild Plum or American Plum
Rosaceae, Rose Family

Habit: Small tree, to 8 m tall. Bark curling in plates.

Leaves: Alternate, deciduous; stalked, often with a gland just below blade; blades elliptic, 5-12 cm long, margins with fine teeth, tips with long narrow points.

Inflorescences: Axillary clusters on older stems.

Flowers: White, about 2 cm wide, with 5 petals, and many stamens.

Fruit: Round, with a central stone, 2-2.5 cm long, red or rarely yellow, smooth-skinned.

Habitat and Distribution: Occasional; along streams, floodplains, fence rows, or moist deciduous woods; central peninsula Florida, west to Alabama, Oklahoma, and Colorado, and north to Montana, Manitoba, and New Hampshire.

Comment: The large spring flowers are very attractive and fragrant. The flesh of the fruit is edible and sweet.

Polypremum procumbens L.

Prunus americana Marshall

White

Prunus angustifolia Marshall
Chickasaw Plum
Rosaceae, Rose Family

Habit: Shrub or small tree, to 4 m tall, forming thickets from root shoots. Branches smooth, some twigs with thorns.

Leaves: Alternate, deciduous; stalked; blades lance-shaped to elliptic, folded upward, 3-8 cm long, margins with fine teeth, tips with short sharp points.

Inflorescences: Axillary clusters.

Flowers: White, about 0.7-1.0 cm wide, with 5 petals, and many stamens.

Fruit: Round, with a stone, 1.5-2.5 cm long, red or yellow, smooth-skinned.

Habitat and Distribution: Occasional, can be locally common; old home sites, old fields, margins of woods, open woods, and disturbed sites; from north peninsula Florida, west to Texas and north to Arkansas, Maryland, and Delaware.

Comment: Chickasaw Plum flowers in early spring. The flowers are not especially fragrant. Thickets have the largest plants towards the center. Flesh of these fruits ranges from sweet to a little less so. This popular shrub has traveled wherever settlers went in late 1800 and early 1900s. It was and still is frequently used for jellies, preserves and various drinks.

Prunus caroliniana (Mill.) Aiton
Cherry-laurel or Laurel Cherry
Rosaceae, Rose Family

Habit: Small to large tree, to 12 m tall.

Leaves: Alternate, evergreen; stalked; blades elliptic, 5-10 cm long, somewhat leathery, often a single glandular spot underneath on each side of the midvein near base, margins smooth or with teeth, tips with short sharp points.

Inflorescences: Spike-like arrays, 1-4 cm long, from leaf axils.

Flowers: White, 2-4 mm wide, with 5 petals, and 10-15 stamens.

Fruit: Round, with a stone, 1-1.3 cm long, dull black.

Habitat and Distribution: Frequent; thickets, fence rows, hammocks, home sites, and farm sites; from central peninsula Florida, west to Texas, and north to North Carolina.

Comment: Cherry-laurel blooms in spring and is very common in cultivation. The fleshy fruits are eaten by birds that spread this species widely. Often used for hedges of specimen plants, this species forms thickets by root shoots.

Prunus angustifolia Marshall

Prunus caroliniana (Mill.) Aiton

White

673

White

Prunus serotina Ehrh.
Wild Cherry or Black Cherry
Rosaceae, Rose Family

Habit: Tree, to 30 m tall.

Leaves: Alternate, deciduous; stalked, with one or two glands just below the blade; blades elliptic, sometimes broadly so, 5-12 cm long, shiny above, with brown woolly hairs along midline beneath, margins with teeth, a few teeth near base with glands.

Inflorescences: Spike-like arrays, 3-10 cm long, from leaf axils.

Flowers: White, about 7 mm wide, with 5 rounded petals, and 20 stamens.

Fruit: Round, with a stone, 0.7-1 cm long, black.

Habitat and Distribution: Common; fence rows, old fields, mesic forests, and pastures; central peninsula Florida, west to Texas, and north to Minnesota and Nova Scotia.

Comment: Occasionally cultivated as a specimen plant because of the multitude of spring flowers. Wild Cherry can be extremely weedy due to the spread of the juicy fruits that are very attractive to birds. The wood is valued for cabinets and interiors.

Prunus umbellata Elliott
Flatwoods Plum
Rosaceae, Rose Family

Habit: Shrub or small tree, to 6 m tall. Twigs rarely thorny.

Leaves: Alternate, deciduous; stalked, with one or two glands just below the blade; blades elliptic to oval, 2-8 cm long, margins with teeth, tips with short sharp points, a few teeth near base with glands.

Inflorescences: Dense clusters from leaf axils.

Flowers: White, 1.2-1.5 cm wide, with 5 petals and many stamens.

Fruit: Round, with a stone, 1-2 cm diameter, purplish black or red.

Habitat and Distribution: Frequent; well-drained soils of flatwoods, sandhills, mixed woods and margins, and pastures; central peninsula Florida, west to Texas, and north to Arkansas and North Carolina.

Comment: As an ornamental, Flatwoods Plum is spectacular when it flowers in the very early spring just before the leaves come out. The fruits are juicy and edible, but quite bitter and sour.

Prunus serotina Ehrh.

Prunus umbellata Elliott

675

White

Psidium guajava L.
Guava
Myrtaceae, Myrtle Family

Habit: Shrub or small tree, usually not over 5 m tall. Branches 4-angled, hairy.

Leaves: Opposite; short-stalked; blades elliptic, 7-14 cm long, leathery, hairy underneath, dotted, veins prominent.

Inflorescences: Flowers solitary from leaf axils.

Flowers: White, rounded, showy, 5 petals, stamens numerous.

Fruit: A berry, rounded, 2-6 cm wide and long, pink or yellow flesh.

Habitat and Distribution: Frequent to common; hammocks, roadsides, canal banks, disturbed sites; central and south peninsula Florida; a native of Tropical America, now cultivated and escaped throughout the tropics and subtropics of the world.

Comment: The attractive flowers appear all year. Guava is a popular cultivated plant. The edible fruit is most frequently used for jelly and juice.

Ptelea trifoliata L.
Wafer-ash, Hop Tree, Skunk-bush
Rutaceae, Citrus Family

Habit: Shrub or small tree, to 8 m tall.

Leaves: Alternate; stalked; blades with 3 broadly elliptic leaflets, each 2-12 cm long, gland-dotted.

Inflorescences: Branched terminal clusters, usually only one sex in each cluster, both sexes on the same plant.

Flowers: White or greenish yellow, 4-5 petals, 4-6 mm long.

Fruit: Round, flat, winged, 1.5-2.5 cm wide, whitish, 2 seeds in each of two cavities.

Habitat and Distribution: Occasional; rich woods, hammocks, hillsides, rocky sites; central peninsula Florida, west to Arizona, and north to Minnesota, Ontario, and New York.

Comment: Wafer-ash blooms in spring. The flowers are attractive, but the fruits in late spring through summer are eye-catching. Several cultivated forms and varieties are available for cultivation and it can be used for hedges. The distinctive fruits can be used as a substitute for Hops in brewing. This species can be a cause of photodermatitis.

Psidium guajava L.

Ptelea trifoliata L.

White

Pterocaulon pycnostachyum (Michx.) Elliott
Blackroot
Asteraceae (Compositae), Aster or Sunflower Family

Habit: Herbaceous perennial, to 80 cm tall, from thick, black roots. Stems erect, winged due to extended leaf bases, covered with white woolly hair.

Leaves: Alternate; stalkless; blades linear-oblong or narrowly elliptic, to 11 cm long, hairless above with white stripe along center of leaf, woolly below, margins smooth to toothed.

Inflorescences: Numerous heads in very dense, terminal spikes, only disc flowers present, heads 4-5 mm long, densely hairy.

Flowers: White to pinkish to yellow, tiny.

Fruit: An achene, 1-1.2 mm long, reddish brown, hairy, cluster of bristles on top.

Habitat and Distribution: Frequent to common; open pinelands, sandhills, and sandy disturbed areas; throughout Florida, west to southeast Alabama, and north to North Carolina.

Comment: Winged stems and the dense cottony heads flowering from spring into fall get Blackroot frequently noticed as something different. The thick black roots give the plant its common name.

Ptilimnium capillaceum (Michx.) Raf.
Mock Bishop's-weed
Apiaceae (Umbelliferae), Carrot Family

Habit: Herbaceous annual, 10-80 cm tall. Stem slender, branched.

Leaves: Alternate; stalked; blades very narrow, threadlike, much divided, whorled in 3s on the leaf axis.

Inflorescences: Spreading, branched, umbrella-like clusters, 2-5 cm wide, from leaf axils.

Flowers: White, with 5 petals less than 1 mm long, with minute sepals.

Fruit: To 2 mm long, ribbed, somewhat rounded, sections semicircular.

Habitat and Distribution: Common; wet ditches, shores, woods, marshes, pastures, and swamps; throughout Florida, west to Texas, and north to Missouri, Kansas, Illinois, Kentucky, and Massachusetts.

Comment: Flowering mostly in warm weather, Mock Bishop's-weed can bloom anytime during the year. It is often considered weedy and is a pest in wetter crops. The masses of blooming plants with the fine foliage in low open wet environments provide a delicate show.

Pterocaulon pycnostachyum
(Michx.) Elliott

Ptilimnium capillaceum
(Michx.) Raf.

White

Pyracantha koidzumii (Hayata) Rehder
Formosa Firethorn
Rosaceae, Rose Family

Habit: Shrub, to 4 m tall. Many branched, lateral branches sharply thorn-tipped, young twigs hairy, older twigs smooth.

Leaves: Alternate, evergreen; short-stalked; blades broader at tip, 1-2.5 cm long, margins smooth, tip rounded.

Inflorescences: Terminal, hairy clusters on branchlets.

Flowers: White, 5 small petals, 20 stamens.

Fruit: A pome, about 5 mm in diameter, orange-red.

Habitat and Distribution: Occasional; disturbed sites; throughout Florida; a native of Taiwan.

Comment: The noticeable flowers can be seen in spring, but the bright evergreen leaves and abundant colorful fruit are spectacular. Several cultivated forms are listed. Formosa Firethorn is frequently cultivated.

Raphanus raphanistrum L.
Wild Radish or Jointed Charlock
Brasicaceae (Cruciferae), Mustard Family

Habit: Herbaceous winter annual, 30-80 cm tall, from a taproot. Stems covered with stiff hairs.

Leaves: Alternate; stalked; blades larger on lower parts of plant, smaller and narrower on upper parts, lower leaves often dissected, upper leaves lobed, margins of all blades toothed.

Inflorescences: Stalked, in an open elongated assortment at stem and branch tips, flowering progressively upward.

Flowers: White or yellow, less commonly pink or purple, 2-3 cm wide, with 4 petals.

Fruit: Pods narrow, 2-6 cm long and 3-5 mm diameter, with constrictions between seeds.

Habitat and Distribution: Frequent, locally common; disturbed sites, particularly in cultivated fields; throughout Florida and throughout the temperate regions of the world; a native of the Mediterranean region of Europe.

Comment: Flowering in spring, Wild Radish is familiar to many as a weed of roadsides and gardens. In mass it makes a showy wildflower.

Pyracantha koidzumii (Hayata) Rehder

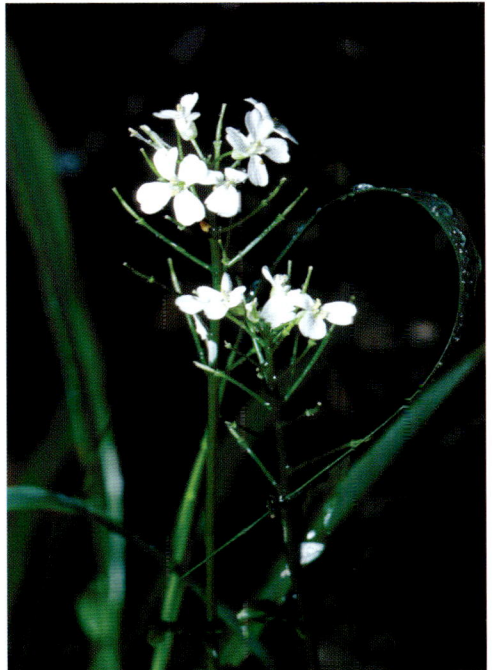

Raphanus raphanistrum L.

White

Rhexia mariana L.
Pale Meadow-beauty or Dull Meadow-beauty
Melastomataceae, Melastome Family

Habit: Herbaceous perennial, 20-80 cm tall, spreading by horizontal roots. Stems hairy, branched or unbranched.

Leaves: Opposite; short-stalked; blades narrow to elliptic to somewhat rounded, to 6 cm long, sparsely hairy, 3-nerved, margins toothed and hairy.

Inflorescences: Few-flowered, branched clusters at stem tips.

Flowers: White, pale lavender, or pink, 2-5 cm wide, with 4 uneven petals.

Fruit: A capsule, urn-shaped, to 1 cm long, usually hairy, sometimes smooth.

Habitat and Distribution: Common; low, wet open sites, ditches, bogs, flatwoods, marshes; south central peninsula Florida, west to Texas, and north to Missouri, Illinois, Indiana, and Massachusetts.

Comment: Flowering all warm months, Pale Meadow-beauty is a great wildflower. Quite common in all moist to wet sites, large colonies are spread in most of the openings. The large flowers, while not spectacular, are easily seen throughout the sites. They can be easily propagated by seeds and roots.

Rhexia parviflora Chapm.
White Meadow-beauty or Apalachicola Meadow-beauty
Melastomataceae, Melastome Family

Habit: Herbaceous perennial, 10-40 cm tall, from short underground runners. Stems 4-sided, joints hairy.

Leaves: Opposite; stalked; blades rounded to elliptic, to 3 cm long, slightly hairy, margins toothed, teeth hair-tipped.

Inflorescences: Wide-branched arrangements at stem tips.

Flowers: White, to 2 cm wide, with 4 uneven petals.

Fruit: A capsule, urn-shaped, to 3-7 mm long, few hairs near tip.

Habitat and Distribution: Rare; wet sands of shallow ponds, bog borders, Hypericum ponds; from Liberty and Franklin Counties west to Santa Rosa County in the Florida panhandle, endemic.

Comment: Flowers can be seen in the summer, June through August. The flowers and numbers of plants at a site are large enough to make a good wildflower display, although White Meadow-beauty is rare and listed as Endangered by the State of Florida.

White

Rhexia mariana L.

Rhexia parviflora Chapm.

683

White

Rhododendron viscosum (L.) Torr.
Swamp-honeysuckle or Swamp Azalea
[*Rhododendron serrulatum* (Small) Millais; *Rhododendron viscosum* (L.) Torr. var.
***serrulatum* (Small) Ahles]**
Ericaceae, Heath Family

Habit: Shrub, to 5 m tall. Branches hairy.

Leaves: Alternate, deciduous; short-stalked; blades somewhat elliptic, 1-7 cm long, hairy, margins with long hairs, no teeth.

Inflorescences: Teminal clusters.

Flowers: White or pinkish, funnel-shaped, 1.5-3.5 cm long, with 5 pointed lobes arching outward and stamens extending past floral tube, sticky, hairy, fragrant.

Fruit: A capsule, 1-2 cm long, with sticky hairs.

Habitat and Distribution: Frequent; wet woods and swamps, mostly in seepage areas; from central peninsula Florida, west to Mississippi, and north to Ohio and Maine.

Comment: Flowering in summer, the very sweet fragrance of the flowers reminds people of Japanese Honeysuckle (*Lonicera japonica*) and blooms about the same time. Easily cultivated, especially with irrigation, it is an attractive shrub flowering at a time when many other shrubs do not. Swamp-honeysuckle will not endure flooding, although in nature it often grows where its roots are always wet.

Rhus copallinum L.
Winged Sumac
Anacardiaceae, Sumac or Cashew Family

Habit: Shrub, to 9 m tall, from underground runners. Younger stems hairy, older stems smooth.

Leaves: Alternate, deciduous; stalked; blades once-divided, with 9-23 elliptical leaflets, always with one leaflet at the tip making an odd number, each 2-10 cm long, hairy underneath.

Inflorescences: Large sprays at tips of young branches, spray as wide as long, male and female flowers separate and on separate plants.

Flowers: White, yellowish or greenish, usually with 5 petals, each 2-3 mm long.

Fruit: A drupe, round, 3-5 mm diameter, reddish pink, hairy.

Habitat and Distribution: Common; dry woods, often along margins, sandhills, flatwoods, hammocks, old fields, fence rows, rights-of-way, and disturbed sites; throughout Florida, west to Texas, and north to Missouri, Michigan, and New Hampshire.

Comment: Winged Sumac can be a prolific weed due to the propensity of shoots from the underground runners of a mature plant. Flowering occurs in late spring through the summer. Fruits can be seen from August into the following winter. The very large sprays of flowers are quite showy. The reddish pink fruit clusters from the female flowers are also attractive. Fall leaf colors, however, are spectacular. The orange to red autumn leaves are the primary reason this shrub is cultivated. As a specimen plant it needs to be put where the runners can be contained. Perhaps it is best placed along a fence or border where the runners can produce additional plants to provide a bigger fall show. Winged Sumac plant parts and its fruit are not poisonous. Propagation is by root cuttings or seeds.

Rhododendron viscosum (L.) Torr.

Rhus copallinum L.

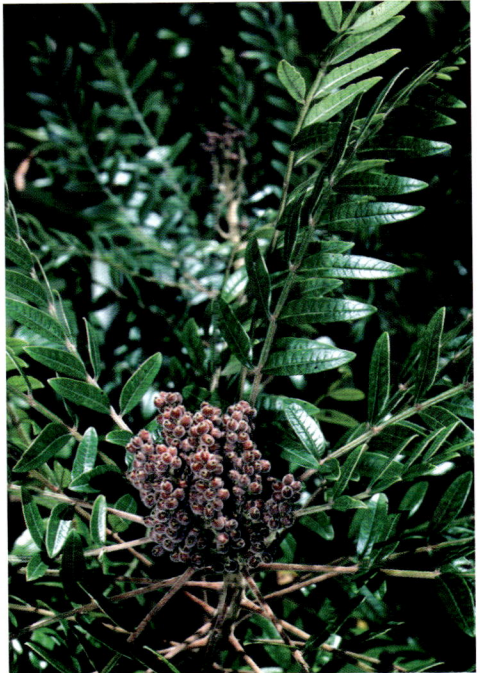

White

Rhynchospora colorata (L.) Pfeiff.
Whitetop Sedge
[*Dichromena colorata* (L.) Hitchc.]
Cyperaceae, Sedge Family

Habit: Herbaceous perennial, to 60 cm tall, from extensive thin underground runners.

Leaves: Alternate, basal; stalkless; blades slender, grass-like, shorter than flowering stem.

Inflorescences: Terminal, composed of several spikelets, above 4-6 bracts.

Flowers: Whitish, tiny; bracts long, slender, to 3 mm wide at base, tapering to tip, arching outward, white basal portion 2.5 cm long or less, green toward tip.

Fruit: A 1.2-1.5 mm long nutlet-like achene.

Habitat and Distribution: Common; wet marshes, prairies, swales, ditches, shores, shallow water, and wet flatwoods; throughout Florida, north to Virginia, and west to Texas; eastern Mexico, British Honduras, West Indies.

Comment: Resembling grass, the distinctive white color of the bracts of the heads can be seen in most wet open habitats in all the warm months. Sometimes it is used in dried flower arrangements.

Rhynchospora latifolia (Baldwin) W.W. Thomas
Giant Whitetop Sedge
[*Dichromena latifolia* Baldwin]
Cyperaceae, Sedge Family

Habit: Herbaceous perennial, 0.5-1.2 m tall, from underground runners.

Leaves: Alternate, basal; stalkless; blades slender, grass-like, shorter than flowering stem.

Inflorescences: Terminal, composed of several spikelets, above 7-10 bracts.

Flowers: Whitish, tiny; bracts long, narrow, 8-10 mm wide at base, tapering to tip, white basal portion usually longer than 2.5 cm, green toward tip.

Fruit: A 1.2-1.5 mm long, nutlet-like achene.

Habitat and Distribution: Frequent; marshes, bogs, ditches, swales, shores, shallow water, and wet flatwoods; throughout Florida, west to Texas, and north to Virginia.

Comment: Resembling grass, the large distinctive white color of the bracts of the heads is always noticed during all the warm months. Like Whitetop Sedge this species is also used in dried flower arrangements.

White

Rhynchospora colorata (L.) Pfeiff.

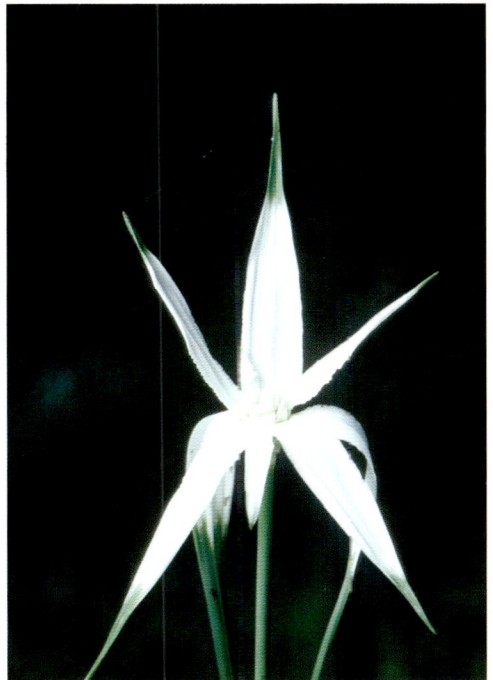

Rhynchospora latifolia
(Baldwin) W.W. Thomas

687

White

Richardia brasiliensis Gomes
Brazil Pusley or Tropical Mexican-clover
Rubiaceae, Madder Family

Habit: Herbaceous perennial, 10-70 cm long, from a long, thickened, fleshy, sometimes branched tap root. Stems spreading, lying flat, hairy, rough.

Leaves: Opposite; short-stalked; blades ovate to elliptic, 1.5-4 cm long, densely hairy on both surfaces.

Inflorescences: Very dense, terminal clusters.

Flowers: White, funnel-shaped, 3-5 mm long, with 6 spreading, pointed lobes.

Fruit: Dry, separating into 4 nutlets, 2.5 mm long, short stiff hairs.

Habitat and Distribution: Common; dry soils of lawns, old fields, rights-of-way, dunes, prairies, pastures, disturbed sites; throughout Florida, west to Texas and Mexico, and north to Virginia; a native of Tropical America.

Comment: Found flowering in any month lacking frost, Brazil Pusley is a common weed. The bright white flowers in tight clusters, in the right light, can be attractive in areas with very low vegetation. The bumpy tap roots are known to contain root-knot nematodes. The presence of light will increase seed germination which means that soil disturbance will probably mean more Brazil Pusley plants.

Richardia grandiflora (Cham. & Schltdl.) Schult. & Schult. f.
Large-flower Pusley or Large-flower Mexican-clover
Rubiaceae, Madder Family

Habit: Herbaceous perennial, to 80 cm long, from a tap root. Stems spreading, mostly lying flat, hairy.

Leaves: Opposite; short-stalked; blades slightly broader at tips, hairy on both surfaces.

Inflorescences: Dense terminal clusters.

Flowers: White, blue, pink, or violet, funnel-shaped, 2 cm long, with 6 spreading lobes.

Fruit: Dry, separating into 4 nutlets.

Habitat and Distribution: Occasional, locally common; dry soils of lawns, rights-of-way, pastures, disturbed sites; central and southern peninsula Florida; a native of South America.

Comment: Large-flower Pusley flowers during all warm months. The flowers are very noticeable, particularly so in weedy turf areas. Mowing does not seem to affect the plant and soil disturbance increases the numbers of plants. As a wildflower Large-flower Pusley is rapidly moving within Florida. This ornamental weedy flower is being used to decorate sandy, nearly bare spots in turf.

Richardia brasiliensis Gomes

Richardia grandiflora (Cham. & Schltdl.) Schult. & Schult. f.

White

Ricinus communis L.
Castor-bean
Euphorbiaceae, Spurge Family

Habit: Herbaceous annual or perennial, shrub, or tree, 15 m tall. Branches and stems smooth, often waxy gray, hollow between joints.

Leaves: Alternate; long-stalked; blades attached to blade inside of margin (peltate); blades 20 to 60 cm wide, with 5-11 pointed radiating lobes, margins with teeth, often reddish.

Inflorescences: Terminal, spike-like cylinder, male and female flowers separate, but in the same spike, male flowers below the female.

Flowers: Male yellow with many stamens; female flowers at top of spike, green with feathery red styles; flowers with 5 sepals and no petals.

Fruit: A capsule, round, 3-lobed, 1.2-1.6 cm diameter, covered with soft spines.

Habitat and Distribution: Frequent; disturbed sites; throughout Florida, widespread in the warm and tropical parts of the world; a native of Africa.

Comment: Flowering in summer and fall, Castor-bean is popular as an ornamental. The unusual leaves and fruits are attractive. There are many named cultivated forms with color variations and size modifications. The plant is tropical in nature and freezes easily. It readily escapes from cultivation by seeds. Oil from the seeds produces the Castor Oil of commerce. The oil is highly valued for machines. Seeds are VERY POISONOUS if ingested.

Rivina humilis L.
Rouge Plant, Bloodberry, Coralito, Pigeon-berry
Phytolaccaceae, Pokeweed Family (Petiveriaceae, Guinea Hen Weed Family)

Habit: Herbaceous perennial, 0.3-1.5 cm tall, from a thick rootstock. Stems erect to clambering and vine-like, sometimes woody below, branches spreading.

Leaves: Alternate; stalked; blades ovate to somewhat lance-shaped, 3-15 cm long, margins smooth.

Inflorescences: Elongated, spike-like, terminal or axillary arrangements.

Flowers: White or pink, 3-6 mm wide, 4-parted.

Fruit: A berry, round, 2-3.5 mm in diameter, bright red.

Habitat and Distribution: Frequent; woods, thickets, and shady disturbed sites, often on rocky soils; peninsula Florida, west to Arkansas and Texas, throughout tropical America; naturalized in tropical Asia.

Comment: Rouge Plant will bloom during all warm months. The flowers are attractive, but the bright, shiny red fruits are quite showy and conspicuous. Sometimes Rouge Plant is cultivated for the fruits. Fruits are used for a red dye. Leaves and roots are POISONOUS, and the fruits are also thought to be TOXIC.

Ricinus communis L.

Rivina humilis L.

White

Robinia pseudoacacia L.
Black Locust
Fabaceae (Leguminosae), Pea or Bean Family

Habit: Tree to 25 m tall. Bark thick, deeply furrowed.

Leaves: Alternate; stalked; blades with 7-19 oval leaflets, leaflets 2-5 cm long, with a pair of woody spines 0.3-2.5 cm long at the base of the leaf stalk.

Inflorescences: Numerous flowers, in drooping, spike-like arrangements, from leaf axils.

Flowers: White with a yellow spot, fragrant, typically pea-shaped, 1.5-2 cm long.

Fruit: A flat pod, 5-10 cm long, about 1 cm wide, hairless, upper surface with a narrow wing.

Habitat and Distribution: Occasional to locally frequent; woods, thickets, margins, pastures, roadsides, windrows, canyons, water course banks, cultivated and disturbed areas; northern peninsula and panhandle Florida, throughout the United States, north to Nova Scotia and Quebec, west to Washington and California; native from Pennsylvania, south to northern Georgia, west to northern Arkansas; introduced into Europe.

Comment: Black Locust flowers in spring. The flowers are showy and quite fragrant. It has been extensively planted for ornament and wind brakes. The wood is resistant to decay and has been used for fence posts. Leaves, bark, and seeds are TOXIC if ingested.

Rorippa nasturtium-aquaticum (L.) Hayek
Water-cress
[*Nasturtium officinale* R. Brown]
Brassicaceae (Cruciferae), Mustard Family

Habit: Herbaceous aquatic perennial, 10-80 cm long. Stems smooth, floating to ascending, rooting at nodes.

Leaves: Alternate; stalked; blades 2-15 cm long and 2-5 cm broad, divided into 3-11 leaflets, leaflets mostly rounded to lance-shaped, margins often with blunt teeth.

Inflorescences: Long, terminal or axillary cluters.

Flowers: White, with 4 petals each 4-5 mm long.

Fruit: In slender pods, curved, 1-2 cm long; seeds reddish, about 1 mm long, net-veined.

Habitat and Distribution: Occasional to infrequent; streams and springs; central peninsula Florida, throughout the United States and southern Canada; a native of Europe.

Comment: Flowering in late spring through the summer, the flowers are not especially showy. This plant is the Water-cress of commerce. The pungent leaves are used for salads.

White

Robinia pseudoacacia L.

Rorippa nasturtium-aquaticum (L.) Hayek

693

White

Rosa laevigata Michx.
Cherokee Rose
Rosaceae, Rose Family

Habit: High-climbing shrub, to 5 m tall. Stems thorny, with flattened, recurved, thick prickles.

Leaves: Alternate, evergreen; stalked; blades once-divided with 3 or 5, elliptic-ovate to lance-shaped leaflets, each 1-8 cm long, margins with teeth.

Inflorescences: Solitary, long-stalked flowers from leaf axils.

Flowers: White, 6-8 cm across, with 5 broad petals and many stamens, stalks and floral tubes with stiff glandular-tipped hairs.

Fruit: A hip, orange to red, pear-shaped, 3-4 cm long.

Habitat and Distribution: Occasional; hammocks, thickets, margins of woods, and along roadsides; central peninsula Florida, west to Texas, and north to South Carolina; West Indies; a native of China.

Comment: The large showy flowers of Cherokee Rose appear in spring. This attractive rose is an impressive wildflower. As an escape from cultivation, it is frequently seen climbing over shrubs along roadsides.

Rubus cuneifolius Pursh
Sand Blackberry
Rosaceae, Rose Family

Habit: Herbaceous perennial shrub, 0.3-1.5 m tall, from underground runners. Aboveground stems biennial, usually erect, flowering in the second year, producing short flower bearing branches in the second year, covered with straight or curved prickles.

Leaves: Alternate, deciduous, thorny; stalked; blades once-divided, with 3, 4 or 5 elliptic to diamond shaped leaflets, each to 5 cm long, with dense white or gray hairs beneath, margins toothed, curved prickles on midveins underneath.

Inflorescences: Solitary to many flowers, terminal on branchlets.

Flowers: White, to 3 cm across, with 5 petals, and many stamens.

Fruit: An aggregate of many small berries, usually 1-1.5 cm long, black when ripe, juicy.

Habitat and Distribution: Common; sandhills, open woods, thickets, clearings, flatwoods, margins of wetlands, and along roadsides; throughout Florida, west to Mississippi, and north to Connecticut.

Comment: Spring blooming, this aggressive wildflower also can provide a juicy edible fruit. Sand Blackberry requires good conditions during the development of the fruit for it to be juicy and not dry and mostly tasteless. Old dried stems can make walking through colonies and harvesting fruit extremely difficult due to the multiple stiff thorns.

Rosa laevigata Michx.

Rubus cuneifolius Pursh

White

Rubus trivialis Michx.
Southern Dewberry
Rosaceae, Rose Family

Habit: Herbaceous perennial shrub, to 2 m long, from a rootstock. Aboveground stems biennial, trailing or lying flat, flowering in the second year, producing short flower bearing branches in the second year, covered with straight or curved prickles, and straight, stiff, simple and glandular-tipped hairs.

Leaves: Alternate, deciduous, thorny; stalked; blades once-divided, with 3 or 5 elliptic to lance-shaped leaflets, each to 6 cm long, smooth beneath, margins toothed, and midveins with curved prickles underneath.

Inflorescences: Flowers 1-3, terminal on branchlets.

Flowers: White, to 5 cm across, with 5 petals, and many stamens.

Fruit: An aggregate of many small berries, usually 1.5-3 cm long, black when ripe, juicy.

Habitat and Distribution: Common; fields, rights-of-way, thickets, fence rows, open woods, and open disturbed wetland margins; throughout Florida, west to Texas, and north to Missouri, Illinois, and Virginia.

Comment: Flowering in spring, Southern Dewberry forms low mats, often over low vegetation. The fruit is juicy and edible.

Sabatia angularis (L.) Pursh
Rose-pink Sabatia, Common Marsh-pink, Bitter-bloom, Square-stem Rose-gentian
Gentianaceae, Gentian Family

Habit: Herbaceous annual, 20-90 cm tall. Stems 4-sided, wing-margined, freely branching.

Leaves: Opposite; stalkless; blades rounded, 1.5-4 cm long, 3- to 7-veined.

Inflorescences: Terminal, in a wide-branching arrangement.

Flowers: White, usually pink, with 5 petals, each 1-2 cm long and half as wide.

Fruit: A capsule, oval or conical, to 8 mm long, to 5 mm wide.

Habitat and Distribution: Frequent (rare in Florida); moist meadows, open woods, calcareous hammocks, marshes, pastures, prairies, ditches, and fields; central panhandle Florida, west to eastern Texas, and north to Kansas, Illinois, Michigan, and New York.

Comment: Flowering during summer, Rose-pink Sabatia is a frequently encountered wildflower. With several flowers and large petals, Rose-pink Sabatia is quite lavish.

Rubus trivialis Michx.

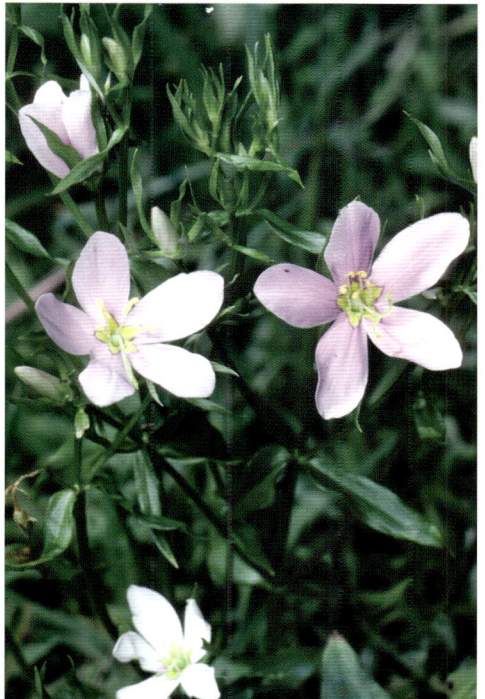

Sabatia angularis (L.) Pursh

White

Sabatia bartramii Wilbur
Large Marsh-pink, Bartram's Sabatia, Bartram's Rose-gentian, Ten-petal Marsh-pink
[*Sabatia decandra* (Walter) R.M. Harper, misapplied; *Sabatia dodecandra* (L.) BSP.
var. *coriacea* (Elliott) Ahles]
Gentianaceae, Gentian Family

Habit: Herbaceous perennial, to 1 m tall, from short underground runners. Stem rounded, smooth, branched, branches usually alternate.

Leaves: Opposite; stalkless; blades succulent, basal broadest at tip, stem lance-shaped to narrowly elliptic to linear, 2-8 cm long.

Inflorescences: Flowers terminal, usually only one appearing at a time.

Flowers: Rarely white, usually pink with yellow eye, 10-12 petals, each 2-3 cm long.

Fruit: A somewhat rounded capsule, 6-7 mm long.

Habitat and Distribution: Frequent; wet pinelands, ditches, and pond margins; throughout Florida except southern tip, west through southern Alabama to southeastern Mississippi, and north through southern Georgia into southern South Carolina.

Comment: Flowering from summer into fall, the large many-petaled flowers are easily seen, although not very many flowers at a time. Large Marsh-pink is rare northward in the range.

Sabatia brevifolia Raf.
Narrow-leaved White Sabatia or Short-leaf Rose-gentian
[*Sabatia elliottii* Steud.]
Gentianaceae, Gentian Family

Habit: Herbaceous annual, to 70 cm tall. Stems with alternate branching, smooth, round or with slight ridges.

Leaves: Opposite; stalkless; lower blades often broader at tip, stem blades very narrow, 1-4 cm long.

Inflorescences: A wide, branched arrangement, at stem tips.

Flowers: White with yellow center, 5 petals, 1-1.3 cm long.

Fruit: A capsule, 4-5 mm long.

Habitat and Distribution: Common; wet pinelands, pond margins, seepage bogs; throughout Florida, west to southern Alabama, and north to southern South Carolina.

Comment: Flowers can be seen from spring into summer southward in the range and in fall northward. The large white flowers are easy to see in the large clusters.

Sabatia bartramii Wilbur

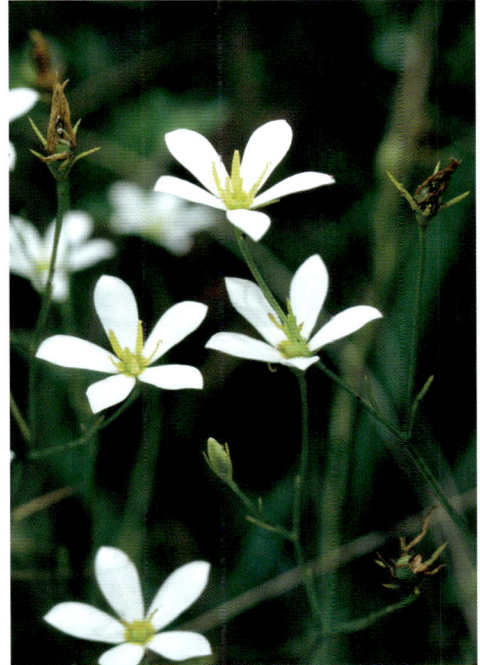

Sabatia brevifolia Raf.

White

Sabatia macrophylla Hook.
Large-leaf Rose-gentian
[*Sabatia recurvans* Small]
Gentianaceae, Gentian Family

Habit: Herbaceous perennial, to 1.2 m tall, from a thick underground runner. Stems erect, round, waxy-gray upwards, branches opposite.

Leaves: Opposite; stalkless; blades lance-shaped to oblong, thick, 3-6 cm long.

Inflorescences: Terminal, flat-topped to rounded arrangements.

Flowers: White, 5 petals, each 5-7 mm long.

Fruit: A capsule, 3-4.5 mm long.

Habitat and Distribution: Occasional; wet flatwoods, seepage bogs, pond margins, pits, and ditches; north peninsula Florida, west to eastern Louisiana, and north to southern Georgia.

Comment: Summer blooming Large-leaf Rose-gentian can quickly be distinguished by the clusters of bright white flowers and the waxy-gray stem.

Sagittaria filiformis J.G. Sm.
Standing Water Arrowhead
[*Sagittaria stagnorum* Small]
Alismataceae, Water-plantain Family

Habit: Herbaceous aquatic perennial, seldom upright, from thin underground runners and bulbs.

Leaves: Alternate; narrowly strap-like, thin and floating when water is high, when water is down, plants ascending with a stalk and rounded blades.

Inflorescences: All flowers floating, in 2-6 whorls, at top of leafless flowering stem, male and female flowers separate, in separate whorls in the same inflorescence, lowest whorl female, upper whorls male.

Flowers: White, petals 3, rounded, each to 6 mm long.

Fruit: A nutlet-like achene, to 1.5 mm long; fruiting heads to 8 mm wide.

Habitat and Distribution: Common; shallow water of ponds, lakes, and ditches; throughout peninsula Florida, west to central panhandle Florida and Alabama, and north to North Carolina.

Comment: The few small flowers of Standing Water Arrowhead seen floating in shallow water in all the warm months are not very noticeable. However, frequently, hundreds to thousands of plants will be in flower at the same time covering the water surface with a dramatic white display.

Sabatia macrophylla Hook.

Sagittaria filiformis J.G. Sm.

White

Sagittaria graminea Michx.
Grassleaf Arrowhead
Alismataceae, Water-plantain Family

Habit: Herbaceous aquatic perennial, to 60 cm tall, from short thick underground runners.

Leaves: Above water leaves alternate; stalked or stalkless; blades grass-like, to 50 cm long.

Inflorescences: Flowers in 2-12 whorls, at top of leafless flowering stem, male and female flowers separate, in separate whorls in the same inflorescence, lowest 1-2 whorls female, upper whorls male.

Flowers: White, petals 3, round, each 0.6-2 cm long, male flowers with less than 20 stamens.

Fruit: A nutlet-like achene, 1.2-3 mm long, winged; fruiting heads to 1 cm wide.

Habitat and Distribution: Common; ponds, prairies, swamps, slow streams, and ditches; throughout Florida, north to Newfoundland and Labrador, west to the Great Plains and Texas; Cuba.

Comment: The whorled white flowers seen in all the warm months are not particularly showy, but sometimes in mass they can provide an eye-catching display.

Sagittaria lancifolia L.
Lanceleaf Arrowhead or Bulltongue Arrowhead
Alismataceae, Water-plantain Family

Habit: Herbaceous aquatic perennial, to 1.5 m tall, from thick stout underground runners.

Leaves: Above water leaves alternate; stalks to 1.2 m long; blades elliptic, to 40 cm long and 10 cm wide, leathery, thick.

Inflorescences: Flowers in 5-12 whorls, at top of leafless flowering stem, male and female flowers separate, in separate whorls in the same inflorescence, lowest 3-4 whorls female, upper whorls male.

Flowers: White, petals 3, round, each 1-2 cm long, male flowers with 28-30 stamens.

Fruit: A nutlet-like achene, 1.5-2.3 mm long, winged; fruiting heads to 1.5 cm wide.

Habitat and Distribution: Common; ponds, lakes, streams, rivers, canals, ditches, and swamps; throughout Florida, west to Oklahoma and Texas, and north to Maryland and Delaware.

Comment: The large white whorled flowers of Lanceleaf Arrowhead are showy and can be seen in all warm months. It is easy to transplant into wet areas and is frequently used as an ornamental in ponds.

Sagittaria graminea Michx.

Sagittaria lancifolia L.

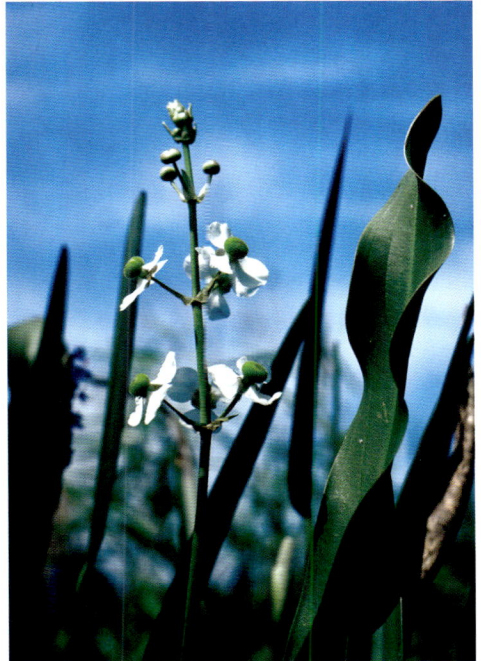

White

Sambucus canadensis L.
Elderberry
[*Sambucus nigra* L. subsp. *canadensis* (L.) R. Bolli; *Sambucus simpsonii* Rehder]
Caprifoliaceae, Honeysuckle Family (Adoxaceae, Moschatel Family)

Habit: Shrub, 2-4 m tall, with underground runners. Stems woody, easily broken, with white pith.

Leaves: Opposite; stalked; blades with 3-11 narrow elliptical leaflets, each 5-18 cm long, margins toothed.

Inflorescences: Terminal, dense, flat-topped, long-stalked clusters.

Flowers: White, 3-5 mm across, 5-lobed.

Fruit: A berry-like drupe, round, 4-6 mm in diameter, purplish black.

Habitat and Distribution: Common; moist to wet habitats, thickets, roadsides, ditches, banks, clearings, floodplains, and other moist disturbed sites; throughout Florida, south to the West Indies, west to Texas and Mexico, and north to South Dakota, Manitoba, Quebec, and Nova Scotia.

Comment: Elderberry flowering occurs from spring into fall. The very large flat clusters of flowers are easily seen in moist to wet areas along most roads. The large clusters of the dark purplish fruits are also attractive. Many animals eat the fruits and consequently carry the seeds to a great many habitats, including upland dry areas. Elderberry is a common weed in pastures. The underground runners frequently produce large colonies. The cooked flowers and ripe fruits are edible and used for several foods and drinks, including: pies, wines, jellies, preserves. The unripe fruits and all other parts of the plant are POISONOUS. Dried leaves have been used as an insecticide. Pith is used in laboratories for holding specimens to make sections for microscopic examination.

Samolus ebracteatus Kunth
Water Pimpernel or Limerock Brookweed
Primulaceae, Primrose Family

Habit: Herbaceous perennial, 10-60 cm tall. Stem smooth, seldom branched, sometimes in tufts.

Leaves: Alternate, basal rosette when young; stalkless or winged at base; blades oblong, wider towards ends, 3-10 cm long.

Inflorescences: Long, spike-like stalks from the leaf axils.

Flowers: White to pinkish, 5-9 mm wide, with 5 squarish lobes.

Fruit: A capsule, 3-4 mm wide.

Habitat and Distribution: Frequent; brackish and fresh wetlands, dunes, wet limestone, streams; throughout Florida, and west along the Gulf coast to Texas, Oklahoma, Nevada, Mexico, and the West Indies.

Comment: Blooming in all the warm months, this grayish green wildflower is not easy to see. The flowers are small but several appear at the same time. When several tufted plants occur in a bare area Water Pimpernel never fails to provoke comment.

Sambucus canadensis L.

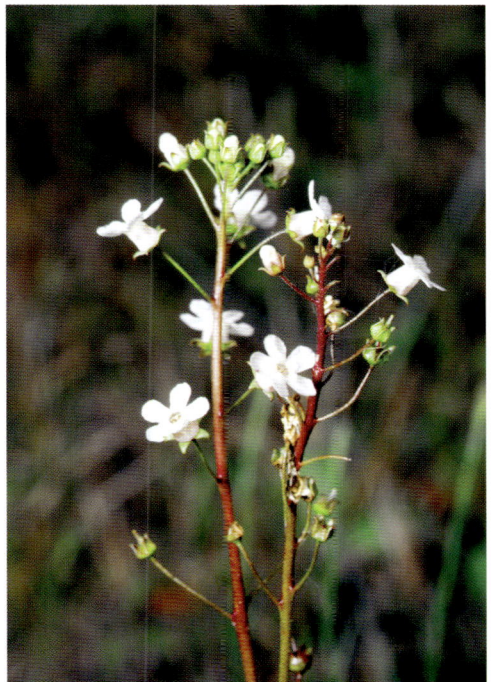

Samolus ebracteatus Kunth

White

Sanguinaria canadensis L.
Bloodroot or Red Puccoon
Papaveraceae, Poppy Family

Habit: Herbaceous perennial, 8-40 cm tall, from thick, underground runner, runner with red juice.

Leaves: Leaf usually 1; stalked; blade rounded, 5-18 cm wide, shallowly- or deeply-lobed, with undulating edges.

Inflorescences: Solitary flower on a long stalk.

Flowers: White, to 6 cm diameter, with 8-12 elliptic petals, 1.5-3 cm long.

Fruit: A capsule, slender, 3-6 cm long and 0.5-1 cm diameter.

Habitat and Distribution: Infrequent to occasional; rich woods, mesic hammocks, deciduous forests, wooded slopes; panhandle Florida, west to Texas, and north to Manitoba, Ontario, Quebec, and Nova Scotia.

Comment: The striking single flower pops up in early spring. The bright white blooms stand out against the dark background of the forest. Bloodroot can easily be transplanted. Occasionally, it is offered in nurseries. One cultivated form is listed. The dried underground runners have been used medicinally.

Sarcostemma clausum (Jacq.) Roem. & Schult.
White Vine, Milk Vine, White Twine Vine
[*Funastrum clausum* (Jacq.) Schltr.]
Asclepiadaceae, Milkweed Family (Apocynaceae, Dogbane Family)

Habit: Perennial vine, to 4 m long, forming mats over other vegetation. Stems sometimes semi-woody.

Leaves: Opposite; short-stalked; blades somewhat rounded, to oblong, to nearly linear, 2-8 cm long, somewhat succulent.

Inflorescences: Stalked clusters from the leaf axils.

Flowers: White, fragrant, to 1.3 cm wide, with 5 pointed petals.

Fruit: Dry, capsule-like, opening on one side, slender, tapered, to 8 cm long.

Habitat and Distribution: Frequent; hammocks, low wet coastal habitats; central and southern peninsula Florida; the West Indies, continental tropical America.

Comment: White Vine can bloom in any warm month. This vine is most often noticed when the clusters of the fragrant flowers appear. Sometimes regarded as a weed when forming mats over hedges and other landscaping.

Sanguinaria canadensis L.

Sarcostemma clausum
(Jacq.) Roem. & Schult.

White

Saururus cernuus L.
Lizard's-tail
Saururaceae, Lizard's-tail Family

Habit: Herbaceous perennial, 0.3-1.2 m tall, from underground runners. Stems erect, branched, hairy.

Leaves: Alternate; stalked; blades heart-shaped, 5-15 cm long, and 2-9 cm wide, margins smooth.

Inflorescences: Long, tapering spikes from leaf axils, spikes droop toward tips.

Flowers: White, comprised of 6-8 filaments.

Fruit: Fleshy, wrinkled, oval, 2-3 mm long.

Habitat and Distribution: Common; wet habitats, swamps, marshes, ditches, hammocks, and shores; throughout Florida, west to Oklahoma and Texas, and north to Kansas, Minnesota, Ontario, and Quebec.

Comment: Summer blooming, but very noticeable anytime because of the heart-shaped leaves. The plants grow in profusion and are seemingly in all shallow wet habitats. The showy long, tapering spikes of flowers resemble a lizard's tail for which the plant is named.

Scaevola plumieri (L.) Vahl
Beachberry, Inkberry, Gull-feed, Black Soap, Mad Moll
Goodeniaceae, Goodenia Family

Habit: Herbaceous perennial or shrub, 0.3-1.8 m tall, succulent, from underground runners.

Leaves: Alternate, spiraled, dense, with a tuft of silky hairs in the axils; stalkless; blades spatulate, to 10 long, thick.

Inflorescences: Loose arrangements in leaf axils.

Flowers: White to pinkish, 2-3 cm long, tubular, tube open to base on one side, hairy inside tube.

Fruit: A berry-like drupe, round, 1-1.5 cm in diameter, black, juicy.

Habitat and Distribution: Frequent; coastal dunes and strands; coastal central and south peninsula Florida; south into the West Indies and continental tropical America.

Comment: Beachberry can be found blooming in any warm month. The peculiar flowers, open on one side, are sometimes called half-flowers. The rich green succulent leaves suffused with white flowers peeping out along the stem make this a good native ornamental to plant along the shore. The colonial nature of the plants helps to secure sandy soils.

Saururus cernuus L.

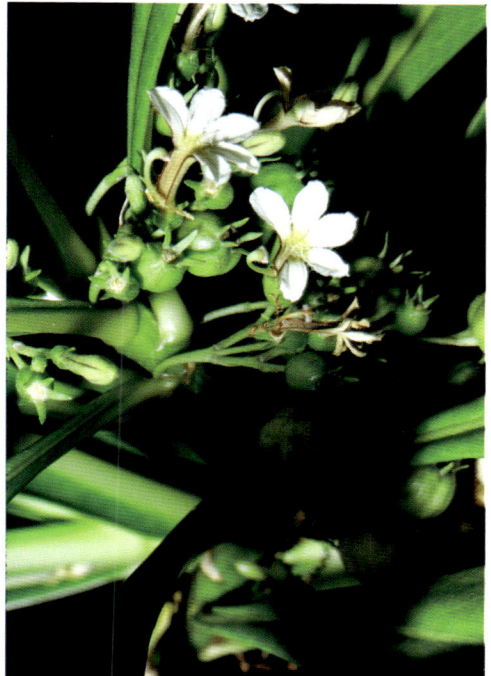

Scaevola plumieri (L.) Vahl

White

Schinus terebinthifolius Raddi
Brazilian Pepper Tree or Christmas-berry Tree
Anacardiaceae, Cashew Family

Habit: Shrub, to 10 m tall. Stems and branches usually arching, young branches hairy.

Leaves: Alternate; stalked; blades divided into 3-11 (usually 7) leaflets, leaflets stalkless, leaflets elliptic to lance-shaped, each to 5 cm long, margins smooth or toothed.

Inflorescences: Terminal, large, branched arrangements, male and female flowers separate and on separate plants.

Flowers: White, 5 petals, each about 1.5 mm long.

Fruit: A drupe, red, round, about 6 mm in diameter.

Habitat and Distribution: Common; virtually any disturbed area, invades all natural communities, washes, canyons; scattered counties in the north peninsula and throughout central and southern peninsula Florida, southern California; widely introduced into warm temperate climates within the New World; a native of South America.

Comment: The large clusters of flowers can be seen in spring and summer. The bright, holly-like, fruits are very showy. As an ornamental Brazilian Pepper Tree is highly desirable, but is so weedy and invasive that it is prohibited from sale by the State of Florida. Several birds and other wildlife eat and spread the seeds. The plant is cold sensitive which is preventing its spread northward. The dried fruits are ground and used as pepper for seasoning. Some individuals are allergic to the flowers and fruits that can cause dermatitis. The sap, like that of other plants (Poison-ivy, Poison-oak, Poison-wood, Poison Sumac, Mango) in this family, can cause a severe contact DERMATITIS. This plant is listed by the Florida Exotic Plant Pest Council as a Category I Invasive Species.

Schoenolirion albiflorum (Raf.) R.R. Gates
White Sunnybell
[*Schoenolirion elliottii* Feay ex A. Gray]
Liliaceae, Lily Family

Habit: Herbaceous perennial, to 70 cm tall, from a bulb-like base.

Leaves: Alternate, basal; stalkless; blades long, grass-like, narrow.

Inflorescences: Spike-like, usually branched arrangements at stem tips.

Flowers: White, petals 5-6 mm long.

Fruit: A rounded, 3-lobed and 3-seeded capsule, 5-6 mm wide.

Habitat and Distribution: Occasional, can be locally common; cypress ponds, cypress swamps, cypress prairies, and wet flatwoods; peninsula to central panhandle Florida, adjacent Alabama and Georgia.

Comment: Spring and summer, White Sunnybell is a noteworthy wildflower in wet habitats, especially where Cypress is present and deep water does not stand.

Schinus terebinthifolius Raddi

Schoenolirion albiflorum
(Raf.) R.R. Gates

White

Serenoa repens (W. Bartram) Small
Saw Palmetto
Arecaceae (Palmae), Palm Family

Habit: Perennial shrub, normally prostrate or to 7 m tall. Stems horizontal, leaning, or erect, branching.

Leaves: Alternate; long-stalked, stalks with spiny teeth on margins; blades fan-like, to 1 m broad, green or waxy blue-green.

Inflorescences: Short, thick, branched spikes from leaf bases.

Flowers: White, 5-6 mm long, with 3 petals.

Fruit: Oblong to ellipsoid, to 2 cm long, black, juicy, with a hard elongated seed.

Habitat and Distribution: Common; flatwoods, scrub, coastal dunes, prairies, upland islands in marshes, and sandy hammocks; throughout Florida, west to Louisiana, and north to South Carolina.

Comment: Flowering in spring and summer the fragrant attractive blooms are popular with bees. Fruit is edible. The taste of various plants and can be described as barely edible to almost good. Saw Palmetto is one of if not the oldest plant in the Southeast U.S. The stems growing along the ground are perhaps hundreds of years old. Stems grow very slowly and commonly the oldest are nearly to over one hundred feet long. It is very difficult to trace older stems. Stems of various plants grow in a tangle over and under each other. Also, as stems begin decaying they often have gaps because portions do not decay at the same rate. The low habit, slow growth, and clusters of flowers and fruit make this a highly desirable shrub for landscaping. The green or waxy bluish-green color forms are massed together, but can be easily separated by color form in the wild.

Sida acuta Burm. f.
Southern Sida or Wire-weed
[*Sida carpinifolia* L. f.]
Malvaceae, Mallow Family

Habit: Herbaceous annual or short-lived perennial, 0.3-1 m tall, from a taproot. Stems branching, often low-spreading.

Leaves: Alternate; short-stalked; blades lance-shaped to linear, 2-10 cm long, wider toward base, green on both surfaces, margins with teeth.

Inflorescences: Flowers solitary, short-stalked, in leaf axils.

Flowers: White or usually yellow, with 5 petals, each 0.8-1.5 cm long.

Fruit: 5-8 mm broad, comprised of a ring of 7-12 capsules, each 2-toothed.

Habitat and Distribution: Common; disturbed sites, lawns, roadsides, fields; throughout Florida, west to Mississippi, and north to southern coastal South Carolina; West Indies, Central and South America, Africa.

Comment: Normally thought of as a weed, Southern Sida flowers during warm weather. When several plants are aggregated in a lawn or pasture the effect is very attractive. Sometimes called Wire-weed because the deep taproot is as difficult to pull out of the ground as buried rusted wire.

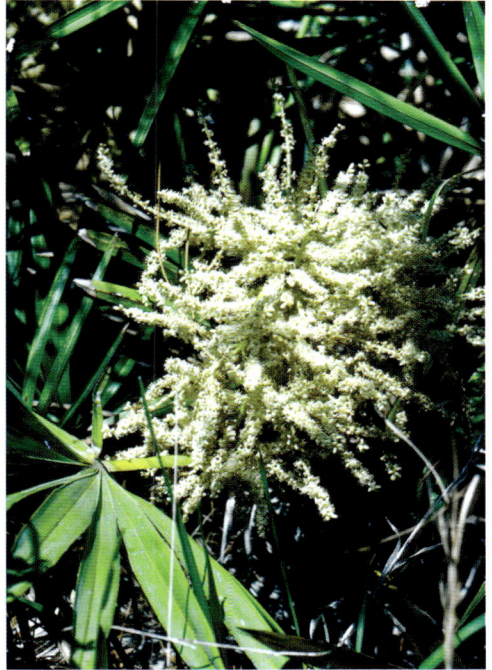

Serenoa repens
(W. Bartram) Small

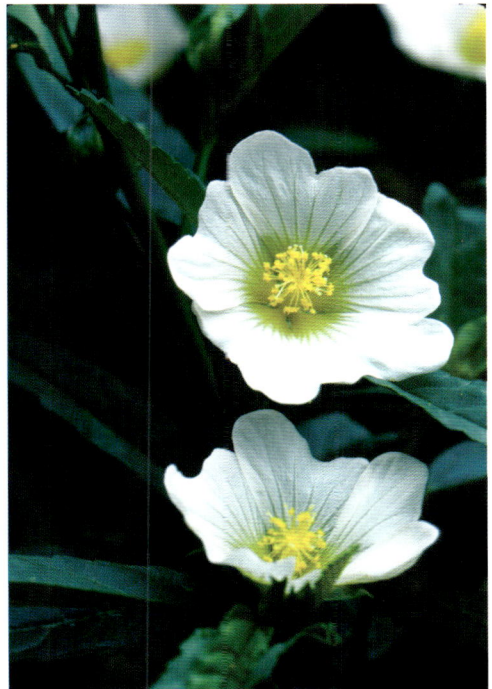

Sida acuta Burm. f.

White

Silene antirrhina L.
Sleepy Catchfly
Caryophyllaceae, Pink Family

Habit: Herbaceous annual, 30-90 cm tall, from a taproot. Stems erect, branched, often with sticky areas just below the joints.

Leaves: Opposite; stalkless; blades basal and along stems, spatulate to lance-shaped to linear, 2-7 cm long.

Inflorescences: Open, terminal, branched arrangements.

Flowers: White or pink, tubular, 6-10 mm long, with 5 petals.

Fruit: A capsule, rounded, to 8 mm long.

Habitat and Distribution: Occasional; sandy soils of open, disturbed sites, fields; central peninsula Florida northward throughout the United States and southern Canada, south to Mexico.

Comment: Sleepy Catchfly is a weed that blooms in spring and summer. The petals barely exceed the sepals. It is not much of a wildflower and more of a curiosity.

Solanum americanum Mill.
American Black Nightshade
[*Solanum nigrum* L., misapplied]
Solanaceae, Nightshade Family

Habit: Herbaceous annual, to 1 m tall. Stems smooth, branched.

Leaves: Alternate; stalked; blades broadly elliptical to somewhat lance-shaped, 2-10 cm long, margins smooth to wavy and coarsely toothed.

Inflorescences: Stalked, nodding, somewhat flat-topped clusters, from leaf axils upwards on the plant.

Flowers: White, star-shaped, 6-8 mm wide, with 5 petals curving out and backwards.

Fruit: A berry, round, 5-9 mm diameter, shiny black when ripe.

Habitat and Distribution: Common; dry habitats, disturbed sites, margins, fields, roadsides, gardens, openings; throughout Florida, west to Texas, and north to North Dakota and Maine.

Comment: The numerous white flowers found on a mature American Black Nightshade plant are sparkly. The plants are often larger than the vegetation surrounding them so that as wildflowers they stick out and are very attractive. The plant is very weedy. Many animals and birds eat the fruit and distribute the seeds. The entire plant and unripe fruits should be considered POISONOUS.

Silene antirrhina L.

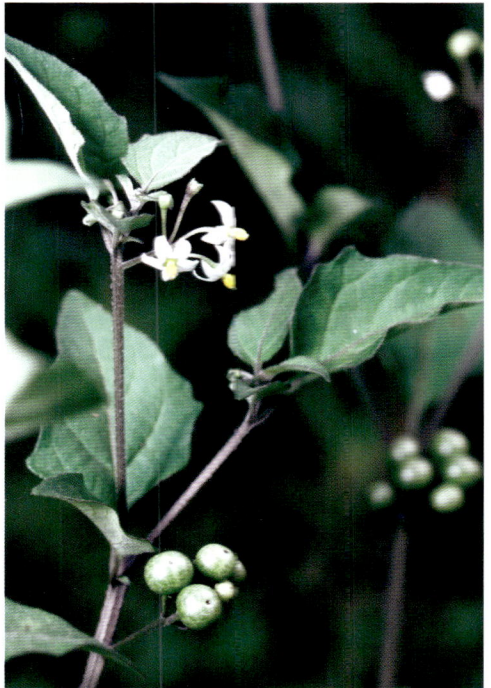

Solanum americanum Mill.

White

Solanum donianum Walp.
Blodgett's Nightshade or Mullein Nightshade
[*Solanum blodgettii* Chapm.; *Solanum verbascifolium* L.]
Solanaceae, Nightshade Family

Habit: Shrub or small tree, to 3.5 m tall. Branches spreading, young branches densely star-shaped hairy with short yellowish or grayish hairs.

Leaves: Alternate; distinctly stalked; blades elliptic-oblong to oblong, to 25 cm long and 7 cm wide, sandpapery, densely star-shaped hairy, tips sharp-pointed, bases rounded.

Inflorescences: At stem tips, terminal or sometimes opposite the uppermost leaves, in several- to many-flowered rounded clusters, branches forking in pairs.

Flowers: White or slightly bluish, united at the base, 5-7.5 mm long, densely star-shaped hairy on the outside; anthers prominent, 4.5-5.5 mm long; sepals united, parted to the middle, rounded, 2-2.5 mm long hairy.

Fruit: A round, shiny, red, smooth berry, 5-7.5 mm in diameter; seeds flattened, 2-2.5 mm long.

Habitat and Distribution: Occasional; hammocks, open marshy areas, scrub lands, thickets; southern peninsula Florida, Bahamas, British Honduras, southern Mexico.

Comment: Flowers, though irregular, can be found at any time of year. The green leaves are not especially noticeable, but Blodgett's Nightshade flowers make this an attractive shrub. The plant is even more handsome because the fruit ripens to a shiny, bright red while the younger flowers are still opening . If ingested, as with all members of the nightshades, foliage and unripe fruits are regarded as POISONOUS.

Solanum pseudogracile Heiser
Large-flowered Black Nightshade
Solanaceae, Nightshade Family

Habit: Herbaceous annual, to 70 cm tall. Stems trailing, climbing.

Leaves: Alternate; stalked; blades ovate to diamond-shaped, to 11 cm long, margins with rough teeth.

Inflorescences: Stalked, nodding, somewhat round-topped clusters, from leaf axils on the upper plant.

Flowers: White or purplish, star-shaped, to 1 cm wide, with 5 petals.

Fruit: A berry, round, to 1 cm diameter, black when ripe.

Habitat and Distribution: Infrequent; disturbed habitats; from central peninsula and panhandle Florida north into the northeast states.

Comment: The flowers of Large-flowered Black Nightshade are easily seen and identified by the large flowers occurring from spring into fall. The larger fruit of this species is also noteworthy. Like most of the Black Nightshades this plant is also quite weedy, but less frequent. The Black Nightshades have been widely used as medicinal and food plants in various parts of the world. Ingested foliage and fruits are regarded as POISONOUS.

Solanum donianum Walp.

Solanum pseudogracile Heiser

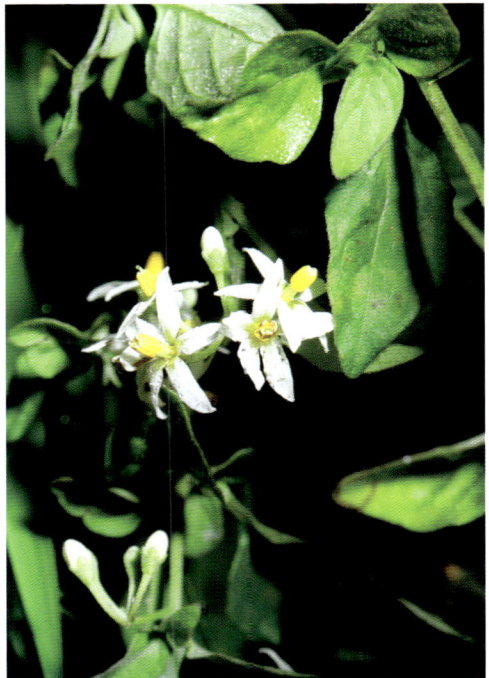

White

Solanum sisymbriifolium Lam.
Sticky Nightshade
Solanaceae, Nightshade Family

Habit: Herbaceous annual, 0.3-1 m tall. Stems with shaggy and sticky hairs and yellow spines.

Leaves: Alternate; stalked; blades lance-shaped, 6-22 cm long, with multiple often deep lobes, top and bottom with sticky and star-shaped hairs, veins with yellow spines.

Inflorescences: 1 to several flowers in elongated terminal arrangements or along stem.

Flowers: White to blue, to 3 cm across, with 5 pointed petals.

Fruit: A berry, round, to 2 cm diameter, red with yellow spines.

Habitat and Distribution: Rare to infrequent; disturbed habitats, roadsides; central peninsula Florida, west to Texas, and north to Ohio and New England; native to Argentina.

Comment: The large flowers bloom from spring into fall followed by the large red fruits covered with yellow spines. The plant is not very attractive. As always, the foliage and unripe fruits are regarded as TOXIC if ingested.

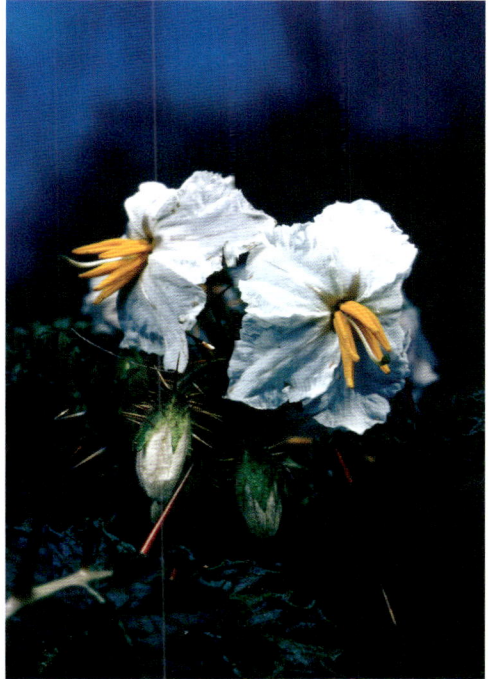

Solanum sisymbriifolium Lam.

White

Spermacoce verticillata L.
White Head Broom or Boton Blanco
[*Borreria terminalis* Small; *Borreria verticillata* (L.) Meyer]
Rubiaceae, Madder Family

Habit: Herbaceous semi-woody perennial, 5-50 cm tall, sometimes rooting at nodes.

Leaves: Opposite; stalkless; blades linear to linear-lance-shaped, 1-5 cm long, sometimes several smaller leaves clustered in the axils.

Inflorescences: Round, dense, terminal and upper axillary clusters.

Flowers: White, tubular, to 3 mm long, with 2 lobes.

Fruit: A tiny, smooth, leathery capsule, about 1 mm in diameter.

Habitat and Distribution: Frequent to common; dry disturbed habitats, lawns, pastures, roadsides, and other open areas; peninsula Florida, coastal Texas; throughout the West Indies, warm temperate and tropical America, south Pacific, tropical Africa; a native of tropical America.

Comment: This small shrubby plant is more of a weed than a wildflower, but the dense clusters of very small flowers bloom in all warm months and provide a bit of wildflower value to otherwise green open spaces.

Spermolepis divaricata (Walter) Raf.
Roughfruit Scaleseed or Southern Spermolepis
Apiaceae (Umbelliferae), Carrot Family

Habit: Herbaceous annual, 10-70 cm tall. Stems slender, solitary or branching.

Leaves: Alternate; short-stalked or lacking stalks; blades 1-8 cm long, finely divided into very narrow or threadlike segments, about 0.5 mm wide, divisions opposite on leaf axis.

Inflorescences: Wide-spreading, thin, long-stalked clusters in upper leaf axils.

Flowers: White, minute, with 5 petals, each to 1 mm long; no sepals.

Fruit: Two-parted, rounded, 1.5-2 mm long, about 1.5 mm wide, warty with many small bumps.

Habitat and Distribution: Frequent; moist to wet disturbed sands, exposed pond bottoms, moist woodlands, prairies; south peninsula Florida, west to Texas, and north to Kansas, Missouri, and Virginia.

Comment: Roughfruit Scaleseed is a thin weedy-looking plant flowering in spring. In open wet sandy disturbed areas it is very difficult to separate this plant from Mock Bishop's-weed [*Ptilimnium capillaceum* - minute sepals, leaf axis with whorled divisions, fruit smooth] and Marsh Parsley [*Cyclospermum leptophyllum* (*Apium leptophyllum*), not pictured - no sepals, leaf axis with opposite divisions, fruit smooth].

Spermacoce verticillata L.

Spermolepis divaricata (Walter) Raf.

White

Spiranthes odorata (Nutt.) Lindl.
Fragrant Ladies'-tresses
[*Spiranthes cernua* (L.) Rich. var. *odorata* (Nutt.) Correll]
Orchidaceae, Orchid Family

Habit: Herbaceous perennial, 0.1-1.1 m tall, from above ground runners. Stems erect, smooth on the lower stem, hairy on the upper stem.

Leaves: Alternate, 3-5; stalkless; blades lance-shaped and broader near tip, 5-52 cm long and 0.6-4 cm wide.

Inflorescences: 10-30 flowers at stem tips, in several tight spirals around stalk.

Flowers: White, tubular, 1-1.8 cm long, lip oblong and tapering to tip, fragrant.

Fruit: A capsule, ellipsoid, 1 cm long.

Habitat and Distribution: Frequent; swamps, wet grassy habitats, wet margins, floodplains, often in shallow standing water; throughout Florida, west to Texas, Oklahoma, and Arkansas, and north to Delaware.

Comment: This Ladies'-tresses blooms in fall and winter. The large flowers, tall stem, and wet habitat help identify this orchid. Even though the plant is quite thin the large spiraled flowers are noticeable. An added attraction is the very pleasant fragrance from the flowers.

Spiranthes longilabris Lindl.
Long-lipped Ladies's-tresses
Orchidaceae, Orchid Family

Habit: Herbaceous perennial, 12-60 cm tall. Stems erect, sparsely hairy.

Leaves: Alternate, 3-5, basal; stalkless; blades lance-shaped, 3-15 cm long, and less than 5 mm wide, sometimes withering before flowering.

Inflorescences: 10-30 flowers at stem tips, in a single spiral around stalk.

Flowers: White or cream, tubular, 0.9-1.5 cm long, lip oblong with a wavy and lacerate tip.

Fruit: A capsule, ellipsoid, 8 mm long.

Habitat and Distribution: Occasional; wet habitats, swamps, wet flatwoods, prairies, marshes, and meadows; throughout Florida, west to Texas, and north to Virginia.

Comment: Like most Ladies'-tresses, this long-lipped species will catch the eye due to the spiral arrangement at stem tips. Identification features are the longer flowers and single spiral.

Spiranthes odorata (Nutt.) Lindl.

Spiranthes longilabris Lindl.

White

Stellaria pubera Michx.
Giant Chickweed, Star Chickweed, Great Chickweed
[*Alsine pubera* (Michx.) Britton]
Caryophyllaceae, Pink Family

Habit: Herbaceous perennial, 10-40 cm tall. Stems erect to spreading, hairy.

Leaves: Opposite; stalkless; blades elliptic to lance-shaped, 1-9 cm long.

Inflorescences: Few to many flowers at stem tips in a widely branched arrangement.

Flowers: White, 0.7-1.2 cm across, with 5 deeply lobed petals resembling 10 slender petals.

Fruit: A capsule, 3.5-5.5 mm long.

Habitat and Distribution: Frequent; dry habitats, hammocks, rich woods; Gadsden and Liberty Counties in panhandle Florida (rare), west to Alabama, and north to Illinois and New Jersey, seldom on the Coastal Plain.

Comment: The lobed petals of Giant Chickweed are so deeply lobed as to resemble ten instead of five petals. When flowering in spring the large starry blooms are conspicuous. They are sometimes planted in wild gardens.

Stylisma patens (Desr.) Myint
Trailing Stylisma
[*Bonamia patens* (Desr.) Shinners; *Bonamia patens* var. *angustifolia* (Nash) Shinners; *Stylisma angustifolia* (Nash) House; *Stylisma patens* subsp. *angustifolia* (Nash) Myint; *Stylisma trichosanthes* House, misapplied]
Convolvulaceae, Morning-glory Family

Habit: Herbaceous perennial vine, to 1.5 m long. Stems trailing, branched.

Leaves: Alternate; essentially stalkless; blades linear to narrowly lance-shaped, to 3 cm long, hairy.

Inflorescences: Flowers solitary on long stalks from leaf axils.

Flowers: White, bell-shaped, 1.3-2.5 cm long and 1.2 cm wide.

Fruit: A papery, 4-seeded capsule, 4-8 mm long.

Habitat and Distribution: Frequent; dry, sandy, open habitats - scrub, sandhills, pinelands, and roadsides; central peninsula Florida, west to Alabama, and north to North Carolina.

Comment: Trailing Stylisma blooms from spring into fall. The large white flowers, standing upright, on this inconspicuous prostrate vine are almost startling and seem so out of place for such a small, almost indistinct plant.

Stellaria pubera Michx.

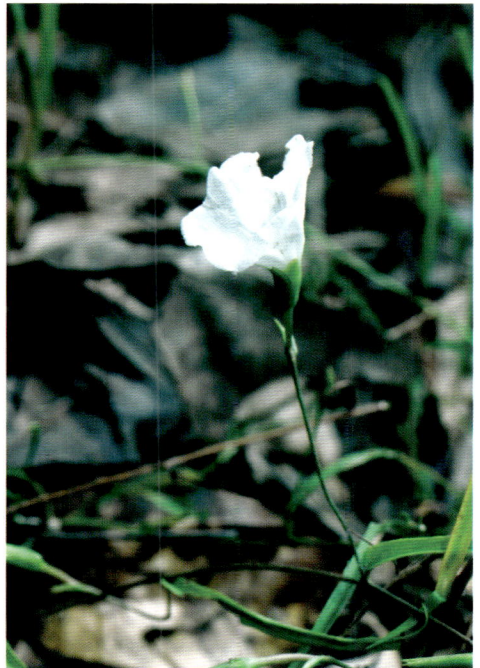

Stylisma patens (Desr.) Myint

White

Styrax americanus Lam.
Storax, American Snowbell, Mock-orange
[*Styrax pulverulentum* Michx.]
Styracaceae, Storax Family

Habit: Shrub, to 6 m tall.

Leaves: Alternate; stalked; blades elliptic to somewhat oblong, 3-8 cm long, margins smooth to toothed, smooth to hairy beneath.

Inflorescences: Flowers solitary or in groups of 2-4 in leaf axils.

Flowers: White, 8-12 mm long, hairy, with 5 slender lobes, curling back.

Fruit: A capsule, round, 6-8 mm in diameter.

Habitat and Distribution: Occasional to locally frequent; moist woods, swamps, shores, ponds in forests, forested depressions, mixed deciduous forests, along streams; central peninsula Florida, west to Texas, and north to Missouri, Illinois, Indiana, and Tennessee.

Comment: Storax flowers in spring. The flowers are large and attractive. Many blooms occur on the branches making for a good show as a shrubby ornamental wildflower.

Syngonanthus flavidulus (Michx.) Ruhland
Shoe-button or Yellow Hatpins
Eriocaulaceae, Pipewort Family

Habit: Herbaceous perennial, to 30 cm tall, from pale brown, spongy roots. Stems erect, thin, hairy.

Leaves: Alternate, in a basal rosette; stalkless; blades 1-7 cm long, very narrow.

Inflorescences: Terminal, in domed heads.

Flowers: White, grayish or yellowish, to 1 cm wide, bracts yellow.

Fruit: A capsule, 3-parted; seeds 0.4-0.5 mm long, smooth.

Habitat and Distribution: Common; wet habitats - margins, flatwoods, and marshes, often in shallow sandy depressions where water stands for very short periods; throughout Florida, west to Alabama, and north to North Carolina.

Comment: Shoe-button flowers in spring and the old heads can persist for months. Frequently the plants can be found in mass.

Styrax americanus Lam.

Syngonanthus flavidulus
(Michx.) Ruhland

White

Tephrosia chrysophylla Pursh
Golden Hoary-pea
Fabaceae (Leguminosae), Pea or Bean Family

Habit: Herbaceous perennial, 30-60 cm long. Stems lying flat, branched, spreading, hairy.

Leaves: Alternate; short-stalked; blades with 5-7 ovate leaflets, wider toward tips, tips notched, hairless and shiny above, hairy and with visible veins below.

Inflorescences: Small clusters, opposite the leaves.

Flowers: White changing to pink then reddish, typically pea-shaped, to 1-1.4 mm long.

Fruit: In flat, 3-4 cm long, slender, pods.

Habitat and Distribution: Frequent; dry sandy soils, pinelands, sandhills, and roadsides, from central peninsula Florida, west to Mississippi, and north to southeastern Georgia.

Comment: Golden Hoary-pea flowers from spring into fall. The flowers are easy to see on the open ground among the sparse vegetation. The distinctive leaves are often golden hairy underneath.

Thalictrum thalictroides (L.) A.J. Eames & B. Boivin
Rue Anemone or Windflower
[*Anemonella thalictroides* (L.) Spach]
Ranunculaceae, Buttercup Family

Habit: Herbaceous perennial, 5-25 cm tall, from thickened roots. Stems thin, smooth, erect.

Leaves: Alternate; basal long-stalked, stem stalkless; blades with 3 leaflets, each rounded, 3-toothed, 1-3 cm wide.

Inflorescences: Flowers solitary or in groups of 2-5 on the ends of stems.

Flowers: White, with 5-10 spreading, petal-like sepals, each 0.5-1.5 cm long, stamens numerous and yellow, 5-15 separate female carpels.

Fruit: Oblong, 3.5-4.5 mm long, 1-1.5 mm broad, ribbed.

Habitat and Distribution: Rare southward to frequent northward; limestone bluffs and woods, rich woods; Leon and Gadsden Counties in panhandle Florida, west to Mississippi and Arkansas, and north to Kansas, Minnesota, Ontario, and New Hampshire.

Comment: This delicate pretty flower is seen in spring. Rue Anemone is listed by the State of Florida as Endangered.

Tephrosia chrysophylla Pursh

Thalictrum thalictroides
(L.) A.J. Eames & B. Boivin

White

Thunbergia fragrans Roxb.
White Thunbergia or White-lady
Acanthaceae, Acanthus Family

Habit: Perennial vine, to 2 m long. Stems trailing or climbing.

Leaves: Opposite; stalked; blades broadly lance-shaped, 4-12 cm long, margins irregularly toothed.

Inflorescences: Flowers solitary, stalked, in leaf axils.

Flowers: White to purple or lavender, tubular, 3-5 cm wide, 5 lobed.

Fruit: A capsule, to 2 cm long, to 8 mm in diameter, a thick beak on top.

Habitat and Distribution: Occasional; disturbed habitats; central and south peninsula Florida; a native of India and Ceylon, now widely dispersed in warm regions.

Comment: The large flowers seen in most warm months, but mostly in late spring, are quite showy. White Thunbergia is very widely cultivated. It can be trained over a trellis or fence.

Tofieldia racemosa (Walt.) BSP.
Asphodel
Liliaceae, Lily Family

Habit: Herbaceous perennial, to 70 cm tall, from underground runners. Stems with sandpapery, rough, sticky hairs.

Leaves: Alternate, mostly basal; stalkless; blades grass-like, 7-40 cm long, 1-6 mm wide.

Inflorescences: 1-7 small clusters of 2 or 3 flowers along the upper stem.

Flowers: White, petals 3, sepals 3, 3-5 mm long, spreading.

Fruit: A capsule, to 3 mm long, rounded.

Habitat and Distribution: Occasional; low, wet pine flatwoods, pine savannas, seeps, and bogs; from northern peninsula Florida, west to Texas, and north to New Jersey.

Comment: The white spires of flowers can be seen poking out of the grasses and sedges in open wetlands during the summer and fall. The uppermost flowers bloom first.

Thunbergia fragrans Roxb.

Tofieldia racemosa (Walt.) BSP.

White

Toxicodendron radicans (L.) Kuntze
Poison-ivy
Anacardiaceae, Cashew Family

Habit: Perennial vine or small shrub, high climbing to 30 m or more. Stems with abundant aerial roots.

Leaves: Alternate; stalked; blades with 3 ovate to elliptic leaflets, each to 20 cm long, often wider at base, margins smooth or few toothed.

Inflorescences: Loose axillary clusters, male and female flowers separate and on separate plants.

Flowers: White or greenish, small, about 2 mm long, with 5 petals.

Fruit: A drupe, roundish, 4-7 mm broad, white or ivory.

Habitat and Distribution: Common; virtually all moist to wet, open to wooded habitats; throughout Florida, west to Arizona, south into Mexico, and north to British Columbia and Nova Scotia; Bermuda, Bahama Islands, Asia.

Comment: Flowering occurs from spring into early summer. The small flowers are not easily seen, but the fruit is somewhat ornamental. The entire plant, including the roots, contains a TOXIC sap that causes severe dermatitis on most individuals. The sap leaks out to all surfaces. If ingested the sap is quite POISONOUS. Animals seem to be immune to the toxicity. If plants are consumed by fire, the tiny microscopic particles within the smoke will be covered with the TOXIC sap. Very serious problems can be caused by this TOXIC smoke getting in eyes, nose, or mouth.

Tradescantia fluminensis Vell.
White-flowered Spiderwort or Small-leaf Spiderwort
Commelinaceae, Spiderwort Family

Habit: Herbaceous perennial, to 1 m or more long. Stems prostrate, smooth, somewhat succulent, flowering stems ascending.

Leaves: Alternate; sheathing at base, sheaths with long hairs on margin, stalkless; blades ovate, 2-9 cm long, somewhat succulent, with parallel veins.

Inflorescences: Small, terminal clusters.

Flowers: White, 3 petals, each 8-10 mm long.

Fruit: A capsule, rounded, to 3 mm long.

Habitat and Distribution: Occasional, locally common; moist to wet hammocks, and floodplains; central peninsula, west to the central panhandle in Florida; California; a native of South America.

Comment: Flowering in most warm months, White-flowered Spiderwort is an attractive wildflower. The shiny green leaves and stems are very appealing. It can weakly ascend over and into low vegetation. However, it can become dense as a blanket and smother all low native vegetation. This plant is listed by the Florida Exotic Plant Pest Council as a Category I Invasive Species.

Toxicodendron radicans
(L.) Kuntze

Tradescantia fluminensis Vell.

733

White

Tradescantia ohiensis Raf.
Common Spiderwort
Commelinaceae, Spiderwort Family

Habit: Herbaceous perennial, 20-80 cm tall, from fleshy roots. Stems usually single, smooth, erect or ascending.

Leaves: Alternate; sheathing at base, sheaths with hairs on margin, stalkless; blades linear, firm, 15-45 cm long, and 0.6-4 cm broad, with parallel veins.

Inflorescences: Terminal and axillary clusters on the upper stems.

Flowers: White or blue, 2-3 cm wide, petals 3, 1-3 leaflike bracts immediately below the clusters.

Fruit: A capsule, somewhat rounded, 4-6 mm long.

Habitat and Distribution: Frequent, locally common; woods, sandy roadsides, meadows, prairies, and dry disturbed sites; central peninsula Florida, west to Texas, and north to Minnesota and Massachusetts.

Comment: The white or blue flowers in spring are showy. There are usually many in each cluster. Dry roadsides and adjacent fields often provide the best displays. Common Spiderwort is being used more frequently in cultivation.

Trifolium nigrescens Viv.
Ball Clover
Fabaceae (Leguminosae), Pea or Bean Family

Habit: Herbaceous annual, to 40 cm tall. Stems clustered, sprawling, trailing and mat-forming or ascending.

Leaves: Alternate; stalked; blades with 3 leaflets, leaflets rounded, broader at tip, 0.6-2 cm long.

Inflorescences: Terminal or axillary rounded clusters, clusters about 1-1.5 cm in diameter.

Flowers: White to pink, 6-9 mm long.

Fruit: A flattened, small, ovoid, 1-4 seeded pod.

Habitat and Distribution: Occasional; vacant lots, fields, lawns, roadsides and other disturbed sites; central and western panhandle Florida, Alabama, Mississippi, Louisiana, and Tennessee.

Comment: Flowering in spring and summer, this small clover has dirty white to pinkish flowers in round clusters. It is sometimes cultivated.

Tradescantia ohiensis Raf.

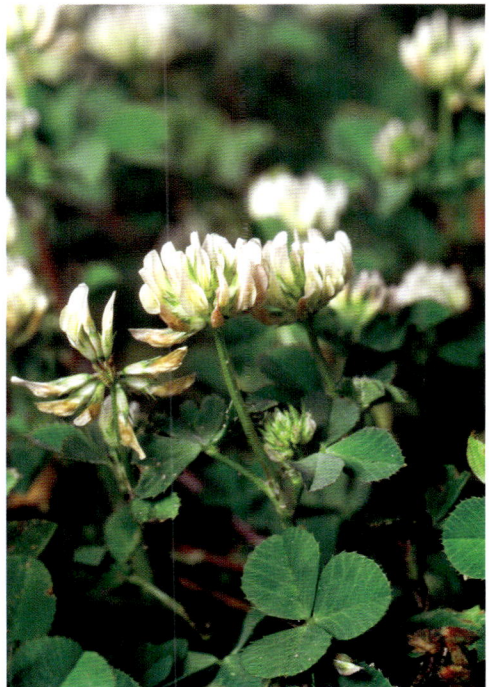

Trifolium nigrescens Viv.

735

White

Trifolium repens L.
White Clover, Dutch Clover, Ladino Clover
Fabaceae (Leguminosae), Pea or Bean Family

Habit: Herbaceous perennial, 10-40 cm long, with above ground runners. Stems creeping, branching, smooth or slightly hairy, sometimes rooting at nodes.

Leaves: Alternate; long-stalked; blades with 3 roundish leaflets, broader at tip, each 1-3 cm long, with finely toothed edges.

Inflorescences: Terminal or axillary rounded clusters, clusters 1-3 cm in diameter.

Flowers: White or pink, 6-12 mm long.

Fruit: A slender, flattened, 4-5 mm long, 3-4-seeded pod.

Habitat and Distribution: Common; turf, fields, meadows, agricultural, and other disturbed sites; throughout Florida, and most of the United States; a native of Eurasia.

Comment: Flowering from spring into fall, White Clover is a common wildflower. It is an extremely common forage and turf-forming plant, frequently found as a weed in lawns. It is thought to have been cultivated in North America in the early 1700's.

Typha domingensis Pers.
Southern Cat-tail
Typhaceae, Cat-tail Family

Habit: Herbaceous perennial, 2-3 m tall, from underground runners. Stems solitary, erect.

Leaves: 6-10, alternate, but twist to appear as if arranged on opposite sides of the stem, from a basal sheath; stalkless; blades linear, tapering at tip, 2-3 m long, 0.6-1.2 cm wide, slightly rounded on back surface toward tip.

Inflorescences: Terminal cylindrical spikes, male and female flowers separate, but on the same plant, male spikes located above female spikes, male and female spikes separated by a bare-stemmed constriction 0.7-6 cm long, female spikes to 45 cm long and 2.2 cm in diameter.

Flowers: Whitish, becoming brown, male flowers quickly falling, female flowers persisting.

Fruit: A tiny cylindrical nutlet with a persistent style.

Habitat and Distribution: Common; shallow fresh and brackish water of ditches, marshes, and ponds; throughout Florida, west to California, and north to Utah, Nevada, Kansas, Maryland, and Delaware.

Comment: Flowering in spring and summer, these usually tall, quite distinctive plants are frequently weedy and can cover large disturbed wet areas. In deeper nutrient rich water they often reach their maximum height of about 3 meters. The rootstock was used for food. The leaves can be used for making mats.

Trifolium repens L.

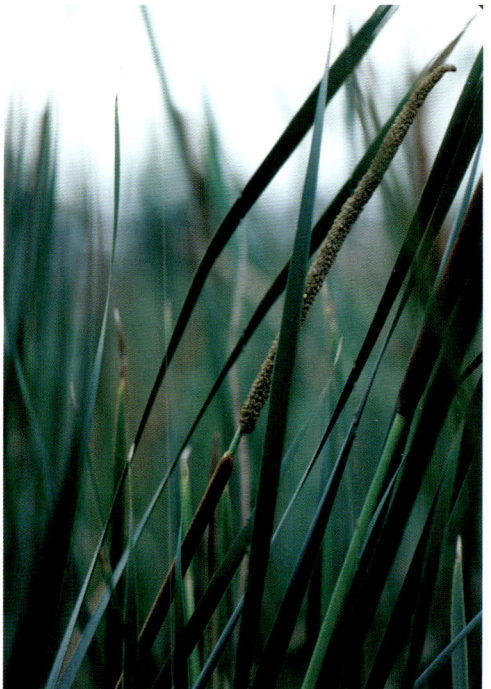

Typha domingensis Pers.

White

Typha latifolia L.
Common Cat-tail
Typhaceae, Cat-tail Family

Habit: Herbaceous perennial, 1-2.5 m tall, from underground runners. Stems solitary, erect.

Leaves: Alternate, but twist to appear as if arranged on opposite sides of the stem, from a basal sheath; stalkless; blades linear, tapering at tip, 1-2.5 m long, 0.8-1.5 cm wide, flat on back surface toward tip.

Inflorescences: Terminal cylindrical spikes, male and female flowers separate, but on the same plant, male spikes located above female spikes, male and female spikes contiguous, not separated, female spikes to 20 cm long and 3 cm in diameter.

Flowers: Whitish, becoming brown, male flowers quickly falling, female flowers persisting.

Fruit: A tiny cylindrical nutlet with a persistent style.

Habitat and Distribution: Common; shallow fresh water of ditches, marshes, ponds, lakes, rivers, and streams; throughout Florida, west to Mexico and California, and north to Alaska and Newfoundland.

Comment: Like Southern Cat-tail, Common Cat-tail blooms in spring and summer and is seemingly in all wet disturbed sites. As for Southern Cat-tail, the Common Cat-tail rootstock was used for food and the leaves for making mats.

Vaccinium arboreum Marshall
Sparkleberry
Ericaceae, Heath Family

Habit: Shrub or small tree, to 10 m tall. Stems crooked, bark grayish, flaking to reveal a reddish inner bark, twigs round, young twigs hairy, older twigs smooth.

Leaves: Alternate, usually evergreen; short-stalked or stalkless; blades oval, 2-4 cm long and 1-3 cm wide, leathery, with minute golden glands along the margins.

Inflorescences: Each flower stalked and along a branch-like axis from leaf axils, on last year's branches.

Flowers: White to pinkish, bell-shaped, 5-8 mm long.

Fruit: Berries round, 5-8 mm diameter, black.

Habitat and Distribution: Frequent; sandy, very, dry wooded sites; from central peninsula Florida, west to Texas, and north to Kansas, Missouri, Illinois, and Virginia.

Comment: This shrub is a remarkable cultivated plant. The many flowers in spring are showy. Black berries mature and persist into the winter. The edible, but not good, fruits are somewhat dry, and do provide some food for birds. The semi-evergreen shiny almost round leaves are attractive. An added bonus is the crooked stem with peeling bark giving the stem a mottled grayish and reddish appearance.

Typha latifolia L.

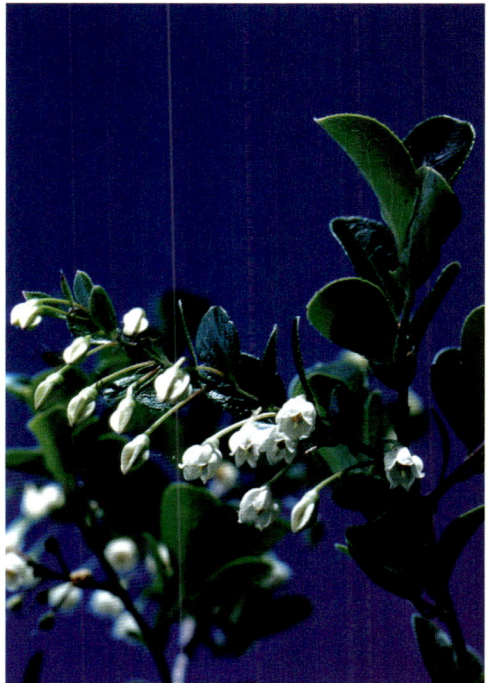

Vaccinium arboreum Marshall

White

Vaccinium corymbosum L.
Highbush Blueberry
Ericaceae, Heath Family

Habit: Shrub, 1-3 m tall. Stems erect, with 1 to many branches from base, twigs hairless, angled.

Leaves: Alternate; short-stalked; blades elliptical, 4-8 cm long and 1.5-4 cm broad, with usually smooth margins.

Inflorescences: Clusters on last year's shoots, just before or as new leaves appear.

Flowers: White, urn-shaped, 5-11 mm long, usually drooping.

Fruit: Berries round, 5-12 mm diameter, blue to black, smooth.

Habitat and Distribution: Frequent; usually in wet pinelands and other low wet wooded habitats; throughout Florida, west to Texas and Arkansas, and north to Virginia.

Comment: Attractive flowers in spring precede the sweet edible fruits. The green angled twigs help with identification during the late summer and winter when no flowers or fruit are present.

Vaccinium myrsinites Lam.
Shiny Blueberry
Ericaceae, Heath Family

Habit: Shrub, 20-60 cm tall, from underground runners. Stems branched, twigs slightly angled, smooth to hairy.

Leaves: Alternate, evergreen; short-stalked; blades elliptic, 0.8-1.5 cm long, shiny, hairy beneath, margins toothed.

Inflorescences: Clusters of 2-8 flowers on last year's shoots, just before or as new leaves appear.

Flowers: White, urn-shaped, white to pink, 6-8 mm long, drooping.

Fruit: Berries round, 6-8 mm diameter, blue to black.

Habitat and Distribution: Common; sandhills and flatwoods; throughout Florida, west to Alabama, and north to South Carolina.

Comment: Shiny Blueberry is a small colonial shrub spreading extensively by means of underground runners. When blooming the white flowers in spring are attractive, especially when combined with the shiny leaves. The sweet edible fruit is an added bonus. This shrub ought to be considered for low borders.

Vaccinium corymbosum L.

Vaccinium myrsinites Lam.

White

Vaccinium stamineum L.
Deerberry
Ericaceae, Heath Family

Habit: Perennial shrub, to 4 m tall, when damaged often forming underground runners. Stems commonly numerous, branched, twigs round, hairy or smooth.

Leaves: Alternate; short-stalked; blades broadly elliptic, 3-8 cm long, thin, deciduous, pale waxy gray underneath, margins often turned down, usually smooth.

Inflorescences: Drooping flowers on long stalks from upper leaf axils.

Flowers: White, bell-shaped, 5-10 mm long, with 5 lobes, anthers and stigma exserted.

Fruit: Berries round, to 1.6 cm diameter, green, pink, or purple, often waxy gray.

Habitat and Distribution: Common; upland, dry, sandy hammocks, sandhills, and flatwoods; central peninsula Florida, west to Texas, Oklahoma, Missouri, and north to Ohio, Ontario, and Massachusetts.

Comment: Mostly spring flowering, the waxy gray leaves of this many-branched shrub can be attractive, but appear somewhat less so in fall and winter. The fruits are sweet and edible, but very seedy.

Valerianella radiata (L.) Dufr.
Corn Salad
Valerianaceae, Valerian Family (Caprifoliaceae, Honeysuckle Family)

Habit: Herbaceous annual, 15-70 cm tall.

Leaves: Opposite; stalkless; blades oblong, broader at tip, 2-6 cm long, margins often hairy.

Inflorescences: Terminal clusters at tips of forked branches, clusters enclosed by small, leaf-like bracts.

Flowers: White, funnel-shaped, 1.5-2 mm long.

Fruit: Rounded, to 2 mm long, 1 nutlet develops.

Habitat and Distribution: Occasional or infrequent; moist to wet habitats, fields, woodland margins, roadsides, and other areas with disturbed soils; north peninsula Florida, west to Texas, and north to Kansas, Illinois, Ohio, and Connecticut.

Comment: As a wildflower, Corn Salad is not particularly showy when it blooms in the spring. The plant itself is used as a pot herb or in salads.

Vaccinium stamineum L.

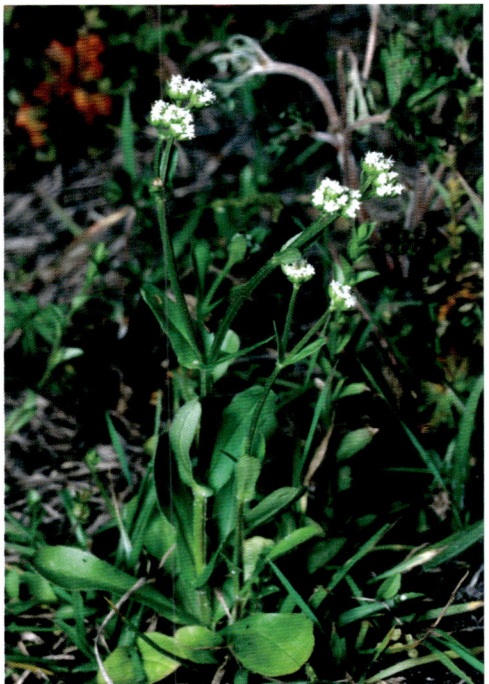

Valerianella radiata (L.) Dufr.

White

Verbesina virginica L.
Frostweed
Asteraceae (Compositae), Aster or Sunflower Family

Habit: Herbaceous perennial, 0.6-2.8 m tall, from fleshy roots. Stems single, densely hairy, winged.

Leaves: Alternate; stalked, stalks winged; blades oval or lance-ovate, lobed, 9-22 cm long, hairy, rough, pale green below, with toothed margins.

Inflorescences: Heads in compact clusters on long stalks from leaf axils.

Flowers: White, with 1-5 rays, each 5-10 mm long and 2-4 mm broad.

Fruit: Nutlets to 5 mm long, winged, with 2 bristles at top.

Habitat and Distribution: Common; wooded and open sites, hammocks, fields, margins, stream banks, bottom lands; throughout Florida, west to Texas, Kansas, and Missouri, and north to Virginia.

Comment: This quite showy summer and fall blooming wildflower is noticeable when poking out of wooded areas. It also can be a significant weed, but could be utilized along wooded margins.

Viburnum nudum L.
Possum Haw or Withe-rod
[*Viburnum cassinoides* L., misapplied]
Caprifoliaceae, Honeysuckle Family (Adoxaceae, Moschatel Family)

Habit: Shrub, to 5 m tall. Twigs covered with scurfy scales.

Leaves: Opposite; stalked; blades elliptic, oblong-elliptic, or lance-ovate, 5-15 cm long, with smooth or scalloped margins, undersides with scurfy scales.

Inflorescences: Dense, flat-topped terminal clusters.

Flowers: White, 5-8 mm broad, with 5 petals.

Fruit: A drupe, ellipsoid, blue-black, 6-10 mm long.

Habitat and Distribution: Frequent; bogs, swamps, bays, and wet woods; central peninsula Florida, west to Texas, and north to Manitoba and Newfoundland.

Comment: In spring the large flat clusters of flowers are quite showy, especially among the large dark green shiny leaves. This shrub can be a great specimen plant.

Verbesina virginica L.

Viburnum nudum L.

White

Viburnum obovatum Walter
Walter Viburnum or Small-leaf Viburnum
Caprifoliaceae, Honeysuckle Family (Adoxaceae, Moschatel Family)

Habit: Shrub, to 4 m tall. Twigs covered with scurfy scales.

Leaves: Opposite; short-stalked; blades spatulate, 2-6 cm long, margins smooth to toothed, undersides with scurfy scales.

Inflorescences: Dense, flat-topped, terminal clusters.

Flowers: White, 4-7 mm broad, with 5 petals.

Fruit: A drupe, black, ellipsoid, 6-9 mm long.

Habitat and Distribution: Common; floodplains, hammocks, moist woods, banks, wet flatwoods, and swamps; coastal plain, throughout Florida, west to Alabama, and north to South Carolina.

Comment: Walter Viburnum, when blooming in spring, can be an outstanding spectacular flowering shrub. When in the open the plant is densely covered with small branches, each with a cluster of bright white flowers. This shrub is best cultivated against a border as it looks somewhat unkempt when not in flower.

Viburnum rufidulum Raf.
Rusty Black Haw, Southern Black Haw, Blue Haw
Caprifoliaceae, Honeysuckle Family (Adoxaceae, Moschatel Family)

Habit: Shrub or small tree, to 9 m tall. Twigs densely covered with scurfy hairs.

Leaves: Opposite; stalked; blades broadly elliptic to oval, 3-9 cm long, shiny, leathery, margins with small teeth, undersides with rust-colored hairs.

Inflorescences: Dense, flat-topped, terminal clusters.

Flowers: White, 5-10 mm broad, with 5 petals.

Fruit: A drupe, ellipsoid, dark blue to purplish, 9-14 mm long.

Habitat and Distribution: Frequent to occasional; moist to dry often rocky woods, pine and oak woodlands, margins, banks; central peninsula Florida, west to Texas, and north to Missouri, Illinois, Indiana, and Virginia.

Comment: Rusty Black Haw is an outstanding ornamental. The blooms occur in spring. When covered in flowers the shrub can be spectacular. When not blooming the shiny, leathery leaves are also quite attractive.

Viburnum obovatum Walter

Viburnum rufidulum Raf.

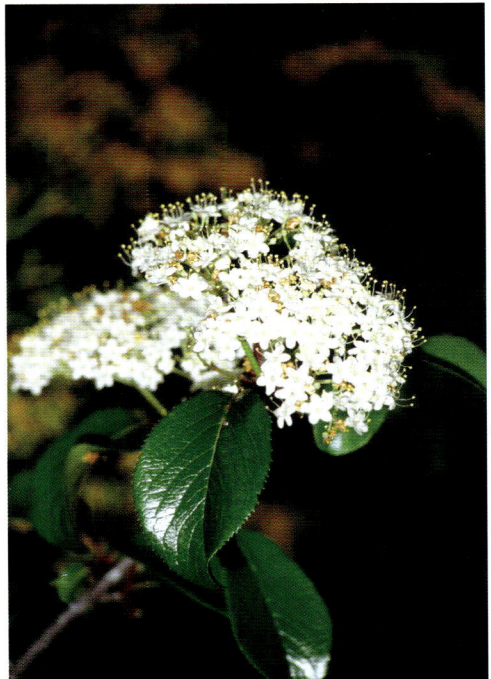

White

Viola lanceolata L.
Longleaf Violet, Strap-leaved Violet, Bog White Violet
Violaceae, Violet Family

Habit: Herbaceous perennial, 2 to 18 cm tall, from a thick horizontal underground runner.

Leaves: Alternate; stalked; blades linear to somewhat lance-shaped, 2-18 cm long, some with shallow blunt-toothed margins.

Inflorescences: Solitary flowers on long stalks from leaf axils.

Flowers: White, 1-2 cm broad, with 5 petals.

Fruit: A capsule, green, oval, 5-9 mm long.

Habitat and Distribution: Frequent to common; moist to wet sites, bogs, swamps, wet margins, and ditches; throughout Florida, west to Texas, and north to Minnesota, Ontario, and Nova Scotia.

Comment: The flowers in spring can occur on plants covering open wet meadows. The many plants, each with a flower or two, can provide quite a wildflower display.

Viola primulifolia L.
Primrose-leaved Violet
Violaceae, Violet Family

Habit: Herbaceous perennial, to 12 cm tall, from short thick rootstock, forming thin above ground runners.

Leaves: Alternate; stalked; blades narrow to broadly oval to somewhat lance-shaped, 1-10 cm long, variously hairy, with finely toothed margins.

Inflorescences: Solitary flowers on long stalks from leaf axils.

Flowers: White, 1-2 cm wide, with 5 petals.

Fruit: A capsule, green, ellipsoid, 7-10 mm long.

Habitat and Distribution: Frequent; moist to wet woods, bogs, margins, flatwoods, ditches, and marshes; throughout Florida, west to Texas, and north to Indiana, New Brunswick, and Maine.

Comment: Usually the spring flowers and plants are both small. When a number of the plants are blooming at the same time the flowers appear as small white dots in the meadow.

Viola lanceolata L.

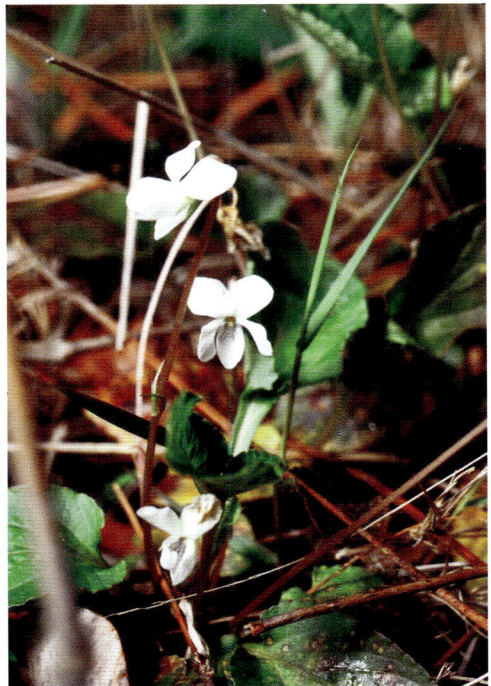

Viola primulifolia L.

White

Viola sororia Willd.
Florida Violet or Dooryard Violet
[*Viola affinis* Le Conte]
Violaceae, Violet Family

Habit: Herbaceous perennial, 7 to 19 cm tall, from a thick horizontal underground runner.

Leaves: Alternate, bunched at tip of runner; stalked; blades 2-8 cm long, broadly rounded to triangular or heart-shaped, hairy, margins smooth.

Inflorescences: Solitary flowers on long stalks from leaf axils.

Flowers: White to greenish-white to violet, 1.5-3.5 cm wide, with 5 petals.

Fruit: A capsule, rounded, purple to purple-spotted, 0.6-1.6 cm long.

Habitat and Distribution: Frequent to common; open, moist to wet woody sites and clearings; throughout Florida, west to Texas, and north to Wisconsin, Quebec, and Vermont.

Comment: Spring flowering like most violets, Dooryard Violet can be somewhat weedy. The larger flowers of this plant are quite attractive and often seen in wooded areas with an open ground cover. This violet is frequently the first wildflower noticed in the very early spring.

Warea carteri Small
Carter's Warea
Brassicaceae (Cruciferae), Mustard Family

Habit: Herbaceous annual, 0.4-1.5 m tall.

Leaves: Alternate; stalkless; blades very narrow, 1-3 cm long, tapering to base.

Inflorescences: Terminal clusters.

Flowers: White, on stalks to 1.2 cm long, petals 4, to 6 mm long, broad at tip, very narrow at base, densely covered with glandular hairs.

Fruit: A pod-like capsule, slender, 5-6 cm long.

Habitat and Distribution: Occasional; dry soils of sandhills and scrub; southern and central peninsula Florida.

Comment: When seen flowering, occurring all year, the long stalks and glandular hairy petals are very distinctive.

Viola sororia Willd.

Warea carteri Small

White

Yucca aloifolia L.
Spanish Bayonet or Spanish Dagger
Agavaceae, Agave Family

Habit: Shrub, 1-3 m tall. Trunk often leaning, thick, stocky, sometimes branching.

Leaves: Alternate; stalkless; blades rigid, narrow, tapering from midleaf to tip, 20-60 cm long, tip a sharply pointed spine, margins sharp, with fine teeth.

Inflorescences: Narrow, short-stalked clusters from stem tips.

Flowers: White, 4-6 cm long, bell-shaped, parts in 3s.

Fruit: A berry, oval, 7-10 mm long, and to 3 cm diameter, leathery.

Habitat and Distribution: Frequent; coastal or less frequently inland, sandhills, shell mounds, sand dunes, brackish marsh margins; throughout Florida, west to Louisiana, and north to North Carolina; a native of Mexico and the West Indies.

Comment: The showy flowers appear in any warm season. The thick stem is not strong and the very heavy leaves often force the plant to lean or fall to the ground. Buds near or at the tip create branches and a terminal branch will become the main stem and again grow upright. This plant was heavily cultivated until recently as a hedge and to discourage prying eyes and burglars.

Yucca flaccida Haw.
Weakleaf Yucca or Bear-grass
Agavaceae, Agave Family

Habit: Shrub, to 80 cm tall. Stem absent.

Leaves: Alternate, in a basal rosette; stalkless; blades tapering from midleaf to tip, 30-80 cm long, 1-4 cm wide, flexible, tip a sharply pointed spine, margins with curling threads.

Inflorescences: Long-stalked, terminal cluster, with wide-spreading, hairy branches, stalk 2-4 m long.

Flowers: White or greenish, 3-6 cm long, bell-shaped, parts in 3s.

Fruit: A capsule, oblong, 5-7 cm long, and 1.5-2 cm wide.

Habitat and Distribution: Common; dry sandy habitats - sandhills, scrub, hammocks, and flatwoods; central peninsula Florida, west to Texas, and north into Georgia.

Comment: Sometimes cultivated, Weakleaf Yucca is spring blooming. The short round plants with "threads" hanging from the leaf margins are attractive. The very long stalks with bright white showy flowers are quite magnificent.

Yucca aloifolia L.

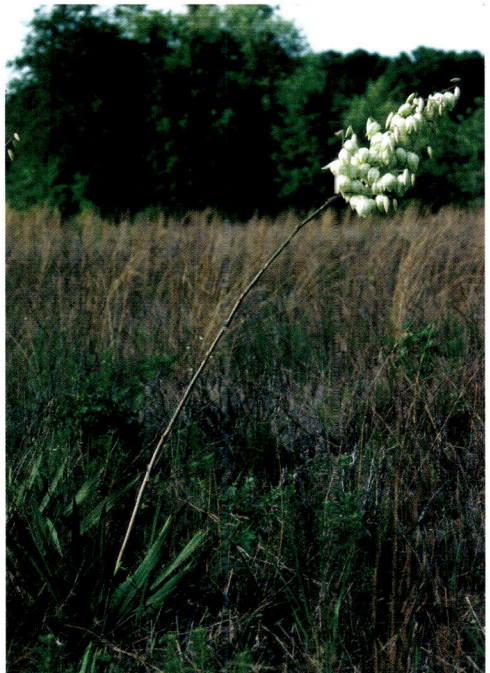

Yucca flaccida Haw.

White

Zephyranthes atamasco (L.) Herb.
Atamasco-lily
Amaryllidaceae, Amaryllis Family

Habit: Herbaceous perennial, to 35 cm tall, from sheathing bulbs.

Leaves: Alternate, basal; stalkless; blades slender, grass-like, 15-45 cm long and 3-4.5 mm wide, upper surface curved inward, margins sharp, dried blades flat.

Inflorescences: Flowers solitary, terminal on a leafless flowering stalk.

Flowers: White, turning pinkish with age, 5-10 cm long, bell-shaped, 6-parted, female part of flower (stigma) longer than anthers.

Fruit: A capsule, rounded, 3-parted, with many flat, shiny black seeds.

Habitat and Distribution: Frequent; wet, often grassy, meadows and woods, flatwoods, floodplains, pastures, ditches, and limestone outcrops; from central peninsula Florida, west to Mississippi, and north to Virginia.

Comment: The spectacularly large spring flowers of Atamasco-lily are easily seen among the low grass-like vegetation. When blooming in forested floodplains and hammocks the flowers often seem to have appeared like magic along the ground. When not blooming the leaves are almost impossible to find. Although quite frequent, Atamasco-lily is listed as Threatened by the State of Florida.

Zephyranthes simpsonii Chapm.
Simpson Rain-lily
Amaryllidaceae, Amaryllis Family

Habit: Herbaceous perennial, to 20 cm tall, from sheathing bulbs.

Leaves: Alternate, basal; stalkless; blades very slender, grass-like, 15-45 cm long and 3-4.5 mm wide, upper surface with longitudinal grooves, margins rounded, dried blades wrinkled.

Inflorescences: Flowers solitary, terminal on a leafless flowering stalk.

Flowers: White with tinges of pink or purple, 6-7.5 cm long, bell-shaped, 6-parted, female part of flower (stigma) at same level as anthers.

Fruit: A capsule, rounded, 3-parted, with many flat, shiny, black seeds.

Habitat and Distribution: Frequent; wet, often grassy, meadows and woods, flatwoods, floodplains, pastures, and ditches; central and south peninsula Florida.

Comment: Spring flowers are easily seen among pasture and roadside grasses. These flowers are very showy and scarcely distinguishable from Atamasco-lily. Although frequent in distribution, Simpson Rain-lily is listed as Threatened by the State of Florida.

Zephyranthes atamasco (L.) Herb.

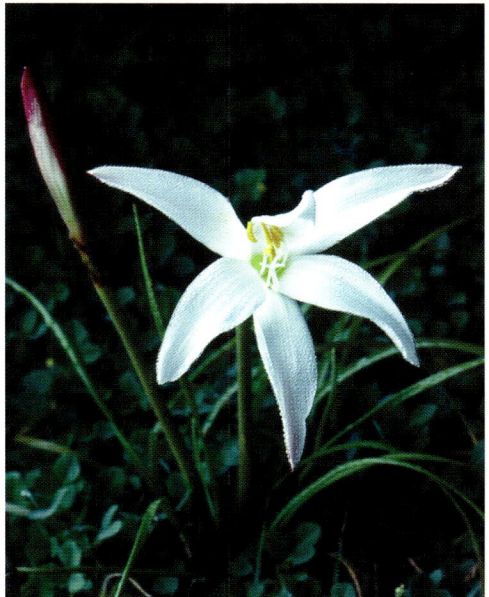

Zephyranthes simpsonii Chapm.

White

Zigadenus densus (Desr.) Fernald
Crow-poison
Liliaceae, Lily Family

Habit: Herbaceous perennial, 0.5-1.5 m tall, from a somewhat bulbous base.

Leaves: Alternate, 1-3, basal; stalkless; blades very slender, grass-like, to 50 cm long, 2-7 mm broad.

Inflorescences: Dense, cylindrical clusters at the tip of a branched, leafless, flowering stalk.

Flowers: White becoming pink, 8-10 mm wide, 6-parted.

Fruit: A capsule, to 12 mm long, and 3-4 mm wide.

Habitat and Distribution: Common; wet flatwoods, pinelands, bogs, and swamps; from central peninsula Florida, west to Texas and north to Virginia.

Comment: The cylinders of delicate white flowers bloom in spring on thin stalks. The masses of these small blooms are pretty. Bulbs of Crow-poison are very POISONOUS if ingested.

Zigadenus densus (Desr.) Fernald

Green

Ambrosia artemisiifolia L.
Common Ragweed, Short Ragweed, Altamisa
Asteraceae (Compositae), Aster Family

Habit: Herbaceous annual, 0.2-2 m tall, from a taproot. Stems erect, branched, hairy.

Leaves: Opposite on lower stem, alternate on upper stem; stalked; blades hairy, narrowly winged, once or twice divided into several pointed segments with toothed edges, 4-15 cm long.

Inflorescences: In elongated, spike-like clusters of heads.

Flowers: Yellowish to green, cup shaped heads, male and female flowers separate, but on the same plant, female flower clusters 2.5-3.5 mm long and 1.5-2.5 mm wide, male flower clusters 2-4 mm wide.

Fruit: A nutlet, brown, 3-4 mm long, 1-beaked, with 5-7 short spines.

Habitat and Distribution: Infrequent to common; disturbed sites, fields, pastures, roadsides, and gardens; throughout Florida, throughout the United States and into southern Canada.

Comment: One of the most prolific weeds in disturbed soils throughout the United States. Blooms in the summer and fall are within small green clusters, proving very difficult to see. This plant is the prolific source for allergenic pollen.

Ambrosia trifida L.
Giant Ragweed
Asteraceae (Compositae), Aster Family

Habit: Perennial erect herb, to 0.9 m tall, from a tuberous rootstock. Stem single, finely hairy.

Leaves: Opposite; stalkless; blades oblong, lance-shaped, or oval, to 7 cm long.

Inflorescences: Male flowers in terminal spikes; female flowers in 3- to 6-flowered round-topped clusters in leaf axils at stem tips.

Flowers: Male and female flowers separate, but on the same plant, greenish yellow, 5 petals, each 1.2-1.5 cm long, spreading downward, each female flower with a crown 7-9 mm long.

Fruit: A woody, beaked nutlet, 8-9 mm long, with several short spines.

Habitat and Distribution: Occasional; moist to wet soils of bogs, flatwoods, and pine savannahs; central peninsula and panhandle Florida, north to coastal plain Georgia.

Comment: Flowers are attractive in the summer, but not especially showy. Fragrant Milkweed is used as a wildflower in home gardens.

Ambrosia artemisiifolia L.

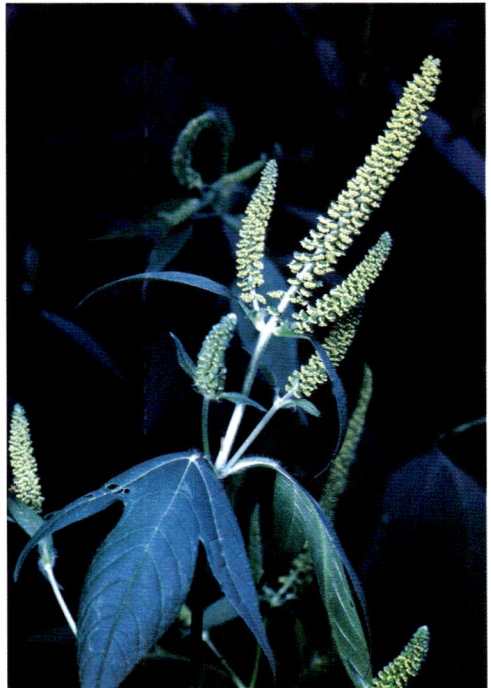

Ambrosia trifida L.

Green

Arisaema triphyllum (L.) Schott
Jack-in-the-pulpit
Araceae, Arum Family

Habit: Perennial herb, 20-80 cm tall, from a subterranean corm.

Leaves: 1-2, or usually 1 on young nonflowering plants; alternate; blades with 3 leaflets, 8-30 cm long, broadly elliptic to oval.

Inflorescences: Greenish spikes hidden within a green, purplish, or striped upright tubular structure 4-8 cm long, with a pointed hood arching over the mouth.

Flowers: Lacking petals, male flowers composed of 2-5 stamens, female flowers composed of a single pistil.

Fruit: Berries red, in large, elongate clusters, 2.5-15 cm long.

Habitat and Distribution: Frequently occurs in hammocks, moist woods, and swamps throughout Florida, west to Texas and Arkansas, and north to Minnesota and Nova Scotia.

Comment: The flowers, blooming in spring and summer, are interesting looking, but not particularly showy. However, the fruit, seen from summer into fall, is quite eye-catching.

Asclepias connivens Baldw.
Fragrant Milkweed
Asclepiadaceae, Milkweed Family

Habit: Perennial erect herb, to 0.9 m tall, from a tuberous rootstock. Stem single, finely hairy.

Leaves: Opposite; stalkless; blades oblong, lance-shaped, or oval, to 7 cm long.

Inflorescences: Round-topped clusters of 3 to 6 flowers, from leaf axils at stem tips.

Flowers: Greenish yellow, 5 petals, each 1.2-1.5 cm long, spreading downward, crown 7-9 mm long.

Fruit: A capsule, 8-9 mm long.

Habitat and Distribution: Occasional; moist to wet soils of bogs, flatwoods, and pine savannahs; central peninsula and panhandle Florida, north to coastal plain Georgia.

Comment: Flowers are attractive in the summer, but not especially showy. Fragrant Milkweed is used as a wildflower in home gardens.

Arisaema triphyllum (L.) Schott

Asclepias connivens Baldw.

Green

Asclepias longifolia Michx.
Florida Milkweed
Asclepiadaceae, Milkweed Family

Habit: Perennial herb, 20-80 cm tall. The clustered stems are purple.

Leaves: Opposite or whorled, very narrow, 8-18 cm long, 2-12 mm wide.

Inflorescences: Three to several flat-topped clusters of flowers, each 3-5 cm across, towards ends of branches.

Flowers: The greenish-white petals with purple tips are curved down and outwards and 5-6 mm long. The hoods in the center of the flower are 2-3 mm long, without outward curving horns.

Fruit: Lance like pods, 9-11 cm long and finely hairy.

Habitat and Distribution: Moist and wet sites, from southern Florida west to eastern Texas and north to Delaware.

Comment: Florida Milkweed is in a group termed the green milkweeds. The greenish white flowers can be found all the warm months especially in the spring. This attractive plant should be considered for gardens. The milky sap is thought to be POISONOUS.

Asclepias obovata Ell.
Green Milkweed
Asclepiadaceae, Milkweed Family

Habit: Perennial erect herb, 0.4-0.7 m tall, from a deep slender rootstock. Stem single, unbranching, hairy.

Leaves: Opposite, oval or wider toward tips, 5-9 cm long and 2-3.5 cm broad, densely hairy beneath.

Inflorescences: Compact domed clusters of flowers, each 3-3.5 cm wide.

Flowers: Petals 5, greenish yellow, 8-10 mm long, spreading downward, with an upright crown in center, crown 5-7 mm diameter.

Fruit: A slender, erect capsule, 8-12 cm long.

Habitat and Distribution: Occurs in sandy pinelands from the Florida panhandle, west to Texas, and north to South Carolina.

Comment: Green Milkweed, blooming from June into September, has large flowers that are pretty but not especially showy. This milkweed is used occasionally in gardens.

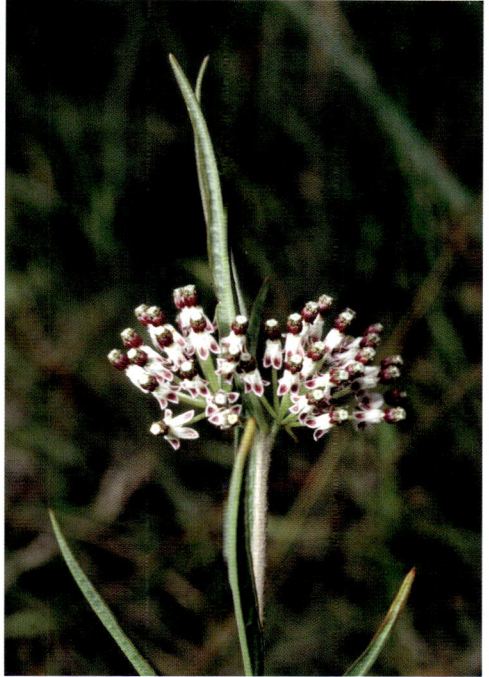

Asclepias longifolia Michx.

Asclepias obovata Ell.

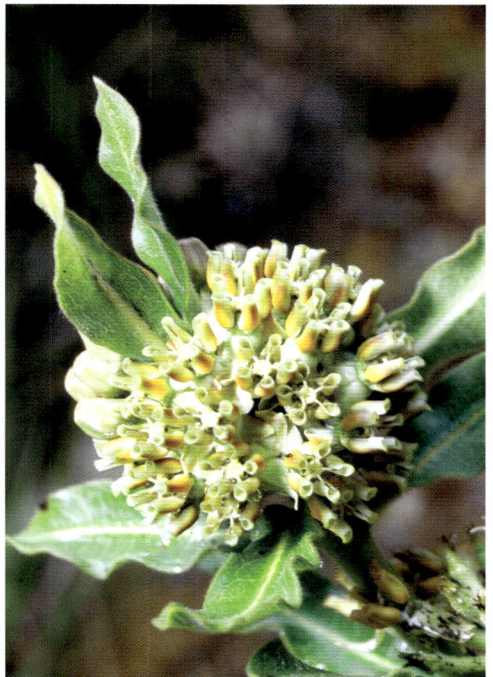

765

Green

Chenopodium album L.
Lamb's-quarters
Chenopodiaceae, Goosefoot Family

Habit: Herbaceous annual, 0.3-1.5 m tall. Stems smooth, branched, grooved, entire plant covered with mealy powder.

Leaves: Alternate; stalked; blades mostly diamond shaped, 3-10 cm long, thickish, 3-veined, larger lower blades with roughly toothed margins, upper leaves smaller, often with smooth margins.

Inflorescences: Congested or interrupted, terminal or axillary spikes.

Flowers: Tiny, lobes green with white margins.

Fruit: A black, glossy nutlet enclosed in a papery covering, about 1 mm in diameter.

Habitat and Distribution: Frequent; disturbed sites, including cultivated fields, yards, and vacant areas; throughout Florida, throughout most of the temperate and subtropical world.

Comment: The very weedy plant blooms in warm months. Even though the flowers are not attractive, the mealy powder covering the entire plant gives it an unusual look. Some plants have purple lines and/or joints that are also quite appealing. Young leaves are edible.

Cladium jamaicense Crantz
Sawgrass
Cyperaceae, Sedge Family

Habit: Perennial herb, to 3 m tall, from large stolons.

Leaves: Narrow, blade- shaped, to 1 m long and 0.8-1.5 cm broad, with cutting, saw-like teeth along edges and midvein, borne in large, dense tuft like clumps.

Inflorescences: Large, terminal, loose, branched clusters.

Flowers: Green to reddish, elliptical, 3.5-4 mm long, borne in loose, branched clusters to 80 cm long.

Fruit: A green to purple, roundish nutlet, 2-3 mm long.

Habitat and Distribution: Common in shallow water of brackish and fresh water marshes and swamps throughout Florida, south into the West Indies, west to Texas and north to southeast Virginia.

Comment: Blooming in summer and fall, this tall, sharp leaved sedge is a common component of marshes. Extensive sawgrass marshes occur in the warmer parts of the coastal plain. This species is not attractive as a wildflower, but is so prominent it deserves recognition.

Chenopodium album L.

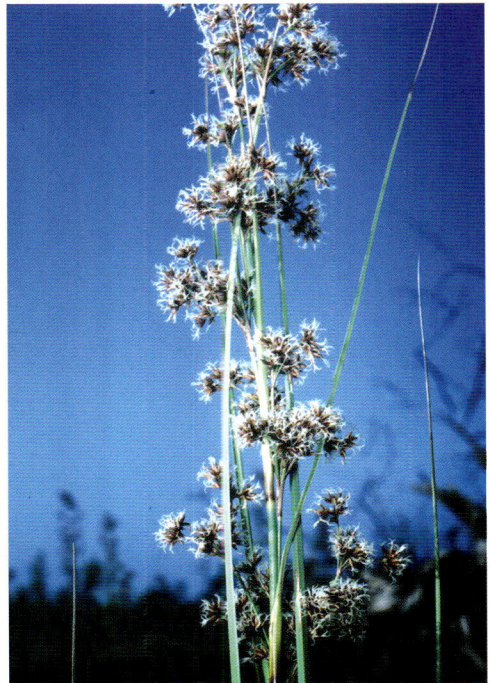

Cladium jamaicense Crantz

Green

Conocarpus erectus L.
Button Wood
Combretaceae, White Mangrove Family

Habit: Perennial shrub or tree, usually about 3 m tall, but can be to 20 m tall.

Leaves: Alternate, short-stalked, with a pair of glands on the stalks just below the blade; blades pointed, elliptic to ovate, 2-10 cm long, evergreen, smooth or with silvery hairs, lower surface with small pits along midvein at junction with lateral veins.

Inflorescences: Dense, round heads, 9-14 mm in diameter, clustered at the tips of branches.

Flowers: Greenish, minute.

Fruit: Cone-like, rounded, purplish green, seeds 4-7 mm long, winged.

Habitat and Distribution: Occurs frequently in sandy coastal areas of central peninsula Florida south through the Florida Keys into the West Indies, and Mexico south through Central America into South America.

Comment: Button Wood can flower all year. The flowers are hardly noticeable, but the foliage is prized for hedges and occasionally for specimen plants. The silver form, Silver Button Wood, with abundant silver gray hairs is *Conocarpus erectus* L. var. *sericeus* Griseb. and is a valued foliage plant. The bark can be used for medicine and tanning and the wood is valued for charcoal.

Cyperus surinamensis Rottb.
Surinam Sedge
Cyperaceae, Sedge Family

Habit: Perennial herb, to 1 m tall, from fibrous roots. Stems 0.2-1 m tall, usually 0.2-0.6 m tall, triangular, with short, stiff, downward pointing hairs.

Leaves: Narrow, 0.1-1.5 cm broad, basal.

Inflorescences: Terminal, branched clusters of spikes above a few narrow, leaf-like, unequal radiating bracts.

Flowers: Greenish, not overlapping, in 2 rows on short spikes 4.5-5 mm long and 1.5-2 mm broad.

Fruit: Slender and nearly oblong, reddish brown, granular nutlet 0.6-0.8 mm long, with a short stalk.

Habitat and Distribution: Found in sandy and clay soils of marshes, shores, wet flatwoods, wet clearings, ditches, and wet swales from South Carolina, west to Texas, south into the West Indies, and Mexico south through Central America into South America.

Comment: Flowering all the warm parts of the year Surinam Sedge is not particularly attractive, but is quite common and frequently encountered in roadside ditches and swales.

Conocarpus erectus L.

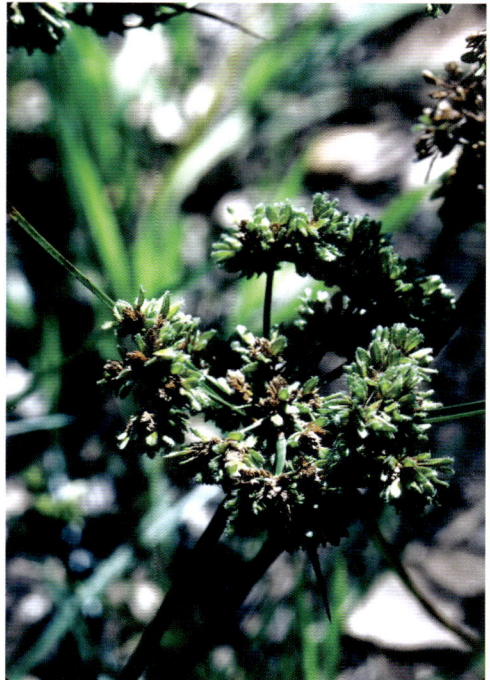

Cyperus surinamensis Rottb.

Green

Encyclia tampensis (Lindl.) Small
Florida Butterfly Orchid
Orchidaceae, Orchid Family

Habit: Perennial, epiphytic herb, 15-60 cm tall, growing above ground on trees. Stems round, bulb-like.

Leaves: 1-3, narrow, thickish, 5-30 cm long.

Inflorescences: Loose groups of few to many flowers on a thin flowering stalk.

Flowers: Orange, yellow, brown, or most frequently, greenish, to 4 cm across, 6 parted, with strap-shaped to spoon-shaped petals and white and purple 3-lobed lip.

Fruit: A drooping capsule, to 3 cm long.

Habitat and Distribution: Found frequently on trees in hammocks and swamps, from north peninsula Florida, south into and through the West Indies.

Comment: Florida Butterfly Orchid flowers in summer. As a cultivated or native wildflower the delicate varied colors can be very noticeable, particularly when found or placed on lower branches.

Epidendrum conopseum Jacq.
Green-fly Orchid
Orchidaceae, Orchid Family

Habit: Perennial, epiphytic herb, 5-30 cm tall, growing above ground on trees. Stems slender, smooth, erect or spreading upward.

Leaves: 2-3, elliptical, 3-9 cm long and 0.4-1.4 cm broad, leathery.

Inflorescences: Loose few-flowered clusters, erect or drooping, on a thin flowering stalk.

Flowers: Pale green or cream, to 2 cm across, 6-parted, with strap-shaped or spoon-shaped petals and 3-lobed lip.

Fruit: A drooping capsule, 1.5-1.8 cm long.

Habitat and Distribution: Frequently occurs on Live Oaks and Magnolias in hammocks and near swamps from Lake Okeechobee in south central peninsula Florida, west to Louisiana, and north to North Carolina, and in Mexico.

Comment: Flowering in summer and into the winter in warmer climates, Green-fly Orchid is a delicate wildflower best seen in cultivation or on lower branches.

Encyclia tampensis (Lindl.) Small

Epidendrum conopseum Jacq.

Green

Eryngium yuccifolium Michx.
Rattlesnake Master
Apiaceae, Carrot Family

Habit: Perennial herb, 0.3-1.6 m tall. Stem solitary, erect, and branched above.

Leaves: Narrow, 15-80 cm long and 1-3 cm broad, stiff, leathery, with distinct spiny edges, in a basal rosette, smaller upwards along the stem.

Inflorescences: Many-headed, branched, terminal clusters.

Flowers: Heads whitish green, roundish, 0.8-2.5 cm long and almost as broad, compact, not spiny.

Fruit: A club-shaped, scaly, 2-parted nutlet, 2.5 mm wide.

Habitat and Distribution: Frequent in dry sandy soil of pinelands, prairies, and open woods throughout Florida, west to Texas, and north to Kansas, Indiana, Minnesota, and New Jersey.

Comment: Blooming in the summer months, this distinctive plant always provokes comment. The button-like flower heads provide one of the common names, Button Rattlesnake Master, for this species. The Yucca-like leaves give it the specific scientific name. As a garden plant this species is underutilized.

Eulophia alata (L.) Fawc. & Rendle
Wild Coco Orchid
Orchidaceae, Orchid Family

Habit: Perennial, terrestrial herb, 0.7-1.5 m tall, from a large corm and fibrous root system.

Leaves: 4-6, slender, folded, 20-120 cm long, sheathing at base.

Inflorescences: Terminal, erect, many-flowered.

Flowers: Showy, dull green- purple to dark reddish-brown, with basal portion 1.5-2.6 cm long, petals broader at the tip.

Fruit: An oblong, dangling capsule, 3.5-4 cm long.

Habitat and Distribution: Found infrequently in low damp sites of pastures, swamps, and roadsides from central peninsula Florida south through the West Indies and southern Mexico into South America, also in Africa.

Comment: Colonies of this very showy wildflower are seen in late summer and fall. A single plant can be attractive and many are much more so.

Eryngium yuccifolium Michx.

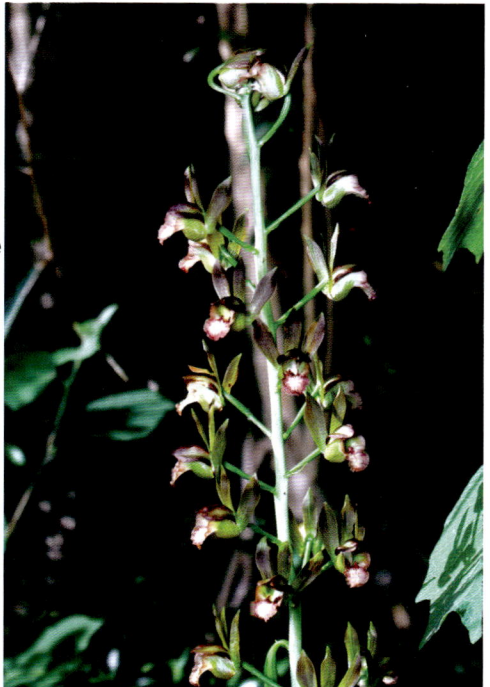

Eulophia alata (L.) Fawc. & Rendle

Green

Euonymus americana L.
Hearts-a-bustin' or Strawberry Bush
Celastraceae, Staff Tree Family

Habit: Perennial shrub, to 2 m tall. Stems erect or straggling, green. Small branches stiffly erect and 4-angled.

Leaves: Opposite, deciduous, oval or wider at base, 3-9 cm long and 1-3 cm broad, thin, with scalloped or toothed edges.

Inflorescences: Clusters of 1-3 flowers borne in leaf axils, along or toward tips of branches.

Flowers: About 1 cm wide, light green to greenish-purple, with 5 broad, oval petals 2.5-4 mm broad, with fl eshy central disc.

Fruit: A red, warty capsule, 1-2 cm in diameter, splitting when mature; seeds bright orange to scarlet.

Habitat and Distribution: Occurs frequently in hammocks and rich woods from central peninsula Florida west to eastern Texas and north to southern Ohio, Missouri, and New York.

Comment: This highly recommended native ornamental flowers and fruits in late spring and early summer. The red, warty fruits are especially showy, particularly when mature and splitting, showing the bright seeds. The shiny green leaves turn red and orange in the autumn.

Lepuropetalon spathulatum Ell.
Little People
Saxifragaceae, Saxifrage Family

Habit: Small annual herb, 1-3 cm across, forming a rosette.

Leaves: Alternate, 3-10 mm long, 1-2 mm broad, stalked and wider toward tip, usually red-spotted and with smooth margins.

Inflorescences: Flattish clusters at stem tips.

Flowers: Greenish, minute, with 5 green sepals and 5 smaller white petals.

Fruit: Angular capsules, wider at top, about 1.5 mm long, with numerous, minute, red seeds.

Habitat and Distribution: Found in moist clay soils along ditches, pond margins, and in depressions. In Florida only in Gadsden county in the panhandle, then west to eastern Texas and Mexico, north to Georgia and the Carolinas.

Comment: This very tiny plant is easily overlooked and is listed as an Endangered species by the State of Florida.

Euonymus americana L.

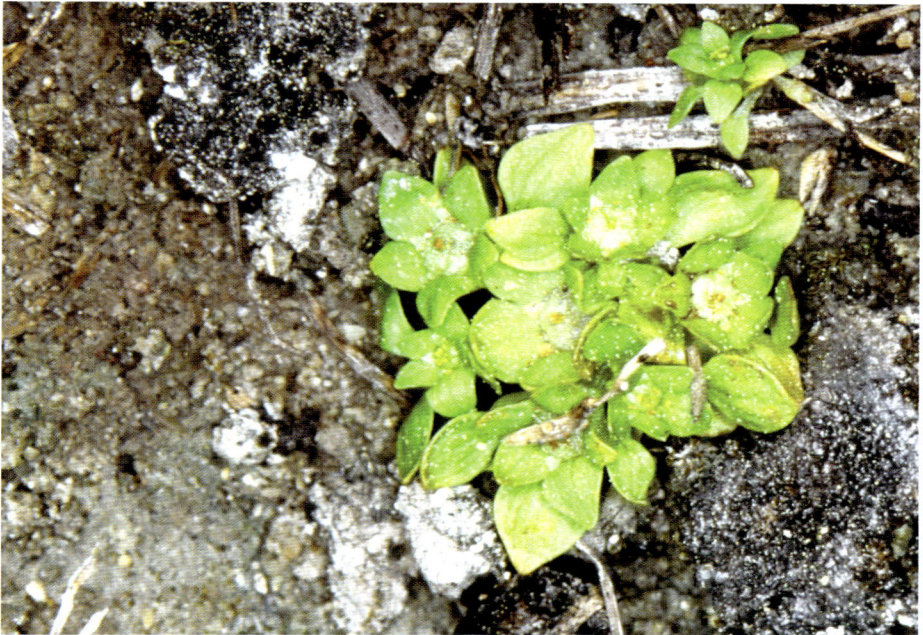

Lepuropetalon spathulatum Ell.

Green

Ludwigia microcarpa Michx.
Tiny Seedbox
Onagraceae, Evening primrose Family

Habit: Perennial herb, 10-50 cm tall. Stems erect or sprawling, simple or branched, hairless, usually winged.

Leaves: Alternate, oval or wider toward tips, to 12 mm long and 5 mm broad, on very short stalks or attached directly to stems.

Inflorescences: Single flowers borne along stems in leaf axils.

Flowers: Greenish, with 4 triangular sepals 1.5 mm long and no petals.

Fruit: A broad 4-angled, wider toward the top, 1.5-2 mm long and wide.

Habitat and Distribution: Common in ditches, marshes, and other moist to wet sites throughout Florida, west to Louisiana, and north to Missouri, Tennessee, and North Carolina.

Comment: This difficult to notice plant flowers in most of the warm months. In the autumn it produces leafy runners from which grow additional plants.

Myrica cerifera L.
Wax-myrtle or Southern Bayberry
Myricaceae, Wax-myrtle Family

Habit: Perennial shrub to small tree, usually 3 to 6 m tall, but occasionally to 13 m. Bark white. Stems often several, sometimes with under ground runners forming colonies.

Leaves: Alternate, elliptical, to 10 cm long and 2.5 cm broad, evergreen, with rusty glandular dots on both surfaces, leaves shorter toward branch tips. Blade margins with a few teeth toward the tip.

Inflorescences: Male and female flowers on separate plants (dioecious), in cylindrical or conical catkins, catkins 0.6-2 cm long and 4-6 mm in diameter.

Flowers: Small, greenish or slightly pink.

Fruit: Round, 2.5-3.5 mm diameter, grayish-white, heavily coated with a gray wax.

Habitat and Distribution: Common in dry to wet soils of pinelands, woods, swamps, bogs, ponds, and old fields, fence rows, and thickets throughout Florida, west to Texas and north to Oklahoma, Arkansas, and New Jersey. Also, Mexico, West Indies, Bermuda, and Central America.

Comment: Wax myrtle blooms in the early spring before shoot production. This very versatile plant can be used as a specimen shrub or tree. It is easily pruned making it practical for use as a hedge or street shrub. The flowers are not especially showy, but the bark, rusty leaves, attractive fruit, and nice shape make it a handsome ornamental. A dwarf form usually less than 1 m tall known as *Myrica cerifera* var. *pumila* Michx., Dwarf Wax-myrtle, is sometimes cultivated. The roots have nitrogen producing nodules. Leaves when crushed are distinctively aromatic.

Ludwigia microcarpa Michx.

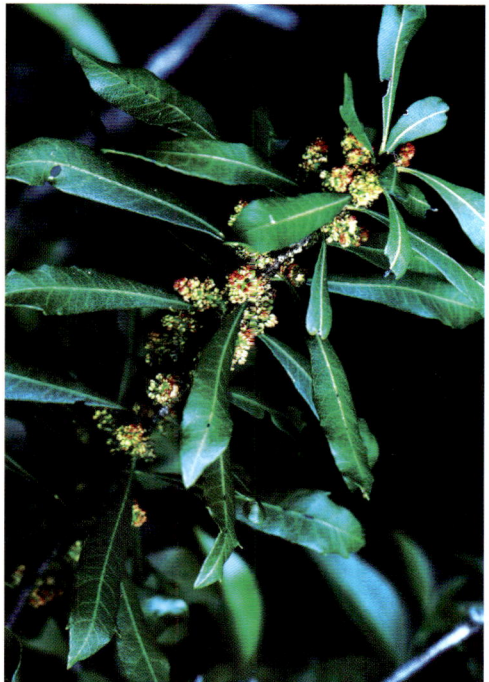

Myrica cerifera L.

Green

Onosmodium virginianum (L.) A. DC.
False Gromwell
Boraginaceae, Borage Family

Habit: Perennial herb, 20-80 cm tall. Stems slender, branched above, stiffly appressed hairy.

Leaves: Alternate, evergreen, elliptic to rounded, often very narrow, sometimes broader at the tip, 2-13 cm long.

Inflorescences: Terminal, long, coiled, leafy spikes.

Flowers: Greenish-yellow to orange, tubular, 7-10 mm long, with 5 narrowly triangular lobes to 3 mm long, and a threadlike style extending past mouth of tube.

Fruit: White nutlets, usually only 1 or 2 maturing out of 4, 2-2.8 mm long.

Habitat and Distribution: Found infrequently in dry woods and pinelands from central peninsula Florida west to Louisiana and north to New York and Massachusetts.

Comment: Flowering is usually in late spring and early summer. The greenish-yellow flowers are not as easily noticed as the orange-flowered forms.

Phoradendron leucarpum (Raf.) Rev. & M. C. Johnston
Mistletoe
[*Phoradendron serotinum* (Raf.) M. C. Johnston]
Viscaceae, Mistletoe Family

Habit: Woody parasite

Leaves: Opposite, leathery, entire, elliptic to oblanceolate, 2-13 cm long and 1-4 cm wide.

Inflorescences: Male and female flowers grow on thick, somewhat fleshy spikes in the leaf axils of separate plants (dioecious).

Flowers: Small, greenish; several tiny flowers are embedded in the tissue of each spike.

Fruit: White, globose, single seeded berry. Seeds white, ellipsoid, flattened, and about 3 mm long.

Habitat and Distribution: On many kinds of, usually deciduous, trees; West Virginia and New Jersey, south to Florida, and west to Oklahoma and east Texas.

Comment: This shrubby, tree living parasite is commonly identified by its leathery leaves and white berries. The profusion of green branches are woody and brittle. Branches are characteristically swollen at each leaf node. The white berries contain a sticky substance that serves to stick the seeds to branches of woody species where they may germinate.

It is more easily seen in the winter in deciduous trees due to the loss of leaves from the host. The fruits are commonly eaten by several species of birds, especially Cedar Waxwings. Caterpillars of the Great Purple Hairstreak Butterfly utilize Mistletoe leaves as their sole source of food. For centuries Mistletoe had been used medicinally, however, it is quite POISONOUS. Ingestion of berries and teas has caused deaths of humans and livestock!

Onosmodium virginianum (L.) A. DC.

Male

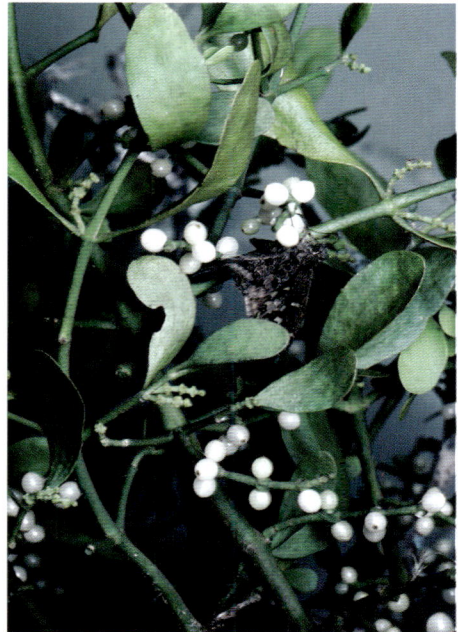

Female

Phoradendron leucarpum (Raf.) Rev. & M.C.Johnston

Green

Rhus copallinum L.
Winged Sumac or Shining Sumac
Anacardiaceae, Cashew Family

Habit: Perennial shrub, to 4 m tall or small tree to 8 m. Stems hairy.

Leaves: Alternate, deciduous, divided, with 9-23 narrow oval leaflets, each 2-10 cm long and 1-4 cm broad, glossy, sometimes with toothed edges, leaflets borne on winged rachis.

Inflorescences: Large, terminal, dense, branched, spray-like clusters, 5-30 cm long. Male and female flowers are on separate plants (dioecious).

Flowers: Greenish, yellow, or whitish, to 3 mm wide, with 5 tiny petals.

Fruit: Drupe round, to 4 mm diameter, pink to red, hairy.

Habitat and Distribution: Common in open dry soils of thickets, woods, old fields, fence rows, and other disturbed areas throughout Florida, west to Texas, and north to Nebraska, Illinois, Indiana, Minnesota, Wisconsin, and Maine.

Comment: This very widespread eastern species blooms in the spring. Flowering is pretty, although brief. The female plants with their clusters of red fruits are attractive for some months until birds strip the fruit. Fall foliage is very showy producing reds and oranges. Winged Sumac can be clonal producing additional plants from the roots. Due to this clonal nature the shrub is best cultivated in sandy soils against a border such as along a fence or in a line of native vegetation.

Rumex hastatulus Baldw.
Sour Dock or Heartwing Sorrel
Polygonaceae, Buckwheat Family

Habit: Winter annual or short-lived perennial herb, 0.2-1.3 m tall, from a taproot.

Leaves: With a sheath, lower leaves with a stalk, blades to 8 cm long, 3-lobed, with the center lobe being long and sword-shaped, and the outer lobes being slender, smaller, emerging from the base of the leaf, and perpendicular to the center lobe.

Inflorescences: Dense terminal clusters, 30-40 cm long, on long, slender stems. Male and female flower are on separate plants (dioecious).

Flowers: Flowers very small, about 3 mm across, sometimes greenish, pink, or purple, becoming red as fruits mature.

Fruit: Dry, red to reddish brown, with a membranous wing, approximately 3 mm wide. Male and female flowers on separate plants (dioecious). Seeds to about 1.5 mm long.

Habitat and Distribution: Occurs in sandy soils of old fields and along roadsides, from central peninsula Florida north to Massachusetts and west to Illinois, Kansas, and Texas.

Comment: Flowers occur in the spring. Masses of female plants when in fruit make a spectacular red show particularly when seen on a hillside field with no competing vegetation.

Rhus copallinum L.

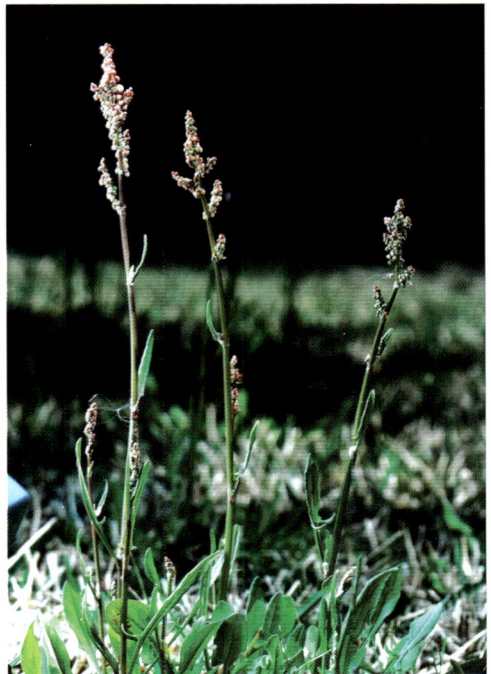

Rumex hastatulus Baldw.

Green

Rumex obtusifolius L.
Bitter Dock
Polygonaceae, Buckwheat Family

Habit: Perennial herb, to 1.2 m tall, from a taproot, with coarse, leafy, unbranched stems.

Leaves: With a sheath, lower leaves with a stalk, blades elliptical to heart-shaped, 15-35 cm long and 6-15 cm broad at base, smaller on upper stems.

Inflorescences: Many, dense, leafy, cylindrical or cone-shaped clusters.

Flowers: Green, 3-5 mm long, lacking petals; flower stalks longer than the fruits.

Fruit: Dry, winged, 3.5-5 mm long and wide with 2 to 4 teeth on each margin. Seeds to 2 mm long.

Habitat and Distribution: Infrequently found in moist to wet soils of ditches, marshes, floodplains, and disturbed areas from central peninsula Florida northward throughout the United States into Quebec, Nova Scotia, and British Columbia; Native to Europe.

Comment: Flowering is during the warm months. This plant is not showy and is frequently overlooked, but, when fruiting, the masses of green, long-stalked, papery fruits are a curiosity.

Sassafras albidum (Nutt.) Nees
Sassafras
Lauraceae, Laurel Family

Habit: Perennial shrub or small tree to 30 m tall. Bark green on young stems.

Leaves: Alternate, deciduous, short stalked; blades elliptical, 6-12 cm long, 2-8 cm broad, with 0-2 lobes, when lobed, lobes mitten like.

Inflorescences: Few- to many-flowered clusters along branches. Male and female flowers are on separate plants (dioecious).

Flowers: Greenish yellow, to 1 cm across, 6-parted.

Fruit: Drupe dark blue, oblong, 8-10 mm long.

Habitat and Distribution: Common in sandy soils along woodland margins, fencerows, and old fields from central peninsula Florida north to Maine and west to Iowa, Kansas, and Texas.

Comment: This weedy shrub flowers from spring in the south into early summer northward. The mitten like lobed leaves are quite distinctive. Bark from the roots is used for flavoring and to make a tea. The plant itself is noticeable because of the leaf shape, but is otherwise not noteworthy as an ornamental.

Rumex obtusifolius L.

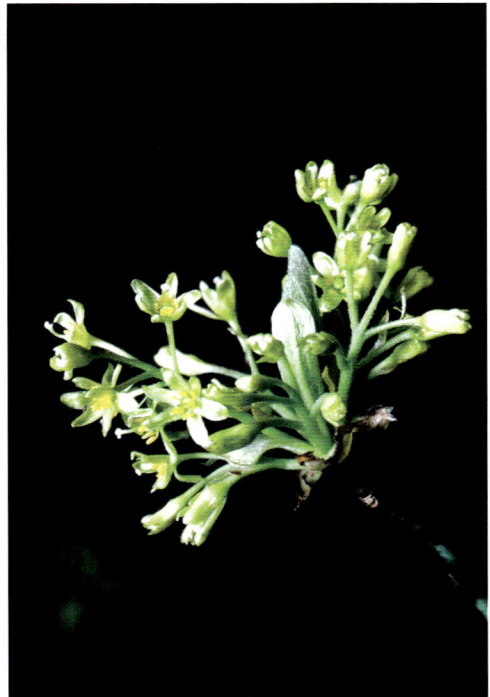

Sassafras albidum (Nutt.) Nees

Green

Smilax auriculata Walt.
Wild-bamboo
Smilacaceae, Greenbrier or Catbrier Family

Habit: Perennial vine from a tuberous under ground runner, climbing by tendrils. Stems green, smooth, coarse, prickly, forming low thickets, rarely climbing high.

Leaves: Alternate, evergreen, with stalks; blades oval to triangular or 3-lobed, 2-12 cm long, leathery.

Inflorescences: Roundish or domed clusters along stems. Male and female flowers occur on separate plants (dioecious).

Flowers: Green, bell shaped, 3-8 mm long, 6-parted.

Fruit: Berries round, black, 5-8 m diameter.

Habitat and Distribution: Commonly occurs in sandy soils of pinelands, hammocks, sandhills, and scrub from Florida, west to Louisiana and Arkansas, and north to North Carolina; also Bahama Islands.

Comment: Wild-bamboo flowers throughout the year in the southern part of its range and during the cooler months of spring and fall northward. This vine with its evergreen leaves can serve as an ornamental especially on an arbor.

Smilax auriculata Walt.

Green

Stillingia sylvatica L.
Queen's Delight
Euphorbiaceae, Spurge Family

Habit: Herbaceous perennial, 0.2-1.2 m tall. Stems few to several from base, branching only below flower clusters.

Leaves: Alternate, elliptic, 3.5-9 cm long and 1-4.5 cm broad, borne on stalks to 4 mm long or directly on stems, margins with incurved glandular teeth.

Inflorescences: Dense spikes, 5-12 cm long, at tips of branches.

Flowers: Green, minute.

Fruit: A smooth, triangular capsule, 8-10 mm long and 5-15 mm broad.

Habitat and Distribution: Common in sandy soils of sandhills and flatwoods throughout Florida, west to Texas and New Mexico and north to Kansas and Virginia.

Comment: Queen's Delight blooms all year in the warmer parts of Florida and during the warm months elsewhere. This unusual looking plant is sometimes called Queen's Root.

Zanthoxylum clava-herculis L.
Hercules'-club
Rutaceae, Citrus Family

Habit: Perennial shrub or tree, to 10 m tall. Stems very prickly, older prickles on corky growths. Prickles often on larger twigs and on leaves.

Leaves: Alternate, compound with 5-19 pointed, lanceolate to oval leaflets, some wider at base, 3-7 cm long, leathery, with scalloped edges having glandular teeth, leaflets borne along thorny stalk.

Inflorescences: Branched clusters at ends of branches.

Flowers: Greenish-white, small, with 5 petals.

Fruit: A dry, red pod, 4-5 mm long.

Habitat and Distribution: Occurs frequently in hammocks, woods, and coastal sites throughout Florida, west to Texas, Arkansas, and Oklahoma, and north to Virginia.

Comment: Flowers occur in summer. The inner bark is used to ease the pain of toothaches leading to the common name, Toothache Tree. The tree also has other medicinal uses.

Stillingia sylvatica L.

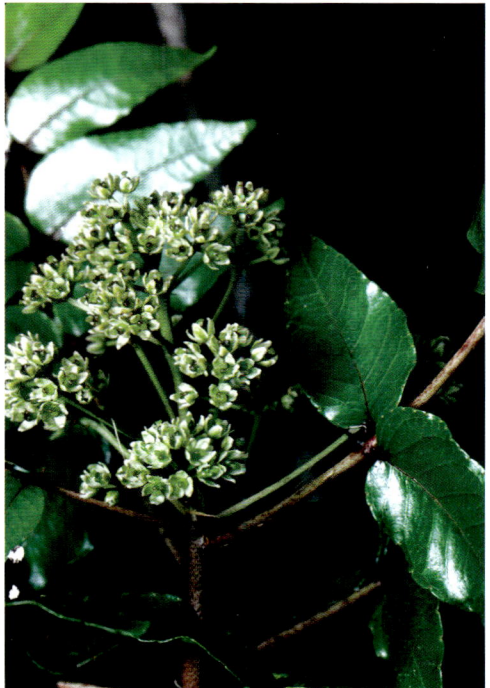

Zanthoxylum clava-herculis L.

Glossary

Achene: a small, dry, one-seeded fruit that does not open or split at maturity, such as found in a sunflower or dandelion.

Aggregate: crowded into a cluster.

Alternate: an arrangement of a single leaf, bud, or branch attached singly at different points of the stem; appearing to alternate.

Annual: a plant starting from seed, completing its life cycle, and dying within one year.

Anther: the saclike portion of the male part of a flower (stamen) that bears pollen.

Apex: the tip of a stem, root, or leaf.

Aquatic: living in water.

Ascending: sloping or leaning upward or outward.

Awn: a slender or stiff bristle.

Axil: the upper angle between the leaf and stem.

Axis: the main stem of a flower cluster.

Basal rosette: a cluster of leaves radiating at ground level at the base of a plant.

Beak: a hard point or projection, seen frequently on seeds and fruits.

Bearded: having long hairs.

Berry: a fleshy or pulpy fruit that does not break open, as the tomato; in general any pulpy or juicy fruit.

Biennial: a plant that completes its life cycle and dies in two years. The first year, seeds germinate and vegetative growth forms. The second year, flowering, seed set, and death occur.

Bisexual: flowers with male (stamens) and female elements; having both sexes.

Blade: the expanded, usually flat, portion of a leaf.

Brackish: somewhat salty.

Bract: a modified, usually reduced leaf associated with a flower or flower cluster.

Branch: a lateral stem.

Bristle: a short, coarse, stiff hairlike part.

Bud: a usually tightly bunched, undeveloped shoot or flower often located at the tip of a stem or branch or located in the axil of a stem or branch.

Bulbil: a small bulb; a small plant produced vegetatively in the axils of the inflorescence.

Bulblet: a small bulb, as one in a leaf axil or in the inflorescence.

Bur: a structure with spines or prickles that is frequently hooked or barbed.

Calyx: the outer parts of a flower composed of leaflike parts called sepals; sometimes colored like petals.

Capsule: a simple, many-seeded, dry fruit, splitting upon drying into two or more parts.

Glossary

Carpel: a basal juvenile seed bearing structure of the female part of the flower.

Ciliate: fringed with hairs on the margin.

Clasping: a type of leaf attachment where the leaf base partly or completely encircles the stem.

Claw: the narrow base of some petals and sepals.

Clonal: vegetative produced from a single individual.

Clump-forming: tufted, grows in a compact cluster.

cm (centimeter): equals approx. 3/8 or 0.4 inches.

Collar: outer side of the grass leaf at the junction of sheath and blade.

Compound: composed of two or more distinct similar parts.

Compressed: flattened laterally.

Cone: an inflorescence with overlapping scales; as in pine.

Conical: cone-shaped.

Cordate: heart-shaped.

Corm: a stout, short, vertical, bulblike underground food storage stem.

Corolla: the flower petals.

Cosmopolitan: common to all or many parts of the world.

Cultivar: form or type of a plant originating as a result of cultivation.

Cotyledon: the primary leaf of the embryo; seed leaf.

Cylindrical: cylinder shaped.

Deciduous: not persistent or evergreen.

Decumbent: lying on the ground but rising at the tip.

Dicotyledon, dicot: broadleaf plants with two seed embryos (leaves) or cotyledons when it emerges from the soil; these also have netted leaf veins, showy flowers, flower parts in fours or fives, and often have yearly growth.

Diffuse: loose and widely spreading.

Dioecious: male and female flowers separate and on separate plants.

Disc flower: a type of flower with a tubular-shaped corolla (united petals) that is found in a head as all or part of the complete flower of many members of the Compositae (or Asteraceae) family.

Dissected: divided into numerous narrow segments of lobes.

Distinct: separate.

Divided: cut to the base or midrib.

Drupe: fleshy fruit with a single seed enclosed in a hardened inner fruit wall.

Glossary

Elliptic: a narrow shape with relatively rounded ends that is widest at the middle; football-shaped.

Embryo: seed portion that develops into a juvenile plant.

Endemic: confined to a very limited geographic area, such as endemic to Florida.

Entire: a type of leaf margin without teeth, lobes, or divisions; smooth.

Erect: upright, often perpendicular to the surface.

Evergreen: having green leaves throughout the winter.

Eye: the center of a flower.

Fascicle: a bunch or cluster.

Fibrous roots: slender, branched roots of similar size arising from a similar point.

Filament: anther-bearing stalk of a stamen (male part) of a flower; threadlike.

Filiform: threadlike, long and very slender.

Flaccid: without rigidity; limp or weak.

Fleshy: thick, juicy; succulent.

Flora: a work which contains the enumeration of plants within an area.

Floret: a small flower.

Folded: arrangement of the youngest leaf in the bud shoot where the leaf buds are folded together lengthwise with the upper surface inside the fold.

Foliage: leaves; mass of leaves.

Fruit: a matured ovary with its enclosed seeds; the ripened female part of the flower.

Genus (plural genera): a group of related species.

Glabrous: smooth, without hairs or bristles.

Gland: an appendage which usually secrets a fluid.

Glandular hair: a small hair terminated by a small pinhead-like gland.

Glume: one of a pair of bracts at the base of a grass flower.

Grasslike: leaves long and narrow, usually more than 10 times as long as broad.

Gymnosperm: plant that produces seeds but not fruits. The seeds are not borne within an ovary and are said to be naked.

Habit: growth form of the plant.

Habitat: natural environment where plant grows.

Head: a dense cluster of stalkless flowers.

Hemispheric: shaped like half a sphere.

Herbaceous: nonwoody plant; can be annual or perennial.

Glossary

Hip: the rose fruit with a fleshy cup containing achenes.

Hybrid: cross between two species.

Immersed: growing under water; submerged.

Inflorescence: the flowering portion of a plant.

Internode: the section of stem between two successive nodes or joints.

Involute: rolled inward.

Joint: node of a stem.

Keel: a prominent ridge, often comprised of tissue on both sides of a midrib which has grown together.

Lanceolate: a shape longer than wide and broadest below the middle.

Leaf: the main organ borne by the stem or axis; usually comprised of a broad flat blade and may have a stalk or not.

Leaflet: one of the divisions of a compound leaf.

Legume: member of the pea or bean family having a dry fruit (pod) that splits open along two longitudinal sutures.

Ligule: projection at the inside junction of the grass leaf blade, which may be membrane-like or a row of hairs.

Linear: a long and narrow shape with parallel margins.

Lobe: a segment of a simple leaf cut rather deeply into curved or angular segments.

m (meter): equals approx. 3.3 feet.

Membranous: thin, transparent, and flexible; membrane-like.

Midrib: the main or central vein or rib of a leaf or leaflet.

Midvein: the primary vein.

mm (millimeter): equals approx. 1/32 of an inch.

Monoecious: male and female flowers separate, but on the same plant.

Monocotyledon, monocot: grass and grasslike plants in which embryos (seedlings) have one cotyledon (seed leaf), parallel-veined leaves, inconspicuous flowers, flower parts in multiples of threes, and no secondary growth.

Naturalized: originally cultivated, now living in the wild.

Nerve: a simple vein or slender rib of a leaf or bract.

Node: the joint of a stem.

Nodding: hanging down.

Node: the point or level of a stem at which one or more leaves and roots are attached.

Nutlet: any small and dry nutlike fruit or seed.

Glossary

Oblanceolate: a shape longer than wide and broader at the tip.

Oblong: an elongate shape with approximately parallel sides, more or less rectangular.

Ocrea: a sheath or tube around the stem at a node; common in the Polygonaceae family.

Once-divided: one row of leaflets along a single axis.

Open: loose.

Opposite: an arrangement of paired leaves or leaflets attached oppositely from each other at the same node.

Orbicular: circular or round in shape.

Oval: rounded; broadly elliptic.

Ovule: an immature seed.

Ovate: a shape similar to a hen's egg; widest below the middle.

Ovary: lower part of the female structure of a flower containing the ovules or later the seed.

Palmate: a type of leaf where leaflets or lobes originate from a common point and diverge like the fingers from the palm of the hand.

Palmate venation: three or more nearly equal veins extending out into the blade from the leaf stalk like the fingers from the palm of the hand.

Palmately divided: a type of arrangement where leaflets arise from finger-like divisions originating from a common point of attachment.

Panicle: a flower cluster composed of branched branches.

Peltate: a type of leaf attachment where the stalk is attached inside the blade margin; umbrella-like.

Pendulent: drooping, hanging downward.

Perennial: a plant that normally lives for more than two years.

Persistent: remaining attached.

Petal: one member of the inner whorl of a flower, usually colored.

Petiole: a leaf stalk.

Pistil: female part of the flower composed of stigma, style, and ovary, formed from one or more carpels.

Pith: soft central tissue of a stem.

Plumose: having fine hairs.

Pod: a dry fruit that opens to disperse seeds. **Pollen:** the male spores borne by the anther.

Pome: a fleshy fruit with the sepals remaining on the upper end, like an apple.

Prickle: sharp outgrowth from a surface.

Glossary

Prostrate: parallel to or lying flat on the ground.

Raceme: an elongated flower cluster with each flower on an individual stalk.

Rachis: the axis of a flower spike or a divided leaf.

Ray flower: a type of flower with a strap-shaped petal, located around the margin of the flowering head found in many members of the Compositae (Asteraceae) family.

Reclining: sprawling or lying down.

Recurved: curved downward or backward.

Reticulate: a network pattern, netted.

Rhizome: a creeping, horizontal under ground stem, producing shoots above ground and roots below; distinguished from a root by the presence of joints (nodes), buds, or scalelike leaves.

Rosette: circular cluster of leaves usually appressed to or located near the ground level.

Rudimentary: small, often incompletely developed.

Runner: a slender stolon or horizontal stem.

Scabrous: rough to the touch (sandpapery).

Scale: any small, thin, dry membrane-like leaf or bract.

Scapose: a naked, flowering stem.

Scurfy: covered with minute, membranous scales.

Seed: a ripened ovule.

Seedhead: a collection of flowers clustered on a main stem.

Sepal: an outer part of a flower that is usually petal-like in appearance and often green in color.

Serrated: toothed; with sharp teeth.

Sessile: without a stalk.

Shoot: a general term for the above ground portion of a plant. **Shrub:** a woody perennial, usually branching from the base with several main stems.

Smooth: lacking hairs, divisions, or teeth; not rough to the touch.

Solitary: alone.

sp. (plural spp.) species (singular and plural).

Species: a group of individuals having certain distinctive characteristics in common.

Sphere: a round object.

Spike: an unbranched flower cluster (arrangement) with stalkless flowers.

Spur: a tubular projection.

Stalk: any slender supporting structure.

Stamen: the male or pollen-bearing organ of a flower consisting of a filament (stalk) and the anther.

Star-shaped: branched (applies to hairs).

Stem: plant organ; functions for support, leaf production, food storage, and limited food production.

Sterile: without seeds or pollen.

Stigma: upper, often feathery, part of the female flower that receives pollen.

Stipule: a bractlike appendage at the base of some leaves.

Stolon: a creeping, above ground stem that roots at the joints (nodes).

Style: the stalk of a pistil that connects the stigma to the ovary.

Submerged: growing under water.

Succulent: soft and fleshy.

Sucker: a plant shoot that arises from an adventitious bud on a root.

Summer annual: a plant that germinates in spring, grows and flowers in summer, sets seed in fall, dying after setting seed.

Synchronous: happening at the same time.

Tapering: gradually becoming smaller toward one end; not abrupt.

Taproot: a single enlarged vertical main root lacking major divisions. **Tendril:** a slender, twisting, threadlike structure of a leaf or stem that allows plants to climb.

Terminal bud: bud located at the end or apex of a stem or branch.

Thorn: a sharp-pointed stiff woody projection, can be curved.

Tooth: any marginal projection.

Translucent: transparent to light.

Tree: a woody perennial, usually with a single trunk or stem.

Trailing: prostrate, but not rooting.

Trunk: the main stem of a tree, often solitary.

Tuber: thickened storage portion of a rhizome or stolon, bearing noces (joints) and buds.

Tufted: in compact clusters, forming clumps.

Twice divided: two rows of lateral branches along an axis that are again divided into two rows of leaflets.

Undulate: wavy.

Vein: ribs of a leaf.

Vine: annual or perennial, herbaceous or woody, climbing, creeping, or trailing.

Whorled: three or more attached in a circular arrangement at the same place.

Glossary

Winter annual: a plant that begins germination in late summer, grows vegetatively during the winter, flowers and sets seed in the later spring to early summer, dying afterwards.

Woody: consisting or composed of wood or woodlike tissue.

Index

Index

Index

Index

Index

Index

Index

Index

Index

Index

Index

Index

Index

Index

Index

Index

Wildflowers of Florida and the Southeast

This book has images of over 750 wildflowers with comprehensive descriptions to aid in identifying these colorful gems of this region. This book will be of value to enthusiastic amateurs who wish to learn about the wildflowers they see in the area as well as to the trained botanist. Wildflowers are separated by color for identification because that is what you see first. The written description gives the geographic range as well as the type of habitat where the bloom will be found. The seasons of flowering, size of the plants, type and shape of leaves, and many more details are included to give you more information on each plant.

David W. Hall, Ph.D.

Is a botanist who currently owns and operates an environmental consulting firm in Gainesville. He was previously employed as a Senior Scientist for KBN/Golder Associates from 1991-1997 following his role at the University of Florida as the Director of the Plant Identification and Information Services. He is a recognized expert in the field of plant identification and forensics and has published 11 books and over 140 articles. He has accrued numerous awards for his botanic, forensic, and agricultural activities.

William J. Weber D.V.M.

Was a practicing veterinarian in Leesburg, FL. His love of nature photography became another career. His articles and photography appeared in many magazines such as National Wildlife Magazine, Natural History Magazine, and Florida Wildlife Magazine. His photographic images have appeared as covers on over 100 national magazines. He is the author of six books.

Jason H. Byrd, Ph.D., D-ABFE

Is a board certified forensic entomologist and Associate Director of the William R. Maples Center for Forensic Medicine at the University of Florida. He was the first individual to be elected as President and Chairperson to the only two professional associations for forensic entomolgists in North America, and has twice served as President of the American Board of Forensic Entomology.

While this book focuses on the state of Florida and the states of the Southeast it will be useful in many areas of the country since many of these flowers bloom west into Texas and into some of the more northern states. The glossary of plants and their parts is useful everywhere.

$40.00
ISBN 978-0-615-39502-9
54000>

9 780615 395029